D1684603

Maxwell on Heat and Statistical Mechanics

Maxwell on Heat and Statistical Mechanics

On "Avoiding All Personal Enquiries" of Molecules

Edited by
Elizabeth Garber,
Stephen G. Brush,
and C. W. F. Everitt

Bethlehem: Lehigh University Press
London: Associated University Presses

Associated University Presses
440 Forsgate Drive
Cranbury, NJ 08512

Associated University Presses
25 Sicilian Avenue
London WC1A 2QH, England

Associated University Presses
P.O. Box 338, Port Credit
Mississauga, Ontario
Canada L5G 4L8

The paper used in this publication meets the requirements of the American National Standard for Permanence of Paper for Printed Library Materials Z39.48–1984.

This book was produced using plain TeX. The typeface is Computer Modern.

Library of Congress Cataloging-in-Publication Data

Maxwell, James Clerk, 1831–1879.
Maxwell on heat and statistical mechanics : on "avoiding all personal enquiries" of molecules / edited by Elizabeth Garber, Stephen G. Brush & C.W.F. Everitt.
p. cm.
Includes bibliographical references and index.
ISBN 0-934223-34-3 (alk. paper)
1. Maxwell, James Clerk, 1831–1879—Correspondence. 2. Thermodynamics. 3. Virial theorem. 4. Equations of state. 5. Rarefied gas dynamics. I. Garber, Elizabeth. II. Brush, Stephen G. III. Everitt, C. W. F. (C. W. Francis), 1934– . IV. Title.
QC310.2.M39 1995
530.4'3'092—dc20 94-25234
CIP

PRINTED IN THE UNITED STATES OF AMERICA

Contents

List of Serial Abbreviations

Abh. Boehm. Ges. Wiss.	*Abhandlungen der Königlich Boehmischen Gesellschaft der Wissenschaften, Prague*
Abh. Math.-Phys. Cl. Gesell. Wissen. ,Leipzig	*Abhandlungen der Mathematische-Physiche Classe der Königlichen Sachsischen Gesellschsft der Wissenschaften, Leipzig*
Acta. Hist. Sci. Nat. Med.	*Acta Historium Scientiarum Naturalum et Medicinalium*
Amer. J. Phys.	*American Journal of Physics*
Amer. J. Sci.	*American Journal of Science*
Amer. Sci.	*American Scientist*
Ann. Chim.	*Annales de Chimie*
Ann. Chim. Phys.	*Annales de Chimie et de Physique*
Ann. Phil.	*Annals of Philosophy*
Ann. Phys.	*Annalen der Physik*
Ann. Sci.	*Annals of Science*
Arch. Hist. Exact Sci.	*Archives for History of Exact Sciences*
Arch. Int. Hist. Sci.	*Archives Internationales d'Histoire des Sciences*
Ber. Leipzig	*Bericht über der Verhandlungen der Königlichen Sachsischen Gesellschaft der Wissenschaften zur Leipzig, Mathematisch-Physiche Classe*
Berlin Akad. Monats.	*Monatsberichte der Königlich Preussischen Akademie der Wissenschaften zu Berlin*

British J. Hist. Sci.	*British Journal for the History of Science*
British J. Philos. Sci.	*British Journal for the Philosophy of Science*
Cambridge Dublin Math. J.	*Cambridge and Dublin Mathematical Journal*
Cambridge Math. J.	*Cambridge Mathematical Journal*
Chem. Rev.	*Chemical Reviews*
Edinburgh J. Sci.	*Edinburgh Journal of Science*
Edinburgh Rev.	*Edinburgh Review*
Fort. Phys.	*Fortschritte der Physik*
Hist. Sci.	*History of Science*
Hist. Studies Phys. Sci.	*Historical Studies in the Physical Sciences*
J. Chem. Educ.	*Journal of Chemical Education*
J. Chem. Soc. London	*Journal of the Chemical Society, London*
J. Hist. Ideas	*Journal for the History of Ideas*
J. Math. Pures Appl.	*Journal de Mathématiques Pures et Appliquées*
J. Nat. Phil. Chem. Arts	*Journal of Natural Philosophy, Chemistry and the Arts* (Nicholson's *Journal)*
J. Rational Mech. Anal.	*Journal of Rational Mechanics and Analysis*
J. Reine Angew. Math.	*Journal für die reine und angewandte Mathematik*
J. Roy. Inst.	*Journal of the Royal Institution*
J. Scient. Instrum.	*Journal of Scientific Instruments*
Math. Ann.	*Mathematischen Annalen*
Mem. Acad. Sci. Paris	*Mémoires de l'Académie des Sciences, Paris*
Mem. Manchester Lit. Phil. Soc.	*Memoirs of the Manchester Literary and Philosophical Society*
Mon. Not. Astron. Soc.	*Monthly Notices of the Astronomical Society*
Monats. Preuss. Akad.,	*Monatsberichte der Königlichen Preussischen Akademie, Berlin*
NTM	*NTM: Zeitschrift für Geschichte der Naturwissenschaften, Technik und Medizin*

Notes Records R. Soc. London	*Notes and Records of the Royal Society of London*
Phil. J.	*Philosophical Journal*
Phil. Mag.	*Philosophical Magazine*
Phil. Trans. R. Soc. London	*Philosophical Transactions of the Royal Society of London*
Phys. Blätter	*Physikalische Blätter*
Phys. Rev.	*Physical Review*
Phys. Zt.	*Physikalische Zeitschrift*
Proc. American Acad. Arts Sci.	*Proceedings of the American Academy of Arts and Sciences*
Proc. Cambridge Phil. Soc.	*Proceeding of the Cambridge Philosophical Society*
Proc. London Math. Soc.	*Proceedings of the London Mathematical Society*
Proc. Manchester Lit. Phil. Soc.	*Proceedings of the Manchester Literary and Philosophical Society*
Proc. R. Irish Acad.	*Proceedings of the Royal Irish Academy*
Proc. R. Inst.	*Proceedings of the Royal Institution*
Proc. R. Soc. Edinburgh	*Proceedings of the Royal Society of Edinburgh*
Proc. R. Soc. London	*Proceedings of the Royal Society of London*
Rep. (no.) Meeting BAAS	*Report of the (no.) Meeting of the British Association for the Advancement of Science*
Sitz., Berlin Chem. Gesell.	*Berlin Chemische Gesellschaft, Sitzungsberichte*
Sitz., Bonn	*Sitzungsberichte der Niederrheinischen Gesellschaft für Natur- und Heilkunde zu Bonn*
Sitz. Math.-Naturwiss. Cl. Akad. Wiss. Wien	*Sitzungsberichte der Mathematische-Naturwissenschaftliche Classe der kaiserlichen Akademie der Wissenschaften, Wien*
Studies in Hist. Phil. Sci.	*Studies in History and Philosophy of Science*
Trans. Cambridge Phil. Soc.	*Transactions of the Cambridge Philosophical Society*

Trans. Acad. Sci., Connecticut	*Transactions of the Connecticut Academey of Sciences*
Trans. R. Soc. Edinburgh	*Transactions of the Royal Society of Edinburgh*
Zt. Phys.	*Zeitschrift für Physik*
Zt. Phys. und Math.	*Zeitschrift für Physik und Mathematik*

Preface

The title to the third and final volume[1] of James Clerk Maxwell's work on heat and matter could have been "Maxwell on Gas Theory, Thermodynamics, the Virial Theorem and the Equation of State, and Rarefied Gas Dynamics." On the surface this seems to be a patchwork of what was left from his very focused work on gases from the 1860s. But what the twentieth century regards as separate topics were joined in Maxwell's physics by his interest in fundamental questions which arose from the foundations of his physics.

The letters, notes, and papers collected here were composed, in the main, during the last decade of Maxwell's life. They range from his concern about the compatibility of kinetic theory and thermodynamics to the mechanics of rarefied gases. His notes begin with the relationship between kinetic theory and thermodynamics, a problem that arose from his 1867 paper. He was eventually provoked into investigating the issue formally, rather than intuitively. The investigation, in turn, led him to examine more carefully and appreciate more deeply Boltzmann's methods and finally to use them himself. This, together with Peter Guthrie Tait's request for help with his text on thermodynamics, nudged Maxwell into more systematic thinking on the theory of heat. It also persuaded him to publish his own text on the subject.[2] The *Theory of Heat* was published in 1871 and according to his biographers, was written at Glenlair before he returned to Cambridge in 1871.[3] The book was popular and went through five editions before Maxwell's death; it was translated into German in 1877.[4] Maxwell used the volume "for the masses" at Cambridge in his lectures to ordinary degree candidates. He had envisioned it as a text for workingmen and students with no advanced mathematics capability. Yet in reading it one would need a well-developed spatial sense. The important changes Maxwell made to this text were in his understanding of the second law drawn from Josiah Willard Gibbs. His respect for Gibbs's work is reflected in the space it occupied in the remaining editions he supervised. Here Maxwell used the theory of heat to tease out the properties of matter and develop an appreciation of the idea of irreversibility. It is not on the engineering topics of work, or, the efficiencies of heat engines.

His interest in the properties of matter also led Maxwell to bring Clausius's virial theorem and the subsequent work of Van der Waals to his colleague's attention. His review of Van der Waals's published dissertation developed into a short paper on how Maxwell would approach the problem. Thermodynamic arguments, the virial, and Van der Waals allowed glimpses of the inner structure of matter that otherwise could only be speculated about.

Maxwell also continued to develop his own ideas on gas theory in the direction of a more general treatment in which no molecular models were necessary. As in the case of his electromagnetism, Maxwell finally developed a form of his gas theory that depended only on the basic principles of mechanics. The particularities of molecular models would not interfere with its claims to validity. And those claims were crucial, as Maxwell well knew, to the mechanical foundation of his physics. Hence the importance of the experiment to investigate the compatibility of gas theory and thermodynamics that he suggested in his paper of 1879.[5] The agreement of gas theory and thermodynamics reinforced his commitment to a mechanical world picture.

From the twentieth century this second direction to his research, of developing a form of statistical mechanics, is far more important. On looking at the reactions of his contemporaries, however, we see no impact. The research of his last decade which seemed immediately important, was on rarefied gas dynamics. This massive paper seemed to solve the radiometer problem—one that had caused so much stir within and beyond the scientific community through the 1870s. It is a fine example of science as theater. The radiometer was a phenomenon made for public consumption and exploitation: forces acting at a distance seemingly made visible. The invisible world made manifest dropped into a society some of whose members were enmeshed in spiritualism already. What editor could resist it!

Maxwell's sober, complicated mathematics put to rest, at least in the scientific community, such speculations. His life ends then with an intellectual triumph but also in a rather shabby professional squabble exacerbated by his rapidly declining health, pain that clouded his judgment, and friends ready to defend his reputation at any cost.

In this decade the one piece of Maxwell's work that overshadows all others is the publication of his *Treatise on Electricity and Magnetism*. Much of the first edition was written during his years as a private scholar at Glenlair.[6]

He was unable to finish correcting the proofs of a second edition. His other publications of this decade include the short, but important *Matter in Motion*.[7] Like *Heat* it is a text without complex mathematics (except in the Appendices) but one whose merits rested upon clarity of conception and language. His other papers of this decade include many reviews for *Nature*, which usually end as small papers in and of themselves, contributions to the ninth edition of the *Encyclopaedia Brittanica*, and addresses to both popular and scientific audiences, and last but not least his referee reports, mainly for the Royal Society of London. All of his papers, long and short, bear careful reading, and his output is far more than one would expect from an administrator burdened with the design, building, and definition of a new institution in the hostile territory of Cambridge.

The actual title of this volume on gases is taken from his referee report on Osborne Reynolds' paper on rarefied gas dynamics published in 1880.[8] The point of Maxwell's remark was that Reynolds could not have detected what he claimed in Maxwell's 1867 paper on gases.[9] He had been very careful not to deal with where molecules have come from within the element of gas under consideration. Maxwell merely counted them and grouped them using their mean velocities only. He continued by explaining that he had always been careful to examine the properties of gases and not the gas-surface interface, the problem Reynolds had embarked upon. The latter involved molecules both approaching and rebounding off the surface that introduced discontinuities into the mathematics. Within the method of his 1867 paper, Maxwell explained, "I carefully abstain from asking the molecules which enter [the volume under consideration] where they last started from. I only count them and register their mean velocities, avoiding all personal enquiries which would only get me into trouble."[10]

Clearly Osborne Reynolds' misreading of his paper irritated Maxwell. Yet this flippant remark, made in the middle of a detailed, if negative appraisal of Reynolds's theory, illustrates some of the essential qualities of Maxwell's physics as it developed in the 1870s. Earlier in his career, Maxwell previously had thought through the approach to gas theory that Reynolds now embarked upon and already had detected its pitfalls.

As well as this reaction to a particular theory, within this critique and in his own papers, Maxwell also brought to bear a new sense of the proper relationship within physics between speculation, theory, and experiment. These three elements were to be integrated as never before in Maxwell's work. No longer tolerable were the self-contained speculative systems of the eighteenth and the first half of the nineteenth century which were connected only tentatively, if at all, to physical reality as revealed by experiment. His caustic remarks in the 1870s on the deductive system of James Challis are an expression of these new standards.[11] While Challis's theory was self-consistent it bore no illustrated relationship to reality, the reality in which Maxwell grounded his own physics. Expressed in the vernacular or in one of the languages of mathematics, theory and experiment must be integrated. In the earlier, deductive schemes of mathematicians such as Challis, particular extensions of the general mathematics under restrictions dictated by the mathematics, infrequently led to results seemingly compatible with experiment.[12] These compatibilities were, earlier, taken as validating the whole deductive system. This was no longer good enough. Mathematical generalities were tempered by physical imagery, and how precisely experiments were related to mathematical speculations required much careful thought and analysis. Maxwell required that any extension of a theory into the experientially unknown be accompanied by experiments that drew his theoretical speculations back to earth.

In his kinetic theory this grounding in reality included his own experiments on gaseous friction, devised and completed soon after his first gas theory paper indicated that friction in gases would be independent of the pressure. It ended in his proposal to investigate the compatibility of gas theory and thermodynamics.[13] In the development of his theory of electromagnetism, experiments play a different, yet key, role. His speculative leap from the experimental results on the ratio of electromagnetic units to electrostatic units and the wave motions sustainable by his solid mechanical model of the aether and the speed of light were followed by experiments at the Cavendish Laboratory on both of these aspects of his argument.[14] And in this connection perhaps we should begin to see his extension of and admiration for Michael Faraday's ideas as based on Maxwell's recognition of a theory for which experiment served as the logic of its development.

In his own physics, experiment was integrated into the body of the development of the theory. And when Maxwell devised any experiments they pointed to the logical structure of the whole theory, not one particular aspect of it. Devising experiments was, for Maxwell, an important facet of the theoretical enterprise. Experiment played a key role even in his earliest papers in electromagnetism and kinetic theory both based on particular models.

The terrain Maxwell had long since abandoned was precisely that proposed by Osborne Reynolds. To follow their impacts with the walls of their container Reynolds had to speculate on the internal structure of his molecules. For his part Maxwell suggested how Reynolds might accomplish his goals without the necessity of such experientially unsupportable assumptions.

In his own, early papers on electromagnetism and kinetic theory, the particulars of the mechanical models were prominent. However, Maxwell was always sensitive to the speculative nature of such models and was sceptical of their reality. They reflected phenomena, that is they replicated the results of known experiments and even suggested new direction for investigation, but they did not necessarily constitute an explanation of them. They certainly did not represent the actual structure of either gases or the aether.[15] Maxwell sidestepped many of the problems that colleagues had to struggle with later in the century. The mechanical models of both molecules and the aether became more and more elaborate as physicists increased their efforts to cope with the growing body of experiment on molecules.

Yet Maxwell's approach had its limitations. In the gas theory case, he could not see the most important aspect of Reynolds' paper that lead Reynolds to develop dimensional analysis. We can only speculate that perhaps Maxwell's illness already limited his judgment because this aspect of Reynolds's paper seems to give to a mathematically intractable problem a visual quality that Maxwell so appreciated elsewhere. Further, Maxwell could not appreciate that for the solution of some problems the structure of both molecules and the aether need speculating about.

This understanding of boundaries was also tied to Maxwell's growing appreciation of the logical limitations of the most general ideas on which he based most of his physics—the ideas of mechanics. His notions of a general mechanical system, its instability and unpredictability, seem remarkably modern, but with this, Maxwell also knew that no mechanical model was sufficient to account for the character required of either electricity or molecules. His own sense of a mechanical system as one that obeyed the conservation of energy freed Maxwell to explore the scope of mechanics as a descriptive framework for the natural world.

Through these papers Maxwell brought into question the very foundations of much of his own physics—the mechanical worldview. In gas theory this is apparent in his papers on molecular models and statistical mechanics and in electromagnetism in his later "electrical formulation" of his theory.[16] None of this was apparent to his contemporaries for some years after his death. Indeed in 1879 his scientific legacy was seen in terms of his work in molecular science and his mechanical formulation of electromagnetism. His later work presented in this volume, in statistical mechanics and gas dynamics, was simply ignored.

In the broader, social context Maxwell's life changed drastically in 1871 with his appointment as the first professor of experimental philosophy at Cambridge. His appointment was an important stage in the struggle to reform the curriculum of Cambridge, begun long before 1871 and destined to continue into this century.[17] As significant as his position as professor in this renewal was Maxwell's part in the reform of the Mathematics and the Natural Science Tripos. As Examiners and Moderators, Maxwell and William Thomson, in collusion with like-minded dons, had begun to change the nature of the Tripos examinations in the late 1860s. And they needed changing. It had long been understood that mathematics at Cambridge was unsatisfactory and "unproductive."

> There were many complaints that Cambridge was behind the rest of the scientific world, and that, whereas the students of so many other Universities were introduced to the splendid discoveries of such subjects as Electricity and Heat, the Wranglers of Cambridge spent their time upon mathematical trifles and problems, so-called, barren alike of practical results and scientific interest. Maxwell's questions (as Moderator in 1866) infused fresh life into the Cambridge Tripos, and, therefore, into the University studies by the number of original ideas and new lines of thought opened up by them, thus preparing for the change of system in 1873, when so many interesting subjects were added to the Examination.[18]

The Mathematics Tripos consisted of questions largely based in mechanics. The expected answers were exercises in mathematical acrobatics: nineteenth-century methods pursuing eighteenth-century goals. But mathematicians on the continent now focussed upon foundational issues, not the display of technical virtuosity.

In a system geared to processing students for an examination the examiners were crucial in setting both the questions for the ultimate test and the training students underwent. In the early decades of the nineteenth-century, Charles Babbage, John Herschel, and William Whewell realized the importance of examiners in establishing a renewed Tripos. At this later date, to reform the edifice they had built, similar tactics were effective.

Yet more was required. As a body the University might agree to any number of changes, but conservatives were certain of their hold on the system. The wealth of the institution lay in the colleges not the university. The wherewithal was simply not available. Under such circumstances political subtlety and manipulation were at a premium. Independent wealth also made the problem easier to solve. In the case of physics, the duke of Devonshire was prepared to underwrite the cost of building and equipping a physics laboratory at Cambridge. Yet we must wonder whether, given the direction in which Maxwell propelled the laboratory, this patrician industrialist got what he had envisioned.

The Cavendish professor was expected: "to teach and illustrate the laws of Heat, Electricity and Magnetism; to apply himself to the advancement of knowledge of such subjects; and to promote their study in the University."[19] Such a broad definition left a great deal of room for negotiation by the appointee. Maxwell was well aware of this when he agreed, after some persuasion, to accept the position. Maxwell's reluctance, if his letters are to be believed, lay in his acknowledged ignorance of how such institutions functioned and whether Cambridge students could be induced to endure the necessary "dull labour." Yet there were few of these institutions in Britain to guide him. He also understood that interpretations of the position ranged from the establishment of a research institute to the performance of popular lectures to add a gloss of modernity to Cambridge education. Maxwell hoped to cater in some fashion to three groups of students: "the masses," students seriously interested in physics, and those capable of research. As the reforms evolved, the Cavendish Laboratory was quickly established as a research laboratory. As such it became important in the scientific community; its prestige within Cambridge was another matter. However, with such low esteem Maxwell did not have to face the problem of teaching large numbers of students even as the university grew.

With his acceptance of the appointment, Maxwell ended his private life pursuing his research at his own pace and was thrust into the center of Cambridge politics as well as the life of an administrator. In addition he had to help design, then oversee, the building and equipping of the laboratory before he could begin to establish any program of research. All of this consumed large amounts of his time and energy. It is no wonder that his own research output declined. Yet the papers he did publish were important, some profoundly so. While many contain little or no mathematics, the ideas and arguments are clear and significant though many were not reprinted in his collected papers.

While this essay gives us the opportunity to explore the character of Maxwell's science, its focus precludes a discussion of the delights of his humor. Some examples of this are included in the form of poems on matters molecular, gaseous and on forces. At just the time when deadly serious matters are under discussion Maxwell's wit leavens the argument without dulling its point. At times these can be very pointed and aimed at the recipient of the piece. A measure of their forbearance and Maxwell's charm is that the poems were preserved and some were even published. Maxwell's wit is apparent throughout the earliest biography of him where the leaden Victorian seriousness of the tale is bedeviled by his ability to see the ridiculous in himself, his colleagues, and in science and its institutions. He seemed to sense that in the broad scheme of things academics and their concerns were not the most basic, except to academics. This ability to step outside of himself has made the long study of his work, sometimes in minute detail, a delight, just as his intelligence has remained a constant stimulus. He also reminds us that the goals of nineteenth-century physicists were not those fulfilled by twentieth-century physics. Their study of nature has to be understood in their terms and we must try to forget just how the whole tale turned out in the next hundred years. To do so is to begin to understand the complexity and depth of our heritage in physics and to gain an appreciation of just what nineteenth-century physicists thought they were doing.

In the documents < > indicates a word that was deleted in the original, ? a scribble that is indecipherable to all three editors, and [ ] our own insertions (usually a suggested title for some unpublished notes). A [space] means that there is a gap equivalent to several lines of text in the manuscript, perhaps indicating that Maxwell wanted to leave room to insert material. Editorial notes to a document, indicated by superscript letters, appear at the end of the document. These notes identify the correspondent and his or her relation to Maxwell, contemporaries of historical interest, and references to any of Maxwell's own published papers or works in progress. For papers reproduced from *The Scientific Papers of James Clerk Maxwell,* we have retained the original page numbers in the book to facilitate cross-references.

We would like to thank Brigadier Wedderburn-Maxwell, N. R. Thorp, Special Collections Librarian, University of Glasgow, N. H. Robinson of the Royal Society of London, and A. E. B. Owen of Cambridge for giving us permission to publish this material and for helping us to obtain copies of it.

Notes

1. The first volume in the series is Stephen G. Brush, C. W. F. Everitt, and Elizabeth Garber (eds.) *Maxwell on Saturn's Rings* (Cambridge, Mass.: MIT Press, 1983). It contains Maxwell's letters as he composed his prize essay on Saturn's Rings, published in 1859, its review by George Biddel Airy and Maxwell's reworking of the problem by applying the methods of his kinetic theory in the early 1860s. The second volume, Elizabeth Garber, Stephen G. Brush, and C. W. F. Everitt (eds.) *Maxwell on Molecules and Gases,* (Cambridge, Mass.: MIT Press, 1986) contains his letters and notes on his two kinetic theory papers published in 1860 and 1867. These are collected together with his ideas on the nature of molecules and matter.

2. Maxwell, *Theory of Heat* (London: Longmans Green and Co., 1871),2d and 3d eds., 1872, 4th ed. 1875, and 5th ed. 1877. Other editions appeared in 1880 and 1883. New editions, with corrections and additions by Rayleigh appeared in 1897 and 1902.

3. Lewis Campbell and William Garnett, *The Life of James Clerk Maxwell* (Johnson Reprint of 1882 edition; New York: 1969), p. 374.

4. The German edition appeared as Maxwell, *Theorie der Wärme,* trans. Dr. F. Auerbach (Breslau: Maruske & Berendt, 1877).

5. For his experiment on gases under gravity see Campbell and Garnett, *Life,* p. 570; Maxwell, "On Boltzmann's Theorem on the average Distribution of Energy in a System of Material Points," *Trans. Cambridge Phil. Soc.,* 12 (1879): 547–570, reprinted in *The Scientific Papers of James Clerk Maxwell,* vol. 2, ed. W. D. Niven (Dover reprint of 1890 edition; New York: 1965), 713–741.

6. Maxwell, *Treatise on Electricity and Magnetism* 2 vols. (Oxford: Clarendon Press, 1873).

7. Maxwell, *Matter and Motion* (London: Society for the Promotion of Christian Knowledge, 1876).

8. Osborne Reynolds, "On certain Dimensional Properties of Matter in the Gaseous State. Part I and Part II," read to the Royal Society February 5, 1879, and published in *Proc. R. Soc., London* 28 (1879): 304–321 and *Nature* 19 (1879): 435. It was printed in full in *Phil. Trans. R. Soc., London* 170 (1880): 727–845 and reprinted in Reynolds, *Papers on Mechanical and Physical Subjects* vol. 1 (Cambridge: At the University Press, 1900–1903), 257.

9. The paper in question is Maxwell, "On the Dynamical Theory of Gases," *Phil. Trans. R. Soc., London* 157 (1867): 49–88 and *Phil. Mag.,* 35 (1868): 129–145, 185–217, reprinted in *Scientific Papers* vol. 2, 26–78 and Garber, Brush, and Everitt, *Maxwell on Molecules and Gases,* 419–472.

10. Maxwell, "Report on a Paper by Prof. Osborne Reynolds 'On Certain Dimensional properties of Matter in the Gaseous State'," document 15.

11. See Maxwell, '"Challis's 'Mathematical Physics'," *Nature* 8 (1873): 279–280, reprinted in *Scientific Papers,* vol. 2, 338–342 and Garber, Brush, and Everitt, *Maxwell on Molecules and Gases,* 126–128 and the subsequent correspondence in *Nature,* 9 (1873): 131–132. On Maxwell's critique of such systems see Robert Kargon, "Model and Analogy in Victorian Science: Maxwell's Critique of French Physicists," *J. Hist. Ideas* 30 (1969): 423–436.

12. For a case study on the historiographic issues surrounding such mathematical physics, see Elizabeth Garber, "Siméon-Denis Poisson: Mathematics versus Physics in early nineteenth-century France," in Garber (ed.), *Beyond History of Science: Essays in Honor of Robert E. Schofield* (Bethlehem Pa.: Lehigh University Press, 1990). These issues and the case of the eighteenth century will be discussed in Garber, *Mathematics as Language* (forthcoming). The case of nineteenth century German physicists is covered in Christa Jungnickel and Russell McCormmach, *Intellectual Mastery of Nature*, 2 vols. (Chicago: University of Chicago Press, 1986). Volume 1 is pertinent to the period of Maxwell's life. For William Thomson's commitment to a physics-based mathematics, see Crosbie Smith, and M. Norton Wise, *Energy and Empire: A Biographical Study of Lord Kelvin* (New York: Cambridge University Press, 1989). Whether William Thomson had such a commitment that was as clear-cut or as well thought out in his early career as the authors state is debatable.

13. For the experiments on friction, see Maxwell, "On the Viscosity or internal Friction of Air and other Gases," *Phil. Trans. R. Soc., London* 156 (1866): 249–268, reprinted in *Scientific Papers,* vol. 2, 1–25. For the experiment on gas theory and thermodynamics, see op. cit., n. 5.

14. For an indication of the role of experiment in the development of Maxwell's electromagnetic theories and his work on the velocity of light and units see C. W. F. Everitt, *James Clerk Maxwell Physicist and Natural Philosopher* (New York: Charles Scribners, 1975), 111–130, 98–110. However, Daniel Siegel, *Innovation in Maxwell's Electromagnetic Theory of Light* (Cambridge: Cambridge University Press, 1991) only deals with these connections in chap. 5.

15. See his remarks at the beginning of Maxwell, "On Faraday's Lines of Force," *Trans. Cambridge Phil. Soc.* 10 (1856): 27–83, reprinted in *Scientific Papers,* vol. 1, 155–229, and at the end of his first paper on gases, Maxwell, "Illustrations of the Dynamical Theory of Gases," *Phil. Mag.* 19 (1860): 19–32; 20 (1860), 21–37, reprinted in *Scientific Papers,* vol. 1, 377–409, and Garber, Brush, and Everitt, *Maxwell on Molecules and Gases,* 285–318. The assessment of Maxwell's commitment to a mechanical world view may never be settled, but his scepticism in his papers on molecules is striking. This is not obvious if one reads primarily the electromagnetic papers. For Maxwell as mechanist see Daniel Siegel, op. cit., n. 14.

16. Maxwell, "On Boltzmann's Theorem,"op. cit., n. 5. and for electromagnetism see Everitt, op. cit., n. 14, 108–109.

17. The details of Maxwell's appointment and the politics of this stage in the reform of the Cambridge curriculum are in Romualdas Sviedrys, "The Rise of Physical Science at Cambridge," *Hist. Studies Phys. Sci.* 2 (1970): 127–145, followed by a commentary on Cambridge politics by Arnold Thackray, 145–149, and further remarks by the author, 149–151. Technical and economic factors and the belief that true science was "practical" are discussed in Graeme Gooday, "Precision Measurement and the Genesis of Physics Teaching Laboratories in Victorian Britain," *British J. Hist. Sci.* 23 (1990): 25–52, and Smith and Wise, op. cit., n. 12, chap. 5.

18. Campbell and Garnett, *Life*, p. 357.

19. Ibid., p. 350.

Maxwell on Heat and Statistical Mechanics

I Introduction

At first glance the papers on thermodynamics and gas theory Maxwell published in the last decade of his life appear scattered and address topics isolated from one another. Historians usually see this decade of his life as dominated by the production of his *Treatise on Electricity and Magnetism* and the building then establishing of the Cavendish Laboratory. The piecemeal appearance of his other papers is not, therefore, surprising. If we look closer, search out the papers not included in his *Scientific Papers*, and remember Maxwell's reputation in the 1870s, his work begins to take on more coherence. He pursued certain problems with a method that came to maturity in this period.[1] To see the connections Maxwell made between gas theory, thermodynamics, and his electro-magnetic theory, we have to look at his textbook on thermodynamics and his unpublished papers, letters, and notes. Maxwell worked on these problems in overlapping periods of time but in this order. Initially, he faced the problem of the compatibility of kinetic theory with the second law of thermodynamics. He then turned to the nature of the second law, statistical mechanics and the equipartition theorem. His final interest was in the dynamics of rarefied gases.

The compatibility of kinetic theory with the second law of thermodynamics surfaced during the publication of Maxwell's 1867 paper on gases.[2] Maxwell wrote to Sir William Thomson who later refereed that paper.[3] Maxwell concluded that the extension of his theory to a gas under external forces led to perpetual motion, directly contradicting the second law, which he quoted from William Thomson. Yet neither Maxwell nor Thomson wanted to accept that conclusion. This reaction can tell us something about the position of thermodynamics and kinetic theory in physics in the late 1860s. Both were necessary. The second law, variously and obscurely interpreted, was an integral part of physics. Kinetic theory was heuristic and already had led to intriguing results about the the properties of molecules. The two had to be compatible.

The history of the first solution to the paradox is given in Maxwell's addendum to his 1867 paper.[4] Having discovered one mistake in his original calculations he deduced an increase in temperature with height, still contrary to the second law. A second set of corrections was necessary before Maxwell concluded that a gas in mechanical equilibrium under the influence of gravity was also in thermal equilibrium.

This result clearly satisfied the needs of the second law. But it contradicted both observations of the temperature of the atmosphere at various heights and Thomson's thermodynamic analysis of an isolated column of the atmosphere under gravity.[5] Thomson had examined the expansion of a mass of gas as it rose through the column, arguing that the decrease in temperature was due to adiabatic processes. Something still seemed amiss.[6] Thermodynamics appeared to offer the growing science of meteorology a foundation for understanding the laws of thermal change for the atmosphere.[7] Kinetic theory apparently contradicted these early results.

Using the methods of his 1867 paper Maxwell returned to the problem with an argument that proved unsatisfactory. He got the conditions he wanted but only if no energy was transported into or out of the stratum. If the temperature was constant, there was no heat conduction. After the fact his gas theory was compatible with the second law. But compatibility was not apparent in the structure of his kinetic theory; as yet he had not developed methods to address the distribution of energy in a gas. However, the parts of his 1867 paper he was using treated a gas without specifying a particular interaction between its molecules.

There the matter rested until Maxwell published his *Theory of Heat* and elicited Francis Guthrie's puzzled reaction (document II-7). We can sympathize with Guthrie if we consider what Maxwell had published by 1873. Maxwell's 1867 statement extending his results to gases under gravity was cryptic and descriptive and those in his thermodynamics text were assertions about the thermal equilibrium of a gas under external forces. In his text he separated this problem from that of the thermal behavior of the atmosphere.[8]

Guthrie's prodding goaded Maxwell into reconsidering the problem. At this time Maxwell probably began to reread Boltzmann with new attention. Before 1873 Maxwell's arguments about gases under external forces were intuitive, not analytical. He had difficulties following Boltzmann's mathematical path but such analyses were necessary to establish the conditions for thermal equilibrium. Since thermodynamics was now a foundation stone for physics, kinetic theory was at stake as was the growing field of molecular physics.[9] The importance of Boltzmann's early papers on kinetic theory for Maxwell is clear. They gave Maxwell the physical approach and the mathematical methods to attack the problem head on even though Boltzmann had a different perspective on the second law.

Boltzmann's initial approach was to "reduce the two laws [of thermodynamics] to such a form that will show the mechanical relationship between the two."[10] The mechanical foundations for thermodynamics should be based on the actual motions of the atoms of a gas under the influence of external forces. Boltzmann quickly realized that a purely mechanical model could not include irreversible changes encompassed in the second law. In 1868 he turned to Maxwell's kinetic theory published the previous year. Boltzmann's goals were twofold: to understand the laws governing the changes in motion of a set of mass points from one fixed state to another while he explored the theory of heat whose "doctrines are liable to deficiencies and are so removed from their analytical foundations."[11]

This mathematical thrust drove Boltzmann to correct the logical deficiencies in Maxwell's kinetic theory papers but along his own lines.[12] By examining the collision of elastic spheres in mathematical detail, Boltzmann deduced the necessary velocity distribution function for particles free of external forces. His function was the same as Maxwell's. From this Boltzmann constructed the distribution function for the kinetic energies of his molecules. Through a series of particular cases he extended his arguments to molecules under external, central forces. He then jumped to the general case, and it is this section of the paper that Maxwell found so enlightening.

It was clear that the quantity to trace was energy, not velocity; equally useful was Boltzmann's deduction that in thermal equilibrium, during collisions, as many molecules changed their position and velocity coordinates from A and B to A' and B', as from A' and B' to A and B. In this section Boltzmann also looked at the variations in velocity and position that Maxwell exploited later.

Maxwell used the above ideas and methods. However, he ignored Boltzmann's definition of probability in terms of the time spent in a particular energy state as a ratio to a very long time and restated it in terms of position and velocity. Presumably both men took their definitions of probability as equivalent.[13] Maxwell assumed the other results he needed were already established by Boltzmann and did not give any independent demonstrations.

Guthrie's second letter (document II-10) gives some indication of just how far the community of physicists could follow Maxwell and Boltzmann through their discussion of gases. His misunderstanding comes from difficulties with visualizing random motions and their consequences. The results of Maxwell's probability argument and the idea of a distribution of velocities eluded him. In his final letter to *Nature* Guthrie pinpointed the crucial criterion on which the validity of Boltzmann's and Maxwell's arguments about thermal equilibrium rested. How could such a perfectly reversible mechanical process as molecular motion lead to the irreversibility inherent in the second law? In this he also saw the physical consequences of picturing a gas made up of molecules in ordered motion, and he put his finger on an important physical problem. Kinetic theory seemed to imply the possibility of a perpetual motion machine, and Maxwell's comments in his text on thermodynamics did not dispel his suspicions.

Maxwell was thus drawn into considering the problem in more detail. Yet even here Maxwell did not demonstrate the validity of his crucial assertion, the number of molecules going from states A and B to A' and B' equaled those going from A' and B' to A and B. He merely referred to the relevant parts of Boltzmann's paper. One wonders how useful that was to Guthrie in Cape Town. This did not end the debate. Reconciling irreversibility, the second law and kinetic theory was discussed in many guises throughout the late nineteenth century, to which we will return later.

Maxwell returned to the problem of the thermal equilibrium of a column of gas under gravity in his 1875 lecture to the Chemical Society on molecules.[14] His remarks led R. C. Nichols to question the validity of kinetic theory. From Maxwell's discussion Nichols concluded that the temperature of a gas was different at different heights.[15] His arguments were quickly countered by J. J. Murphy and put to rest by S. H. Burbury.[16] Nichol finally withdrew his objections. Nevertheless, important issues were raised, and Maxwell encouraged Henry William Watson to publish a formal treatise on gas theory and its problems.[17]

In his small treatise (fifty-one pages of text), Watson discussed only those aspects of kinetic theory that were especially problematic, trying to establish these results as mathematically sound and physically necessary. The book is a series of mathematical propositions in a style now familiar in the mathematical physics adopted by Cambridge from the French.[18] The mathematical rigor required in German mathematical physics was not apparent here. He made no claims to originality, and the derivative nature of his methods is obvious. If the content of this treatise is any indication, kinetic theory was sorely misunderstood even by specialists in 1876. Watson pointed out that the statistical method was necessary, illustrating his claim with Maxwell's example of the complexity of the phenomena under examination. Watson repeated Maxwell's observation that even if molecules began with the same velocities, they would distribute themselves very quickly throughout the velocity spectrum.

More importantly, Watson used Maxwell's methods of 1867 to deduce the velocity distribution function but more carefully like Boltzmann specifying the conditions of the problem.[19] Watson deduced self-consistent expressions for the known properties of gases but admitted some problems remained, in particular those associated with the specific heats ratio. He also extended Maxwell's methods to gases under external, central forces and to gases made up of molecules between which were intermolecular forces obeying Clausius's virial theorem. In these two cases Watson used generalized coordinates and momenta, following Boltzmann. But Watson went beyond Boltzmann's use by showing the uniform distribution of momentum, in the stationary state. This was a requisite step in the derivation of the second law since the assumption of equal probabilities for kinetic energies for all molecules is inadequate for its proof.[20]

All these results set the stage for Watson's penultimate proposition, the derivation of the second law of thermodynamics from the results of kinetic theory. The proposition is synthetic not analytic, drawing upon previous demonstrations even if only for particular cases. All previous propositions lead up to this one, and we can see why Watson needed to derive the velocity distribution function for several cases separately. Following Boltzmann, Watson traced the changes in the virial, energy, and work done by a gas when a small amount of heat δQ is added to a gas otherwise in thermal equilibrium at temperature, T. The problem reduces, mathematically, to showing that the expression for $\delta Q/T$ in equivalent kinetic theory terms is a perfect differential for all changes in the average energy and volume for the gas. Watson could replicate Boltzmann's results in his 1871 paper for material point molecules under external forces.[21] Now, kinetic theory was not merely compatible with, but contained, the second law of thermodynamics. There are shortcomings to both Boltzmann's and Watson's derivations. $\delta Q/T$ must be constructed anew for every particular molecular interaction and is only valid for intermolecular forces that obey the virial theorem. The derivations are based on a series of special cases. There still was no general proof for any gas and all molecular interactions.[22]

In his review of Watson's *Treatise* Maxwell did not emphasize Watson's derivation of the second law directly but looked at the consequences of another of Watson's results. He considered the constancy of the temperature of a mass of gas under gravity, his own way of approaching the connections between thermodynamics and kinetic theory. This time Maxwell saw the atmosphere explicitly as a connected system and Watson's as an isolated column. He then suggested experiments to demonstrate Watson's result. The first would require a vertical tube about one hundred meters tall, kept in a constant thermal enclosure for several years, the density differences only showing up slowly. The other, almost as demanding of contemporary technology, consisted of rotating a mixture of gases at high speeds in a closed tube. Within hours this would produce the same differences in density along its length that would take years in the static case.

Both the thermal equilibrium of a gas column and the experiment to demonstrate it resurfaced in Maxwell's later papers where the experiment is explored in more detail. Maxwell had in mind not just the problems kinetic theory solved but those it generated—in particular, the differences between the ratio of the experimental values of specific heats of gases and those derived from kinetic theory and the indications of internal motions of molecules deduced from spectral data.[23] The ratio from kinetic theory was simply too small and all the mechanical, molecular models available (and all the devices suggested to avoid the problem) were simply inconsistent or made the disagreement more glaring.

Overall Maxwell simply improved Watson's derivation of his proposition on the distribution of momenta and praised the volume in general. But Maxwell leaves the reader on a disconcerting, negative note. The early triumphs of kinetic theory had been replaced by a concern with issues surrounding the second law and the construction of an adequate molecular model for matter.

Thermodynamics

Maxwell was drawn into protracted work in thermodynamics by accepting Peter Guthrie Tait's invitation to read his manuscript on the theory of heat and its history. Maxwell was faced with an interpretation of the second law which conflicted with his own sense of its implications for physics. In his own work he focused on the irreversibility inherent in the second law, and, indeed, many of his papers on gas theory in the 1870s were on irreversible phenomena. As he wrote to Simon Newcombe, these phenomena were the most likely to reward the labor of research.[24] To see how he would reach such a conclusion and become so involved with the theory of heat, we need to review some of the history of thermodynamics.

The development of thermodynamics and the principle of the conservation of energy are oft told tales which still fascinate historians and physicists. Much controversy remains. Who did what, precisely, as well as the historical maxims that lie within these cognitive deeds continues to be debated intensely, if only on scholarly grounds.[25]

Some of the initial statements of the first law of thermodynamics as well as the later clarifications of both laws depended on a particular, mechanical concept of the nature of heat.[26] In addition, the first "histories" were published in the 1860s. This was just over a decade after the statements of the first law of thermodynamics, the establishment of the concept of energy (disengaged from that of force), and the acceptance of energy conservation as a foundation for physics. In the 1860s there were several murky and incomplete statements and interpretations of the second law and it is here where Maxwell's efforts as a physicist were brought to bear.

The early histories were just enough removed from the events for the physical significance of the first law and conservation to be appreciated yet still close enough for those involved to quarrel over priorities and meanings encoded in experiments and papers written in the 1840s and 1850s. The scientific stakes were high; reputations and careers still hung on the assignment of credit or priority and the importance given to various ideas. Also, Tait's text was published in the middle of the first controversy over Britain losing its economic and scientific leadership to Germany.[27] Tait's account was also molded by one of the combatants, William Thomson. After his initial publications Thomson was simply unable to give any credit to Clausius for his part in establishing and investigating the meaning of the second law. Combined with a patriotism that deteriorated into chauvinism, Tait narrated an Anglo-centric and distorted narrative.[28]

Maxwell was not directly involved in the early development of the mechanical theory of heat or conservation of energy. However, he quickly understood their significance and the difficulties not yet solved by the early papers on thermodynamics (document III-2).[29] This sensitivity was honed in facing the problems of teaching in the late 1850s and with his concerns about kinetic theory and thermodynamics ten years later (document III-3). He was primed for Tait's invitation in 1867 to proofread the book-length manuscript he had developed from two previously published papers. The negative reactions from Rudolph Clausius and others to his history of the subject led Tait to ask for help.[30] His version, in turn, depended on deciphering the physical meaning of the two laws of thermodynamics and hence deciding who had stated them first and in the most general form. These judgments affected the social standing of the various protagonists in a newly established profession where creativity was tied closely to promotion and positions of authority. Writing such a history in these circumstances was bound to offend someone, but Tait's openly chauvinistic "history" led to one of the earliest, open wars over the "discovery" of the laws of conservation of energy and the second law of thermodynamics. Tait, as did his British colleagues, emphasized the conservation of energy as the key conceptual development. He thus downgraded the discovery of the first law to a special case of conservation. This was correct physically, but Tait destroyed Clausius's claim to originality.

Maxwell's initial response to Tait's request was to interpret the physical concepts. Circumstances had changed when Tait approached him several years later for help with a second edition. By this date Maxwell had published his own text on thermodynamics and struggled to understand the second law for himself. He had also read the early papers without Tait's interpolations. Indeed, Maxwell was remarkably sane throughout this unpleasant episode, being able to both criticize Tait, praise German physicists and their champion in Britain, John Tyndall, Tait's nemesis. Yet Maxwell and Tait remained friends.

In the middle decades of the nineteenth century both a clear, unequivocal statement of the law of conservation of energy and the two laws of thermodynamics emerged from the confluence of three historically connected streams of ideas in physics. The first stream came from the idea that the great forces of nature could be transformed into one another.[31] The idea that none of the force was lost in the transformation process was the source of the second stream, joined to the first, speculated and experimented upon by several men through the 1840s. And it is here that the first priority disputes erupt, the main protagonists being James Prescott Joule and J. R. Mayer.[32] From very different starting points Joule and Mayer established the conversion of various forces of nature into heat. In addition, they both saw conservation in the numerical constancy of the quantity of a force required to produce a unit of heat. Joule then focused on inanimate phenomena and experimentally probed the interconversion of electricity then mechanical work into heat.[33]

Mayer's discussion was more philosophical and centered on living processes. He derived his approach from Justus Leibig's chemical theory of animal heat, while he objected strongly to the latter's vitalism.[34] However, both Joule and Mayer understood the need and estimated a conversion factor for force into heat.[35] Both also still thought in terms of the "great forces of nature" and did not see what they were measuring as a manifestation of some other quality, energy. And neither (pace Elkana) did Hermann von Helmholtz.

To better understand both British and German reactions to the research of physicists in other national contexts, we need to examine the uses and contentions over the concept of force in the nineteenth century. Force had two meanings, only one of which was acceptable in nineteenth-century Britain. As used by Michael Faraday or James Joule force meant the agents of change in nature, such as heat, light, electricity, gravitation, and chemical forces, without speculating about the processes by which they brought about observed changes. Force, meaning action at a distance, was written out of the nexus of physical explanation in Britain. For both philosophical and physical reasons force became an explicandum rather than the explicans.[36]

The theory of heat was also expressed in mathematics and must be joined to the development of Cambridge mathematical physics, derived from French mathematical physics, as mathematics of the 1820s.[37] The French mathematics that was absorbed was divorced from physics, although mathematicians may have needed physical models to construct their mathematics.[38] With the French model the grounds for explaining the actions of nature were the facts of experiment and the truths of the calculus only. Speculations about internal mechanisms of change and the internal structure of matter were frowned upon in Britain in the middle decades of the nineteenth-century. There was no evidence to justify them. Rankine's vortical molecules could be ignored, as were Laplace's central forces. Their mathematical deductions were all that counted.

Such modes of explanation have their limitations, and these were partially broken through by James Clerk Maxwell. His models of gases and ether could be sidestepped for the mathematical content of his work. Physically his models were criticized severely by his contemporaries, though exploited by his successors. Maxwell shared his colleagues' dislike for the severely deductive systems of German physicists based on inverse forces though he understood them and appreciated them more than most. Indeed, after Tait's lecture on force at the British Association in 1876 Maxwell wrote a two-edged poem on that lecture.[39]

In contrast, German physicists of the 1840s looked back to the natural philosophy of Immanuel Kant and his concept of matter as Newtonian centers of force and saw such inverse forces as the source of all change. [40] In 1846 Wilhelm Weber had scored a theoretical triumph with his electrodynamics. Based upon the idea of electric currents as centers of oppositely charged forces moving in opposite directions, Weber joined experiment to the direct measurement of electrical quantities and developed an explanatory net for known phenomena.[41]

Hermann von Helmholtz was committed to Kant's philosophy and to action-at-a-distance forces. For Helmholtz all the forces of nature were different manifestations of the action of central forces.[42] In his *Erhaltung der Kraft* Helmholtz examined force in both the senses used here. He provided a new, tentatively mathematical explanation of the agents of heat, electricity, and chemistry. Helmholtz came to the study of the conservation of force from his studies in physiology and was as concerned as Mayer to rid this science of vitalism. His 1847 essay is an exercise in following a particular hypothesis in a simple mathematical form through to its eventual experimental verification.[43]

In this paper he coined a series of names for the various forces that could be transformed one into the other without any loss in the quantity of the force. Helmholtz stated that one could deduce the conservation of force from the impossibility of perpetual motion, or, alternatively from the principle that all the forces of nature were reducible to central forces. Importantly, Helmholtz was then able to express the force due to the position of his centers of force with respect to one another in mathematical form by examining the effect of the change of *vis viva* on this new force, *Spannkraft.* He argued that the change in *vis viva* equalled the sum of the *Spannkraft*, corresponding to the change in the distances between the centers of force.

Helmholtz knew that his conservation law was valid only in the case of central forces, but he failed to examine his basic conjecture underlying this particular case that all forces—agents of natural change—can be reduced to such central forces. The particularity of his derivation was lost on most commentators. While the ambiguities in his usage of the term "Kraft" are obvious, Helmholtz made no attempt to develop a different nomenclature, nor to see his work as anything more than an extension of the conservation of *vis viva* for a particular law of force.[44] The strict reductionism of Helmholtz's paper needs to be emphasized against the almost universal acceptance by his contemporaries that his law of conservation of force was the law of conservation of energy. And here we have the third candidate for priority in discovering the principle of the conservation of energy.

In later sections of his paper Helmholtz tried to connect his conservation law to the phenomena of electricity and heat. In his section on heat Helmholtz joins the third stream in physics that converged with the other two during this decade and the next. He was not the first, nor the most successful, in seeing heat as a type of motion, nor in critiquing Carnot's caloric theory and his analysis of the heat engine.

Under the above circumstances it is perhaps understandable that the British did not respond to Helmholtz's work until it was explained to them. Yet, through the efforts of John Tyndall, Clausius's papers were translated immediately into English. Clausius's work on the theory of heat shared certain principles of explanation with the British. It was mathematical and independent of any theory of matter, particularly of action at a distance forces. All that Clausius assumed in his early papers was that heat was motion.

That heat was a form of molecular motion had driven Joule's investigation into the conversion of mechanical effect into work.[45] However, these investigations depended upon molecular models. Joule's early molecular models were of nuclei surrounded by atmospheres in motion. He did not explore which characteristic of this whirling atmosphere was heat and how this molecular characteristic manifested itself macroscopically. Nonetheless, this model led Joule to perform delicate experiments and, at least to his satisfaction, to demonstrate the validity of his identity of heat with motion and the falsity of Carnot's theory.[46]

Helmholtz's ideas on heat as molecular motion were more vague than Joule's. He also made no attempt to connect molecular motions to their macroscopic manifestations, merely insisting heat was motion.[47] Given the German context of Helmholtz's work he did not need to. All that was required was the statement that the "force equivalence" of heat was the "work produced" by the heat as the system passes through a fall in temperature.[48] Helmholtz could show this idea was compatible with experimental results, meeting the basic requirements of current mathematical physics. While he used Joule's direct measurement of the mechanical equivalent of heat, Helmholtz put more credence in values deduced less directly from experiments on the velocity of sound and the latent heat of water vapor. And there, at least in Helmholtz's research publications the matter rested. His research interests returned to physiology.[49] He chose not to pursue the ideas of his conservation of force paper even after the appearance of the first papers of Rankine, Rudolph Clausius, and William Thomson on the "mechanical theory of heat."[50] It was not until thirty years later that Helmholtz would focus his research on thermodynamics and only then would he use the term energy.

Before his own work in thermodynamics and its history, Maxwell described Helmholtz as "one of the first, and is one of the most active, preachers of the doctrine that since all kinds of energy are convertible, the first aim of science at this time should be to ascertain in what way particular forms of energy can be converted into each other, and what are the equivalent quantities of the two forms of energy."[51] Maxwell's later assessment of Helmholtz's place in the history of thermodynamics is interesting. He did not see Helmholtz as the originator of the first law or the conservation of energy. Rather he cited the impact of Helmholtz's 1847 paper on others. Maxwell judged his work in physiology and optics as more important.[52]

The development of the principle of the conservation of energy was not one of sudden insight but of painful stages beset by conceptual muddle, pregnant hints, but only partial statements. However, before 1860, conservation of energy had become an obvious foundation for physics. The process of clarifying the meaning of this principle was through the development of the new, mechanical theory of heat explored, mathematically and conceptually by Clausius, Rankine, and Thomson.[53]

Clausius and Rankine almost simultaneously published papers on the mechanical theory of heat.[54] Rankine was the first to describe mathematically the motions of his molecules that were the microscopic underpinnings of the macroscopic phenomena of heat. His papers in thermodynamics are usually glossed over and he has recently been placed on the fringes of the conceptual development of thermodynamics.[55] We must, however, recognize that his unique method led to valid results.

Rankine developed his arguments simultaneously on three levels–mathematical, phenomenological and micro-molecular—but none of these levels is completely integrated with any other. Most obviously Rankine never connects the motion of the atmospheres of his whirling vortices of ether with the measurables of thermodynamics. He developed his mathematical argument fully, then tried to visualize it in terms of his molecular model. However, his definitions and interpretations of measurable, thermal quantities do not help much in deciphering the physical content of his mathematical analysis.

Rankine stated the first law of thermodynamics mathematically and introduced a function that mathematically was entropy.[56] But he did not decipher the latter physically. In his later interpretation of thermodynamics Rankine's universe was pure mechanism, and he chose to explore the universal implications of the first law and its generalization using the conservation of energy.[57] This was, for Rankine, an inviolable law of nature, and he could not accept Thomson's idea of energy dissipation. He postulated a cosmic conservation of energy with only local pockets of energy loss.[58] This may be because his universe was literally mechanical. The *vis viva* of the motions of his molecular atmospheres represented the "elastic power" of heat, and his mathematics referred to elastic properties discerned by the senses as thermal effects. Elasticity for him was the primary concern.

For physical clarity we must turn to Clausius. Although he would later explore the "motion that we call heat," Clausius understood that to establish the "mechanical theory of heat" in general it had to stand apart from any particular model of matter. He also pinpointed precisely where the mechanical theory of heat had to part conceptual company with that of Sadi Carnot. Thus he used a great deal of Carnot's (or rather Emil Clapyeron's mathematical version of Carnot's) theory that was independent of caloric theory.

Clausius, like Rankine, was well aware that if heat was a mode of motion all the mathematical apparatus of mechanics was available to him. However, he used mechanics on a more abstract level than Rankine. Clausius assumed only that the particles of matter were in motion and that, "heat is a measure of their *vis viva*." He took as axiomatic that with the production of work by heat a certain quantity of heat, proportional to the work done, is consumed and vice versa. However, when he got down to the mathematical, rather than the physical statement of the implications of this axiom, Clausius fell back on the case explored by Carnot, the reversible thermal cycle of an ideal gas, stating the first law as,

$$dQ = dU + AR\frac{a+t}{V}\,dV.$$

Where U is an arbitrary function of temperature and volume, V, dQ is the amount of heat added to the system, $A = 1/J$ and R is the gas constant. The term $a+t$ is the temperature on the absolute scale. Clausius recognized U as a state function made up of the sensible heat and the heat necessary for internal work. He then explained the results of his mathematical analysis physically after obtaining significant physical, as against mathematical, results.[59] Further Clausius could and did lead the mathematical development to the point where theory joined known, experimental results. The main point of his first paper was to establish that when mechanical work is done heat is consumed. Clausius was at pains to demonstrate that his theory differed only in this one respect from Carnot's and that it led to results that agreed more closely with experiment than Carnot's theory.[60]

Clausius continued the task of establishing this basic axiom through a series of particular cases. He demonstrated that his assumption worked for nonideal gases and vapors, for which he could gain suitable experimental data. The theory clearly offered better explanations for other conjectured phenomena.[61] He followed the implications of the mechanical theory of heat into electrical phenomena in several other papers. In this first decade physicists extended the new theory of heat in a dozen different directions including meteorology, chemistry, and cosmology.[62]

For Clausius, it seems that by working through special cases, particularly on the thermal properties of real gases, he gained more control over the content of his 1850 paper. Yet, in these early papers one can feel a presence, not referred to much but nevertheless guiding the shape of the argument. That presence is William Thomson and his masterful development of Carnot's caloric theory of heat. Clausius's thermodynamic paper of 1850 addressed Thomson's caloric papers directly, and he took pains to counter the arguments of those papers.

From the beginning Clausius was regarded as an also ran. In Germany credit for the "discovery" of the first law was given either to Mayer or Helmholtz despite Clausius's clearer and deeper understanding of the physical issues and his mathematical development of the first law. In Britain his work was largely ignored in favor of that of Joule experimentally and Helmholtz theoretically, for the first law, and William Thomson for the second.[63] Thomson's intellectual position with respect to this new theory was somewhat different from that of Joule, Helmholtz, Rankine or Clausius. Thomson alone had accepted—then developed—Carnot's theory. He had worked in Regnault's laboratory and then he had explored the mathematics of caloric theory. Thomson's training in mathematical physics at Cambridge made this mathematical extension almost inevitable given his recent work in electricity.[64] Generally, Thomson's thermodynamics papers follow the same pattern as his papers on caloric theory: the masterful mathematical development along analytically generalized lines in directions mapped out by others.[65] Thomson was not out to replace one theory completely but to emend in the body of a highly developed mathematical theory of heat only those parts that absolutely required correction.

Three of the papers were something of a salvage operation. Thomson indicated this as he presented the object of his first paper on the dynamical theory of heat as showing, "what modifications of the conclusions arrived at by Carnot, and by others who have followed his peculiar mode of reasoning regarding the motive power of heat, must be made when the hypothesis of the dynamical theory, contrary as it is to Carnot's fundamental hypothesis, is adopted."[66] His next objective was to show that the numerical results he had previously deduced calorically from Regnault's experiments on gases could be reinterpreted with the new theory of heat. These results, together with Joule's mechanical equivalent would give a "complete theory of the motive power of heat." The third objective was to extend the theory of heat to explain universal physical properties "established by reasoning analogous to that of Carnot, but founded in part on the contrary principle of the dynamical theory." (These properties actually are independent of either theory of heat.) The main point of these papers is to create a generalized, mathematical development of Rankine's and Clausius's work and connect it with his Carnot papers. This is very possible because many of the results of thermodynamics and caloric theory depend on the form of the equations they share, not on any physical assumptions. As if to emphasize this, Thomson refers to his 1849 paper on Carnot's theory to remind his readers of valid conclusions already reached.[67]

In 1849 Thomson had followed his initial paper on caloric theory with a general mathematical account of Carnot's ideas, as yet unknown to an English speaking audience. Much of this 1849 paper, besides presenting Carnot's ideas, consists of the manipulation of partial differentials without much physical point to all the mathematics. While voicing his doubts in footnotes as to the validity of caloric theory, Thomson had found no definitive reason for abandoning the theory. In the phenomenon of the lowering of the freezing point of water he had verification of its correctness.[68] His mathematically elegant extension of Carnot's ideas had confronted, but not solved, the vexing issue of irreversible phenomena. By the time of his first mechanical heat theory paper, Thomson also wanted to establish some experimental way of choosing definitively between Joule and Carnot. Experimentally determining Carnot's function would serve this purpose. The implications of Carnot's expression for the efficiency of a heat engine had been brought home to him in his correspondence with Joule on the absolute scale of temperature.[69] What Thomson failed to see was that Carnot's function was not dependent on any theory of the nature of heat. By focusing on Carnot's function as the definitive test, Thomson missed its theoretical significance. In 1851 Thomson criticized Clausius for assuming in his opinion, arbitrarily, a form of Carnot's function which could only be established experimentally. In this he tried to use Regnault's experiments on gases.

Thomson tried to minimize his losses retaining as much of his previous papers, mathematically and physically, as possible and repudiating very little. Only those parts of his papers that now depended on the conservation of heat in cyclic, reversible heat engines were under sustained discussion. In Thomson there is no clean break with the past, rather a patching of cracks and filling conceptual lacunae.

Fundamentally then, Thomson was hampered in his approach to heat theory. He never did write down an expression for the first law but constructed expressions for the efficiency of the reversible, cyclic heat engine. He did not, therefore, see the physical significance of internal energy. And indeed Thomson's approach to the heat engine appears to place engineers' issues before physicists' concerns. One of Thomson's major preoccupations during the late 1840s was that the universe was progressive. No former state could reoccur. We can therefore see his fourth paper on the dynamical theory of heat as an important personal resolution of a fundamental philosophical dilemma that both the caloric and the mechanical theory of heat presented to him.[70] However, this is hidden from his readers, and the paper appears as unrelated to either his previous papers in the series or the one following on reversibility.

The origins of this paper lay in a dispute with Clausius over the explanation of the lowered temperature of a gas issuing from a jet.[71] Clausius attributed all the temperature loss to thermal causes, using arguments from his newly developed mechanical theory of heat. Thomson saw the loss in the work done against friction forces along the inside surfaces of the nozzle and jet. The phenomenon was mechanical not thermal.[72]

In the middle of his considerations of cyclic phenomena Thomson suddenly understood that the streaming of gas from a jet was an "un-reversible" phenomenon. It was accompanied by energy dissipation that was nonrecoverable and, more importantly, so were the workings of real heat engines.[73] The third of four propositions he stated was that "when heat is diffused by conduction there is a *dissipation* of mechanical energy, and a full *restoration* of it to its primitive condition is impossible." From this proposition Thomson deduced the extractable mechanical work from an unequally heated body by equalizing the temperature of that body. This was simply the mechanical equivalent of heat "put out of existence." From this followed an integral expression for the final temperature and the statement of his conclusion on the heat death of the universe. He gave no evidence for his initial assumption or the details leading to his conclusions.[74] For Thomson, during every cycle of a heat engine some energy becomes "unavailable" for work and is ejected as heat. This loss is inevitable and irreversible.

After this foray into the second law, Thomson turned to thermoelectricity to finish his papers on reversible phenomena.[75] Thomson's thermoelectric studies are important if one is focusing on the interrelationships between the forces of nature, exploring them as manifestations of various forms of energy. He chose not to explore the physical implications of the idea of unavailable energy in the depth that he studied thermoelectricity. With his approach he could regard the thermal system as a black box and examine the heat input and output along with the work produced. What went on in the box was uninteresting and unexplored. Precisely because of this Thomson's statement of the second law was ultimately less heuristic in developing an understanding of the thermal properties of physical systems. Far more important was the work of both Rankine and Clausius. By 1853 Rankine had a mathematical statement of the second law, and he subsequently named it, he did not explore its physical significance.

Thomson also was unable to see Clausius's contribution to the development of the second law. His vision of the situation seems more clouded as time progressed.[76] And it was Thomson's version of the laws of thermodynamics and its history that Tait published in the 1860s. Tait solicited Thomson's help as he wrote his history, and he also became immersed in the cosmic implications of his physical theory.[77] For Tait, Thomson's statement of the second law was unequivocal and clear. In contrast, Rankine's thermodynamic function was hidden in the mathematics. Tait saw Clausius's development of the second principle of the mechanical theory of heat stated at the end of his 1850 paper as arbitrary, obscure, and, in 1864 when Tait's papers on the history of thermodynamics were first published, incomplete.

Clausius's path to understanding the second law and his naming and exploration of entropy seems almost perversely convoluted. His attention centered on the actual molecular motions that constituted heat. Thus he explored what happened to those motions when heat was added to a body. Clausius tried various approaches to this problem, all of them permanently useful in some ways to the development of either thermodynamics, kinetic theory, or theories of matter in the nineteenth century.

Clausius's initial introduction of dQ/T was physically arbitrary, only vaguely justified and never wholly explained. That is, unless one can assume that Clausius already understood that dQ/T was a state function (complete differential) and as significant as U, the internal energy. Clausius began his exploration of the second principle of the mechanical theory of heat by restating it as "the theorem of the equivalence of transformations."[78] In any reversible heat cycle where mechanical work is produced there are two types of transformations. The first is the transformation of heat into mechanical work at a constant temperature. The other is the transformation of heat at a higher temperature into heat at a lower temperature. The equivalence value of the first transformation is Q/T where Q is the heat transformed into work at the temperature T. The equivalence value of the second transformation is,

$$Q(\frac{1}{T_1} - \frac{1}{T_2}),$$

where Q is again the heat transformed at T_1 to heat at temperature T_2.[79] In a cyclic reversible process the transformations cancel one another, and "the analytical statement of the second fundamental theorem of the mechanical theory of heat is $\int dQ/T = 0$."[80] For irreversible processes $\int dQ/T > 0$, always as there are uncompensated transformations.[81] Whether Clausius reached his expression dQ/T using his molecular model or not is immaterial here, though there are hints that he did.[82]

However, at this point Clausius does not pursue the objective of exploring what will eventually become entropy but delves into "The Nature of the Motion that we call Heat," along two lines of enquiry. The first led into his papers on kinetic theory.[83] The second line of research was to explore what happened when the configuration of the molecules making up a substance was changed. Clausius already had focused specifically on the change a body undergoes when taken around a reversible cycle, not merely on the input and outcome from a Carnot cycle. He now explored what occurred when heat was added to a body, and both internal and external work was done. He expressed the first law as

$$dQ = dU + dW,$$

where dQ was the heat added to the system and dW was the external work performed by the body, and dU is the change in its internal energy, which he changed into,

$$dQ = dH + dL,$$

where dH was the change in the kinetic energy of the molecules by the addition of heat dQ, and dL was the work done, both internal (in changing the configurations of the molecules) and external. Clausius also assumed that this total work, dL, could be written as

$$dL = TdZ,$$

where T is the temperature of the body and Z its "disgregation" which Clausius defined as "the degree of dispersion of a body."

Disgregation was a state function and for Clausius had equal, if not more physical significance than the as yet unnamed dQ/T. Both H and Z were functions of the temperature only.[84] He arrived at the above forms for these significant physical entities by examining the mathematical properties of the equations which were part of a complete "analytical development" of the fundamental principles of the mechanical theory of heat.[85] This enabled him to naturally integrate into one theoretical point of view a number of previously isolated experimental results and to express a number of the partial differentials later developed by Maxwell.[86]

The exploration of the mathematical properties of the internal changes of bodies was for Clausius a natural continuation of the analysis begun with the more familiar thermodynamical quantities. And it was through the physical consideration of H and Z that Clausius explained why uncompensated transformations are always positive. In nonreversible cases a quantity of H and Z existed in the body above that which can be transformed either into work or heat at a lower temperature. On writing,

$$S = \frac{dQ}{T} = \frac{dH}{T} + dZ,$$

arithmetical manipulation allowed Clausius to conclude

$$\frac{dQ}{T} = S - S_0,$$

and S is the "transformation content of the body." This quantity was renamed entropy, from the Greek for transformation, so that it "may be adopted unchanged in all modern languages." It is here that Clausius puts the two laws of thermodynamics into the couplet,[87]

Die Energie der Welt ist constant.
Die Entropie der Welt strebt ein Maximum zu.

For Clausius, however, entropy had molecular as well as thermodynamic significance, but it was the latter that Clausius saw as more important and that gave the mathematical quantity its physical meaning.

Josiah Willard Gibbs was the first to isolate Clausius's mathematical explorations from the physical ones that motivated them and to disconnect the thermodynamically significant from the molecularly particular. And it was not until 1873 when he read Gibbs that Maxwell saw the significance of this aspect of Clausius's work.[88] Maxwell saw Clausius's molecular explorations as speculative and undermining the foundations of thermodynamics. Clausius had taken the measurable quantities of thermodynamics and extracted from them quantities that could not in principle be measured.[89]

When first consulted by Tait in 1867, Maxwell had available to him in the published literature two competing forms for the second law, one of which was tied to speculations which he explicitly criticized and tried to avoid in his own work.[90] Maxwell, therefore, accepted Tait's interpretation of entropy: It was the same quantity as William Thomson's "Unavailable Energy." He even used Tait's interpretation of the second law in the first editions of his textbook on heat.[91] As far as Tait was concerned, Clausius had added nothing to the development of the theory of heat. He even traced the expression dQ/T to Thomson's first paper on thermodynamics. Thus any claim to priority was taken from Clausius and laid at Thomson's feet.[92]

After the publication of his disgregation and entropy papers, Clausius's next task was to explore disgregation and entropy on a molecular level—the kind of motion that actually was heat. This led to a series of papers on the relationship between Hamilton's principle of Least Action and the second law. In dealing with disgregation on the molecular level, Clausius developed the virial theorem. Both were important to Maxwell, and we will return to them later. However, when it was published, Clausius's strictly mechanical model of molecular motions ran counter to Maxwell's own work in kinetic theory and Maxwell's own understanding of the second law.

Tait's account was clearly driven by chauvinism and clouded by friendship.[93] By the time Tait wrote the initial article conservation of energy was accepted as a principle by which nature operated and the first law of thermodynamics was merely a special application of this general principle. The historical role of the mechanical theory of heat in the emergence and understanding of this principle was lost. Helmholtz became the originator of the principle, and Rankine, Clausius, and Thomson became users of it.[94] The real problems of his physics began with his statement of the second law. Tait was committed to Thomson's interpretation and the notion of the dissipation of energy and its cosmic implications. He thus belittled the content of Clausius's 1851 and 1854 papers. He underestimated all of Clausius's efforts, including his work in kinetic theory and on the virial theorem.

Maxwell was drawn into this mesh of history and physics when he agreed to proofread Tait's manuscript. He then wrote his own text on heat in a domain whose fundamental ideas were not clearly differentiated and would only become so in the decade in which the various editions of the text appeared.[95] Although Maxwell accepted some of Tait's physics, his approach to heat theory comes from the context of his own work and interests—the theory of matter. Tait analyzed a Carnot cycle for an engineering class; Maxwell used it to explore what the cycle revealed about the nature of matter. And this is the thread through the text itself. It pulled together and gave the reader a consistent point of view from which to consider the diverse and confusing sets of phenomena that a theory of heat explained.

The text is grounded in experience—the known properties of matter—and the Carnot cycle was discussed in this context. Maxwell kept many of the diverse points together by simply following the mechanical effects produced by heat, isolating each of the parameters of change in turn, then discussing what they disclosed about the nature of matter.

Maxwell began by defining both thermal and mechanical concepts, explaining their use, and then putting the two together by examining the mechanical effects of heat. He investigated this by looking at $P-V$ diagrams as a means of measuring these effects, and this is followed by a discussion of work and heat as forms of energy. Description of the indicator diagram led Maxwell into Thomas Andrews's experiments on carbon dioxide and the relationship between the different states of matter. He considered Andrews's experiments as extremely important.[96] Following Andrews's experiments Maxwell discussed isothermals and adiabats. Only then did he embark on a description of the Carnot cycle and its efficiency. This is followed by a discussion of available and unavailable energy in heat engines and two statements (Thomson's and Clausius's) of the second law.[97] Maxwell, however, analyzed what happened to the working substance in a heat engine with the concept of available energy, then equated entropy with unavailable energy.[98] Included in the properties of matter are evaporation, condensation, radiation, and conduction. In the last section of the book using his kinetic theory, Maxwell gave a molecular account of the thermal properties of matter.

Perhaps nothing illustrates better Maxwell's difficulties in understanding heat theory and his approach to it than his correspondence with James Thomson (documents III-19, 20, 21). As Maxwell states in his second letter he used the correspondence as a opportunity to clarify his own ideas on Andrews's experiments and their significance. It is quite clear that James Thomson aided Maxwell crucially in his initial understanding in this regard, and in this case, the act of writing was necessary for Maxwell to sort out his own ideas. The correspondence is the richest source we have for following the initial problems Maxwell had within a new field of physics and to document his critical attitude and growing ability to distill essential physical ideas from mere mathematical expressions. Independently, he had begun to perceive thermodynamics in ways that were similar to those of Josiah Willard Gibbs, that is, thinking in terms of volume, temperature, and intrinsic energy, rather than in the engineering terms of heat and work done. It becomes clear why Maxwell, almost alone, could appreciate Gibbs's papers and why he proceeded to publicize their worth.

Following the first, the next two editions of his text appeared without many changes. Meanwhile, Maxwell appears to have read Clausius's papers on Hamilton's Principle and continued to read his other papers as well. He also seems uncertain about his own understanding of the second law even though he was unimpressed with Clausius's efforts to do the same. The crucial influence on Maxwell in changing his understanding of entropy, and in seeing the significance of Clausius's work, was Josiah Willard Gibbs.

Gibbs sent Maxwell reprints of his first two papers on thermodynamics in which he discussed the graphical representation of the thermodynamic properties of homogeneous substances, and he used Andrews's experiments on carbon dioxide as a illustration of his method.[99] Gibbs reversed the usual discussion of thermodynamics. The second law was the starting point for his thermodynamics. He began by defining Clausius's entropy then treated it as a property of matter. For the first time entropy was given new thermodynamic meaning. Gibbs then turned to thermodynamic diagrams. Usually pressure and volume were the variables. Gibbs proposed generalizing this graphical method by considering the energy and entropy of a substance in the same way pressure and volume had been used previously.[100]

The diagram he found most convenient for displaying thermodynamic properties was the volume-energy-entropy one. Each thermodynamic state of the substance was associated with a point in this space. The points could be connected together to form a surface of the equilibrium states for the substance. Gibbs called this the "thermodynamic surface."

Gibbs, his graphical method, and the new point of view of thermodynamics that it offered him brought the science alive for Maxwell. Its possibilities seemed to open up anew. He immediately set about trying to convince Tait and everyone else of its importance. He even constructed a series of thermodynamic surfaces for some very hypothetical substances. He corrected the fourth edition of his text. Entropy became a property of matter, related to, but not identical with, unavailable energy. He also included a long discussion of Gibbs's graphical methods in the text.[101]

The geometry of the three-dimensional entropy-energy-volume surface gave Maxwell a deeper understanding of the thermodynamic properties of homogeneous and even of heterogeneous substances. The geometrical properties of the surfaces directly represented important thermodynamic characteristics and the thermal changes available to the substance. It is the geometry that led Maxwell into the discussion of the physics, rather than the physics leading to a sense of the surfaces that may represent those properties.

A measure of the importance Maxwell attached to this geometrical exploration of physics is the time and skill he expended in constructing plaster models of thermodynamical surfaces for hypothetical substances. He drew on these surfaces geometrically derived features that were also physically significant.[102] He then distributed them to everyone he could persuade to accept one, including Thomas Andrews, James Thomson, and Peter Guthrie Tait. There are signs that at least Andrews found following Maxwell's geometrical discussion hard going. Clearly geometry allowed Maxwell to penetrate the physics of thermodynamics in new ways.

Maxwell's insight into the significance of Gibbs's work lead him to anticipate some of Gibbs's later work on heterogeneous substances. The geometry of the thermodynamic surface confirmed the rightness of his physical argument.[103] Maxwell's own researches into the thermodynamics of heterogeneous substance lead him to develop a concept, "reaction," analogous to Gibbs's concept of the potential. Gibbs and Maxwell derived the same conditions of equilibrium for a mixture and for the effects of chemical combination on equilibrium and finally for the effects of a mixture of physical states on the conditions of equilibrium. Here Maxwell's thermodynamic researches ended. He had begun, but Gibbs explored in far greater depth, all the possible variations and conditions and the implications for chemistry of those thermodynamic conditions. The ideas Maxwell touched upon represent only a small part of Gibbs's theory.[104]

The goal of Gibbs's work in thermodynamics was a general theory of thermodynamic equilibrium chemically as well as physically. Maxwell's vision seemed limited to understanding Andrews's experiments as physically significant rather than chemically important. All the examples Maxwell used in both his letter to Stokes and James Thomson and in his notes revolve around the generalized experiment of a gas-liquid mixture. He was concerned with connecting together the three physical states of matter, which allowed him to further explore intermolecular forces.

All of Maxwell's work on thermodynamics were superseded by Gibbs's paper on the equilibrium of heterogeneous substances. In his published paper on the problem Maxwell called his "reaction" by Gibbs's name-potential and publicized Gibbs's work. And he continued to bring Gibbs's work to the attention of everyone he thought ought to listen.[105] This included members of the Chemical Society, whom he addressed in his paper on the molecular structure of matter, Peter Guthrie Tait, on more than one occasion, and individual chemists such as M. M. Pattison Muir.[106]

Finally Maxwell taught a course on thermodynamics at Cambridge at least once, in the October term of 1878, when he was already quite ill. Ambrose Fleming attended those lectures and left his notes on them to the Cavendish Laboratory in 1931. The substance and level of these notes was that of the *Theory of Heat* but with none of the detailed discussion of entropy or of Gibbs's thermodynamic surface. At least Fleming did not record such discussions. Perhaps Maxwell had reached the limit of his strength as he grew progressively more ill or perhaps he had run into the familiar limitation of trying to lecture on the basics of a new and (for students) difficult subject and of having no time left to discuss its deeper physical significance.[107]

The Virial Theorem

Maxwell's work in thermodynamics expanded in two directions. Initially he investigated how thermodynamics impacted the theory of matter. Second, Maxwell developed his own molecular interpretation of the second law. In both areas of research he avoided constructing molecular models dependent on mechanical particularities that could not be tested experimentally. He considered such models speculative, that is, logically self-consistent but untouched by any connection to the measurable quantities.[108] In addition Maxwell wanted a theory of the three states of matter that, again, reflected the findings of continuity at the heart of Andrews's experiments.

It is in this connection that he became interested in Clausius's Virial Theorem and J. D. Van der Waals's thesis. This interest did not necessarily reflect that of Clausius or of Van der Waals. In his mechanical exploration of the nature of heat, Clausius used the theory of gases. His statement of the Virial Theorem was, for him, an important step to a mechanical equivalent of the second law. On the molecular level Clausius looked at the molecular equivalents of the work produced and the heat added during a Carnot cycle.[109]

Simply stated the theorem is that "the mean vis viva of the system is equal to its virial." The virial was defined as $-\sum\overline{(Xx+Yy+Zz)}$, and the summation extends over all pairs of particles in a system of masses m, m', m'' etc., located at x, y, z, x', y', z', and x'', y'', z'' on which forces X, Y, Z, X', Y', Z', X'', Y'', Z'' act. In his proof of the theorem Clausius wrote the component of the force $X = md^2x/dt^2$ and rewrote this as

$$\frac{d^2x}{dt^2} = 2\frac{d}{dt}(x\frac{dx}{dt}) = 2\left\{\frac{dx}{dt}\right\}^2 + 2x\,\frac{d^2x}{dt^2},$$

multiplied through by $m/4$ and substituted the above for X and arrived at

$$\frac{m}{2}\left\{\frac{dx}{dt}\right\}^2 = -\frac{xX}{2} + \frac{m}{4}\,\frac{d^2}{dt^2}(x^2).$$

Taking the time average of this expression for periodic motion, Clausius reduced this expression further by assuming that $\frac{1}{t}\int_0^t (dx/dt)^2$ and $\frac{1}{t}\int_0^t xX\,dt$ represented the mean values of $(dx/dt)^2$ and xX, respectively. He had to assume that for nonperiodic motions the terms cancelled as in the case for periodic motion. At least the terms "can only fluctuate within certain limits; and the divisor t, by which the time is affected, must accordingly, cause the term to become vanishingly small with very great values of t." [110] The same is true for the other components of the forces and *vis viva.* For one particle the theorem becomes

$$\frac{m}{2}\overline{v^2} = -\frac{1}{2}\overline{(xX + yY + zZ)}.$$

At this point Clausius introduced the example of point masses interacting through attractions and repulsions which were functions $\phi(r)$ of the distances between the mass points. He then constructed

$$Xx + X'x' = -\phi(x)\,\frac{(x'-x)^2}{r}$$

for a pair of point masses. When generalized to three coordinates the virial for the pair becomes

$$-\frac{1}{2}(Xx + Yy + Zz + X'x' + Y'y' + Z'z') = \frac{1}{2}\,r\,\phi(r).$$

Summing over all pairs of point masses the virial became $\frac{1}{2}\sum \overline{r}\,\phi(r)$, which Clausius named the ergal of the system. Clausius then distinguished the internal and external virial, and took the case in which the external force on the gas was its pressure. In this case the external virial became $\frac{3}{2}PV$. The internal forces and virial were denoted by $\phi(r)$ and $r\,\phi(r)$. Clausius assumed that the virial did not change much over the "course of the individual molecules." He then wrote the mean *vis viva* of the internal motions of the molecules of a gas as

$$\frac{1}{2}\sum m\overline{v^2} = \frac{1}{2}\sum \overline{r\,\phi(r)} + \frac{3}{2}PV.$$

It was in this form that Maxwell found the theorem so suggestive.

For Clausius this was but one step, although an important one that he clearly announced as such, in finding a mechanical equivalent to the second law.[111] Eventually, taking a particular form of the virial, the ergal, Clausius established connections with his thermodynamical concept of disgregation, changes in the heat in a body, and finally entropy.[112] All this led to the derivation of the second law from Hamilton's Principle of Least Action.[113]

Clausius's attempts to find a mechanical base for the second law were not the first, or the last, in the nineteenth century. Ludwig Boltzmann had begun his own work on thermodynamics in just this fashion before he turned to kinetic theory.[114] He would return to this approach again in the 1880s, following the lead of Helmholtz. To explore the second law Helmholtz's Monocyclic systems were more subtle and sophisticated than the mechanical systems developed in the 1870s.[115] But all mechanical systems obeyed conservation laws and in Maxwell's opinion were flawed as models of the microscopic foundation for thermodynamics.

While Maxwell would not follow Clausius in these later extensions of the virial, the virial theorem did open up the possibility of investigating the nature of the forces between molecules. Maxwell related the virial theorem to known deviations from the gas laws and pointed out that these deviations could come only from the motions of the gas molecules. He then extended these remarks into a discussion of Andrews's experiments on carbon dioxide, the continuity of the different states of matter, and the recently published dissertation of Johannes Diderik van der Waals.[116] His remarks were brief, but he emphasized the importance of the work even while noting its "mathematical errors."

Johannes Diderik van der Waals (1837–1929) began his examination of the gaseous and liquid states of matter with Clausius's virial theorem, building into a general gas law terms based in the consideration of the forces between molecules.[117] The resulting gas law is

$$\left(P + \frac{a}{V^2}\right)(V - b) = R(1 + \alpha T),$$

where P is the pressure applied externally and a/V^2 is a pressure arising from the the forces between the molecules where V is the volume of the gas. The factor $V - b$ represents the actual volume the gas molecules have to move around inside the externally defined volume; V. The factor b was deduced from considering an effective size of a molecule, taking into account its finite size (which can be used even if the molecules are centers of force) and the mean free path. Van der Waals even took into account both oblique and direct impacts and estimated that b was four times the size of the molecules themselves. R is the gas constant and $1 + \alpha T$ the temperature.

Maxwell's published critique of van der Waals (document IV-3) was not simply based on intuitive generalizations but on detailed calculations (documents IV-6, 7, 8, 9, 10, 11).[118] Maxwell rejected the idea that externally imposed quantities, the measurables, pressure and volume were "subject to correction." They were givens and also were treated as such in Clausius's virial theorem which was the supposed starting point for van der Waals's work. Van der Waals's derivation was not correct, and Maxwell offered his own. Maxwell's corrections were in terms of purely molecular quantities. However, van der Waals's thesis determined the pattern of his discussion. Each of the corrections is taken up in the same order, beginning with the same concerns as van der Waals but leading to quite different conclusions.

Pressure and volume remain untouched. His corrections were to the changes in the molecules' motions by the forces acting between them. His revisions were in the form of a series in terms of ρ/σ the ratio of the density of the gas, ρ, to the density of the molecules, σ.

$$P = \frac{1}{3}\rho \overline{v^2}\left\{1 - 2\,log(1 - 8\frac{\rho}{\sigma})\right\}.$$

Although σ seems in principle unmeasurable, in his notes Maxwell expressed it in terms of the size and mass of the molecules themselves, i.e., it was related directly to the molecular characteristics that form his long-term interest and quantities that could be estimated using kinetic theory. Even Maxwell's remark in his review that investigating the attractive forces was very difficult because of the interactions of three or more molecules was based on his attempt to work out such interactions.

The importance that Maxwell attached to van der Waals's work, because it permitted feeling "our way into the minute structure of bodies," is reflected in his referee report on a paper by Thomas Andrews (document IV-4).[119] Andrews's paper was an extension of the research he had reported in the 1860s. In this phase of his research Andrews concentrated on gases only. Maxwell's review is straightforward enough, but he cannot help detailing how Andrews's corrected expression for the gas law was consistent with kinetic theory and the virial theorem and comparing it with other generalized gas laws. There is no indication in these remarks what he expected Stokes, as secretary of the Royal Society, or Andrews, the author, to do with them. They seem to be an exposition of his opinion of the importance of the experiments themselves, but he never recommended any of his remarks be included in the final paper. These speculations end here with these notes but do give us some feel for the kinds of theories about molecular structure Maxwell took seriously.[120]

Maxwell's Demon and Irreversibility

Although Gibbs gave Maxwell a full comprehension of the second law within thermodynamics, Maxwell already had worked out, in principle, his own understanding of the law's implications in molecular terms.[121] In his early response to Tait's request for help in writing a thermodynamics text, Maxwell had stated his own statistical interpretation of the second law (document III-4), where he introduced his "demon" to Tait. This interpretation emphasized irreversibility, a point about heat phenomena that Maxwell insisted on very early in a letter to William Thomson. It was built into Thomson's version—the one that Maxwell accepted initially.[122] He expanded and explored the statistical nature of the second law and its implications for a mechanical worldview both formally and informally without finally resolving any issues.

At first Maxwell saw the second law as only statistically true, but he envisioned molecules and gases as mechanical systems. The statistical nature of the law arose from our inability, unlike his demon, to sort individual molecules. The statistical nature of the second law depended on the nonexistence of such a sharp-witted "being." The function of his demon was merely to sort molecules by velocity and thus defeat irreversibility. Irreversibility was guaranteed through the mixing of molecules of different velocities. The demon undid this mixing.[123]

Boltzmann responded to the dilemma of an irreversible system composed of reversible, mechanical systems by exploring the randomness of molecular motions. His deeper scrutiny of the issues were in response to physicists and mathematicians such as Josef Loschmidt, Györö Zermelo, and Henri Poincaré.[124] Maxwell's demon, however, resonated with other concerns of his contemporaries. He had introduced his demon at a time when the cosmological and terrestrial consequences of the second law of thermodynamics were being widely discussed. Thomson had begun this line of argument in 1852, declaring that "within a finite period of time past, the earth must have been, and within a finite period of time to come the earth must again be, unfit for the habitation of man as at present constituted, unless operations have been, or are to be performed, which are impossible under the laws to which the known operations going on at present in the material world are subject."[125]

The consequences of Thomson's dissipation principle were elaborated further by Hermann von Helmholtz in a lecture two years later, in which he described the final state of the universe: All energy will eventually be transformed into heat at uniform temperature, and all natural processes must cease; "the universe from that time forward would be condemned to a state of eternal rest."[126]

Thomson reiterated this opinion at the British Association meeting in 1854, "the end of this world as a habitation for man, or of any living creature or plant at present existing in it, is *mechanically inevitable.*"[127] Even at this early date others did not share Thomson's gloomy predictions. Rankine thought it "conceivable that, at some indefinitely distant period, an opposite condition of the world may take place, in which the energy which is now being diffused may be reconcentrated into foci, and stores of chemical power again produced from the inert compounds which are now being continually formed." He imagined that the interstellar ether which transmits light and radiant heat has a finite boundary beyond which there is empty space; the heat radiated from stars and other bodies will be reflected at this boundary and reconcentrated into foci. Dead stars that arrive at these foci would be vaporized, leading to a renewal of activity and life.[128] Arguing that the properties of radiation are compatible with the second law, Clausius quickly pointed out that Rankine's scheme to reconcentrate energy violated that same law.[129] Although his own understanding of the second law was to evolve, Clausius did not change this aspect of his thought. Clausius insisted that the second law explicitly refuted the common idea that "everything is a circuit" so that "the world may go on in the same way for ever." However, as entropy approached its maximum "the occasions of further changes diminish; and supposing this condition to be at last completely attained, no further change could evermore take place, and the universe would be in a state of unchanging death."[130]

By this time the idea of a universal heat death had already penetrated popular philosophy in England. Herbert Spencer wrote that the sun cannot shine forever but must eventually dissipate its energy by radiation into space, a state must eventually be reached in which all molecular motion is equilibrated, and this equilibration "must end in complete rest." "If Man and Society are similarly dependent on this supply of force that is gradually coming to an end; are we not manifestly progressing towards omnipresent death?"[131] This idea was too attractive to drop even after Maxwell pointed out that Spencer's views on molecular motion were not consistent with the kinetic theory of gases.[132] Despite Spencer's attempt to work this idea into a general theory of evolution, scientists were quick to see a contradiction between dissipation and the evolution of organic structures.[133] Indeed scientists' reactions to the idea of a heat death were mixed. Acceptance or rejection of the idea followed no disciplinary or national boundaries.

Clausius's heat death theory was enthusiastically endorsed by the German physiologist Adolf Fick. Yet Robert Mayer rejected the view that if the sun derives its heat from mechanical energy, "the entire machine of creation must eventually come to a standstill."[134] Balfour Stewart agreed with Thomson that the sun, like all other stars, must eventually die, forcing us to "contemplate the death of the universe." Eardley Maitland likewise expected all matter to collect eventually into "a cold and naked ball."[135] But two Americans, H. F. Walling and Pliny Earle Chase, were more optimistic: they looked to the reconcentration of energy or rotational breakup of dead stars to go on forever creating new worlds, thus preserving the idea of cosmic evolution against the alleged tendency to degradation.[136]

Although the demon was born in a letter in response to Tait's request for "some hints" in connection with his forthcoming *Sketch of Thermodynamics*, he is not mentioned in that book.[137] Nor did Tait take the opportunity to discuss Maxwell's demon when he discussed Thomson's form of the second law of thermodynamics in his address to the Mathematics and Physics Section of the British Association in 1871. He was more interested in using the law for deductions about the beginning of the universe rather than its end or about any possible ways of evading that end. Tait's first published reference to Maxwell's demon is in 1876 where he concluded that since we are not demons in Maxwell's sense, the second law is valid in practice, and the ultimate fate of the universe is that all bodies will be at the same temperature.[138]

Before Maxwell introduced his demon to the public in his *Theory of Heat* (1871), a similar idea had been published by the Austrian scientist Joseph Loschmidt.[139] As Edward Daub points out, Loschmidt did not (as Boltzmann later incorrectly recalled) propose a tiny intelligent being to do the job of violating the second law, but he did imagine that a small part of a wall separating two compartments could be opened and closed in such a way as to allow faster molecules to move from one side to the other. As Maxwell himself noted (documents III-6 and 15), the demon is really no more than a valve.[140] When *Theory of Heat* first appeared, reviewers praised it but did not mention Maxwell's introduction of the demon.[141] (The book went through ten editions in thirty years, and it probably was read by a substantial number of teachers and students in Britain.)

The demon became unavoidable after Thomson's 1874 paper where he introduced the "reversibility paradox." Using his demon Maxwell had already raised this issue in a letter to John William Strutt, document III-15. How can a system of particles obeying Newton's laws of motion, which are time reversible, behave irreversibly? Thomson considered a bar of metal, perfectly insulated from its surroundings, with one half initially at one uniform temperature, the other half at another uniform temperature. Diffusion of heat would tend to equalize the temperatures in the two halves. Yet, "this process of diffusion could be perfectly prevented by an army of Maxwell's intelligent demons," stationed at the surface, "separating the hot from the cold part of the bar." He replaced the metal (in which heat conduction occurs primarily by radiation of energy from one molecule to another) by a gas (in which heat is conducted mostly by diffusion of molecules). Each demon was to be armed with a "molecular cricket bat" with which he must strike each molecule coming into his territory so as to send it back with the same energy or else (according to orders) allow it to pass through. In this way the original temperature difference can be maintained while allowing the molecules themselves to diffuse. But "if no selective influence, such as that of the ideal 'demon,' guides individual molecules, the average result of their free motions and collisions must be to equalize the distribution of energy among them." The probability of a spontaneous fluctuation that upsets the temperature equilibrium is extremely small. This indeed could be achieved by reversing the motion of each particle, but it would be followed immediately by equalization. So, for Thomson, Maxwell's demon belongs to the ideal world of reversible "abstract dynamics"; the real world of "physical dynamics," lacking demons, is almost always irreversible.[142]

Loschmidt used the reversibility paradox to draw the opposite conclusion. He rejected the opinion of Thomson and Clausius that the universe must come to an end and tried to "destroy the terroristic nimbus of the second law, which has made it appear to be an annihilating principle for all living beings of the universe; and at the same time open up the comforting prospect that mankind is not dependent on mineral coal or the sun for transforming heat into work, but rather may have available forever an inexhaustible supply of transformable heat." In particular he believed that the solidifying sun would absorb radiant energy and reconcentrate it, reviving the solar system.[143] Although Loschmidt's interpretation of the second law was not generally accepted, his critique did force Boltzmann to revise his ideas and formulate a more explicitly statistical definition of entropy.[144]

In 1878 Maxwell pointed out that the dissipation of energy and the distinction between work and heat depend on our inability to trace and manipulate molecular motions.[145] In a correspondence in *Nature* stimulated by S. Tolver Preston's suggestion that the second law of thermodynamics could be violated by allowing one gas to diffuse into another, A. S. Herschel and Preston discussed possible physical mechanisms equivalent to what Herschel called Maxwell's "sprite."[146] The astronomer Friedrich Zöllner linked "Thomson's demons" to Plato's shadows and LeSage's gravific corpuscles. Tait, in response, complained that Zöllner seemed to think Maxwell, Thomson, and Tait really believed in the existence of Maxwell's demons.[147] Clausius also failed to see the point of Maxwell's demon as invoked by Tait in an argument on the second law that Tait considered "absolutely fatal to Clausius' reasoning." Clausius replied, "My law is concerned, not with what heat can do with the help of demons, but rather with what it can do by itself."[148]

For the first generations of physicists confronted with irreversibility in the context of a mechanical worldview Maxwell's demon made the incompatibility between these two facets of their physics superficial.[149] For William Thomson, the demon ultimately mediated the finite abilities of man to know a world that in principle was not completely knowable. Maxwell's demon differed from other animals only in its very small size. As finite beings we cannot encompass the effectively infinite number of mechanical molecules in a system, and we see irreversibility. In their mediating role Maxwell's demons were required to serve any number of functions beyond the mere sorting demanded by Maxwell.[150] For Maxwell the source of irreversibility lay in the inherent nature of mechanical systems themselves despite the inference others might draw from his demon example.[151]

After Maxwell's death his demon took on a life of its own. It continued to play a pedagogical role in discussions of the statistical nature of thermodynamics, and it began to acquire new powers.[152] It might be responsible for the escape of fast molecules from the earth's atmosphere or for the action of bacilli or for radioactivity. It might acquire the ability to see backward in time.[153] For Henry Adams and other observers of science, it symbolized the power of science in the modern world, pushing its neighbors around like Germany, and becoming a presidential candidate.[154] For modern science, the demon has become the "minimal communicator," painfully aware of the quantized cost of obtaining the information ("negentropy") needed to do his job, never quite sure whether that cost exceeds the thermodynamic benefits gained by producing a more orderly world.[155]

In view of the great popularity of his demon, it may seem surprising that Maxwell did not develop a quantitative treatment of irreversibility based on this idea. Unlike Boltzmann, he did not try to formally demonstrate his statistical interpretation of the second law, rather he explored its general physical and even theological implications.

Maxwell was committed to a mechanical view of nature but his understanding of what mechanics consisted of and how much of a particle's motion could be predicted were changing rapidly. These were not intuitive ideas to keep scientific and theological worldviews together; they came from a commitment to an understanding of the power and generality of energy conservation as a principle in physics and Hamiltonian mechanics. In general terms Maxwell expressed these ideas to Francis Galton. In sending his subscription to the Philosophical Club for 1879 to Galton, Maxwell asked in a postscript,

> Do you take any interest in Fiat, Fate, Free Will etc. If as Boussinesq (of hydrodynamic reputation) "Conciliation du léxistence de la vie at de la liberté morale" (Paris, 1878) does the whole business by the theory of the singular solutions of the differential equations of motion. Two other Frenchmen have been working on the same or a similar track. Cournot (now dead) and de St. Vincent (of elastic reputation Torsion of Prisms etc.) Another, also

> in the engineering line of research Philippe Breton seems to me to be somewhat like minded with these. There are certain cases in which a material system, when it comes to a phase in which the particular path which it is describing coincides with the envelope of all such paths may either continue in the particular path or take to the envelope (which in these cases is also a possible path), and which course it takes is not determined by the forces of the system (which are the same for both cases) but when the bifurcation of path occurs, the system, ipso facto, involves some determining principle which is extra physical (but not extra material) to determine which of the two paths it is to follow.
>
> When it is on the enveloping path it may at any instant, at its own sweet will, without exerting any force or spending any energy, go off along that one of the particular paths which happens to coincide with the actual condition of the system at that instant.
>
> In most of the former methods, Dr Balfour Stewart etc. there was a certain small but finite amount of travail decrachany or trigger-work for the Will to do. Boussinesq has managed to reduce this to mathematical zero, but at the expense of having to restrict certain arbitrary constants of the motion to mathematically definite values, and this I think will be found in the long run, very expensive. But I think Boussinesq's method is a very powerful one against metaphysical arguments about cause and effect and much better than the insinuation that there is something loose about the laws of nature, not of sensible magnitude but enough to bring her around in time.[156]

In his reply Galton clearly followed Maxwell's argument regarding free will, "after all, does the question not coincide in the final principles with that of unstable equilibrium?" Galton then went into his own research, "experiments on the workings of my own mind," advertised an article of his own about to appear, and finished with the remark that he was looking forward to hearing William Thomson on Maxwell's demon at the Royal Institution the following day.[157]

Maxwell had considered the significance of such a mechanical universe for free will earlier in a paper read to the Eranus Club members at Cambridge, made up of former members of the Apostles, who were intellectuals but not physicists. There is much in this essay besides Maxwell's discussion of free will and statistical versus dynamical knowledge. In the part of the essay of interest here Maxwell focused on the limitations of our knowledge of the motions of the molecule.[158] By the time this essay was written, Maxwell already had developed doubts about specific molecular models.[159]

Their behavior was not only unknowable on an individual level but largely unpredictable as well. To illustrate his point Maxwell turned to the stability and instability of systems where a small alteration in one variable brought about rapid changes in many others. In all the cases he examined,

> the system has a quantity of potential energy, which is capable of being transformed into motion, but which cannot begin to be so transformed till the system has reached a certain configuration, to attain which requires an expenditure of work, which in certain cases may be infinitesimally small, and in general bears no definite proportion to the energy developed in consequence thereof.[160]

Maxwell's argument rested on illustration and analogy—where evidence was ambiguous he was ambivalent. Unlike many of his colleagues, Maxwell did not commit himself to developing any particular molecular model. As in the case of his mature electromagnetism, he focused on what we know of physical systems rather than speculating about their internal characteristics. He was, therefore, not addressing the same questions in physics and chemistry as his contemporaries.

His concept of a dynamical system was taken from Hamilton's mechanics whose implications for physical explanation he explored in some detail.[161] Unlike William Thomson or Peter Guthrie Tait who included Hamilton's mechanics in their text, then returned happily to speculating about molecular structure, Maxwell followed the implications of Hamilton's work for mechanical explanations. These implications appear in several places and culminate in his paper on statistical mechanics of 1879.

In addition to his understanding of Hamiltonian mechanics, Maxwell's interest in what we can know of systems, in light of thermodynamics and irreversibility, led him to conclusions that were the opposite of determinism. Statistical knowledge of matter is all that is available to us. This involved unexpected consequences from an "exact science." One such consequence was the asymmetry of functions of time. In some cases we can prophesy about the behavior of systems, but its history is unknown. Unlike astronomy,

> in the theory of the diffusion of matter, heat, or motion, the prophetical problem is always capable of solution; but the historical one, except in singular cases, is insoluble. There may be other cases in which the past, but not the future, may be deducible from the present. Perhaps the process by which we remember past events, by submitting our memory to analysis, may be a case of this kind.[162]

The consequences of irreversibility were now part of his understanding of the nature of mechanics itself.

Whether Maxwell had read Boltzmann's paper on the H-theorem of 1872 by the time he delivered this essay, this paper had an impact on him that we have to explore to understand the development of his statistical mechanics.[163] Although Maxwell understood the second law as statistical, he had not made of this intuitive insight a systematic, defensible argument. This was done by Boltzmann. In examining the gas as a mechanical system Boltzmann concluded that the behavior of its molecules was intrinsically statistical.

Boltzmann based his derivation on the idea that if a molecule passed through all possible states of motion, the fraction of the time it spent in a certain state was the probability that the molecule would be found in that state. He had to assume that the behavior of this molecule was typical of all molecules of the gas. He then examined the specifics that would govern the behavior of the gas in thermal equilibrium. Under these conditions the states of the molecules changed from internal motions and collisions between the molecules. However, the velocity distribution function remained the same. Boltzmann then demonstrated that under these conditions the necessary distribution function was Maxwell's provided that the collisions between the molecules were elastic and the number of molecules between the chosen limits before and after any collisions were independent of one another. Boltzmann then constructed a mathematical function that depended on the equilibrium velocity distribution function and behaved as the entropy of the gas. By concentrating on the mathematical foundations of gas theory, Boltzmann fused it to thermodynamics.

While Boltzmann concentrated on the second law and the intrinsically statistical nature of molecular behavior, Maxwell drew other conclusions.[164] By accepting this inexorable linkage Boltzmann had to answer the critiques of Loschmidt, and the 1880s even retreated from his statistical viewpoint. He set out to explore the extent to which purely mechanical derivations of the second law were possible and their limitations.[165] Like the earlier attempts by Clausius and Szily and the later, more sophisticated analysis of Hermann von Helmholtz, Boltzmann's derivation depended on Hamilton's Principle of Least Action. Maxwell had already deduced that this kind of derivation of the second law based upon mechanics, even those as general as Hamilton's principle, could not account for irreversibility. This is reflected in the structure of his paper in 1878 where he spelled out the mechanical and statistical specification necessary for the problem. Maxwell developed his alternative approach based on his own understanding of mechanics that transcended both determinism and specific molecular models.

Maxwell's ideas on dynamics developed through his papers on electromagnetism and the writing of his *Treatise* and by Thomson and Tait's text on mechanics.[166] Their text was firmly based in the concept of the conservation of energy, with dynamics developed independently of statics. Indeed statics became merely a special case of dynamics. They gave a great deal of space to Hamilton's formulation of mechanics.

In the second edition of his *Treatise* and his text on mechanics, *Matter and Motion*, Maxwell defined a mechanical system as one that was subject to the law of conservation of energy. Although mechanical models were important in the construction of his electromagnetic theory and thinking through the implications of his electromagnetic theory of light, a decade later Maxwell's ideas on dynamical systems had broadened, and this had import for his electromagnetism. The electromagnetic field was mechanical because it carried energy and was subject to the law of conservation of energy. In the 1873 edition of the *Treatise* Maxwell added a short chapter on the Hamiltonian dynamics of connected systems. This came from his assumption that "the phenomena of the electric current are those of a moving system, the motion being communicated from one part of the system to another by forces, the nature and laws of which we do not yet even attempt to define."[167] The Langrangian formulation of mechanics, generalized by Hamilton, allowed Maxwell to ignore the mechanisms and forces that connected the parts of the system. It was perfect for the exploration of the physics of systems outside of mechanics but still subject to the conservation of energy.

Maxwell pursued these methods for gases which were connected systems, like his electromagnetic systems where the forces were unknown. In the case of "molecular science" one could proceed by the method of hypothesis about the connections between the parts of the system. The success of this approach depended on the hypothesis. Even if the results were in accord with the phenomena, and there are no other hypotheses, "we must still admit the possibility of our hypothesis being a wrong one." We are more successful with general hypotheses that simply assume that the phenomena depend on the configuration and motion of a material system.[168]

In Maxwell's 1879 paper on statistical mechanics, as in his first paper on kinetic theory, the statistical and mechanical specifications of the system are linked together. Mechanically he specified the system (molecule or gas) at a particular point in phase space. The statistical specification differs only by placing the system within a small, well-defined volume of phase space around a particular point. It is here that he linked the two specifications explicitly.

Hamiltonian dynamics became a means of developing a theory of gases independent of any specific model of molecules or of their interactions without abandoning the dynamical imagery offered by the law of conservation of energy. The system was separated from the problem of determinism by simply assuming that it moved through all the phases consistent with the principle of the conservation of energy. There was no necessity for the radical abandonment of order for randomness.[169] Maxwell considered that he was only avoiding specifying something that could not be specified anyway "when the system is made up of parts which are so small that we cannot observe them, and whose motions are so rapid and so variable that even if we could observe them we could not describe them."[170]

Maxwell, however, was not the first to use Hamilton's mechanics in gas theory. Watson had used generalized coordinates and momenta in his text of 1877, but Watson used them simply to specify the motions of molecules and to take over into gas theory a theorem proved, in generalized coordinates, by Thomson and Tait. For Maxwell the system is the gas, not the molecule.

Maxwell accepted Boltzmann's solution to the equilibrium distribution of kinetic energy among a finite number of material points. He clearly thought his point of view was sufficiently different, and an examination of the assumptions on which it was based was important enough, to look at the problem again especially "when the assumptions relate to the degree of irregularity to be expected in the motion of a system whose motion is not completely known."[171] Both Watson and Boltzmann assumed that the particles of gases interacted with one another at small distances and under external forces such that the time for such interactions was very small compared to the time between encounters. Such an assumption did not seem to enter actually into their analyses. However, for Maxwell such an assumption was obvious for liquids and solids. Clearly he was implying that gas theory methods could be extended to the other states of matter based on his own demonstration of Boltzmann's theorem that depended on the gas passing through every phase consistent with the conservation of energy.

The conceptual simplicity of this approach sets the results of this, one of the last of Maxwell's papers, outside all of the debates on molecules, recurrence theorems, etc. Perhaps the most devastating conclusion for a physics based in mechanics is that reached on the specific heats ratio from the equipartition theorem. In any dynamical system, defined by the conservation of energy, the specific heat of a body "cannot be less than the product of the degrees of freedom into an absolute constant, unless the potential energy diminishes as the temperature rises."[172] Maxwell was already profoundly disturbed by the implications of the equipartition theorem and knew that no molecular model was adequate to avoid them. The Hamiltonian formulation set the place of the theorem within the logical structure of mechanics. It was an inevitability given that the gas was a definable, mechanical system.

Maxwell used the device of representing the state of motion of a large number of particles, n, by the location of a single point in a "phase space" of $2n$ dimensions, the coordinates and momenta of which were the positions and momenta of the particles. Boltzmann had applied a similar method in configuration space, but the Hamiltonian formalism was simpler, more elegant, and general. Maxwell then used the same postulate as Boltzmann that in time the system would pass through every phase consistent with the conservation of energy. This postulate breaks down for certain conditions, some of which Maxwell discussed, but he argued it should be true even for a system of particles with discontinuous changes in direction of motion and velocity from collisions.[173] If this postulate is accepted, then the time-average of the behavior of a single system can be replaced by the average over a large number of systems distributed throughout the available phase space. This was Maxwell's preference.[174] Following and extending Boltzmann, Maxwell introduced averages over one particular ensemble: a collection of similar systems with identical energies but different initial conditions (Gibbs's "microcanonical ensemble"). Maxwell demonstrated that his original velocity distribution can be derived from such an ensemble by taking the limit as the number of degrees of freedom goes to infinity, keeping the average energy per degree of freedom constant.

As his argument unfolded, Maxwell was at pains to refer back to the concerns of his 1873 paper on Boltzmann's theorem, i.e., the compatibility of his gas theory with thermodynamics. He rounded out his paper with an eminently Maxwellian characteristic: the suggestion for an experiment that would indeed demonstrate that his gas theory was compatible with the behavior of real gases. Based on considering the rotational degrees of freedom in a gas, Maxwell suggested a method to show that the densities of the constituent components in a rotating mixture would be the same as if the component was present alone. He even gave estimates on speeds of rotation, gas mixtures, and time for an experimental run. His correspondence reveals his plans to carry out such an experiment at Cambridge shortly before his death.[175]

In the last year of his life, therefore, Maxwell left his colleagues with a paradox. Gases, considered as mechanical systems and dealt with statistically, led to results consistent with the laws of thermodynamics; but they also led to results not confirmed by experiment. He also left his contemporaries with a puzzle. Later discussions of Maxwell's papers criticized both his mathematics and the gas model implied by the theory.[176] Those working on gas theory in the next two decades misunderstood Maxwell's method. In his previous papers they understood his dynamical system to be the molecule, not the gas. Maxwell seemed to be leaving them the task of inquiring into the internal structure of molecules, and this meant the analysis of collisions using gas theory. But this paper was based on a model of noninteracting systems. Even Boltzmann's long commentary on this paper ignored the significance of the specific heats results and the futility of pursuing mechanical molecular models implied in Maxwell's results. Boltzmann's attention was on the Boltzmann's theorem result.[177]

Maxwell's kinetic theory, molecular models, and the specific heats ratio came under the scrutiny of Tait, Kelvin, Boltzmann, and others through the 1880s and into the 1890s. Together with Boltzmann's and Helmholtz's work on monocyclic systems, the distance between Maxwell's intent and his colleagues' understanding becomes visible. This was complicated by Tait's pugnacious defense of Maxwell's kinetic theory which seemed not to include any serious reading of Maxwell's 1879 paper or most of Boltzmann's work. At least in Britain this debate lead to a more conscious understanding of Boltzmann's work. By 1890, even if Tait did not appreciate his work an active group of theorists did.[178]

Having accepted Boltzmann's specification of the conditions for equilibrium, the attack lead by William Thomson (now Lord Kelvin) on the implications of kinetic theory shifted to the equipartition theorem, specific heats, and hence to molecular models. But all of Kelvin's "test cases" and molecular models were demonstrated to be examples of, not exceptions to, equipartition, reinforcing Maxwell's conclusions. More importantly, this controversy lead to a reexamination of the foundations of kinetic theory itself.

In a report presented at the 1891 meeting of the BAAS Bryan concluded that no mechanical system alone, however specified, could lead to the irreversibility demanded by the second law of thermodynamics.[179] It was agreed that the statistical interpretation of the second law was inevitable, however they might debate the meaning of the H-theorem with Boltzmann. A similar report three years later explicitly dealt with the equipartition theorem. From this meeting two opinions emerged. The first, not to last long, was that gas theory was a purely mathematical theory. The other, more significant, was that neither the ether, nor the electromagnetic oscillations of molecules could be investigated by the methods of gas theory. Theories of molecular structure were being separated from theories of gases.

Two theorists saw that the problems that came with the conclusions of Maxwell's 1879 paper lay not in Maxwell's mathematics or his gas model but in his assumptions. If the "continuity of path" condition, in James Jeans's terminology was correct, Maxwell's conclusions were irrefutable.[180] Rayleigh, in more terse prose, could not fault Maxwell's conclusions, even as he corrected his algebra. Equipartition was correct within Maxwell's assumptions. Kelvin's efforts to avoid the theorem were futile, it was a matter of logic, not choice.[181]

The mathematical problem raised by the continuity of path hypothesis was resolved in the review article of Paul and Tatiana Ehrenfest of 1911. They suggested that the "ergodic hypothesis," which they attributed to Maxwell and Boltzmann, implied that the configuration of the system represented by a point in $6n$-dimensional phase space would have to pass through every point of that space (more precisely, every point of the 6n-1 dimensional space corresponding to the states of a given energy). This behavior might be an impossibility. While superficially plausible, in 1913 Arthur Rosenthal and Michel Plancherel proved it was impossible.[182]

Thus the ergodic hypothesis was replaced by the "quasiergodic" hypothesis. But by this time, it did not seem very relevant to physics on the microscopic level. The problems that had motivated Maxwell to introduce it into gas theory were already being solved through the new quantum theory. In their arguments for the necessity of quantum theory, the Ehrenfests then James Jeans pointed to the breakdown in the equipartition theorem in black body radiation and the behavior of the specific heats in solids. All this brings us back into the heart of Maxwell's concerns, the inner structure of molecules and the inadequacies of mechanics to model their behaviors.

Rarified Gas Dynamics

Maxwell's last major paper on any subject was "On Stresses in Rarified Gases Arising from Inequalities in Temperature,"[183] In part it was a theory of the radiometer. Introduced by William Crookes in 1873 the radiometer, a partially evacuated chamber containing a paddle wheel with vanes blackened on one side and silvered on the other which spun rapidly when radiant heat impinged on it, caused a sensation.[184] It seemed to prove a direct pressure or repulsive force of light or heat and thus had a direct bearing on the problem of the nature of both light and heat.[185]

In the eighteenth century proponents of the corpuscular theory of light searched for its confirmation by demonstrating light had momentum. Several experimenters directed powerful beams of light on delicately suspended bodies with varying results. Interpretation was complicated because some versions of the wave theory also predicted a pressure from light. If the waves were longitudinal, it seemed plausible that there should be a pressure. But, if they were transverse, as proposed by Thomas Young and later Augustin Fresnel at the beginning of the nineteenth century, it was not so obvious that a mechanical force could be exerted in a direction perpendicular to the wave motion. In 1873 Maxwell showed that transverse electromagnetic waves would exert a pressure on any surface they strike, and he calculated the magnitude of this pressure. It was this result for the pressure and energy density of electromagnetic radiation that led Boltzmann to derive the T^4 radiation law in 1884.

A somewhat skeptical observer of the radiometer, the physician W. B. Carpenter, reported the initial reaction of scientists in 1873. When the radiometer "was first exhibited at the soiree of the Royal Society, there was probably not one who was not ready to believe with its inventor that the driving round of its vanes was effected by the direct mechanical energy of that mode of radiant force which we call Light."[186]

Carpenter criticized this "readiness to admit a novelty" as unscientific and compared it with the gullibility of scientists who accepted accounts of supernatural phenomena. This was a dig at Crookes and his involvement in spiritualism. According to Carpenter, "two of the most distinguished among British mathematical physicists" agreed with Crookes that the motion of the vanes was a direct effect of radiant energy. If Maxwell was one of those mathematical physicists referred to, he does not record this opinion in his letter to Tait describing the event (document VI-2). However, in that same year in his *Treatise* Maxwell wrote that concentrated radiation from an electric lamp, "falling on a thin metallic disk, delicately suspended in a vacuum, might perhaps produce an observable mechanical effect."[187] He estimated the force due to ordinary sunlight would be only about one tenth of horizontal magnetic force in Britain.

Maxwell reviewed Crookes's written account of his work for the Royal Society. In his report (document VI-3) Maxwell judged that Crookes "has made a great discovery" and that the observed "repulsions from a heat-emitting body" were "due to radiation" and recommended publication. He noted that while he had suggested a repulsion from radiation in his *Treatise*, "the effects observed by Mr. Crookes seem to indicate forces of much larger values."

In this first paper, Crookes suggested that experiments on the radiometer might help reveal something about the repulsive action of the sun on comets. In the following summer Coggia's comet appeared in the skies of Europe. Its tail was visible for weeks, pointing away from the sun. Astronomers had suggested previously some kind of repulsive action of the sun, perhaps light pressure acted on the volatile matter of the comet and pushed it out the side opposite the sun. Maxwell had discussed comets' tails in earlier correspondence with astronomers but had not adopted the light-pressure explanation.[188] Comet's tails were, at least, a subject of conversation in the Maxwell house during the summer of 1874. According to Garnett the repetition of the word "tail" caused Maxwell's terrier to run around in circles chasing his own.[189]

Osborne Reynolds, professor of engineering at Owens College, Manchester, was one of the first to challenge the light-pressure theory of the radiometer. He thought comets' tails were either an electrical effect or a "negative shadow." In 1873 he published a paper on the steam engine in which he discussed the condensation of a mixture of steam and air on cold surfaces. This may have inspired his suggestion, one year later, that the radiometer effect depends upon the condensation and evaporation of gas molecules at the surface of the vanes. Although this form of his theory was not successful, in a general way it prefigured his later explanations based on gas-surface interactions.

At this time, however, the work of the young physicist Arthur Schuster, who came to Manchester in 1875, was more important. With Reynolds's encouragement, Schuster did what was subsequently regarded as the "crucial experiment" on the radiometer. Maxwell also reviewed his paper for the Royal Society (document VI-7). Schuster observed the rotation of the radiometer case in a direction opposite to that of the vanes.[190] This meant that no momentum was brought into the system, only energy. The actual turning of the vanes must then depend on some internal mechanism.

Several observers had already pointed out that light ought to produce a greater force on the reflecting, silvered side of the vanes than on the absorbing blackened side. But the vanes moved in the direction opposite to that expected. A few scientists continued to explain the radiometer by light pressure. Other theories depended on molecular or electrical actions.[191]

By 1876, the British scientists actively working on the radiometer—Crookes, Schuster, Reynolds, G. Johnstone Stoney, James Dewar, and Tait—agreed that the rotation of the vanes was not a direct effect of the radiation striking them. Rather, it depended on the residual gas in the radiometer and the temperature difference between the two sides of the vanes. Gas molecules striking the hot, blackened side would somehow exert more pressure than those striking the cold, silvered side of the vane. However, they disagreed over the reason for this pressure difference. They recognized that the methods of kinetic theory based upon the Clausius mean-free-path concept were inadequate to describe the behavior of rarefied gases. In the radiometer the mean free path of the molecule might be larger than the distance the molecule traveled between successive collisions with solid surfaces. Temperature equilibrium, usually reached through intermolecular collisions, would not be established. An unequally heated vane might establish a large temperature gradient in the surrounding gas. Without arbitrary assumptions, however, no one was able to calculate accurately the net force on the vanes from this temperature gradient.[192]

This was clearly a problem for an expert in kinetic theory, and Maxwell frequently was asked for his opinion. Realizing that the problem was more difficult than it first seemed, Maxwell was reluctant to commit himself until he considered it carefully. As a reviewer for Stokes at the Royal Society he was at the center of work being done on the radiometer. His reports on the papers of Crookes, Reynolds, and Schuster (documents VI-3, 4, 5, 6, 7 and 15) show his skepticism of existing theories. His reticence is obvious in his success at avoiding any explanation of the radiometer to Queen Victoria when summoned to do so (document VI-8).

Maxwell reported to Cambridge University that the Cavendish Laboratory had acquired four radiometers by May 1877 (none were mentioned in the list of apparatus for the previous year).[193] Since light pressure did not seem to be involved, the flurry of excitement about radiometers was dying down. Maxwell was beginning to define the real problem and set to work seriously.

It was not until just before his death in 1879 that Maxwell completed his long paper on rarefied gases. The main part of the paper was devoted to calculating the force resulting from temperature inequalities in the interior of a gas, deliberately leaving aside surface effects.[194] He found that the stress in the gas was proportional to the second spatial derivative of the temperature. Contrary to earlier theories of George Fitzgerald and G. Johnstone Stoney, a constant temperature gradient would not produce an inequality of pressure. Acting as a heat source a small object in the gas could produce a changing temperature gradient. The resulting stress might be sufficient to account for the motion of the radiometer vanes. If the vane itself is the source of the heat flow, the stress would be found only near the edges of the vane. Maxwell was careful to point out that his theory only gave the stress *normal* to the surface of the vane. Further assumptions were necessary to compute the *tangential* stresses. The usual assumption in kinetic theory was that the physical state of the gas next to the surface is the same as that of the surface, with no tangential stress. This involves two assumptions: that there is no "slipping" of the gas relative to the surface, no velocity discontinuity, and there is no temperature discontinuity.[195]

Maxwell now recognized the defect of his original theory. When a number of solid spheres are at different temperatures in a gas, equilibrium will be reached in which there is a steady flow of heat. As long as only normal stresses are considered there are no forces on the spheres. Nevertheless, Crookes's experiments demonstrated that forces do act "between solid bodies immersed in rarified gases, and this, apparently, as long as inequalities of temperature are maintained."[196] Maxwell attributed the effect to "the phenomena discovered in the case of liquids by Helmholtz and Piotrowski and for gases by Kundt and Warburg, that the fluid in contact with the surface of a solid must slide over it with finite velocity in order to produce a finite tangential stress."[197] Kundt and Warburg found that the velocity of slipping of a gas over the surface from a given tangential stress varies inversely as the pressure. Therefore, the effect should be stronger in rarefied gases. The existence of such currents sweeping along the surfaces of solid bodies immersed in a rarefied gas would destroy the simplicity of Maxwell's kinetic-theory solution of the problem.

In an appendix, added to the paper in May 1879, Maxwell stated that while he had been reluctant to give a quantitative treatment to the slipping effect, one of the referees of the papers encouraged him to do so and had suggested various possible hypotheses about the gas-surface interaction. Maxwell proposed to set off the equations that "express both the fact that the gas may slide over the surface with a finite velocity ... and the fact that this velocity and the corresponding tangential stress are affected by inequalities of temperature at the surface of the solid, which give rise to a force tending to make the gas slide along the surface from colder to hotter places."[198]

He assumed that "of every unit of area [of the surface] a portion f absorbs all the incident molecules, and afterwards allows them to evaporate with velocities corresponding to those in still gas at the temperature of the solid, with a portion $1-f$ perfectly reflects all the molecules incident upon it." As a first approximation, he related the velocity of the gas near a flat surface in the yz plane to the temperature θ, to its gradient, and to the viscosity μ and density ρ of the gas:

$$v - G\left(\frac{dv}{dx} - \frac{3}{2}\frac{\mu}{\rho\theta}\frac{d^2\theta}{dxdy}\right) - \frac{1}{4}\frac{\mu}{\rho\theta}\frac{d\theta}{dy} = 0,$$

where G is the "Gleitungs-coefficient" (slip-coefficient) introduced by Helmholtz and Piotrowski. G is related to the mean free path λ of the molecule by[199]

$$G = \frac{2}{3}\left(\frac{2}{f} - 1\right)\lambda.$$

This is the basis for the modern theory of the radiometer and many other phenomena of rarefied gases. Maxwell did not claim the credit for discovering the tendency of a gas to creep along the surface from a colder to a hotter place. He credited it to Osborne Reynolds, who called it "thermal transpiration" and showed that it was a necessary consequence of kinetic theory. Indeed Reynolds had an important influence as Maxwell states,

> it was not till after I had read Professor Reynolds' paper that I began to reconsider the surface conditions of a gas, so that what

> I have done is simply to extend to the surface phenomena the method which I think most suitable for treating the interior of the gas. I think that this method is, in some respects, better than that adopted by Professor Reynolds, while I admit that his method is sufficient to establish the existence of the phenomena, though not to afford an estimate of their amount.[200]

The most important part of Maxwell's paper, the treatment of the gas-surface interactions, owes its origins to the suggestions of an unnamed referee and its success to Maxwell's knowledge of Osborne Reynolds's as yet unpublished paper![201]

Maxwell sent his paper to the Royal Society in March 1878, and it was refereed by William Thomson who submitted his report on 15 June 1878 (document VI-12). Thomson said that he had already discussed the paper with the author. It seems probable that Maxwell knew Thomson was the referee—he could hardly fail to recognize his style.[202]

In his previous report on Maxwell's 1866 paper on viscosity Thomson had noted the importance of slip.[203] He now proposed that the surface of a solid may be covered with bumps that trap, or reflect, the gas molecules in various ways. He suggested that a theory of gas-solid interactions might be developed on this basis. This was the stimulus for Maxwell to work out just such a theory. The early stages of its development can be gleaned from his report on the sixth of Crookes's papers of October 1878 (document VI-14). Crookes submitted the paper to the Royal Society on June 27th, so Maxwell presumably received it only a few weeks after seeing Thomson's report on his own paper. Maxwell was interested in a new instrument constructed by Crookes, "a small fly with clear mica vanes which can be placed in different parts of the radiometer bulb to detect differences of pressure on its vanes." Explorations with this device would permit a direct test of Maxwell's result that if the isothermal surface adjacent to the surface of the solid was not parallel to that surface, there would be a tangential stress of the gas on the solid. The force would be directed from places where consecutive isothermal surfaces were close together to places where they were further apart. In his report Maxwell continued that "at the surface of the solid the gas cannot support this stress but must yield to it. This causes a current sweeping over the surface." The theory was qualitatively confirmed by Crookes's observations.

Three months later, in January 1879, Osborne Reynolds submitted his paper to the Royal Society.[204] In it was a general theory of the flow of rarefied gases, with applications to the radiometer and to the new phenomenon of "thermal transpiration," one of the first, major discoveries stimulated by the radiometer. Maxwell was the referee, and, he made good use of its contents in extending his own theory.

As Reynolds defined it, thermal transpiration was the flow of gas through porous plates caused by temperature differences on opposite sides of the plate. With hydrogen at atmospheric pressure on one side of the plate a temperature difference of $160°F$ would sustain a permanent pressure difference of one inch of mercury, the higher pressure being on the hotter side. The rate of transpiration depends on the porosity of the plate and the density of the gas. Similar results are obtained provided that "the density of the gas is inversely proportional to the lateral dimensions of the passages through the plates." That is, similar effects would occur in rarefied gases with channels of larger dimensions.

According to Reynolds, it was this "scaling" property of the phenomenon that led him to discover it in the first place. He had recognized that the radiometer effect depended on the ratio of the size of the vanes to the mean free path of the molecules in the gas; there would be no effect for very large vanes or very small mean free paths. Since the mean free path decreases as the pressure and density increase,

> it appeared that by using vanes of comparatively small size the force should be perceived at comparatively greater pressures of gas... On considering how this might be experimentally tested, it appeared that to obtain any result at measurable pressures the vanes would have to be very small indeed; too small almost to admit of experiment. And it was while thinking of some means to obviate this difficulty that I came to perceive that if the vanes were fixed, then instead of the movement of the vanes we should have the gas moving past the vanes—a sort of inverse phenomenon—and then instead of having small vanes, small spaces might be allowed for the gas to pass. Whence it was at once obvious that in porous plugs I should have the means of verifying these conclusions.[205]

In his initial report on this paper (document VI-15) Maxwell heartily approved of his experiments but was less than satisfied with the theory. Reynolds had not stated clearly what happened at the surface, and Maxwell suggested that he should consider that "the surface has on it small prominences of various shapes from which the molecules rebound, with a velocity which is greater or less than that of the gas." He even worked out the beginnings of a quantitative theory based on this idea, remarking that "conditions of this kind might perhaps assist Prof. Reynolds in forming a theory of rebounding molecules."

In the published paper Reynolds states that section VII "was revised and somewhat enlarged in August 1879, in accordance with a suggestion made by one of the referees," but he did not adopt the particular formulation given him by Maxwell. Nevertheless there is a clear train of influence running through the referees' reports from Thomson to Maxwell to Reynolds. In the conclusion of his own report on Reynolds's paper, Maxwell gave his own explanation of thermal transpiration and, incidentally of the radiometer.

Relations between Maxwell and Reynolds deteriorated rapidly after this report.[206] Several more letters passed between Reynolds, Maxwell, Stokes, and William Thomson, the other referee. The letter from Maxwell to Stokes of 2 September 1879 was very sharp in its criticism of Reynolds.[207] Horace Lamb wrote of this letter to Joseph Larmor in 1905:

> Maxwell's letter is certainly amazing and characteristic, and hits off very happily and good-naturedly some of O. R.'s peculiarities. The gaiety of the world will lose by its omission, but I cannot urge its publication at present. I remember O. R. long bore a grudge against Maxwell on account of this paper, and he might still be sensitive to the brilliant raillery.[208]

Reynolds's "grudge" was based on more than the normal resentment of authors against referees who criticize and delay the publication of their work. Maxwell made use of Reynolds's discovery of thermal transpiration, with proper acknowledgment in his own paper, cast doubts on other parts of Reynolds's work, before Reynolds was in a position to defend himself. The situation was aggravated by Maxwell's final illness. Maxwell was dying, slowly and painfully of cancer of the abdomen, the disease that killed his mother at the same age.

Illness may also account for Maxwell's erratic behavior during this period. Writing to Stokes in December 1879 Tait suggested that "Maxwell's hereditary malady had begun to affect him some months before his death." Maxwell had made conflicting recommendations of George Chrystal and William Garnett for the chair in mathematics at Edinburgh,

> I began to fear that his mind was affected: but, happily this phase was very transient. I had full details of his malady from my colleague Sandars, who attended him and who was an old fellow-student of his in Edinburgh, but I considered it much too painful and delicate a subject to be even hinted at in public. No man could have behaved more tenderly or nobly than Maxwell did under long trials of the most overwhelming nature. How he could manage to do splendid mathematical work all the time is inconceivable.[209]

Here was the trap waiting for Reynolds, the circle of Tait, Thomson, Stokes and many another of Maxwell's powerful friends who would rush to his defense if anyone criticized the dead hero.

Two weeks before Maxwell's death Reynolds sent a letter to Stokes to be communicated to the Royal Society. In it he protested the slur on his work in Maxwell's paper, and ventured "to request those interested in the subject to withhold their opinion until they have the opportunity of reading my paper. In the meantime I can only express my opinion that Professor Maxwell is mistaken in supposing that the results which are obtained from his method are more definite than those obtained from mine."[210]

Apparently Reynolds wanted the letter read immediately to the Royal Society, but Stokes was reluctant to do so. In a subsequent letter on 5 November 1879 Stokes wrote,

> A LITTLE AFTER 12 O'CLOCK TODAY PROFESSOR MAXWELL PASSED AWAY QUIETLY WITHOUT APPARENT PAIN.
>
> I GOT LAST NIGHT YOUR LETTER IN WHICH YOU SAY THAT YOU STILL WISH YOUR LETTER TO ME TO BE READ BEFORE THE SOCIETY.
>
> IF IT WERE SIMPLY A COMMUNICATION OF YOUR OWN, SENT IN THE USUAL FORM OF SUCH COMMUNICATIONS, IN CASE OF ANY DOUBTS HAVING ARISEN AS TO THE PROPRIETY OF PRINTING IT IN THE SHAPE SENT IN, IT WOULD HAVE BEEN SIMPLY A MATTER BETWEEN YOURSELF AND THE COUNCIL. BUT AS IT IS DRAWN UP IN THE FORM OF A LETTER TO ME, I FEEL THAT ITS PRESENTATION BY ME WITHOUT NOTE OR COMMENT WOULD IMPLY THAT I QUITE APPROVE OF IT, WHICH I MUST SAY IS NOT THE CASE. THERE ARE TWO COURSES OPEN:–THAT YOU SHOULD ALTER IT SO AS TO MAKE IT WHAT I SHOULD QUITE APPROVE OF, OR THAT IN PRESENTING IT I SHOULD ADD COMMENTS OF MY OWN.[211]

Reynolds chose the second alternative. His letter was read the following April, followed by a note by Stokes.[212] In his note Stokes mentioned that some of Maxwell's work had been based partly on a suggestion of William Thomson. He went on to say that when the paper was in press.

> In a letter I received from him at the time, he informed me that he felt very ill, and was hardly fit even to go through his own paper; though a subsequent letter, in which he entered into some scientific matters, was written in his usual cheerful style. No one had, I believe, at that time any notion of the very serious nature of his illness.[213]

Stokes sent a copy of his letter to Thomson for approval and Thomson replied: "I think your note on Osborne Reynold's letter is quite satisfactory. You have I think kept quite free from anything that could make a cause for bad feeling."[214]

In an earlier letter to Stokes, Thomson had mentioned a paper by W. Feddersen containing experiments that apparently anticipated Reynolds's discovery of thermal transpiration.[215] These experiments, in turn, were based upon a suggestion of Carl Neumann.[216] The above letter from Thomson to Stokes continues,

> Did you look up the German reference which I gave? Maxwell cannot have known of it, as it anticipated Reynolds in the discovery of the thing, though so far as I could tell by a brief dipping into it, the author F. Neumann [?] was in a very unadvanced state of mind in respect to the true theory of gases, and as to the relation between what he had discovered and such things as the Peltier effect in thermo-electricity.[217]

Stokes had passed on the reference to Reynolds, for in a note to his paper added in December 1879 Reynolds states,

> Since the reading of this paper I have had my attention called to a paper by W. Feddersen ... Feddersen made some experiments, and seems to have thought that he had discovered some such phenomenon. But the results he obtained were attributed by M. J. Violle to the presence of the vapour of water, against which no precautions appear to have been taken (*Journal de Physique,* 1875, p. 90). That M. J. Violle was right there can be no doubt, for the results obtained are now seen to be much too large for the true results, and are similar to those which I obtained before I succeeded in sufficiently drying the air.[218]

In attempting to repel the threatened German invasion of his priority, Reynolds had hastily grasped at a French straw. He misunderstood Violle's remarks, even the page number of the citation is incorrect. Reynolds's relations with Maxwell's friends are clearly delineated in this dilemma into which he was thrown.

The hostility continued. In a note of 1881 George Francis Fitzgerald criticized Reynolds and complained that "Prof. Reynolds' paper is very elaborate, and necessarily somewhat difficult, not only from the nature of the subject, but also, in parts, owing to the inelegant method Prof. Reynolds has pursued."[219] Reynolds replied,

> With regard to Professor Maxwell's remarks on my paper, and his own work on the same problem, of course, the sad circumstances of his death occurring, so that this was about the last work he did, renders it very difficult to approach the subject; but

> with reference to what I have already said, and in explanation of the apparently imperfect idea at which he arrived as to the scope and purpose of my method, it may be stated that, before writing his own paper, Professor Maxwell had only seen my paper in manuscript in the condition in which it was first sent to the Royal Society, when the preliminary part was very much compressed, and, as I fear, somewhat vaguely stated, besides being founded on different assumptions from the present. Without entering further upon this now, I may refer to a letter which I addressed to Prof. Stokes after seeing an early copy of Prof. Maxwell's paper, and before I was aware of his illness, and which was subsequently published.[220]

Despite this rather ungraceful exchange between Reynolds and Maxwell's friends, his work admirably complemented that of Maxwell. Maxwell's work created the science of rarefied gas dynamics; Reynolds, besides the immediate discovery of thermal transpiration, gained insight into dimensional analysis. This led to his work on turbulent motion in hydrodynamics.

Maxwell contributed to one other method of great beauty in his notes to his paper of May and June 1879—the use of spherical harmonics for expansions of the distribution function. It exemplified what Maxwell called "the cross-fertilization of the sciences." He had used spherical harmonics in the *Treatise* to solve electrical problems. In the course of this work he made one of his rare contributions to pure mathematics, specifying the harmonic by its poles. A working knowledge of the subject made him familiar with the possibility of expressing any rational, homogeneous function of three variables as a sum of solid harmonics and with the standard theorems of reducing the expansions to manageable forms.

The calculations in rarefied gases depended upon evaluating changes in homogeneous functions of the components ξ, η, ζ of molecular velocity. In his original draft Maxwell had proceeded by methods similar to those used in his gas theory paper of 1867. However, while revising the chapter on spherical harmonics for the second edition of his *Treatise,* Maxwell conceived of the idea of applying harmonics to gas theory. A standard theorem on products of surface and zonal harmonics (Legendre polynomials), discussed in the *Treatise*, eliminates odd terms in the expansion of variations on F, greatly simplifying its evaluation.[221] In addition terms described by harmonics of the same order have identical rates of decay and can be specified by a single, generalized relaxation time. With these and other simplifications, Maxwell carried the approximations to higher order and added another term to the equation of motion of a gas subject to variations in temperature. The condition he found for successful use of the approximation was that variations in properties of the gaseous medium should remain small over distances comparable to the mean free path.

It is a tribute to the power of Maxwell's work that on two distinct occasions his papers on transfer theory and rarefied gas dynamics stimulated new investigations by others years beyond the interval at which science ordinarily receives historical embalmment. In 1910 Sidney Chapman, then a student of Larmor's, read them, without realizing how much fruitless effort had been spent on transfer equations, and he began his own investigation that led to their general solution:

> thus with the ignorant hardihood of youth I attempted a problem that Larmor would certainly have thought unfit for such a novice. It seems it is sometimes good not to know too much.[222]

In 1956 Ikenberry and Truesdell again returned to Maxwell, extending the techniques of the 1879 paper in a process of "Maxwellian iteration," which gave exact, analytical solutions to various problems with inverse fifth-power forces and enabled them to expose a number of shortcomings in the approximation methods used previously in the general theory.[223]

Long before these developments Boltzmann had appreciated Maxwell's methods of the 1879 paper. His doglike humility before Maxwell's work may well have embarrassed Maxwell, the dog lover. The passage is well known, where he compared the 1867 paper to a great symphony, with its majestic organization, splendid reach of thought, and dramatic simplification when the maestro utters the magic spell "put $n = 5$."[224]

In 1894 Boltzmann paid his first visit to England for a discussion on statistical mechanics organized by the British Association. A day or so later certain aloof Cambridge gentlemen were amazed to see a lumbering, dishevelled figure arrive breathless at the door of the Cavendish Laboratory. In a squeaky guttural Boltzmann demanded immediate access to Maxwell's unpublished manuscripts, wherein were calculations of immense importance, specifically the notes for the rarefied gases paper of 1879. He had run all the way from the railway station, where he had arrived without a penny in his pocket. The manuscripts were not found, but they turned up years later and are included here (document VI-25). Their publication is, in part, a response to Boltzmann's plea to the British Association

> to make efforts to ascertain if the manuscript of the investigation made by Maxwell on the application of spherical harmonics to the theory of gases is still in existence, and, if this manuscript should be lost, to encourage physicists to repeat these calculations.[225]

Notes

1. Maxwell's ideas on method have been explored by historians but usually in the context of a particular piece of his work. Here we extend them to most of

his mature work where the implications of his ideas on method and "speculation" reveal the catholic structure of his physics. See Peter Harman, "Edinburgh Philosophy and Cambridge Physics: the Natural Philosophy of James Clerk Maxwell," in *Wranglers and Physicists: Studies in Cambridge Physics in the nineteenth Century*, ed. Peter Harman (Manchester: Manchester University Press, 1985), 202–224. See also, Peter Heimann, "Maxwell and Consistent Modes of Representation," *Arch. Hist. Exact Sci.* 6 (1970): 171–213; Jon Dorling, "Maxwell's Attempts to Arrive at a Non-Speculative Foundation for Kinetic Theory," *Studies in Hist. Phil. Sci.* 1 (1970): 229–248; Robert Kargon, "Model and Analogy in Victorian Science: Maxwell's Critique of French Physicists," *J. Hist. Ideas* 30 (1969): 423–436; Joseph Turner, "A Note on Maxwell's Interpretation of Some Attempts at Dynamical Explanation," *Ann. Sci.* 11 (1956): 238–245; "Maxwell on the Method of Physical Analogy," *British J. Philos. Sci.* 6 (1955): 225–238.

2. Maxwell, "On the Dynamical Theory of Gases," *Phil. Trans. R. Soc. London* 157 (1867): 49–88, reprinted in *Scientific Papers*, vol. 2, 26-78, and Garber, Brush, and Everitt, *Maxwell on Molecules and Gases* (Cambridge, Mass.: MIT Press, 1986), 420–472, 469.

3. Documents II-1 and II-2, The latter reads like a note for the end of the 1867 paper, op. cit. (note 2) before Maxwell detected the error in his calculations.

4. Op. cit. (note 2), 469.

5. William Thomson, "On the convective equilibrium of temperature in the atmosphere," *Mem. Manchester Lit. Phil. Soc.* 2 (1865): 125–131.

6. The two cases are actually different. In Thomson's case a mass of gas moves; in Maxwell's case molecules enter and leave a volume but the gas does not move as a mass with a common velocity.

7. For the uses meteorologists made of thermodynamics see E. Garber, "Thermodynamics and Meteorology," *Ann. Sci.* 33 (1976): 51-65. Gisela Kutzbach, *The Thermal Theory of Cyclones. A History of Meteorological Thought in the Nineteenth Century* (Boston: American Meteorological Society, 1979), fits the development of meteorological ideas on thermodynamics into the understanding of the dynamics of storms.

8. Francis Guthrie seems not to have known of Boltzmann's work, hardly surprising given his location, and no one other than Maxwell published on the problem in English. The question had general interest and Guthrie's position and ignorance of Boltzmann's work was probably shared with others. See Stephen G. Brush, *The Kind of Motion We Call Heat* vol. 1 (Amsterdam: North-Holland, 1976), 349–351.

9. For details on the development of this branch of physics and Maxwell's role in that growth, see Garber, Brush, and Everitt, op. cit. (note 2), 29–47, and references cited therein.

10. Boltzmann, "Über die mechanische Bedeutung des zweiten Haupsatzes der Wärmetheorie," *Sitz. Math.-Naturwiss. Cl. Akad. Wiss., Wien* 53 (1866): 195–220, reprinted in, *Wissenschaftliche Abhandlungen*, ed. Fritz Hasenörhl (New York: Chelsea Pub. Co., 1968) vol. 1, 9–33, p. 9.

11. Boltzmann, op. cit. (note 10). Most historians examine Boltzmann's contributions to thermodynamics. For his work in kinetic theory, see Stephen G. Brush, op. cit. (note 8); Brush, *Kinetic Theory* 3 vols. (New York: Pergamon Press, 1966–72) and his translation of Boltzmann, *Lectures on Gas Theory* (Berkeley: University of California Press, 1964).

12. His first paper in kinetic theory was Boltzmann, "Studien über das Gleichgewicht der lebendigen Kraft zwischen bewegten materiellen Punkte," *Sitz. Math-Naturwiss. Cl. Akad. Wiss., Wien* 58 (1868): 517–560.

13. For Maxwell on the probability of a molecule being in a particular state, see Maxwell, "On the final State of a System of Molecules subject to Forces of any Kind," *Rep. of the 42nd Meeting BAAS* (1873), Trans. Sect. A, 29–32, and document II-14. For the implications of these assumptions and later discussions of the ergodic hypothesis, see Brush op. cit. (note 8) vol. 2, 363–386.

14. Maxwell, "On the dynamical Evidence of the molecular Constitution of Bodies," *Nature* 11 (1875): 357–359, 374–377; *J. Chem. Soc. London* 13 (1875): 493–508; *Scientific Papers*, vol. 2, 418–438.

15. R. C. Nichols, "On the dynamical Evidence of Molecular Constitution," *Nature* 11 (1875): 486–487. Nichols' argument showed that he had read Maxwell's paper op. cit. (note 14) carefully and was trying to argue his point using both kinetic theory and the virial theorem.

16. Joseph John Murphy, "Equilibrium of Gases," *Nature* 12 (1876): 26; Samuel. H. Burbury, "Equilibrium of Temperature in a vertical Column of Gas," *Nature* 12 (1876): 107. Nichol, "Equilibrium of Gases," *Nature* 12 (1876): 67. Burbury apparently had written to Nichol and convinced him of the rightness of Maxwell's argument, and his letter was then published.

17. Henry William Watson, *A Treatise on the Kinetic Theory of Gases* (Oxford: Clarendon Press, 1876). In the preface Watson thanked Maxwell for his encouragement and the use of his unpublished notes on gases.

18. See Garber, " Reading Fourier, Constructing Physics: French Mathematical Physics and Its Readers (1820–1850)" in press.

19. The proposition was to demonstrate that a velocity distribution be the permanent one—shades of Boltzmann, but the methods of deduction are Maxwell's.

20. In his review Maxwell improved Watson's derivation of this result but otherwise found the content of the treatise very satisfactory.

21. Boltzmann, "Analytischer Beweis des zweiten Haupsatzes des mechanischen Wärmetheorie aus Sätzen über das Gleichgewicht der lebendigen Kraft," *Sitz. Math.-Naturwiss. Cl. Akad. Wiss., Wien* 63 (1871): 712–732. Reprinted in *Wiss. Abh.*, vol. 1, 288–308.

22. Boltzmann was already deriving such a general derivation—the culmination of this generalized approach. The term "statistical mechanics" was introduced by Josiah Willard Gibbs in the early twentieth century.

23. For a discussion of Maxwell and molecular models, see Garber, "Molecular Science in Late Nineteenth-Century Britain," *Hist. Studies in Phys. Sci.* 9 (1976): 265–297.

24. See Garber, Brush, and Everitt, op. cit. (note 2), 274.

25. For example, see Clifford Truesdell, *The Tragicomical History of Thermodynamics, 1822–1854* (New York: Springer-Verlag, 1980). This is a logical, mathematical reconstruction of how thermodynamics could and should have developed that argues against the principal maxim of most historian's accounts, that the evidence of the documents determines the shape and texture of the analysis.

26. The importance of the mechanical theory of heat can be seen, for example, in the work of Clausius and Rankine. For discussions of their theories, see Keith Hutchison, "W. J. M. Rankine and the Rise of Thermodynamics," *British J. Hist. Sci.* 14 (1981): 1–26; "Rankine, Atomic Vortices and the Entropy Function," *Arch. Int. Hist. Sci.* 31 (1981): 72–134; G. Bierhalter, "The Mechanical Foundation for the second Law," *Arch. Hist. Exact Sci.* 37 (1987): 77–99.

27. In this debate, the decline of science in Britain was seen as a component of its economic decline and became part of a plea for more government funding for science. For the intellectual dimensions of this challenge, see David Knight, *The Age of Science* (Oxford: Basil Blackwell, 1986), chap. 4.

28. For an account of William Thomson's reaction to Clausius's thermodynamics and his part in the construction of Tait's history, see Crosbie Smith and M. Norton Wise, *Energy and Empire: A Biographical Study of Lord Kelvin* (Cambridge: Cambridge University Press, 1989), chap. 10, especially 341–348.

29. See especially the sources listed in note e., document III-2.

30. Tait's book on thermodynamics appeared as Tait, *Sketch of Thermodynamics* (Edinburgh: Edmonston & Douglas, 1868). Clausius protested Tait's history in Clausius, "Zur Geschichte der mechanischen Wärmetheorie," *Ann. Phys.* 145 (1872): 132–146. Tait's reply is on p. 496. Protest also appeared in Clausius, *Phil. Mag.* 43 (1872): 106–115, 117–118, in a letter to the editor. Tait again replied on pp. 338–339. Clausius had also claimed his proper due from Joule. See Clausius, "On the Discovery of the True Form of Carnot's Function," *Phil. Mag.* 11 (1856): 388–390, and 12 (1856): 463. The slights Clausius felt may help

explain his distant, formal response to Maxwell's letter enquiring about entropy, see document III-43.

31. Yehuda Elkana indicates that in Michael Faraday's work, for example, there is transformation without the recognition of conservation. See, Elkana, *The Discovery of the Conservation of Energy* (Cambridge Mass.: Harvard University Press, 1974). In contrast, see Thomas S. Kuhn, "Energy Conservation as an Example of Simultaneous Discovery," in *Critical Problems in the History of Science* ed. Marshall Clagett (Madison Wisconsin: University of Wisconsin Press, 1959), 321–356.

32. Others can be added to this list, including Ludwig Colding, but the main protagonists remain Joule and Mayer. This controversy has recently broken out again. See Eric Mendoza and D. S. L. Cardwell, "On a Suggestion Concerning the Work of Joule," *British J. Hist. Sci.* 14 (1981): 177–180, and D. S. L. Cardwell, *James Joule: A Biography* (Manchester: University of Manchester Press, 1989) where they counter the argument in Henry John Steffens, *James Prescott Joule* (New York: Science History Publications, 1979) that Joule used Mayer without acknowledgment. See also, J. T. Lloyd, "Background to the Joule-Mayer controversy," *Notes Records R. Soc., London* 25 (1970): 211–225. Mayer's lack of support among his peers is traced by Ken Caneva, in *Julius Robert Mayer* (Princeton, N. J.: Princeton University Press, 1993).

33. For a recent discussion of Joule's work, see William H. Cropper, "James Joule's Work in Electrochemistry and the Emergence of the First Law of Thermodynamics," *Hist. Studies Phys. Sci.* 19 (1988): 1–15. See also Steffens, op. cit. (note 32).

34. For full discussion of Mayer's ideas, see Caneva op. cit. (note 32). Also helpful are Peter Heimann,"Mayer's Concept of 'Force,'" *Hist. Studies Phys. Sci.* 7 (1976): 277–296. and Hutchison, "Mayer's Hypothesis: A Study of the Early Years of Thermodynamics," *Centaurus* 20 (1976): 279–304.

35. Joule's work has attracted closer scrutiny by historians of physics than Mayer's. His work is more approachable because it is experimental and lies within the modern boundaries of physics. Mayer's work belongs to a scientific tradition less easily encompassed in past or present disciplinary boundaries—the study of animal heat.

36. Some historians have traced the philosophical foundations of this antipathy to action-at-a-distance forces to Scottish Common Sense Philosophy. See Richard Olsen, *Scottish Philosophy and British Physics, 1750–1880* (Princeton, N. J.: Princeton University Press, 1975); George Elder Davie, *The Democratic Intellect: Scotland and Her Universities in the Nineteenth Century* (Edinburgh: Edinburgh University Press, 1959). For William Thomson in particular, see Smith and Wise, op. cit. (note 28) passim. These discussions do not explain the methodological choices of George Gabriel Stokes, John Herschel, or William Hopkins, for example.

37. For a detailed history of French mathematical physics as mathematics, see Ivor Grattan Guinness, *Convolutions in French Mathematics, 1800–1840: From*

the Calculus and Mechanics to Mathematical Analysis and Mathematical Physics 3 Vols (Boston Mass.: Birkhauser, 1990). For the adoption and form mathematical physics took in Britain, see Garber, op. cit. (note 18) and Garber, *Mathematics as Language* (forthcoming).

38. Garber, "Siméon-Denis Poisson: Mathematics versus Physics in Early Nineteenth-Century France," in *Beyond History of Science: Essays in Honor of Robert E. Schofield* ed. Elizabeth Garber (Bethlehem Penn.: Lehigh University Press, 1990) delineates through an example why French mathematical physics is *not* physics.

39. The poem is was originally published in C. G. Knott, *Life and Scientific Work of Peter Guthrie Tait* (Cambridge: Cambridge University Press, 1911) 253–255. The poem was not for publication. Tait kept it in his scrapbook.

40. For the development of German physics as discipline and profession, see Christa Jungnickel and Russell McCormmach, *Intellectual Mastery of Nature: Theoretical Physics from Ohm to Einstein* (Chicago: University of Chicago Press, 1986) vol. 1 **The Torch of Mathematics 1800–1870**, and Katheryn Olesko, *Physics as a Calling: Discipline and Practice in the Königsberg Seminar for Physics* (Ithaca: Cornell University Press, 1991)

41. Wilhelm Weber,"Elektrodynamische Maasbestimmungen," *Abh. bei Begrundung der König. Sachs. Gesell. der Wissen. am Tage der zweihundertjahrigen Geburtstagfier Leibnizens, Leipzig*, 1846, 211–378, abstracted in *Ann. Phys.* 73 (1848): 193–233: It is this version of the paper that was translated into *Scientific Memoirs*, vol. 5 (1852): 489–529.

42. For a full account of the connections between Helmholtz's ideas and those of Kant see, Peter Heimann, "Helmholtz and Kant: The Metaphysical Foundations of *Über die Erhaltung der Kraft*," *Studies in Hist. Phil. Sci.*, 5 (1974): 205–238. These connections are obscured in Elkana, op. cit. (note 31), which leads him later to find generality where there is only particularity. For an alternative critique, see Truesdell, op. cit. (note 25), who concludes, correctly, that "no one" established the law of conservation of energy before 1850. Helmholtz's Kantianism was lifelong. He never relaxed his vision of nature as causally driven and that matter was composed of centers of force.

43. Hermann von Helmholtz *Über die Erhaltung der Kraft* (Weinheim: Physik-Verlag, 1983). Translated as "On the Conservation of Force: A Physical Memoir," *Taylor's Scientific Memoirs*, vol. 4 (1853). Reprinted and translated in *Selected Writings of Hermann Helmholtz*, ed. Russell Kahl (Middletown, Conn.: Wesleyan, 1971).

44. Both Elkana and Heimann claim that indeed Helmholtz's Kraft and energy are one and the same. See, Elkana, "Helmholtz's Krafte," *Hist. Studies Phys. Sci.* 2 (1970): 263–298 and op. cit. (note 31); Heimann, "Conversion of Forces and the Conservation of Energy," *Centaurus* 18 (1974): 147–161.

45. Joule's first paper on heat as a mode of motion was Joule, "On the Electrical Origin of the Heat of Combustion," *Rep. Ninth Meeting BAAS* (1842), Trans. Sect. A., 31; *Phil. Mag.* 20 (1842): 98–113.

46. See Steffens, op. cit. (note 32) for an account of Joule's experiments. Steffens notes that Joule's results were less than convincing because he measured such small temperature changes. This criticism was repeated through the nineteenth century, even by experimentalists, see Henry Rowland, "On the Mechanical Equivalent of Heat," *Proc. American Acad. Arts Sci.* 15 (1880): 75–200. For a detailed assessment of the accuracies of Joule's thermometers, see J. R. Ashworth, "Joule's Thermometers in the Possession of the Manchester Literary and Philosophical Society," *J. Scient. Instrum.* 7 (1930): 361–363.

47. Helmholtz used the same evidence as Joule to argue against the materiality of heat—friction producing infinite quantities of heat and heat produced by current electricity.

48. Kahl, op. cit. (note 43), 33.

49. Helmholtz began his neurological researches while he was still an army surgeon at Potsdam in 1843. In his research he timed nerve pulses and began his work on the physiology of hearing and sight. For an account of this work see Leo Koenigsberger, *Hermann von Helmholtz* vol. 1 (Braunschweig: Vieweg 1902–03).

50. However, while he did not add to the development of the theory of heat Helmholtz wrote review articles for his colleagues and kept up his priority claims for the discovery of the first law. See, Helmholtz,"Theorie der Wärme," *Fort. Phy.* 6 and 7 (1850–1851): 561-598; 9 (1853):404–405; 12 (1856): 345, 355–356.

51. Maxwell to Lewis Campbell, April 1862, in, Lewis Campbell and William Garnett, *The Life of James Clerk Maxwell* (London: Macmillan, 1882), first edition (1882) reprinted with a new preface by Robert Kargon and appendix with letters (New York: Johnson Reprint Corp., 1969), 335.

52. Maxwell, "Hermann Ludwig Ferdinand von Helmholtz," *Nature*, 15 (1877): 389–391 reprinted in vol. 2, *Scientific Papers*, 592–598.

53. Truesdell, op. cit. (note 25) offers a logical reconstruction of the development of thermodynamics, insisting that the development ought to be one smooth evolution of increasingly more general logic. His is a very useful mathematical look at thermodynamics disengaged from the particular views of nature to which they were attached historically.

54. Rankine presented his ideas on the nature of heat and its mechanical effects to the Royal Society of Edinburgh in 1849. His papers appeared after those of Clausius. Rankine, "On the mechanical Action of Heat Especially in Gases and Vapours," read at the October 1849 meeting, *Trans. R. Soc. Edinburgh* 20 (1850–51) [1853]: 87–120, 147–190, 565–589; *Phil. Mag.* (1854): 3–21, 111–121. See also Rankine, "On the Centrifugal Theory of Elasticity as Applied to Gases and

Vapours," *Phil. Mag.* 2 (1851): 509–542. Rankine's theory is about the mechanics of gases made up of vortices. See Keith Hutchison, "W. J. M. Rankine and the Rise of Thermodynamics," *British J. Hist. Sci.* 14 (1981): 1–26.

55. Hutchison, op. cit. (note 54).

56. Rankine, "Mechanical Action," op. cit. (note 54).

57. Rankine first introduced conservation as a general principle of nature in Rankine, "On the General Law of the Transformation of Energy," *Proc. Philos. Soc., Glasgow* 3 (1853): 276–280; *Phil. Mag.* 5 (1853): 106–117, elaborated and developed in "Outlines of the Science of Energetics," *Edinburgh New Philos. J.* (1855): 120–121, and "The Conservation of Energy," *Phil. Mag.* 17 (1859): 250–253, 347–348, where, he differentiated clearly between force and energy and stated the circumstances of the conservation laws of physics.

58. For an assessment of Rankine's mechanical theory see Hutchison, op. cit. (note 54) and Hutchison, "Der Ursprung der Entropiefunktion bei Rankine und Clausius," *Ann. Sci.* 30 (1973): 341–364.

59. Clausius later gave his readers a lesson in the elements of partial differential equations in answer to a mathematician who critiqued his usage. The disciplines were indeed separating. See Clausius, *On the Mechanical Theory of Heat,* trans. Hirst (London: Taylor and Francis, 1867), 1-13.

60. Clausius, "Über der bewegende Kraft der Wärme ...," *Ann. Phys.* 79 (1850): 368–397, 500–524, translated in *Phil. Mag.* 2 (1851): 1–21, 102–119. Details of this first paper are given in Truesdell, op. cit. (note 25) along with a detailed assessment of its logical limitations. See also Eri Yagi,"Analytical Approach to Clausius's First Memoir on Mechanical Theory of Heat," *Historia Scientiarum* 20 (1981): 77–94.

61. For example, the effect of pressure on the freezing point of water. The correct result had been postulated by James Thomson on the basis of his brother's work in caloric theory. It fell out of the mathematics. Clausius could, however, offer the same results from a theory that integrated the phenomenon physically making a coherent whole. See Clausius, "Über die Einfluss der Druckes auf der Gefrieren der Flussigkeiten," *Ann. Phys.* 81 (1850): 168–172, translated in *Phil. Mag.* 2 (1851): 548–550. The article in the *Annalen* is preceded immediately by a note that an article on the same subject by William Thomson had just appeared. See Thomson, "On the Effect of Pressure on the Freezing Point of Water, Experimentally Demonstrated" *Proc. R. Soc., Edinburgh* 2 (1850): 267–271. This was the experimental verification of his brother's prediction that appeared in James Thomson, "Theoretical Considerations on the Effects of Pressure in Lowering the Freezing Point of Water," read January 1849, published in *Trans. R. Soc., Edinburgh* 16 (1849): 575–80. For a different interpretation of this collaboration, see Smith and Wise, op. cit. (note 28), 298–300. Both Thomsons thought in terms of Carnot's theory. However, the result is independent of both the caloric and the dynamical theories of heat.

62. For the reactions of biologists and geologists to Thomson's cosmological speculations see Joe Burchfield, *Lord Kelvin and the Age of the Earth* (New York: Science History Pub., 1975). The cosmological context of Thomson's thermodynamics and his forays into the question of the age of the earth is examined also in Smith and Wise, op. cit. (note 28), chaps. 4, 15, and 16.

63. In his address of 1853 as president of Section A of the BAAS, William Hopkins traced the development of the new theory of heat through the work of Rumford to Joule who established "the true foundation in experiment" of that theory. See Hopkins, "Address," *Rep. 23rd Meeting BAAS*, Trans. Section A, xli–lvii (1853).

64. See Garber, "Fourier" op. cit. (note 37).

65. Smith and Wise, op. cit. (note 28), chaps 9 and 10, interpret Thomson's work in thermodynamics quite differently. For Smith's assessment of Thomson's place in the history of thermodynamics, see Smith, "William Thomson and the Creation of Thermodynamics: 1847–1855," *Arch. Hist. Exact Sci.* 16 (1976–77): 232–288.

66. Thomson, "On the Dynamical Theory of Heat; with Numerical Results Deduced from Mr. Joule's 'Equivalent of a Thermal Unit' and M. Regnault's 'Observations on Steam'," *Proc. R. Soc.* Edinburgh 3 (1851): 48–52, published in full in *Trans. R. Soc., Edinburgh* 20 (1851): 261–288 and *Phil. Mag.* 4 (1852): "Part I," 8–21, 10. More parts were published in subsequent issues through the 1850s.

67. For example, Thomson, op. cit. (note 66), 20. The reference is to his expression for the efficiency of a heat engine.

68. Thomson "An Account of Carnot's Theory of the Motive Power of Heat with numerical Results Deduced from Regnault's Experiments on Steam," *Proc. R. Soc., Edinburgh* 24 (1849): 198–204, published in full in *Trans. R. Soc., Edinburgh* 16 (1849): 541–574. Thomson's first paper on Carnot was Thomson, "On an Absolute Scale of Temperature, Founded on Carnot's Theory of the Motive Power of Heat, and Calculated from the Results of Regnault's Experiments on the Pressure and Latent Heat of Steam," *Phil. Mag.* 33 (1848): 313–317.

69. Thomson, "On the Dynamical Theory of Heat. Part II. On the Motive Power of Heat through Finite Regions of Temperature," *Phil. Mag.* 4 (1852): 105–117.

70. Smith and Wise, op. cit. (note 28), chap. 4 and passim for Thomson's philosophical problems with the mechanical theory of heat. These center on the philosophical issue of conservation in a world that evolves. However, Carnot's theory was also based on conservation, the conservation of heat, which in their account does not seem so problematical to Thomson.

71. See,Clausius, op. cit. (note 60), 16–17. Thomson's challenge is in Thomson, "On a Remarkable Property of Steam Connected with the Theory of the Steam Engine," *Phil. Mag.* 37 (1850): 386–389, *Ann. Phys.* 81 (1850): 477–480.

72. The argument continued during the following year in Clausius, "Ueber das Verhalten des Dampfes bei der Ausdehnung unter verscheiden Unständen," *Ann. Phys.* 82 (1851): 263–273; *Phil. Mag.* 1 (May 1851): 398–405. Thomson's theory was developed in Thomson, "Note on the Effect of Fluid Friction in Steam Which Issues from a High-Pressure Boiler through a Small Orifice into the Open Air," *Phil. Mag.* 1 (June 1851): 474 and, "II," (October 1851): 273–274. Clausius replied in Clausius, "Reply to a Note from Mr. W. Thomson on the Effects of Fluid Friction," *Phil. Mag.* 11 (1851): 139–142. Thomson reiterated his opinion in, "On the Dynamical Theory of Heat Part III," *Phil. Mag.* 4 (September 1852): 168–176, 176.

73. Thomson, "On a Universal Tendency in Nature to the Dissipation of Mechanical Energy," *Phil. Mag.* 4 (October 1852): 304–306. In this paper Thomson credits Clausius for the statement of the "axiom" on which his paper is based (p. 304).

74. Thomson, op. cit. (note 73). For details of the cosmic and geological implications of Thomson's work, see Burchfield, op. cit. (note 62).

75. Thomson mentioned thermoelectric phenomena in his first paper on thermodynamics. See Thomson, op. cit. (note 66), 15–16. Truesdell op. cit. (note 25) has analyzed these papers from the thermodynamic point of view. Thomson, "Account of Researches in Thermoelectricity," *Proc. R. Soc., London* 7 (1854–55): 49–58, *Phil. Mag.* 8 (1854): 62–69; "Account of Experimental Investigations to Answer Questions Originating in the Mechanical Theory of Thermo-electric Currents," *Proc. R. Soc., Edinburgh* 3 (1854): 255; "On the Dynamical Theory of Heat. Part VI. Thermoelectric Currents," *Proc. R. Soc., Edinburgh* 3 (1857): 255–256, *Trans. R. Soc., Edinburgh* 21 (1857): 123–171, *Phil. Mag.* 11 (1856): 214–225, 281–297, 379–388, 433–446. These are his first papers and the major reports on his work. They were followed with smaller papers on further experimental work through the 1850s. Thermoelectricity later became the focus for Tait's research. Clausius also worked on thermoelectricity in the 1850s.

76. See Smith and Wise, op. cit. (note 28), 343–344.

77. See P. M. Heimann, "*The Unseen Universe*: Physics and the Philosophy of Nature in Victorian Britain," *British J. Hist. Sci.* 6 (1972): 73–79.

78. Clausius, "Ueber ein veränderte Form des zweiten Haupsatzes der mechanischen Wärmetheorie," *Ann. Phys.* 93 (1854): 481–507; *Phil. Mag.* 12 (1856): 81–98. The phrase is introduced as the title to a section of the paper on p. 85.

79. E. E. Daub "Sources for Clausius's Entropy Concept: Reech and Rankine," *Proc. Fifteenth International Congress for the History of Science* (1978): 342–358, traces the origins of this expression to the work of F. Reech, not mentioned

by Clausius until 1863. However, Clausius's argument is not the same as that of Reech. For a discussion of the mathematical generality but flawed logic of Reech's work see Truesdell, op. cit. (note 25), chap. 10 and Appendix and Truesdell and S. Bharatha, *The Concepts and Logic of Classical Thermodynamics* (New York: Springer-Verlag, 1977), see also Daub, *Isis* 70 (1979): 478.

80. Clausius, op. cit. (note 78), 94.

81. Clausius, op. cit. (note 78), 96. In his paper Clausius is careful to mention previous work on irreversibility in both Rankine and Thomson. There is a verbal statement of this expression, but irreversibility is not explored. This expression is explored in Clausius's first paper on disgregation, "Über die Anwendung des Satze von der Aequivalenz der Verwandlungen auf die innere Arbeit," *Ann. Phy.* 116 (1862): 73–112. He discusses the thermodynamic implication of this expression in Clausius, "Über verscheidene für die Anwendung bequeme Formen der Hauptgleichungen der mechanischen Wärmetheorie," *Ann. Phys.* 125 (1865): 353–401. Daub, op. cit. (note 79) sees a major development in Clausius's understanding of irreversibility in his paper on the steam engine, published in light of his reading of Rankine on the same subject. See Clausius,"Ueber die Anwendung der mechanischen Wärmetheorie auf Dampfmaschine," *Ann. Phys.* 97 (1856): 441–476, 513–558; *Phil. Mag.* 12 (1856): 241–265, 338–354, 426–443. Rankine's paper appeared as Rankine, "On the Geometrical Representation of the Expansive Action of Heat, and the Theory of Thermo-dynamic Engines," *Phil. Trans. R. Soc., London* 144 (1855): 115–175.

82. Even in Clausius's early papers on thermodynamics, op. cit. (note 60), 2, the molecular basis of his own thought surfaces. It is explicit in his papers on disgregation begun in 1862 as well as his papers in kinetic theory from 1857 onward.

83. Clausius, "Ueber die Art der Bewegung, welche wir Wärme nennen," *Ann. Phys.* 100 (1857): 353–380; *Phil. Mag.* 14 (1857): 108–127. Clausius's papers on kinetic theory have been discussed in Garber, Brush, and Everitt, op. cit. (note 2), 4–7, and references cited there.

84. Clausius's papers on disgregation include Clausius, op. cit. (note 81), 1862. The expression noted appears in Clausius, "Über der Bestimmung der Disgregation eines Körpers und die wahre Wärmecapacitat," *Ann. Phys.* 127 (1866): 477–483; *Phil. Mag.* 31 (1866): 28–33, 28. See Gunter Bierhalter, "Die mechanischen Entropie- und Disgregations Konzepte aus dem 19 Jahrhundert: Ihre Grundlagen, ihr Versagen und ihr Enstehungshintergrund," *Arch. Hist. Exact Sci.* 32 (1985): 17–41, and M. J. Klein, "Gibbs on Clausius," *Hist. Studies Phys. Sci.* 1 (1969): 127–149.

85. His physical conclusions seemed to go against the tide of experimental evidence in that the heat capacities of a body in all three physical states were the same. This meant arguing that the measured specific heats were not the "true" specific heats. See Clausius, op. cit. (note 81), 1865. Clausius appears to have developed Rankine's argument along these lines.

86. Clausius, "Ueber die Bestimmung der Energie und Entropie eines Kör pers," *Zt. Phys. Math.* 11 (1866): 31–46; *Phil. Mag.* 32 (1866): 1–17. Here he explores internal energy but as a function of volume and temperature only.

87. Clausius, op. cit. (note 81), 1865, 401.

88. See document III-25 where Maxwell announces his discovery of Gibbs to Tait, December 1873.

89. See documents III-14, III-24.

90. See note 30 for details of Tait's publications.

91. Maxwell's statements of the second law, equating it to unavailable energy, appear in document III-21.

92. Tait never retracted this judgment despite Maxwell's efforts to educate him in Gibbs's thermodynamics and the differences between Thomson and Clausius. Maxwell gave an alternative interpretation of the history and the physics in his review of the second edition of Tait's text. See document III-45. If anything in his second edition Tait gave even less credit to Clausius. See Tait, *Sketch of Thermodynamics* (Edinburgh: David Douglas, 1877), chap. 3.

93. By the publication date of his book on thermodynamics in 1867 Tait was already collaborating with William Thomson on their textbook in mechanics. It was planned as a survey of "natural philosophy," but they completed only the first part on mechanics. It was the first textbook successfully to incorporate energy conservation into its conceptual and problem-solving structure. See William Thomson and Peter Guthrie Tait, *Treatise on Natural Philosophy* (Oxford: Clarendon Press, 1867) (second edition 1879). For a discussion of this collaboration and the initial plans for the text and what was actually completed, see Smith and Wise, op. cit. (note 28), chap. 4.

94. The assigning of priority was still a delicate matter. If Helmholtz was granted priority for energy conservation, Mayer's originality was undermined, which improved Joule's priority claims. Tait, however, did not state this in so many words. He merely quoted at length from a carefully constructed letter from Helmholtz, a less than impartial source. Tait, op. cit. (note 30), chap. I. See also Clausius, "Ueber einige Stellen der Schrifte von Helmholtz 'Ueber die Erhaltung der Kraft'," *Ann. Phys.* 89 (1853), 568–579 and 91 (1854): 601–604.

95. Maxwell, *Theory of Heat* (London: Longman, 1871). Editions appeared regularly until the fourth in 1875, the one in which major revisions of his ideas on the second law appear. See document III-27. The text was written for working men as well as students at Cambridge. Maxwell had a lifelong commitment to the education of workingmen, lecturing at Aberdeen to such groups to the horror of other academics. As an undergraduate he was drawn to Christian Socialism and taught workingmen physics as a don at Cambridge. He continued to do so at London after 1860. But addressing these audiences led to limitations. The texts

went only as far as was deemed necessary for artisans to understand the engines that they operated, and the mathematics was rudimentary. However, to follow his geometrical arguments, his readers must have had a well-developed spatial ability.

96. Document III-18 in which Maxwell contacts James Thomson to clarify the meaning of Andrews's experiments.

97. See document III-21 for Maxwell's discussion of available energy.

98. Op. cit. (note 97) Maxwell had misunderstood Clausius on two counts. If we write the first law as

$$dU = TdS + dW,$$

the energy "lost" to the system is $dQ = TdS$. In addition dS depends only on the initial and final states of the system, while dQ depends on the path of the taken.

99. Josiah Willard Gibbs, "Graphical Methods in the Thermodynamics of Fluids," *Trans., Connecticut Acad. Sci.* 2 (1873): 309–342, and "A Method of Geometrical Representation of the Thermodynamic Properties of Substances by Means of Surfaces," 2 (1873): 382–404, reprinted in *The Scientific Papers of J. Willard Gibbs*, vol. 1, ed. Henry Andrews Bumstead and Ralph Gibbs Van Name (New York: Dover reprint of Longmans Green edition 1903, 1961), 1–32, and, 33–54. The discussion of Andrews's work appears on 44–47.

100. M. J. Klein, "The Early Papers of J. Willard Gibbs A Transformation of Thermodynamics," *Proceedings of the Fifteenth International Congress of the History of Science* (Edinburgh: Edinburgh University Press, 1978), 330–341.

101. See Garber, "James Clerk Maxwell on Thermodynamics," *Amer. J. Phys.* 37 (1969): 146–155. Klein, op. cit. (note 100) notes Maxwell's appreciation of Gibbs's geometrical approach to thermodynamics.

102. For a detailed account of Maxwell's construction of these surfaces, see Ian B. Hopley, "Clerk Maxwell's Contribution to Physics," Ph. D. thesis, University of London, 1956, vol.1, 176–189.

103. See document III-33. This letter to Stokes was written almost a year before Gibbs's paper was published. Gibbs, "On the Equilibrium of Heterogeneous Substances," *Trans. Connecticut Acad. Sci.* 3 (1876–1878): 108–248, 343–542, reprinted in *Gibbs's Scientific Papers*, vol. 1, 55–353. According to Lynde Phelps Wheeler, Gibbs's biographer, Maxwell received galley proofs of the first part of this paper. Therefore, he could and did use the language and the terms Gibbs's coined before the paper was published after his letter to Stokes and his notes in document III-34 but before those of document III-35. Wheeler, *Josiah Willard Gibbs,* rev. ed. (New Haven: Yale University Press, 1952), 94.

104. Maxwell's work overlapped only aspects of the first part of Gibbs's paper, Gibbs op. cit. (note 103) (1876–1878), 63–70.

105. Document III-36.

106. Maxwell addressed the chemists in Maxwell, op. cit. (note 14). The reference to Gibbs is on p. 426. See also Garber, Brush, and Everitt, op. cit. (note 2). For Pattison Muir, see L. P. Wheeler, op. cit. (note 103), 85. For Tait see document III-42. Tait included a description of Gibbs's geometrical method in the second edition of his text. See Tait, op. cit. (note 92) (1877), chap I. Maxwell was successful at bringing Gibbs to the attention of American physicists. While building the faculty of Johns Hopkins physics department H. A. Rowland wrote to its president Gilman, May 8th, 1879.

> Some time since I called your attention to the necessity of having the subjects of Mechanics and Mechanical Drawing taught in the University. I would like at the present time to call your attention to Prof. Willard Gibbs of Yale College as an excellent candidate for the chair of Mechanics not only from his eminence but because he will fill a wide gap in the mathematical staff. There are two methods of considering such a subject as mechanics, the one being purely mathematical, and the other a combination of the mathematical and physical, and they must be cultivated by minds of different kinds. Of such men I may mention Cayley and Sylvester on the mathematical side and Helmholtz, Thomson and Maxwell on the mathematical-physical side and in the latter class should also be included Newton who was indeed the beau ideal of this class of minds. Prof. Willard Gibbs belongs preeminently to the latter class; to men who not only grasp the subject from a mathematical standpoint but who see the subject in *all* its bearing, and to whom the problems of nature are something more than targets on which to practice with their mathematics.
>
> As to his eminence there can be no doubt. Maxwell in his small work on the "Theory of Heat" has introduced no less than thirteen pages from the papers of Prof. Gibbs, and he told me personally the new method of Prof. Gibbs allowed problems to be solved which he, Maxwell, had almost concluded were incapable of solution.
>
> In conclusion I may add that Maxwell, who has the best opportunity for judging in the world, states that this class of men is very rare while brilliant men of purely mathematical class are comparatively common,
>
> Yours respectfully,
> H. A. Rowland

The editors wish to thank Bob Kargon for drawing this letter to their attention.

107. Ambrose Fleming, "Notes on Maxwell's Thermodynamic Lectures, 1878–1879," Cambridge University Library, Maxwell Collection.

108. See Garber, op. cit. (note 23) and Maxwell, "Challis's 'Mathematical Principles of Physics'," *Nature* 8 (1873): 279–280, reprinted in *Scientific Papers,* vol. 2, 338-342; Garber, Brush, and Everitt, op. cit. (note 2), 126–130 and the following exchange between Maxwell and Albert J. Mott, 131–132.

109. Clausius, "Ueber einen auf die Wärme anwendbaren mechanischen Satz," *Ann. Phys.* 141 (1870): 124-130; *Phil. Mag.* 40 (1870): 122–127. This paper is

a continuation of his disgregation papers. He later expressed disgregation and ergon in terms of Hamilton's Principle of Least Action. See, Clausius, "Ueber einen neuen mechanischen Satz in Bezug auf stationäre Bewegungen," *Ann. Phys.* 150 (1873): 106–130; *Phil. Mag.* 46 (1873): 236–244, 266-276. Clausius's virial theorem has recently been examined in Bierhalter, "Das Virial-Theorem in seiner Beziechung zu den mechanischen Grundlegungen des zweiten Hauptsatzes der Wärmelehre," *Arch. Hist. Exact Sci.* 27 (1982): 199–211.

110. Clausius, op. cit. (note 109), 127.

111. Clausius, "Über die Anwendung einer von ihm aufgestellten mechanischen Gleichung auf die Bewegung eines materiellen Punkte uneinander," *Nachrichten von der Königlichen Gesellschaft der Wissenschaften* 8 (1871): 245–266; *Phil. Mag.* 42 (1871): 321–332.

112. Clausius, "Über die Zurückfuhrung des zweiten Hauptsatzes der mechanischen Wärmetheorie auf allgemeine mechanische Principien," *Ann. Phys.* 142 (1871): 433–461; *Phil. Mag.* 42 (1871): 161–181.

113. Clausius, "Über den Zusammenhang des zweite Haupsatzes der mechanischen Wärmetheorie mit dem Hamilton'schen Princip," *Ann. Phys.* 146 (1872): 585–591; *Phil. Mag.* 44 (1871): 365–369. Others were involved in similar attempts. See Bierhalter, "Zu Szilys Versuch einer mechanischen Grundlegungen des zweiten Haupsatzes der Thermodynamik," *Arch. Hist. Exact Sci.* 28 (1983): 25–35.

114. Boltzmann, "Über die mechanischen Bedeutung des zweiten Haupsatzes der Wärmetheorie," *Sitz. Math.-Naturwiss. Cl. Akad. Wiss., Wien* 52 (1866): 195–220. See Bierhalter, "Boltzmanns mechanische Grundlegung des zweiten Haupsatzes der Wärmelehre aus dem Jahre 1866," *Arch. Hist. Exact Sci.* 24 (1981): 195–205, 207–220.

115. Helmholtz,"Verallgemeinung der Sätze über die Statik der monocyklischer System," *Sitz., Berlin* 159 (1884): 1197–1201; "Studien zur Statik monocyklischer System," (1884): 159–177, 311–318, 755–759; "Principien der Statik monocyklischen System," *J. reine angew. Math.* 97 (1884): 111–140, 317–336. Boltzmann, "Ueber die Eigenschaften monocyclischer und anderer damit verwandter Systeme," *Sitz. Math.-Naturwiss. Cl. Akad. Wiss., Wien* 90 (1884): 231–245; "Neuer Beweis eines von Helmholtz aufgestellen Theoreme betreffend die Eigenschaften monocyclischer Systeme," *Gottingen Nachr.* (1886): 209. These later attempts to search for a mechanical basis for thermodynamics have recently been examined by Bierhalter,"Die von Helmholtzschen Monozykel- Analogien zur Thermodynamik ...," *Arch. Hist. Exact Sci.* 29 (1983): 95–100.

116. Maxwell, op. cit. (note 14). The section on the virial is on pp. 420–21. Scientists had only recently realized the continuity of the three states of matter. See Brush, op. cit. (note 8), vol. 1, 256–264.

117. J. D. van der Waals thesis appeared as *Over de continuiteit van den gasen vloeistoftoestand* (Leiden, 1873), trans. R. Threlfall and J. Adair titled

"The Continuity of the Liquid and Gaseous States," *Physical Memoirs* (London: Physical Society of London, 1890): vol. 1. The full thesis has recently been translated with a detailed essay on later developments in physics based on van der Waals's equation of state in *On the Continuity of the Gaseous and Liquid States*, ed. and trans. J. S. Rowlinson (Amsterdam, 1988). See also M. J. Klein,"The Historical Origins of the van der Waals Equation," *Physica* 73 (1974): 28–47

118. Maxwell's assessment of van der Waals has recently been critiqued in Kosta Gavroglu, "The Reaction of British Physicists and Chemists to van der Waals' Early Work and to the Law of Corresponding States," *Hist. Stud. Phys. Bio. Sci.* 20 (1990): 199–238. Gavroglu sees the problem in Maxwell's emphasis on the differences between the gaseous and liquid states and the transition from one to the other. van ders Waals's focus was on the continuity of the two states of matter. He suggests that Maxwell's negative attitude led to the neglect of van der Waals work in Britain in the late nineteenth century.

119. Rather than seeing Maxwell not accepting van der Waals's work we have to understand that he saw its significance for the theory of matter in terms we no longer accept as important. His referee reports indicate that he did give considerable thought thought to van der Waals's work.

120. The relationship between kinetic theory and theories of matter have been analyzed in Brush, *Statistical Physics and the Atomic Theory of Matter from Boyle and Newton to Landau and Onsager* (Princeton: Princeton University Press, 1983), 39–78. See also Brush, op. cit. (note 8) vol. 2, 401–416.

121. Brush, op. cit. (note 120), 73–77, discusses the relationship between Gibbs's thermodynamics and statistical mechanics and Maxwell's reaction to it. See also, M. J. Klein, "Maxwell His Demon and the Second Law of Thermodynamics," *Amer. Sci.* 58 (1970): 84–97, and E. E. Daub, "Maxwell's Demon," *Stud. Hist. Phil. Sci.* 1 (1970): 213–227. For the broader, cultural context of Maxwell's demon, see Silvan S. Schweber, "Demons, Angels and Probability: Some Aspects of British Science in the Nineteenth Century," in *Physics as Natural Philosophy*, eds. Abner Shimony and Herman Feshbach (Cambridge Mass.: MIT Press, 1982), 319–363.

122. Smith and Wise, op. cit. (note 28), 86–99 emphasize this orientation in Thomson's physics.

123. Brush, op. cit. (note 11), vol. 2, 590, points out the importance of this argument in Maxwell's understanding of irreversibility.

124. See Brush, op. cit. (note 11), vol. 2, 627–637 for Zermelo and Poincaré and the mechanical worldview.

125. Thomson, op. cit. (note 73), 106

126. Helmholtz's *Popular Scientific Lectures* (New York: Dover, 1962), 74 from *Ueber die Wechselwirkung der Naturkräfte und die darauf bezüglich neusten*

Ermittelungen der Physik, trans. John Tyndall (Königsberg: Gräfe and Unzer, 1853); "On the Interaction of Natural Forces," *Phil. Mag.* 11 (1856): 489–518.

127. Thomson, "On Mechanical Antecedents of Motion, Heat, and Light," *Rep. 24th Meeting BAAS.*, pt. II. (1854) 59–63. Reprinted in Thomson's *Mathematical and Physical Papers,* vol. 2 (Cambridge: Cambridge University Press, 1884), 37. Emphasis is in the original.

128. W. J. M. Rankine, "On the Reconcentration of the Mechanical Energy of the Universe," *Phil. Mag.* 4 (1852): 358–360.

129. Clausius, "Ueber die Concentration von Wärme—und Lichtstrahlen und die Gränzen ihrer Wirkung," *Ann. Physik* 121 (1864): 1–44; translation in Clausius, op. cit. (note 59) 290f.

130. Clausius, "On the Second Fundamental Theorem of the Mechanical Theory of Heat," *Phil. Mag.* 35 (1868): 405–419, 418–419, translated from the author's German text of a lecture given in September 1867.

131. H. Spencer, *First Principles* (reprint of the fourth edition; New York: Dewitt Revolving Fund, 1958), 487–508.

132. See their 1873 correspondence, in Garber, Brush, and Everitt, op. cit. (note 2), 156–162.

133. W. Smyth, "Mr. Spencer and the Dissipation of Energy," *Nature* 5 (1872): 322. J. Croll, "What Determines Molecular Motion?—The Fundamental Problem of Nature," *Phil. Mag.* 44 (1872): 1–25. See S.G. Brush, *The Temperature of History: Phases of Science and Culture in the Nineteenth Century* (New York: Burt Franklin, 1978), chap. V.

134. A. Fick, *Die Naturkräfte in ihrer Wechselsbeziehung* (Würzburg: Stakel, 1869). J. R. Mayer, "Mechanical Theory of Heat," *Nature* 1 (1870): 566–567. Translated extract from his lecture "Ueber nothwendige Consequenzen und Inconsequenzen der Wärmemechanik," 1869.

135. B. Stewart, "What is Energy? IV.—The Dissipation of Energy," *Nature* 2 (1870): 270–271. See also Stewart, *The Conservation of Energy* 2d ed. (London: King 1874), 142, 153, 167; E. Maitland, "The Extinction of Stars," *Nature* 2 (1870): 211-212.

136. H.F. Walling, "The Relation of the Dissipation of Energy to Cosmical Evolution," *Proc. Amer. Assoc. Adv. Sci.* 22 (1873): 46–49. P. E. Chase, "Cosmical Activity of Light," *Phil. Mag.* 50 (1875): 250–253.

137. Knott, op. cit. (note 39), 213.

138. P.G. Tait, "Address to Section A, Mathematical and Physics Section of the British Associaton," *Rept. 41st Meeting BAAS.* (1871): 1–8; *Nature* 4 (1871):

270–273. The demon does appear in Tait, *Lectures on Some Recent Advances in the Physical Sciences* (London: Macmillan, 1876), 188–120, 146. See also the book published anonymously by B. Stewart and P. G. Tait, *The Unseen Universe; of, Physical Speculations on a Future State* 7th ed. (London: Macmillan, 1878). Maxwell's demon is invoked to violate the second law in a spiritual world.

139. J. Loschmidt, "Der zweite Satz der mechanischen Wärmetheorie," *Sitz. Math.-Naturwiss. Cl. Akad. Wiss., Wien*, 59 (1869): 395–418.

140. E. E. Daub, op. cit. (note 121).

141. For example, [Anon.] *Phil. Mag.* 43 (1872): 149–151; B. Stewart, "Maxwell on Heat," *Nature* 5 (1872): 319–320.

142. W. Thomson, "The Kinetic Theory of the Dissipation of Energy," *Proc. R. Soc. Edinburgh* 8 (1874): 325–334; *Nature* 9 (1874): 441–444; *Phil. Mag.* 33 (1892): 291–299, reprinted in Brush, *Kinetic Theory*, vol. 2. In a footnote Thomson defined the demon: "according to the use of this word by Maxwell, is an intelligent being endowed with free-will and fine enough tactile and perceptive organization to give him the faculty of observing and influencing individual molecules of matter." The report on this paper by Neesen in *Fortschr. Physik* 30 (1874): 673–674 mentions "Wessen (Maxwell's Däemonen) die Fähigkeit haben, jedes einzelne Molecül in jedem Augenblick zu fassen."

143. J. Loschmidt, "Über den Zustand des Wärmegleichgewichtes eines Systems von Korpern mit Rucksicht auf die Schwerkraft," *Sitz. Math.-Naturwiss. Cl. Akad. Wiss., Wien*, 73 (1876): 128–142, 366–372; 75 (1877): 287–298; 76 (1877): 209–225, abstract in "Cosmical Results of the Modern Heat Theory," *Nature* 18 (1878): 184–185.

144. L. Boltzmann, "Über die Beziehung eines allgemeine mechanischen Satzes zum zweiten Hauptsatze der Wärmetheorie," *Sitz. Math.-Naturwiss. Cl. Akad. Wiss., Wien*, 75 (1877): 67–73; translation in Brush, op. cit. (note 11), vol. 2. See also Daub, op. cit. (note 121).

145. See the last paragraph of Document III-52 in Garber, Brush, and Everitt, op. cit. (note 2) and Document III-45 in this volume.

146. S. T. Preston, "On the Diffusion of Matter in Relation to the Second Law of Thermodynamics," *Nature* 17 (1877): 31–32; "On a Means for Converting the Heat-Motion Possessed by Matter at Normal Temperature in Work," *Nature* 17 (1878): 202–204. J. Aitken, "On a Means for Converting the Heat Motion Possessed by Matter at Normal Temperatures into Work," *Nature* 17 (1878): 260. A.S. Herschel, "On the Use of the Virial in Thermodynamics," *Nature* 18 (1878): 39–40. S. T. Preston, "On the Availability of Normal-Temperature Heat-Energy," *Nature* 18 (1878): 92–93.

147. F. Zöllner, "Thomson's Däemonen und die Schatten Plato's," in his *Wissenschaftliche Abhandlungen*, Band I (Leipzig: Staackmann, 1878), 710–732. P. G. Tait, "Zöllner's Scientific Papers," *Nature* 17 (1878): 420–422.

148. Translation quoted from Klein, op. cit. (note 121), 93. R. Clausius, "Ueber eine von Hrn. Tait in der mechanischen Wärmetheorie angewandte Schlussweisse," *Ann. Phys.* 2 (1877): 130–133; reprinted with slight modifications in *Die Mechanische Wärmetheorie*, 2d ed., Bd. II, 2 (Braunschweig: Vieweg, 1879), 314–317

149. The problem of irreversibility was also discussed later in the context of the validity of the equipartition theorem as a consequence of Boltzmann's H theorem—a discussion in which the demon played no part. See Brush, op. cit. (note 11), vol. 2, 616–627, and Garber, op cit (note 23).

150. W. Thomson, "The Sorting Demon of Maxwell," *Proc. R. Inst.* 9 (1879): 113–114. For another view of the nature of demons, see J. Bailey, "Demonology at Home and Abroad," *Blackwood's Edinburgh Magazine* 99 (1866): 502–518. Smith and Wise, op. cit. (note 28), 623–628, discuss the varied functions that Maxwell's demon performed for William Thomson.

151. Smith and Wise, op. cit. (note 28), chap. 12, discuss the development of Thomson's ideas also along these lines.

152. S. T. Preston, "Temperature Equilibrium in the Universe in Relation to the Kinetic Theory," *Nature* 20 (1879): 28. G. G. Stokes, "On the Bearings of the Study of Natural Science, and of the Contemplation of the Discoveries to Which the Study Leads, on Our Religious Ideas," *J. Trans. Victoria Inst.* (1880): 226–247. We thank David Wilson for this reference. W. Thomson, "Steps toward a Kinetic Theory of Matter," *Rept. 54th Meeting BAAS.* (1884): 613–622; *Nature* 30 (1884): 417–421. P. G. Tait, *Heat* (London: Macmillan, 1884), 366. H. Poincaré, "Le Mécanisme et l' Experiénce," *Revue de Metaphysique et de Morale* 1 (1893): 532–537, English translation in Brush, op. cit (note 11), vol. 2. K. Pearson, *The Grammar of Science* (London: Scott, 1865), 100–101. J. H. Jeans, *The Dynamical Theory of Gases* (Cambridge: Cambridge University Press, 1904), 168.

153. H. Whiting, "Maxwell's Demons," *Science* 6 (1885): 83. On bacilli see G.J. Stoney, "Suggestion as to a Possible Source of the Energy Required for the Life of Bacilli, and as to the Cause of Their Small Size," *Sci. Proc. R. Dublin Soc.* 8 (1893): 154–156. On radioactivity and the demon see W. Crookes, "Presidential Address," *Rept. 68th Meeting BAAS* (1898): 3–38, 27. For philosophical implications with regard to time see K. Pearson, *The Grammar of Science*, 2d ed. (London: Black, 1900), note VII.

154. H. Adams, letter to Cecil Spring Rice, 11 November 1897, in *Letters of Henry Adams (1892-1918)*, ed. W. C. Ford (Boston: Houghton Mifflin, 1938), 135–136; letter to Brooks Adams, 2 May 1903, in *Henry Adams and his Friends*, ed. H. D. Cater (Boston: Houghton Mifflin, 1947), 545; *Mont-Saint-Michel and Chartres* (Boston: Houghton Mifflin, 1905), 375; "The Rule of Phase Applied to History,"in *The Degradation of the Democratic Dogma* (reprint, New York: Capricorn/Putnam, 1958), 273. R.P. Blackmur, *Henry Adams* (New York: Harcourt Brace Jovanovich, 1980), 234–235.

155. H. S. Leff and A. F. Rex eds., *Maxwell's Demon: Entropy, Information, Computing* (Princeton, N. J.: Princeton University Press, 1990). For a bibliography, see H. S. Leff, "Resource Letter MD-1: Maxwell's Demon," *Am. J. Phys.* 58 (1990): 201–209.

156. James Clerk Maxwell to Francis Galton, February 26th, 1879, University College, London. The editors wish to thank professor Ruth Cowan for bringing this letter and Galton's reply to their attention.

157. Francis Galton to Maxwell, February 27th, 1879. The paper referred to by Thomson is "The sorting Demon of Maxwell," *Proc. R. Inst.* 9 (1879): 113–114, read February 28th, 1879, *Nature* 20 (1879): 123. Quote also in Porter, *The Rise of Statistical Thinking* (Princeton N. J.: Princeton University Press, 1986), 201 n.

158. For recent discussions see Brush, op. cit. (note 120) and Theodore Porter, "A Statistical Survey of Gases: Maxwell's Social Physics," *Hist. Studies Phys. Sci.* 12 (1981): 77–116. Porter sees Maxwell's view of mechanics as deterministic. However, for Maxwell our lack of knowledge of molecules' motions was inherent in the nature of mechanical systems.

159. See Garber, op. cit. (note 23).

160. Maxwell, "Does the Progress of Physical Science Tend to Give any Advantage to the Opinion of Necessity (or Determinism) over That of the Contingency of Events and the Freedom of the Will?" in Campbell and Garnett, op. cit. (note 51), 434–444, the essay is dated February 11th, 1873.

161. Peter Harman considers only the philosophical aspects of Maxwell's ideas on mechanics and only follows his physical arguments in the theory of the electromagnetic field. Harman, "Edinburgh Philosophy and Cambridge Physics: The Natural Philosophy of James Clerk Maxwell," in *Wranglers and Physicists*, ed. Peter Harman (Manchester: University of Manchester Press, 1985), 202–224.

162. Maxwell, op. cit. (note 160), 440.

163. Boltzmann had actually begun this new approach in 1868 by offering a derivation of the distribution law based upon combinatorial theory. Boltzmann, "Studien uber das Gleichgewicht der lebendigen Kraft zwischen bewegten materiellen Punkte," *Sitz. Math.-Naturwiss. Cl. Akad. Wiss., Wien* 58 (1868): 517–560.

164. Brush, op. cit. (note 11), vol. 1 277, argues that the two are not inexorably linked.

165. Boltzmann op. cit. (note 115).

166. Thomson and Tait, op. cit. (note 93). Maxwell had read the proofs of the volumes and generally consulted with both authors, usually through Tait, during

its composition and corrections. The construction and contents of Thomson and Tait's treatise is discussed in Smith and Wise, op. cit. (note 28), chap. 11.

167. Maxwell *Treatise on Electricity and Magnetism*, vol. 2 reprint of 3d ed.; New York: Dover, 1945), 198.

168. Maxwell, *Matter and Motion* (London: Society for the Promotion of Christian Knowledge, 1876), 122. The volume was reviewed by Tait in *Nature* 16 (1877): 119.

169. Brush, op. cit. (note 11), vol. 2, 593 has pointed out that Maxwell's ideas on irreversibility are very close to twentieth-century stochastic explanations rather than to statistical explanations.

170. Maxwell, op. cit. (note 168), 48.

171. Maxwell, "On Boltzmann's Theorem on the Average Distribution of Energy in a System of Material Points," *Trans. Cambridge Phil. Soc.* 13 (1879): 547–570, 548. Reprinted in vol. 2, *Scientific Papers*, 713–741, and document V-7.

172. Maxwell, op cit (note 171), 549–550.

173. The validity of this postulate, the "ergodic" hypothesis, was much discussed afterward with considerable misrepresentation of Maxwell's opinions. For a discussion of the issues and Maxwell's point of view, see Brush, "Foundations of Statistical Mechanics, 1845–1905," *Arch. Hist. Exact Sci.* 4 (1966): 147–183. See also Brush, op. cit. (note 11), vol. 2, 335–421.

174. This was also Gibbs's choice who called such an average the "ensemble" average. It is much easier to compute in most cases.

175. See Campbell and Garnett, op. cit. (note 57), 570–571.

176. For general surveys of the debate, see G. H. Bryan, "Report on the Present State of Our Knowledge of Thermodynamics Part II The Laws of Distribution of Energy and Their Limitations," *Rep. 64th Meeting BAAS* (1894): 64–108. See also Paul and Tatiana Ehrenfest, *The Conceptual Foundations of the Statistical Approach in Mechanics*, trans. M. J. Moravcsik (Ithaca: Cornell University Press, 1959); Brush, op. cit. (note 103), 363–385; M. J. Klein, op. cit. (note 134); and H. Bernhardt, "Über die Entwicklung und Beudeutung der Ergodenhypothese in den Anfangen der statistischen Mechanik," *NTM*, 8 (1971): 13–25.

177. Boltzmann,"On Boltzmann's Theorem on the Average Distribution of Energy in a System of Material Points," *Phil. Mag.* 14 (1882): 299–312. The one physicist who really developed Maxwell's gas theory ideas and coined the language for its use was Josiah Willard Gibbs. For Gibbs's statistical mechanics, see Klein, "Some Historical Remarks on the Statistical Mechanics of J. W. Gibbs," in *From*

Ancient Omens to Statistical Mechanics, ed. J. L. Beggren and B. R. Goldstein *Acta Hist. Sci. Nat. Med.* 39 (1987).

178. For some details of this debate see Brush, op. cit. (note 11), vol. 2, 356–363.

179. G. H. Bryan, "Report of the State of Our Knowledge of Thermodynamics. Part I Researches Related to the Connection of the Second Law with Dynamical Principles," *Rep. 61st Meeting BAAS* (1891): 85–122.

180. James Jeans developed his own hydrodynamical model of the motion of the phase points representing states of the gas. His alternative condition for the validity of the equipartition theorem was that the "streamlines" in this abstract fluid in 6n-dimensional space be closed. By 1904 Jeans could not find any reason for the condition not being valid. Equipartition stayed in the foundations of gas theory. See Jeans, *The Dynamical Theory of Gases* (Cambridge: Cambridge University Press, 1904).

181. John William Strutt, Third Baron Rayleigh, "Remarks on Maxwell's Investigation Respecting Boltzmann's Theorem," *Phil. Mag.* 33 (1892): 356–359 and "The Law of Partition of Energy," *Phil. Mag.* 49 (1900): 98–118.

182. Paul and Tatiana Ehrenfest, op. cit. (note 176). See, M. J. Klein, *Paul Ehrenfest* (New York: Elsevier, 1970). For the history of the ergodic hypothesis beyond the Ehrenfests' article, see Stephen G. Brush, "Proof of the Impossibility of an Ergodic System," *Transport Theory and Statistical Physics* 1 (1971): 287–311.

183. Maxwell, "On Stresses in Rarified Gases Arising from Inequalities in Temperature," abstract in *Proc. R. Soc. London* 27 (1878): 304–308 and *Nature* 18 (1878): 54–55, printed in full in *Phil. Trans. R. Soc. London* 170 (1880): 231–256, reprinted in vol. 2, *Scientific Papers*, 681–712.

184. William Crookes, "On Attraction and Repulsion Accompanying Radiation," *Proc. R. Soc. London* 22 (1873): 37–41; *Phil. Mag.* 48 (1874): 81–95, printed in full in *Phil. Trans. R. Soc. London* 164 (1874): 501-527; "Part II," *Proc. R. Soc. London* 23 (1875): 373–378, printed in full *Phil. Trans. R. Soc. London* 165 (1875): 510–547; "Part III and IV," *Proc. R. Soc. London* 24 (1876): 276–279, 279–283, printed in full in *Phil. Trans. R. Soc. London* 166 (1876): 325-376. The Bakerian Lecture "Part V," was read in December 1878 and abstracted in *Proc. R. Soc. London* 27 (1878): 29–33 and published in full in *Phil. Trans. R. Soc. London* 169 (1878): 243–318. For the role of Crookes in the development of the radiometer, see A. E. Woodruff, "William Crookes and the Radiometer," *Isis* 57 (1966): 188–198, and "The Radiometer and How It does Not Work," *Physics Teacher* 6 (1968): 358–364.

185. It was long thought that light might exert a pressure, although there was no proof until the work of Lebedev in Russia and Nichols and Hull in the United States around 1900. See Petr Lebedev, "Maksvello-Bartolyevskiya

davleniya luchistoi energii," *Zhurnal Russkago FizikoKhimicheskago Obschestva pri Imperatorskom S.-Petersburgskom Universitet* 32 (1900): 211–217. "Untersuchungen uber die Druckkrafte des Lichts," *Ann. Phys.* 6 (1901): 433–458; E. F. Nichols and G. F. Hull, "A Preliminary Communication on the Pressure of Heat and Light Radiation," *Physical Review* 13 (1901): 307–320. For historical surveys and further references, see Edmund Whittaker, *A History of the Theories of Aether and Electricity* (New York: Thomas Nelson, 1951), vol. 1, 273–274, Crookes op. cit. (note 206) (1874) and G. Berthold, "Notizen zur Geschichte des Radiometers," *Ann. Phys.* 158 (1876): 483–487.

186. W. B. Carpenter, *Nineteenth Century* 1 (1877): 242. For William Thomson's fascinated but noncommittal reaction, see Silvanus P. Thompson, *The Life of William Thomson, Baron Kelvin of Largs,* (London: Macmillan, 1910) vol. 2, 1125. Maxwell also witnessed Crookes experiments at the same soiree, see document VI-2; his remarks are descriptive only.

187. Maxwell, op. cit. (note 167), vol. 2, article 792, 441.

188. Maxwell to George Bond, August 1863, in Brush, Everitt, and Garber, *Maxwell on Saturn's Rings* (Cambridge Mass.: MIT Press, 1983), 166–168, 167, and this volume document VI-1.

189. Allan Ferguson, "The Clerk-Maxwell Centenary Celebration," *Nature* 128 (1931): 604–608.

190. Arthur Schuster, "On the Nature of the Force Producing the Motion of a Body Exposed to the Rays of Heat and Light," *Proc. R. Soc. London* 24 (1876): 391–392, printed in full in *Phil. Trans. R. Soc. London* 166 (1877): 715–724. See also Schuster, "The Radiometer and Its Lessons,"*Nature* 17 (1877): 143, and Schuster, *Biographical Fragments* (London: Macmillan, 1932), 230–232.

191. A. Ledieu, "Examen de l'action mécanique possible de la lumière. Étude du radioscope de M. Crookes," *Comptes Rendus* 82 (1876): 1241–1245, 1293–1297; P. Montani,"Sull'azione meccanica esercitata dalla luce," *Atti dell Reale Accademia dei Lincei Roma* 3 (1876): 597–600. W. M. Williams, "The Philosophy of the Radiometer and Its Cosmical Revelations," *Journal of Science* 6 (1876): 517–522. I. Carbonelle and E. Ghysens, "L'action mécanique de la lumière," *Annales de la société scientific de Bruxelles* 1 (1877): 19–74, seem to retract their explanation of the radiometer at the end of their paper. Ether effects were invoked by James Challis, "A Theory of Mr. Crookes's Radiometer," *Phil. Mag.* 1 (1876): 395–397; "Theoretical Explanations of additional Phenomena of the Radiometer," *Phil. Mag.* 2 (1876): 374–379; and "A Theory of the Action of the Cup-shaped Radiometer with Both Sides Bright," *Phil. Mag.* 3 (1877): 278–281, 395–396; E. H. Cook, "The Existence of the Luminiferous Ether,"*Phil. Mag.* 11 (1881): 477–492, and as late as 1905 by Duncan in Robert K. Duncan, *The New Knowledge* (New York: Barnes, 1905). Electrical explanations were suggested by Joseph Desaulx, "The Direct Motion in the Radiometer, an Effect of Electricity," *Nature* 14 (1876): 288–289; "The Inverse Rotation of the

Radiometer, an Effect of Electricity," *Nature* 14 (1876): 449–450; W. G. Hankel, "Ueber das Crookes'sche Radiometer," *Ann. Phys.* 2 (1877): 627–631; W. de Fonvielle, "The Radiometer in France," *Nature* 14 (1876): 296–297; and P. de Heen, "Note sur la théorie du radiomètre, sur la photographie Le Bon et sur la nature de l'électricité," *Bulletins de l'Academie royale des sciences, Bruxelles* 32 (1896): 75–82. Those who thought gas effects more likely than light pressure but did not develop detailed theories include A. Berin, "Sur la radiomètre," *Ann. chim.* 10 (1877): 396–408; H. A. Cunnington, "Heat and Light a Motive Power; or Experiments with Radiometers," *Popular Science Review* 15 (1876): 128–137; A. H. L. Fizeau, discussion remarks, *Comptes Rendus* 82 (1876): 1252, 1413; G-A. Hirn, "Sur le maximum de la puissance repulsive possible des rayons solaires," *Comptes Rendus* 82 (1876): 1472–1476; J. Moutier, "Sur l'évaporation; sur les cycles réversibles; sur les mouvements des corps échauffés," *Les Mondes* 40 (1876): 574–579; J. W. Phelps, "The Radiometer," *Popular Science Monthly* 14 (1878): 76–77.

192. See, J. Puluj, "Ueber das Radiometer," *Sitz. Math.-Naturwiss Cl. Acad. Wiss. Wien* 80 (1880): 132–136. F. Zollner, "Untersuchungen über die Bewegungen strahlender und bestrahleter Korper," *Ann. Phys.* 160 (1877): 296–317, 459–466. E. Pringsheim, "Ueber das Radiometer," *Ann. Phys.* 18 (1883): 1–32.

193. *Cambridge University Reporter*, 15 May 1877, 434, and 20 May 1876, 496.

194. See document VI-22. Maxwell well knew that consideration on the molecular level of inelastic collisions of molecules should be avoided. See document VI-15.

195. Maxwell had specifically rejected the possibility of slip in analyzing his experiments on the viscosity of gases. See Maxwell, "On the Viscosity or Internal Friction of Air and other Gases," *Phil. Trans. R. Soc. London* 154 (1866): 249–268, reprinted in *Scientific Papers*, vol. 2, 1–25, 10.

196. Maxwell, op. cit. (note 183), 685.

197. H. von Helmholtz and G. von Piotrowski, "Ueber Reibung tropfbarer Flussigkeiten," *Sitz. Math.-Naturwiss. Cl. Akad. Wiss. Wien* 40 (1860): 607–658, and A. Kundt and E. Warburg, "Ueber Reibung und Wärmeleitung verdunnte Gase," *Ann. Phys.* 155 (1875): 337–366, 525–550; 156 (1876): 177–211.

198. Maxwell, op. cit. (note 183), 699.

199. Maxwell quoted the results Kundt and Warburg had found for air at different pressures P, flowing through glass capillary tubes from 17° C to 28 ° C, G = 8/P centimetres Maxwell, op. cit. (note 183), 703–712.

200. Maxwell, op. cit. (note 189), 703–704.

201. Reynolds's paper appeared eventually as, Osborne Reynolds, "On Certain Dimensional Properties of Matter in the Gaseous State. Part I Experimental Researches on Thermal Transpiration through Porous Plates, and on the Laws of

Transpiration and Impulsion, including an Experimental Proof that a Gas is not a Continuous Plenum. Part II On an Extension of the Dynamical Theory of Gas which includes the Stresses, Tangential and Normal, Caused by a Varying Condition of the Gas, and Affords an Explanation of the Phenomena of Transpiration and Impulsion," *Proc. R. Soc London* 28 (1879): 304–321, printed in full in *Phil. Trans. R. Soc. London* 170 (1880): 727–845.

202. Thomson seems to have used this method of "refereeing" freely with other authors. See, Smith and Wise, op. cit. (note 28). Stokes did make some attempt at anonymity. As secretary of the Royal Society he sent Maxwell a typed copy of portions of the report rather than the original.

203. For Thomson's report on Maxwell's paper, see Royal Society Archives.

204. This is the paper published as Reynolds, op. cit. (note 201).

205. Reynolds op. cit. (note 201), 305.

206. For details on this see Stephen Brush and Francis Everitt, "Maxwell, Osborne Reynolds and the Radiometer," *Hist. Studies Phys. Sci.* 1 (1969): 105–125.

207. Document VI-22.

208. Letter from Horace Lamb to Joseph Larmor, October 1 1905, in Cambridge University Library, Stokes Papers, Add. 7618, Box 3. The letter was written as Larmor prepared an edition of Stokes's correspondence.

209. Letter from Peter Guthrie Tait to George Gabriel Stokes, December 1879, Cambridge University Library, Stokes Papers, Box 13.

210. Reynolds, op. cit. (note 201).

211. Letter from Stokes to Reynolds, 5 November 1879. Cambridge University Library, Stokes-Kelvin Papers, Add. 7618, Box 1.

212. Reynolds, op. cit. (note 201), read at the April 8th 1880 meeting, with a letter from Stokes of 13 March 1880. The page 300 appears twice in this volume of the *Proc. R. Soc. London*. See also the letters from Reynolds to Stokes, 6 November 1879 and 16 March 1880 in *Memoir and Scientific Correspondence of the Late Sir George Gabriel Stokes*, ed. Joseph Larmor (Cambridge: Cambridge University Press, 1907), vol. 1, 231–232.

213. Stokes, op. cit. (note 212). The letters to which he probably refers are documents VI-19 and 22, respectively.

214. William Thomson to Stokes, 11 April 1880. Cambridge University Library, Stokes-Kelvin papers Add. 7618, Correspondence, Box 1.

215. W. Feddersen, "Über Thermodiffusion von Gasen," *Ann. Phys.* 148 (1873): 302–311; *Phil. Mag.* 46 (1873): 55–62.

216. Carl Neumann, "Vorlaufige Conjectur über die Ursachen der thermoelektrischen Strome," *Ber. Leipzig*, 24 (1872): 49–64.

217. Thomson, op. cit. (note 214).

218. Reynolds, op. cit. (note 201). The Violle paper is J. Violle, "Exposé des recherches entreprises sur la thermodiffucion," *Journal de Physique* 4 (1875): 97–104. In attempting to repel the German invasion Reynolds hastily grasped at a French straw. He misunderstood Violle's remarks.

219. George F. Fitzgerald, "On Professor Osborne Reynolds's Paper 'On Certain Dimensional Properties of Matter in the Gaseous State', " *Phil. Mag.* 11 (1881): 103–109.

220. Reynolds, "Certain Dimensional Properties of Matter in the Gaseous State. An Answer to Mr. F. G. Fitzgerald," *Phil. Mag.* 11 (1881): 335–342.

221. Maxwell, op. cit. (note 167), vol. 2, article 135a, equation 50. For gases, see Maxwell, op. cit. (note 183), 688.

222. Sidney Chapman to Stephen Brush, 11 July 1961, quoted in Brush, op. cit. (note 8), p. 455

223. E. Ikenberry and C. Truesdell, "On the Pressures and the Flux of Energy in a Gas According to Maxwell's Kinetic Theory, I," *J. Rational Mech. Anal.* 5 (1956): 1–54.

224. Boltzmann, "Über das Gleichgewicht der lebendigen Kraft zwischen progressiver und Rotationsbewegung bei Gasmolekülen," *Berlin Akad. Monats.* (1888): 1395–1408, translated in Planck, "Maxwell's Influence on Theoretical Physics in Germany," in J. Thomson, *James Clerk Maxwell: A Commemorative Volume 1831–1931* (Cambridge: Cambridge University Press, 1931), 45–65, 55. The 1867 paper under discussion is Maxwell, op. cit. (note 2).

225. Boltzmann, "On Maxwell's Method of Deriving the Equations of Hydrodynamics from the Kinetic Theory of Gases," *Rep. 64th Meeting BAAS* (1894): 579.

II Documents: From Kinetic Theory to Thermodynamics

1. Letter from Maxwell to William Thomson, February 27, 1866

Glasgow University Library, Kelvin Papers, M 19

8 Palace Garden Terrace
London W.
1866 Feb 27

Dear Thomson

In working at the Dynamical Theory of Gases I have come on the following paradox, which I intend to think about, but I should be obliged to you for the benefit of your views.

1st Suppose $\overline{\xi^2}$ to represent the mean square of the velocity of a molecule in direction of x it is easy to show that

$$p = \rho\,\overline{\xi^2}$$

2nd If $\overline{\eta^2}$ and $\overline{\zeta^2}$ be the mean square of the velocities in the other directions and if β be the ratio of the energy of rotation or other internal motion to that of translation then the mean total energy of a molecule will be

$$\frac{1}{2}M(\overline{\xi^2} + \overline{\eta^2} + \overline{\zeta^2})(1 + \beta)$$

and that of unit volume

$$\frac{3}{2}(1 + \beta)\rho\,\xi^2 = \frac{3}{2}(1 + \beta)p$$

and that of unit of mass

$$\frac{3}{2}(1 + \beta)\frac{p}{\rho}$$

3rd Now let unit of mass be enclosed in volume V and let V become $V+dV$ then work $= p\,dV$ is done and we must have

$$d(\frac{3}{2}(1+\beta)pV) + p\,dV = 0$$

or since $p\,V = \xi^2$ and $V = 1/\rho$

$$\frac{3}{2}(1+\beta)\frac{d\xi^2}{\xi^2} = \frac{d\rho}{\rho}$$

All this is the ordinary theory putting $\frac{3}{2}(1+\beta) = 1-\gamma$ and $\xi^2 = CT$(emperature)

Now comes the difficulty

4th To determine the conditions of equilibrium of temperature in a heavy gas.

The mean energy of a molecule is

$$\frac{1}{2}M(\xi^2+\eta^2+\zeta^2)(1+\beta)$$

If it ascends a distance dx against a force g this is diminished by $g\,M\,dx$. If its mean energy thus diminished is greater than that of the molecules in the stratum into which it has come it will increase the mean energy there, if less it will diminish it therefore for equilibrium of heat

$$g\,M\,dx + d(\frac{1}{2}M(\xi^2+\eta^2+\zeta^2)(1+\beta) = 0$$

or

$$gdx + \frac{3}{2}(1+\beta)d\overline{\xi^2} = 0$$

Now

$$g\,dx = -\frac{dp}{\rho} = \frac{3}{2}(1+\beta)\frac{d\overline{\xi^2}}{<\overline{\xi^2}>} =< \frac{dp}{p} = \frac{5}{2}(1+\beta)>$$

This is the condition of no conduction of heat up or down in a heavy gas.

Now since

$$p = \rho\overline{\xi^2}, \qquad \frac{dp}{p} = \frac{d\rho}{\rho} + \frac{d\xi^2}{\xi^2}$$

If then a mass of gas under gravity is left to itself the law of temperature will be

$$\frac{dT}{T} = \frac{2}{3(1+\beta)}\frac{dp}{p}$$

whereas if the pressure of the gas is changed [by] dp

$$\frac{dT}{T} = \frac{2}{(5+3\beta)}\frac{dp}{p}$$

so that the temperature in a vertical column decreases faster with the <height> pressure than that of a portion of gas carried up the column bodily.

Now for the paradox

5th Take a large mass of gas and let it come into thermic equilibrium. Take a small portion in a cylinder and piston without weight or counterbalanced. Raise it and let it expand so as to have the same pressure with the surrounding gas. Then as it is hotter it will be lighter and will be buoyant, so that, as it goes by it will do work. When it comes to the top let it remain till it cools to the surrounding temperature then lower it keeping the pressure equal to the surrounding pressure then it will be colder and therefore denser than the surrounding gas and will sink with a force which may be made to do work. Thus by means of a material agency mechanical effect is derived from the gas under gravity by cooling it below the temperature of the coldest of the surrounding objects.

See Thomson Dyn. θ of H [Dynamical Theory of Heat] 2nd Law

Whether the dyn.θ. of Gases is good or not 1 2 and 3 are good mechanics and true of gases. 4 is the only difficulty and the only way out of it seems to be that in the case supposed a molecule moving upwards has not the same mean energy as one moving horizontally or downwards, at the same height. This would involve different pressures in different directions and is otherwise objectionable. So there remains as far as I can see a collision between Dynamics and thermodynamics.[a]

I made a new set of experiments on viscosity of air and have come within 1/150 of what I got in summer (corrected for temperature of course). If you want the results for your book or anything else here they are

$$\text{Coefft of friction } \mu \text{ dimensions } \frac{M}{LT}$$

$$\text{at } 62°F\mu = .09362\frac{grains}{feetseconds}$$

$$\text{at } < 32\ 0°C > \mu = .01878(1+\alpha\theta)\frac{grammes}{meterseconds}$$

centigrade scale for θ

μ is quite independent of the pressure and proportional to the absolute temperature from 50° to 185°F. This value is about double Stokes and half Meyer's. Hydrogen is .516 of air by my expts. Graham by transpiration gets .485 but he required less hydrogen and less time for his expts and I think got purer gas. Carbonic acid .859 by me .807 by Graham. All these gases are dry. Damp air is a very little smoother than dry about 1/60 part for pressure 4in temperature 70°F. Hydrogen and air, equal vols. about 15/16 of air.

Results of dynamical theory not yet tested.

1 Coefficient of conductivity for heat (measured as energy) $= \frac{9(1+\beta)}{2}\frac{p}{\rho\,\theta}\mu$ where μ is the coefficient of friction or viscosity.
2 Equations of motion are of the form

$$\rho\frac{\partial u}{\partial t} + \frac{dp}{dx} - \mu\left(\frac{d^2u}{dx^2} + \frac{d^2u}{dy^2} + \frac{d^2u}{dz^2}\right) - \frac{\mu}{3}\frac{d}{dx}\left(\frac{du}{dx} + \frac{dv}{dy} + \frac{dw}{dz}\right)$$

the same as Stokes. For different gases

$$\frac{1}{\mu} \propto \sqrt{\frac{M}{K}}$$

where M = mass of molecules K force at unit distance.
3 Mixed gases diffuse till the law of density for each is the same as if the others were away. The equation of diffusion is

$$u_1 p_1 = \frac{D}{p_1 + p_2}\left(X\rho_1 - \frac{dp_1}{dx}\right)$$

where u_1 is the vely [velocity] and p_1 the pressure of one gas p_2 of the other. D is a coefficient depending on the two gases' mutual action. It is independent of density and varies with [the] square of temperature therefore since it is divided by $p_1 + p_2$ the actual rate of diffusion will be as the temperature and inversely as the density.
B.[alfour] Stewart is busy with his "Chincera bombyglasso in vacuo." I am going to see the new electrical machine at the R I [Royal Institution] which I take to be a foreign development of C F Varleys multiplier. There is no doubt that it is the right thing in electrical machines to have no friction but to work up the electricity by induction

Yours truly

J. Clerk Maxwell

a. One section of the letter on experiments on the elasticity of glass is omitted here.

2. [Paradox of the final equilibrium of temperature in a column gas subject to gravity]

Cambridge University Library, Maxwell Collection, Scientific papers 6.

Since μ is independent of the density and proportional to the pressure the coefficient of conductivity for heat which is $\frac{9}{2}\frac{p}{\rho\,\theta}\mu\,(1+\beta)$ will also be independent of the <pressure> density and will vary directly as the temperature.[a]

The condition of there being no conduction of heat is

$$\frac{d\theta}{dp} = \frac{2}{3(1+\beta)}\frac{\theta}{p}$$

Now when the pressure of a gas is gradually changed no heat being allowed to enter or escape

$$\frac{d\theta}{dp} = \frac{2}{5+3\beta}\frac{\theta}{p}$$

If therefore a mass of gas under the action of gravity were to be left to itself till conduction of heat ceased, the temperature would diminish more rapidly with the height than that of a portion of gas carried up bodily. If the portion of gas were to ascend it would be warmer than the surrounding gas and would therefore tend to ascend still, and if it were to descend below its original position it would be colder than the surrounding gas and would tend to descend further. Hence the condition of final equilibrium of heat in a gas acted on by gravity is one of mechanical instability so that such a mass of gas left to itself will perpetually be converting part of its heat into visible motion or currents and the energy thus developed will be reconverted into heat by friction.

If however the motion were properly regulated the energy thus developed could be transferred to machinery <till> so as to convert the invisible agitation of the gas into any other form of energy and thus form a perpetual motion. For instance if a portion of the gas were carried upwards and made to expand so as always to be of the same temperature with the surrounding gas it would be rarer than the surrounding gas and <there would be> the resultant of pressure and gravity on the portion of gas would act upwards and so do work. If when at the highest point it is allowed to acquire the temperature and pressure of the surrounding gas, and is then lowered and compressed so as to be always of the temperature of the surrounding gas, it will be denser than the surrounding gas and the resultant of gravity and pressure will act down- wards and shall do work. Thus from a mass of gas acted on by gravity energy may be abstracted to any amount and the gas cooled to a corresponding extent. This result is directly opposed to the second law of Thermodynamics which <states> affirms that "it is impossible <to transfer heat into mechanical force by cooling a body below> by means of inanimate material agency to derive mechanical effect from any portion of matter by cooling it below the temperature of the coldest of the surrounding objects."*[b]

I think it necessary to confirm a result so much opposed to so important a doctrine by a mere elementary investigation in which we do not require to consider the precise nature of the action between the molecules or to determine the coefficient of conductivity.

a. These notes were probably penned before Maxwell found the error in his paper of 1867, i.e., Maxwell,"On the Dynamical Theory of Gases," *Phil. Trans. R. Soc. London* 157 (1867): 49–88, reprinted in *Scientific Papers*, vol. 2, 26–78 and Garber, Brush, and Everitt, *Maxwell on Molecules and Gases*, 419–472.

b. Footnote at the bottom of the page, W. Thomson. "On the Dynamical Theory of Heat.[with numerical results deduced from Mr. Joule's 'Equivalent of a thermal Unit' and M. Regnault's 'Observations on Steam'']," Trans. Edin 1851 p.265. [*Trans. R. Soc. Edinburgh* 20 (1851): 261–288, p.265.]

3. Letter from William Thomson to George Gabriel Stokes, October 13, 1866

Royal Society Archives, Referee Reports, vol. 6 no. 179[a]

Blackdale Largs
by Greenock
Oct 13/66

My dear Stokes,[b]

You have no doubt by this time got my letter addressed to Lenifield [sic] Cottage, but if not no matter now.[c] I hope my delay in returning Maxwell's paper (which I posted, addressed to you at Malahide, two days after my return from London, being just 18 days after the Great Eastern's arrival)[d] has caused [no] delay in its publication, although I fear it must have been inconvenient to you. It was with great difficulty I could find time to read it and think about it sufficiently and I had to use my last railway journey to learn a little more of it otherwise I sh$^{\underline{d}}$ have returned it several days sooner.

There can be no doubt as to its suitability for the Transactions, in my opinion. But it might I think be considerably improved by guarding and modifying somewhat some of the statements. For instance throughout the idea explicitly expressed in the beginning, is adhered to, of regarding the molecules as mere centres of force, repelling as D^{-5}. Now this is inconsistent with giving them kinetic energy of rotation round their centres of inertia and if they were really mere centres of force we *must* have $\beta = 0$, and so the proportions of the spec.[ific] heats & could not be made to fit the reality of gases. It should I think be explained in a sentence or two early in the paper that the molecules are regarded *not* as centres of force but as really (according with their name) little heaps of matter, acting on one another with forces *not* in lines through their centres of inertia (in as much as <increases> changes in the energy of motions of their centres is accompanied with increase of their rotatory energies keeping the average proportion $\beta : 1$ constant, and that the average value of the component of the mutual force between each pair in the direction of the line joining their centres is prop.[ortional] to D^{-5}. <This <state> condition <could [?] to this> be achieved for instance by making each molecule a group of magnets arranged to constitute what>

This supposition differs from those previously worked on by Herapath, Clausius & Maxwell only in making D^{-5} the law of average repulsion, instead of $(D/a)^{-n}$ where a is the radius of a hard spherical atom, and n is a very great (infinitely great) number.

There is one very important investigation against the conclusion drawn from which I think the author should warn the reader (as he warned me in correspondence 3/4 of a year ago).[e] It is that in which the law of temperature of air in a vertical column is sought. The conclusion, as Maxwell wrote to me, violates Carnot's principle. He put it to me then, and I thought of it a long time, without being able to see through it. Neither can I now when I have had his paper in my hands. But yet I feel convinced that the conclusion must be wrong.

Imagine a closed vessel of impermeable material– a vertical column (circular or rectangular if you please) with a <symmetri> plane partition EF also impermeable, dividing it into symmetrical parts communicating freely at the top and bottom of the partition but perfectly separated elsewhere. Fix plates of <metal> polished silver KL, K'L' symmetrically in the position shown and let the halves of these towards K, K' be smoked. Leave the whole to itself and the Maxwell equilibrium will be unstable. For let a <a> current <will> ever [so] slight be established in the direction shown by the arrow heads. Then in virtue of the air being (as Maxwell's conclusion makes it) hotter at the top, and the consequent radiation from K to K', the air moving down from K will become cooled below its equilibrium temperature and that moving

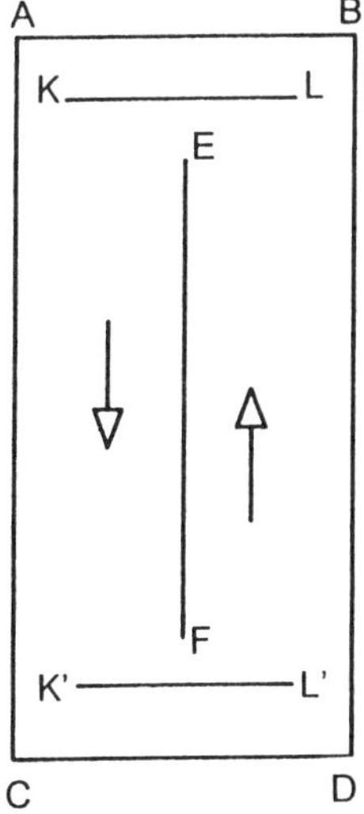

horizontally from K' to <K' an> L' and upwards from L' will become hotter for the same reason. Thus the current will be kept flowing, and by putting a wind mill in any proper position work will be done till all the kinetic energy of the particles is gone; i.e., the whole heat of the gas will convert itself into weights raised. Surely there must be some flaw in a kinetic investigation that leads to such a result. [The following sentence is inserted in the margin.] I am very confident that the bodily convective equilibrium of temperature in a vertical column must agree with the molecular equil$^{\underline{m}}$ [f] of temperature in the true molecular theory. Whatever flaw may be in Maxwell's investigation if any I have not been able to see. But I cannot see it as a rigorous mathematical working out from the premises and therefore it seems to me probable that some of the assumptions used as stepping stones must be in-valid. The only definite suggestion that has occurred to me is that Maxwell does not seem to have introduced the change in energy in the free motions of each particle between successive collisions due to gravity and that possibly by taking this into account a different conclusion may be had without giving up any of the assumptions.[g]

I have made a few pencil marks on the first page or two, where it appeared to me that the author had stated too unguardedly fallacious conclusions of the older molecular attempts. It is obvious (and early proved by experiment as I have done in lecturing on the subject) that for cork Poisson's ratio is nearer 0 than 1/4. Everett in experiments made this summer on glass and brass which will soon be communicated to the R. S. [Royal Society] has found for glass something decidedly less than 1/4 (something about .22 or .23 if I remember right).[h] For brass he finds a considerably larger value than Kirchoff [sic] had found for his brass.[i] Besides, a long time ago I asserted (Phil. Trans. somewhere about 1855) that a solid violating Poisson's condition could be <constructed> built up of small parts (molecules) each fulfilling it.[j] Here for instance is one way of doing this. Make ordinary spiral springs of steel wire, suppose (w$^{\underline{h}}$ <very nearly> somewhat approximately does really fulfill Poisson's condition and may be supposed to do so rigorously for the present) and build up a solid of such molecules, <connect into> connected any way you please. In as much as when you compress or elongate a single spiral spring in the direction of its axis, the lateral expansion or contraction is very small, the Poisson ratio for <such> a <com> solid so composed will be nearer 0 than 1/4. It no doubt is so for a ball of worsted or for a cube of sponge and no doubt cork is composed, so far as first grade of molecular structures somewhat like sponge. This reminds me of one of Maxwell's initial remarks w$^{\underline{h}}$ requires correct$^{\underline{n}}$ — viz. that <molecular> theories of matter are either atomic, or the one that supposes all matter continuous *and homogeneous.* The only views that have ever appeared to me true or natural as to the constitution of matter are those that suppose all space to be full but the properties of known bodies to be due to or necessarily associated with molecular structure as of a sponge or other organic tissues or brick work, i.e., that there are vast variations of density from point to point within spaces of dimensions some small fraction of a wave length (though not inappreciably small).[k]

I am at a loss where to address this but as you say nothing about leaving Malahide let it take its chances there.

Yours always truly,
William Thomson

a. This letter is reprinted in David B. Wilson, *The Correspondence between Sir George Gabriel Stokes and Sir William Thomson, Baron Kelvin of Largs* (Cambridge: Cambridge University Press, 1989), vol. 1, 327–331.

b. This letter is William Thomson's referee report of Maxwell,"On the Dynamical Theory of Gases," *Phil. Trans. R. Soc. London* 157 (1867): 49–88, reprinted in *Scientific Papers*, vol. 2, 26–78 and Garber, Brush, and Everitt, *Maxwell on Molecules and Gases*, 419–472. For details of the refereeing system at the Royal Society, see, Marie Boas Hall, *All Scientists Now: The Royal Society in the 19th Century* (London: Cambridge University Press, 1984).

c. For this letter, which contains nothing on the contents of Maxwell's paper see Wilson, op. cit. (note a), vol. 1, 327.

d. The delay in returning Maxwell's paper was not by procrastination. Thomson was enmeshed in the new communications and electrical industries as well as continuing his career as professor of natural philosophy at Glasgow University. The reference to the return of the Great Eastern also is about the laying of the Atlantic cable. After seven years of successive failures, public quarrels between Thomson and the engineer-in-chief and the near bankruptcy of the Atlantic Telegraph Company William Thomson was knighted for his role in the final successful venture. Had this attempt failed a major political scandal could have erupted as both the British and American governments underwrote some of the enormous costs of what was essentially a private venture. For the latest account of Thomson's career and role in the Atlantic Telegraph Company see, M. Norton Wise and Crosbie Smith, "Measurement, Work and Industry in Lord Kelvin's Britain," *Hist. Studies Phys. Sci.* 17 (1987): 147–173, and *Energy and Empire: A Biographical Study of Lord Kelvin* (Cambridge: Cambridge University Press, 1989), part IV. The slip in this next sentence is interesting, given the rivalry between the two men that began in electromagnetic theory. For which see, Wise and Smith, supra, chap. 13.

e. See Document II–1.

f. equilibrium

g. The result Thomson discusses, that kinetic theory implies that under gravity a column of gas cannot be in thermal equilibrium, was due to a calculational error. See, Maxwell, op. cit. (note a), *Scientific Papers*, vol. 2, 469–470, and Garber, Brush, and Everitt, op. cit. (note b), 75–76.

h. Joseph D. Everett (1831-1901), professor of physics, Queens College, Belfast published several papers on the elastic properties of solids. Everett, "Account of Experiments on the Flexural and Torsional Rigidity of a Glass Rod, Leading to the Determination of the Rigidity of Glass," *Phil. Trans. R. Soc. London* 156 (1866): 185–192, *Phil. Mag.* 31 (1866): 476–477, "Account of Experiments on Torsion and Flexure for the Determination of Rigidities," *Proc. R. Soc. London* 15 (1867): 356; 16 (1868), 248, *Phil. Trans. R. Soc. London* 158 (1868): 363–370 are those to which Thomson refers.

i. Gustav Robert Kirchhoff (1824-1887) published several theoretical papers on elasticity. The paper in which Poisson's ratio is discussed is Kirchhoff, "Über das Verhaltnis der Quercontraction zur Längendilation bei Stäben von federhaten Stahl," *Ann. Phys.* 108 (1859): 369–392, and *Ann. Chim.* 59 (1860): 498–805, and *Phil. Mag.* 23 (1862): 28–46.

j. William Thomson, "Elements of a Mathematical Theory of Elasticity," *Phil. Trans. R. Soc. London* 146 (1856): 481–498 reprinted in *Mathematical and Physical Papers* vol. 3, 84–112. Siméon-Denis Poisson (1781-1840) published a number of theoretical papers on elasticity. The one in which he deduced the ratio in question through some very dubious mathematical manipulations is Poisson, "Mémoire sur l'équilibire et le mouvement des corps élastique," *Mém. Acad. Sci. Paris* 8 (1828): 357–570, 623–627, 360. Poisson examined the tension in a prismatic, homogeneous bar and concluded that the ratio of the axial elongation to lateral contraction was 1 : 4 and hence the relation between the modulus of compression, E, to that of shear, G, was $G = 0.4E$. Thomson's ideas on elasticity were published in William Thomson and Peter Guthrie Tait, *Treatise on Natural Philosophy* (Oxford: Clarendon Press, 1867) and were based on experiments done by his students at Glasgow University. Poisson's ratio is discussed in Stephen P. Timoshenko, *History of the Strength of Materials* (New York: Dover reprint, 1983) chap. viii; his assessment of Thomson's contributions to the theory of elasticity, 260–263. In marked contrast is Isaac Todhunter's *A History of the Theory of Elasticity and the Strength of Materials*, ed. Karl Pearson (New York: Dover reprint, 1960) vol. 1, assessment of Poisson. See also James F. Bell, *The Experimental Foundations of Solid Mechanics* Handbuch der Physik, Band VIa part 1 (New York: Springer-Verlag, 1973) which gives a very thorough discussion of nineteenth- century work. Maxwell discusses this further in a letter to George Gabriel Stokes, September 25th, 1875 reprinted in *Memoir and Scientific Correspondence of the Late George Gabriel Stokes*, ed. Joseph Larmor (Cambridge: At the University Press, 1907) vol. 2, 36–37. For the theory of elasticity in Maxwell's theory of gases see, Garber, Brush, Everitt, op. cit. (note b), 23–27.

k. For Maxwell's reply to Thomson's critique of his molecular ideas see Document II–4.

4. Letter from Maxwell to George Gabriel Stokes,[a] December 18,1866

Memoir and Scientific Correspondence of the Late Sir George Gabriel Stokes, ed. Joseph Larmor (Cambridge: Cambridge University Press, 1909) vol. 2, 27–28.

---

8 Palace Gardens Terrace, W.
1866, Dec. 18

I enclose a report on Mr Tarn's paper and Thomson's report on mine.[b] Thomson's theory of matter being continuous but much more dense at one place than another is a molecular theory, for the dense portions constitute a finite number of molecules: and I do not assert nor does anyone since Lucretius that the space between the molecules is absolutely empty.
I take Statics to be the theory of systems of forces without considering what they act on. Kinetics to be the theory of motions of systems without regard to the forces in action, Dynamics to be the theory of the motion of bodies as the result of given forces, and Energetics to be the theory of the communication of energy from one body to another. I therefore call the theory a dynamical theory because it considers the motion of bodies as produced by certain forces.[c]

I think I have stated plainly enough the difference between my molecules and pure centers of force. I do not profess to have a dynamical theory of the molecule itself.

I have considered the equilibrium of heat in a vertical column and find I made a mistake in equation 143. I now make the temperature the same throughout and have inserted (on two pages) an addition to that effect which I mean to be instead of p.60. I have also amended equation 54 in which the effect of external forces was omitted by mistake in writing out and copying from equation 47.

The result is a corroboration of the theory of the distribution of velocity as the errors are in the theory of "Least Squares"; for we require to have

$$\overline{\xi^4} = 3\overline{\xi^2}\,.\,\overline{\xi^2}$$

which agrees with that theory.[d]

I think something might be done to the statical theory of elasticity with centres of force whose displacement is a function of the index of the molecule as well as of its initial coordinates, the numbers of kinds of indices being small and the groups of n particles being originally of the same form, but it would require an investigation to itself.

I remain, Yours truly,

J. Clerk Maxwell

I shall be at Cambridge from 26 Dec. to 26 Jan. at Trinity.

a. George Gabriel Stokes (1819-1903) was Secretary of the Royal Society from 1854 until 1885 and acted as editor for the *Philosophical Transactions.*

b. The paper Maxwell refereed was E. W. Tarn, "On the Stability of Domes," *Proc. R. Soc. London* 15 (1867): 182–188. His report is in the Royal Society Archives. Maxwell's own paper under discussion was published as Maxwell, "On the Dynamical Theory of Gases," *Phil. Trans. R. Soc. London* 157 (1867): 49–86, reprinted in *Scientific Papers,* vol. 2, 26–78 and Garber, Brush, and Everitt, *Maxwell on Molecules and Gases*, 419–472.

c. In spite of Maxwell's attempts here to justify his nomenclature of "dynamical" theory, by 1873 he had switched to the term "kinetic." Maxwell discusses this further in letters to William Garnett of July 9th and 14th, 1877, reprinted in Lewis Campbell and William Garnett, *The Life of James Clerk Maxwell* (New York: Johnson reprint of 1882 edition, 1969), 397–399.

d. Maxwell goes into more detail over the origins of this error in Maxwell,in Maxwell, op. cit. (note b), "Additional Remarks," Garber, Brush, and Everitt, 469–470.

5. [Distribution of temperature in a vertical column of gas][a]

Cambridge University Library, Maxwell Collection, Scientific Papers 6

It is possible to consider the permanent distribution of temperature in a vertical column of gas independently of any theory as to the law of force between the molecules.

Let M be the mass of a molecule, N the number of molecules in unit of volume, then $M\,N = \rho$ is the density of the gas.

Let ξ, η, ζ be the components of the velocity of a molecule.

The components of momentum of a particle are $M\xi, M\eta, M\zeta$ and its energy is $1/2\,\beta\,M(\xi^2 + \eta^2 + \zeta^2)$, the factor β being the ratio of the whole energy to the energy of translation.

Now consider the transfer of any quantity Q across unit of area in the direction of x in unit of time

For the dN particles whose x-velocity has the same value ξ, the number transferred is $\xi\,dN$. If Q is what each transfers, the quantity transferred is

$$\int \xi\,Q\,dN$$

where the integration is to be extended to all the groups of which N is made up.

1 Let $Q_1 = M$, the mass of the particle, then
Mass transferred $= \int \xi\,M\,dN = \rho\,u$ where u is the <mean> velocity of the gas in mass

(2) Let $Q_2 = \xi\,M$, the momentum of the particle

Momentum transferred $= \int \xi^2\, M\, dN = p$ the pressure of the gas

(3) Let $Q_3 = 1/2\, \beta\, M(\xi^2 + \eta^2 + \zeta^2)$, the energy of the particle

Energy transferred $= 1/2 \int \beta\, M\, \xi(\xi^2 + \eta^2 + \zeta^2)\, dN =$ rate of conduction of heat

(4) Let $Q_4 = 1/2\, \beta\, M\, \xi(\xi^2 + \eta^2 + \zeta^2)$ [?] the part which a single molecule contributes towards conduction of heat, and let us consider the variation of $\int Q_4\, dN$ as arising 1st from transfer of Q_4 into and out of the element considered 2nd from change of the value of Q_4 due to the impressed force whose component in direction of x is X and 3rd from collisions among the molecules.

1st The change of $\int Q_4\, dN$ arising from transfer of Q_4 through the surfaces <which have> x and $x + dx$ is

$$-\frac{d}{dx} \int 1/2\, \beta\, M\, \xi^2(\xi^2 + \eta^2 + \zeta^2)\, dN$$

2nd Since $\frac{\partial \xi}{\partial t} = X$

$$\frac{\partial \xi}{\partial t}(\xi^2 + \eta^2 + \zeta^2) = X(3\xi^2 + \eta^2 + \zeta^2).$$

3rd The effect of collisions is of the form

$$\frac{3}{2} k_1 A_2 \beta\, \rho \int M\, \xi(\xi^2 + \eta^2 + \zeta^2)\, dN = R.$$

This is shown in my paper on Dynamical Theory of Gases Phil. Trans. 1867 but we do not here require to consider it.

We thus obtain the equation

$$\frac{\partial}{\partial t} \int Q_4\, dN + \frac{d}{dx} \int \frac{1}{2} \beta\, M\, \xi^2\, (\xi^2 + \eta^2 + \zeta^2)\, dN$$

$$= X \int \frac{\beta}{2} M\, (3\xi^2 + \eta^2 + \zeta^2)\, dN - \frac{3}{2} k_1 A_2 \beta\, \rho \int M\, \xi(\xi^2 + \eta^2 + \zeta^2)\, dN$$

If the condition of the system is permanent the first term vanishes.

If the distribution of velocities among the molecules is according to the law

$$dN = \frac{N}{\alpha^3 \pi^{3/2}} e^{-\{\frac{\xi^2 + \eta^2 + \zeta^2}{\alpha^2}\}}\, d\xi\, d\eta\, d\zeta$$

then since

$$\int \xi^2 M\, dN = p$$

and

$$\int \eta^2 M\, dN = \int \zeta^2 M\, dn = p$$

and

$$\int \xi^4 M\, dN = 3\frac{p^2}{\rho}$$

$$\int \xi^2(\eta^2 + \zeta^2) M\, dN = 2\frac{p^2}{\rho}$$

and we find

$$\frac{d}{dx}\frac{5}{2}\beta\frac{p^2}{\rho} = \frac{5}{2}\beta\, X\, p - \frac{3}{2}k_1 A_2 \beta\, \rho \int M\, \xi(\xi^2 + \eta^2 + \zeta^2)\, dN$$

But $X = \frac{1}{\rho}\frac{dp}{dx}$ and $\frac{p^2}{\rho} = \frac{p}{\rho\,\theta} p\,\theta$ and $\frac{p}{\rho\,\theta}$ is a constant
Hence

$$\frac{5}{2}\beta\,\frac{p}{\rho\,\theta}(\theta\,\frac{dp}{dx} + p\,\frac{d\theta}{dx}) = \frac{5}{2}\beta\,\frac{p}{\rho}\frac{dp}{dx} - R$$

or

$$\frac{5}{2}\beta\,\frac{p^2}{\rho\,\theta}\,\frac{d\theta}{dx} = -R$$

so that the rate of conduction is zero if the temperature is uniform.[b]

A quantity of gas is in equilibrium in a vessel under the action of gravity, to determine the pressure and temperature at any point when both the pressure and temperature have assumed their final state.

The condition of <equi> mechanical equilibrium is

$$\frac{dp}{dx} = \rho\, g$$

where x is measured in the direction in which g acts. We have also

$$p = \frac{1}{3}\,\rho(\xi^2 + \eta^2 + \zeta^2)$$

and in the state of equilibrium $\xi^2 = \eta^2 = \zeta^2$

The whole energy of a molecule M is

$$\frac{1}{2}\,M(1+\beta)(\xi^2 + \eta^2 + \zeta^2)$$

where $\xi\,\eta\,\zeta$ and β have values peculiar to that molecule.

Now let the molecule move so as to increase x by dx then gravity will do work upon it $= Mgdx$ and its energy will now be increased

$$< \frac{1}{2}\,M(1+\beta)\,(\xi^2 >$$

by this amount. If its energy is now greater than the mean of that of the molecules among which it has arrived it will increase the mean energy of

<these> the molecules in that part of the field, and if it is less it will diminish the mean energy but if the <difference of the> mean energy of these molecules is greater by $M\,g\,dx$ than the original energy of the molecule considered, its arrival will not alter the mean energy of the surrounding molecules.

That this must be the case generally we must have

$$\frac{1+\beta}{2}\frac{d}{dx}(\xi^2+\eta^2+\zeta^2)=g$$

and if this condition be fulfilled, any molecule passing from one position to another will have the excess of its energy above the mean energy of the surrounding molecules the same throughout its path.

a. These notes were penned before Maxwell read Boltzmann's 1872 paper on the thermal equilibrium of gases but after he wrote his paper on the dynamical theory of gases because he is using the methods of the latter and does not refer to the former.

b. The following paragraphs begin on a fresh sheet and may have been penned later, although the subject matter and the notation are the same.

6. Excerpts from Maxwell "Molecular Theory"

Maxwell, *Theory of Heat* (New York: Appleton, 1872), 300–301.[a]

The second result of our theory relates to the thermal equilibrium of a vertical column. We find that if a vertical column of gas were left to itself, till by the conduction of heat it had attained a condition of thermal equilibrium, the temperature would be the same throughout, or, in other words, gravity produces no effect in making the bottom of the column hotter or colder than the top.

The result is important in the theory of thermodynamics, for it proves that gravity has no influence in altering the conditions of thermal equilibrium in any substance, whether gaseous or not. For if two vertical columns of different substances stand on the same perfectly conducting horizontal plate, the temperature of the bottom of each column will be the same; and if each column is in thermal equilibrium of itself, the temperatures at equal heights must be the same. In fact, if the temperatures of the tops of the two columns were different, we might drive an engine with this difference of temperature, and the refuse heat would pass down the colder column, through the conducting plate, and up the warmer column; and this would go on till all the heat was converted into work, contrary to the second law of thermodynamics.

But we know that if one of the columns is gaseous, its temperature is uniform. Hence that of the other must be uniform, whatever its material.

This result is by no means applicable to the case of our atmosphere. Setting aside the enormous direct effect of the sun's radiation in disturbing thermal equilibrium, the effect of winds in carrying large masses of air from one height to another tends to produce a distribution of temperature of a quite different kind, the temperature at any height being such that a mass of air, brought from one height to another without gaining or losing heat, would always find itself at the temperature of the surrounding air. In this condition of what Sir william Thomson has called the Convective equilibrium of heat, it is not the temperature which is constant, but the quantity ϕ, which determines the adiabatic curves.[b]

In the convective equilibrium of temperature, the absolute temperature is proportional to the pressure raised to the power $\frac{\gamma-1}{\gamma}$, or 0.233.

The extreme slowness of the conduction of heat in air, compared with the rapidity with which large masses of air are carried from one height to another by the winds, causes the temperature of the different strata of the atmosphere to depend far more on this condition of convective equilibrium than on true thermal equilibrium.

a. Maxwell devoted one long chapter in his text on thermodynamics to "Molecular Theory" and thus to connecting kinetic theory and thermodynamics. This section appears on p. 320 in later editions. The passage above is the source of Frederick Guthrie's problem in Document II–7.

b. Maxwell defined ϕ as Rankine's Thermodynamic Function, i.e. entropy. However, Maxwell did not mention Clausius's concept nor had he related it to Rankine's or William Thomson's formulations of the second law. Thomson's paper to which Maxwell refers is, William Thomson,"On the Convective Equilibrium of Temperature in the Atmosphere," *Mem. Manchester Lit. Phil. Soc.* 2 (1865): 125–131. For a discussion of this paper see Garber, "Thermodynamics and Meteorology," *Ann. Sci.* 33 (1976): 51–65.

7. Letter from Francis Guthrie "Kinetic Theory of Gases" [a]

Nature 8 (1873): 67.

On page 300 of the second edition of Maxwell's excellent little text-book on the "Theory of Heat," it is stated, as a result of the kinetic theory of gases therein set forth, that "gravity produces no effect in making the bottom of the column" (of gas) "hotter or cooler than the top."

I cannot see how this result follows from the kinetic theory of gases. On the contrary, it seems obvious that thermal equilibrium can only subsist according to the kinetic theory, where the molecules encounter each other with equal average amounts of *work* or *vis viva*, and in order that this may be the case, the velocity of the molecules (and consequent temperature) of any upper layer must be less than that of the molecules in the layer next below; since, in order to encounter each other, the former must descend, and acquire velocity, while the latter must ascend and lose it. This would establish a diminution of temperature from the bottom to the top of a column of air at the rate (in the absence of any counteracting cause) of $1°F$. for 113 ft. of height, as can easily be verified from the fact that on account of the specific heat of air 1 lb. requires 183 foot – pounds to raise its temperature $1°F$. Radiation may diminish this and tend to produce equilibrium, but nevertheless it seems obvious from these two opposing tendencies a residual inequality of thermal condition would result, and that the top of a column would be cooler than the bottom. That this would be the case if the air were in general motion in the form of upward and downward currents, will not, I presume, be disputed; and surely molecular [motion] is on the same footing. If the particles of air are moving in every direction with great absolute velocity, in what respect does this differ from air currents? In fact, all the particles which at any epoch of time are moving in any given direction constitute an air-current in that direction, mingled, it is true, with currents in other directions, but moving with accelerated velocity if descending, and with retarded velocity if ascending, and thus always tending to produce a diminution of temperature with height as a condition of gaseous thermal equilibrium.

J.[b] Guthrie

Graaff Reinet, Cape Colony, April 2

a. Francis Guthrie (1831-1898) was appointed professor of mathematics in the newly established Graaff-Reinet College, Cape Colony, South Africa. Although he explored the botany of South Africa he also published papers on imaginary numbers, thermodynamics as well as the thermal equilibrium of a column of gas. His brother Frederick Guthrie (1833-1886) was a physical chemist and was appointed to the School of Science in South Kensington in 1869. He was also instrumental in establishing the Physical Society of London in 1873.

b. presumably a misprint for F[rancis] Guthrie. All the letters are addressed from South Africa and Maxwell refers to "Principal" Guthrie in a later letter reflecting Francis Guthrie's appointment to the South African College at Cape Town.

8. "Clerk Maxwell's Kinetic Theory of Gases,"

Nature 8 (1873): 84.

Your correspondent, Mr. Guthrie, has pointed out an, at first sight, very obvious and very serious objection to my kinetic theory of a vertical column of gas. According to that theory, a vertical column of gas acted on by gravity would be in thermal equilibrium if it were at a uniform temperature throughout, that is to say, if the mean energy of the molecules were the same at all heights. But if this were the case the molecules in their free paths would be gaining energy if descending, and losing energy if ascending. Hence, Mr. Guthrie argues, at any horizontal section of the column a descending molecule would carry more energy down with it than an ascending molecule would bring up, and since as many molecules descend as ascend through the section, there would on the whole be a transfer of energy, that is, of heat, downwards; and this would be the case unless the energy were so distributed that a molecule in any part of its course finds itself, on the average, among molecules of the same energy as its own. An argument of the same kind, which occurred to me in 1866, nearly upset my belief in calculation, and it was some time before I discovered the weak point in it.

The argument assumes that, of the molecules which have encounters in a given stratum, those projected upward have the same mean energy as those projected downwards. This, however, is not the case, for since the density is greater below than above, a greater *number* of molecules come from below than from above to strike those in the stratum, and therefore a greater number are projected from the stratum downwards than upwards. Hence since the total momentum of the molecules temporarily occupying the stratum remains zero (because, as a whole, it is at rest), the smaller number of molecules projected upwards must have a greater initial velocity than the larger number projected downwards. This much we may gather from general reasoning. It is not quite so easy, without calculation, to show that this difference between the molecules projected upwards and downwards from the same stratum exactly counteracts the tendency to a downward transmission of energy pointed out by Mr. Guthrie. The difficulty lies chiefly in forming exact expressions for the state of the molecules which instantaneously occupy a given stratum in terms of their state when projected from the various strata in which they had their last encounters. In my paper in the *Philosophical Transactions,* for 1867, on the "Dynamical Theory of Gases," I have entirely avoided these difficulties by expressing everything in terms of what passes through the boundary of an element, and what exists or takes place inside it. By this method, which I have lately carefully verified and considerably simplified, Mr. Guthrie's argument is passed by without ever becoming visible.[a] It is well, however, that he has directed attention to it, and challenged the defenders of the kinetic theory to clear up their ideas of the result of those encounters which take place in a given stratum.

J. Clerk Maxwell

a. This probably refers to Maxwell's unpublished notes of Document II–5 where the methods of his paper, "On the Dynamical Theory of Gases," *Phil. Trans. R. Soc. London* 157 (1867): 49–88, reprinted in *Scientific Papers*, vol. 2, 26–78 and Garber, Brush and Everitt, *Maxwell on Molecules and Gases*, 419–472, are used. Maxwell, however, did not avoid the the problem with those methods. These were only solved with his solution published as, Maxwell,"On the Final State of a System of Molecules in Motion Subject to Forces of Any Kind," abstract *Rep. 42nd Meeting BAAS* (1873), Trans. Sect. A, 29–32, and *Nature* 8 (1873): 537–538, reprinted in *Scientific Papers*, vol. 2, 350–354.

9. Letter from Maxwell to Peter Guthrie Tait, August, 1873

Cargill Gilston Knott, *Life and Scientific Work of Peter Guthrie Tait* (Cambridge: Cambridge University Press, 1911), 114.

By the study of Boltzmann I have been unable to understand him. He could not understand me on account of my shortness, and his length was and is an equal stumbling-block to me. Hence I am much inclined to join the glorious company of supplanters and to put the whole business in about six lines.[a]

... In thermal language–Temperature uniform in spite of crowding to one side by force. Molecular volume of all gases equal. Equilibrium of mixed gases follows Dalton's Law of each gas acting as vacuum to the rest (in fact it acts as a vacuum to itself also). In my former treatise I got these results only by way of conclusions. Now they come before any assumption is made as to the law of action between molecules.

a. Unfortunately, Knott only quotes part of this letter and we do not know whether this was ironic or straightforward. The last sentence of the continuation of this letter and Knott's remarks that the letter describes the contents of Maxwell,"On the Final State of a System of Molecules in Motion subject to Forces of any Kind," *Nature* 8 (1873): 537–538, leads us to believe this first paragraph is irony. Maxwell's statement that Boltzmann could not understand him for his shortness is borne out in a statement made by Boltzmann in 1888. See Boltzmann, "Über das Gleichgewicht der lebendigen Kraft zwischen progressiver und Rotationsbewegung bei Gasmolekulen," *Berlin Akad. Monats* (1888): 1395–1408, quoted in Stephen G. Brush, *Kinetic Theory* (New York: Pergamon Press, 1966–72) vol. 3, 274–275.

10. Letter from Francis Guthrie, "On the Equilibrium of Temperature of a Gaseous Column Subject to Gravity"

Nature 8 (1873): 486.

From Mr. Clerk-Maxwell's reply to my note on this subject which appeared in your columns a short time since, it would appear that he does not profess so much fully to explain the difficulty suggested by me as to show that it is capable of explanation, referring your readers to his other works for further information. I would not, therefore, have troubled you further on the subject had it not occurred to me on reading Mr. Maxwell's letter that I could state the case in such a way as to render clearly apparent the grounds for taking different views on this point.

Let a vertical column of gas, subject to gravity and in a state of equilibrium as to pressure and temperature, be divided by a horizontal plane P into two parts, A above and B below.

In the time Δt let a mass M_1 of particles pass in their free course from A to B, and a mass M_2 from B to A.

Let the portion of A from which the particles composing M_1 proceed be called the upper stratum, and the corresponding part of B the lower stratum, then the following consequences may be deduced:–

1. From the equilibrium of density[a]

$$M_1 = M_2.$$

2. From the equilibrium of temperature the amounts of work in M_1 and M_1 while passing through P are equal.

3. From the effect of gravity the work in M while in A reckoning from the commencement of the free course of each particle composing M_1, is less than at P, while that in M_2 is greater.

4. Whence it follows that of the two equal masses M_1 and M_2 in the upper and lower strata respectively M_1 contains less work than M_2.

5. The work in M_1 while in the upper stratum reckoned as before, is the same as that of any other equal average mass in that stratum, and the same is the case also of M_2.

6. The average amounts of work in equal masses in the two strata, and the consequent temperatures of the strata are unequal, the lower stratum having the higher temperature.

I suppose Mr. Maxwell would deny the truth of statement (5). I presume he would argue as follows:–

"Of all the particles in the lower stratum which in the time Δt have at the commencement of their free course a velocity and direction such as would take them through P, gravity in selecting those which compose M_2 excludes those whose velocities are insufficient to overcome the effect of their weights, while in forming M_1 particles of low velocity are selected (included?), which, but for the effects of gravity, would not have cut P in their free courses, consequently the particles in M_1 have an average velocity less than that of the upper stratum from which they come, while the particles of M_2 have a greater average velocity than that of the lower stratum, and consequently the inequality of the average velocity of the particles in the two strata cannot be inferred from the inequality of the average velocities of the particles composing M_1 and M_2 while in those strata."

This argument, therefore, assumes the theory that in a given mass of uniform temperature there are particles moving with every velocity from nothing upwards to a certain limit, and mixed in certain proportions. That this is actually Mr. Maxwell's view I own I might have remembered, but I suppose I overlooked it from an impression in my own mind that the molecular motion was to be regarded as being of a planetary (or in the case of gases a cometary) nature. That in masses of the same temperature velocities were to be regarded as practically uniform, except in so far as affected by the distance of the particles apart, and that the so-called impacts of particles were more properly to be regarded as perihelion passages of bodies moving among each other in hyperbolic orbits. If this view is the more accurate one, then obviously the argument which I have assumed Mr. Maxwell would use, falls to the ground.

Is there no possibility of testing the nature of the thermal equilibrium of a column of still air? The result would at any rate throw an unexpected light on the nature of molecular motion. F. Guthrie

Graaff Reinet College, July 19

a. This condition does not hold. Maxwell addressed this issue in Maxwell, "On the Dynamical Theory of Gases," *Phil. Trans. R. Soc. London* 157 (1867): 49–88, reprinted in *Scientific Papers*, vol. 2, 26–78, 63–64, and in Garber, Brush, and Everitt, *Maxwell on Molecules and Gases*, 419–472, 457–458. However, he had not established it in this published correspondence.

11. Letter from Maxwell, "On the Equilibrium of Temperature of a Gaseous Column subjected to Gravity"

Nature 8 (1873): 527–528.

Since reading Principal Guthrie's first letter on this subject (vol. viii, p. 67), I have thought of several ways of investigating the equilibrium of temperature in a gas acted on by gravity. One of these is to investigate the condition of the column as to density when the temperature is constant, and to show that when this is fulfilled the column also fulfills the condition that there shall be no upward or downward transmission of energy; or, in fact, of any other function of the masses and the velocities of the molecules. But a far more direct and general method was suggested to me by the investigation of Dr. Ludwig Boltzmann*[a] on the final distribution of energy in a finite system of elastic bodies. A sketch of this method as applied to the simpler case of a number of molecules so great that it may be treated as infinite, will be found on p. 535. Principal Guthrie's second letter (vol. viii p. 486) is especially valuable as stating his case in the form of distinct propositions, every one of which, except the fifth is incontrovertible. He has himself pointed out that it is here that we differ, and that this difference may ultimately be traced to a difference in our doctrine as to the distribution of velocity among the molecules in any given portion of the gas. He assumes, as Clausius, at least in his earlier investigations, did, that the velocities of all the molecules are equal, whereas I hold, as I first stated in the *Phil. Mag.* for Jan. 1860,[b] that they are distributed according to the same law as errors of observation are distributed according to the received theory of such errors.

It is easy to show that if the velocities are all equal at any instant they will become unequal as soon as encounters of any kind, whether collisions or "perihelion passages" take place. The demonstration of the actual law of distribution was given by me in an improved form in my paper on the "Dynamical Theory of Gases," *Phil. Trans.* 1866, [sic] and *Phil. Mag*, 1867,[c] and the more elaborate investigation by Boltzmann has led him to the same result. I am greatly indebted to Boltzmann for the method used in the latter part of the sketch of the general investigation (see p. 535) which was communicated in a condensed form to the British Association on Sept, 20. 1873.[d]

J. Clerk Maxwell

a. Footnote at the bottom of the page in [Boltzmann] Studien über das Gleichgewicht der lebendigen Kraft zwischen bewegten materiellen Punkten. Von Dr. Ludwig Boltzmann, Sitzb. d. Akad. d. Wissensch. [58] October 8, 1868 (Vienna) [517–560].

b. Maxwell,"Illustrations of the Dynamical Theory of Gases," Part I *Phil. Mag.* 19 (1860): 19–32, "Part II," 20 (1860): 21–37, reprinted in *Scientific Papers*, vol. 1, 377–409 and Garber, Brush, and Everitt, *Maxwell on Molecules and Gases*, 285–318.

c. Maxwell, "On the Dynamical Theory of Gases," *Phil. Trans. R. Soc. London* 157 (1867): 49–88, reprinted in *Scientific Papers*, vol. 2, 26–78 and Garber, Brush, and Everitt, *Maxwell on Molecules and Gases*, 419–472.

d. Maxwell,"On the Final State of a System of Molecules in Motion Subject to Forces of Any Kind," *Rep. 42nd Meeting BAAS* (1873), Trans. Sect. A, 29–32,

Nature 8 (1873): 537–539, reprinted in *Scientific Papers*, vol. 2, 351–354 and as Document II–13.

12. [Notes on "On the Final State of a System of Molecules in Motion Subject to Forces of Any Kind"]

Cambridge University, Maxwell Collection, Scientific Papers, 5.

Let molecules of several kinds be <contained> in motion within a vessel with elastic sides and let each kind of molecules be acted on by <a> forces which have a potential the form of this potential being in general different for different kinds of molecules. Let the coordinates of a molecule be $x\,y\,z$ and the components of its velocity $\xi\,\eta\,\zeta$ and let it be required to determine the number of molecules of the first kind which, on an average, have their coordinates between x & $x + dx$ y and $y + dy$ and z and $z + dz$ and also their component velocities between ξ and $\xi + d\xi$ η and $\eta + d\eta$ and ζ and $\zeta + d\zeta$. This number must depend on the coordinates $x\,y\,z$ <and> on the component velocities $\xi\,\eta\,\zeta$ and on the limits and we may therefore write it

$$f(x,\,y,\,z,\,\xi,\,\eta,\,\zeta)\,dx\,dy\,dz\,d\xi\,d\eta\,d\zeta$$

We shall begin by <considering the> investigating the laws according to which this number depends on the velocity $\xi\,\eta\,\zeta$ before we proceed to determine in what manner it depends on the position $(x,\,y,\,z)$.

For this purpose we consider the mode in which the velocity of a molecule of the first kind will be changed in consequence of a collision with a molecule, say, of the second kind. The whole number of molecules of the first kind in unit of volume at the given place which have velocities within given limits may be written

$$f_1(\xi_1,\eta_1,\,\zeta_1)\,d\xi_1\,d\eta_1\,d\zeta_1 = n_1$$

The number of those of the second within their limits of velocity are

$$f_2(\xi_2,\eta_2,\,\zeta_2)\,d\xi_2\,d\eta_2\,d\zeta_2 = n_2$$

The number of pairs which can be formed <out of> by taking one molecule of each kind is $n_1\,n_2$.

<The number of these pairs which encounter each other in such a manner that>

We have already shown that when two molecules encounter each other the velocity of their centre of gravity remains unchanged in magnitude and direction but that the line AB representing on the diagram of velocities the relative velocity of the molecule B with respect to A is turned about

G <through an angle 2θ in a plane defined through AB when> as a fixed point without change of magnitude into a new <position> direction which we may suppose to be defined by the angular coordinates θ & ϕ.

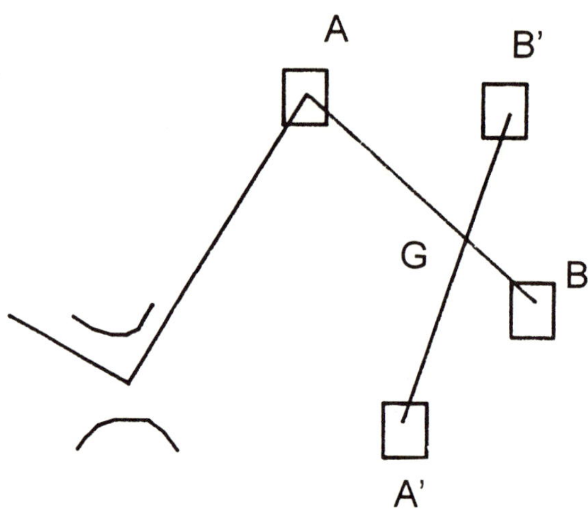

The values of θ and ϕ depend upon the <direction and> magnitude of the <line b which> angular momentum of B with respect to A and on the direction of the axis about which this angular momentum exists, and the number of pairs [of] molecules for which θ lies between θ and $\theta + d\theta$ and ϕ between ϕ and $\phi + d\phi$ will be

$$n_1\, n_2\, F(\theta, \phi,)\, d\theta\, d\phi$$

Now let us suppose < θ and ϕ constant while ξ_2 > the components of relative velocity $(\xi_2 - \xi_1)$, $(\eta_2 - \eta_1)$, $(\zeta_2 - \zeta_1)$ to remain constant while the components of absolute velocity of both molecules are made to vary then

$$d\xi_2 = d\xi_1 \qquad d\eta_2 = d\eta_1 \qquad d\zeta_2 = d\zeta_1$$

Let us also suppose that θ and ϕ remain constant during the variation, then if the values of the component velocities after collision be distinguished by accents, $d\xi_1' = d\xi_2' = d\xi_2 = d\xi_3$ &c. and the same for the other differentials. Hence the number of molecules of the first kind whose velocities as represented in the diagram lie within the element A which encounter molecules of the second kind whose velocities lie within the <equal> element B equal to A and which have encounters such that θ and ϕ lie between given limits is

$$< n_1\, n_2\, F(\theta\, \phi) >$$

$$f_1(\xi_1,\, \eta_1,\, \zeta_1)\, f_2(\xi_2,\, \eta_2,\, \zeta_2)\, F(\theta,\, \phi)(d\xi\, d\eta\, d\zeta)^2\, d\theta\, d\phi$$

and after collisions the velocities are represented by points lying within the elements A' and B', each equal to A.

Let us now consider the number of encounters between pairs of molecules the original velocities of which lie within $d\xi\, d\eta\, d\zeta$ of $-\xi_1' - \eta_1' - \zeta_1'$ and $+\xi_2' + \eta_2' + \zeta_2'$ respectively and for which θ' and ϕ' lie between the same limits as before, except that $\phi' = \phi + \pi$, $\theta' = \pi - \theta$. Their number is evidently

$$f_1(-\xi_1',\, -\eta_1',\, -\zeta_1')\, f_2(+\xi_2',\, +\eta_2',\, \zeta_2')\, F(\theta,\, \phi)(d\xi\, d\eta\, d\zeta)^2\, d\theta\, d\phi$$

<By the principle of the reversibility of motion in any conservative system <since> the velocities after collision will be $-\xi_1 - \eta_1 - \zeta_1$ and $-\xi_2 - \eta_2 - \zeta_2$ will lie between limits of the same magnitude as the original ones at A & B but with reversed sign.>

These pairs of molecules will change their velocities from $OA'\, OB'$ to $OA\, OB$. If the system is in a permanent state <the number of> as many pairs will change from $OA\, OB$ to $OA'\, OB'$ <must be the same as the number of> as change from $OA'\, OB'$ to $OA\, OB$ or dividing out the common factors

$$f_1(\xi_1\, \eta_1\, \zeta_1)\, f_2(\xi_2\, \eta_2\, \zeta_2) = f_1(\xi_1'\, \eta_1'\, \zeta_1')\, f_2(\xi_2'\, \eta_2'\, \zeta_2')$$

One and only one <other> necessary relation exists between the variables before and after collisions namely

$$M_1(\xi_1^2 + \eta_1^2 + \zeta_1^2) + M_2(\xi_2^2 + \eta_2^2 + \zeta_2^2)$$

$$= M_1(\xi_1'^2 + \eta_1'^2 + \zeta_1'^2) + M_2(\xi_2'^2 + \eta_2'^2 + \zeta_2'^2)$$

Writing

$$< \rho^2 = \xi^2 + \eta^2 + \zeta^2 >$$

$$\xi = V\ell \qquad \eta = Vm \qquad \zeta = Vn$$

this becomes

$$M_1\, V_1^2 + M_2\, V_2^2 = \text{constant}.$$

We may also write

$$f(\xi,\, \eta,\, \zeta) = e^{F(V^2, \ell, m, n)}$$

whence on taking the logarithm of ( )

$$F_1(M_1\, V_1^2\, \ell_1\, m_1\, n_1) + F_2(M_2\, V_2^2 i\, \ell_2\, m\, n)$$

is constant provided $M_1 V_1^2 + M_2 V_2^2$ is constant.

Since $\ell_1\, m\, n$ are independent of $\ell_2\, m\, n$ we must have

$$F_1\, M_1\, V_1^2 = A\, M_1\, V_1^2 \qquad \& \qquad F_2\, M_2\, V_2^2 = A\, M_2\, V_2^2 \tag{2}$$

or

$$f_2^a(\xi_1\, \eta_1\, \zeta_1) = C_1\, e^{A M_1 V_1^2} \qquad f_2(\xi_2\, \eta_2\, \zeta_2) = C_2 e^{A M_2 V_2^2}$$

Since the relation between the velocity immediately before and immediately after collision is not affected by the continuous action of finite forces on the molecules the distribution of velocities at any given point of the vessel must be <of the form> such that the number of molecules in the element dx dy dz having velocities between ξ & $\xi + d\xi$ &c is

$$dN = c\, C_1\, e^{AM(\xi_1^2+\eta_1^2+\zeta_1^2)}\, d\xi\, d\eta\, d\zeta\, dx\, dy\, dz$$

where C is a function of $x\, y\, z$ which may be different for different kinds of molecules, while A though it may be a function of the position of the element is the same for the different kinds of molecules.

Let us now suppose that a force whose potential is ψ_1 acts on the molecules of the first kind which we are now alone considering and let us consider the variation of dN <arising> during the time δt the quantities $x\, y\, z\, \xi\, \eta\, \zeta$ varying in the same way as the coordinates and the component velocities of a molecule do, that is, let

$$\delta x = \xi \delta t \qquad \delta y = \eta \delta t \qquad \delta z = \zeta \delta t$$

$$\delta \xi = -\frac{d\psi}{dx}\, \delta t \qquad \delta \eta = -\frac{d\psi}{dy}\, \delta t \qquad \delta \zeta = -\frac{d\psi}{dz}\, \delta t$$

We thus find

$$\begin{aligned} \delta \frac{\log dN}{\delta t} &=< \frac{d \log C_1}{dx} \xi > \\ &= \frac{dC_1}{dx}\xi + \frac{dC_1}{dy}\eta + \frac{dC_1}{dz}\zeta \\ &\quad + 2\, A\, M\Big(\xi\frac{d\psi}{dx} + \eta\frac{d\psi}{dy} + \zeta\frac{d\psi}{dz}\Big) \\ &\quad + M(\xi^2 + \eta^2 + \zeta^2)\Big(\frac{dA}{dx}\xi + \frac{dA}{dy}\eta + \frac{dA}{dz}\zeta\Big) \end{aligned}$$

Now since the same set of molecules have by their motion passed into this new condition the variation of N must be zero and this must be true whatever the particular values of $\xi\, \eta\, \zeta$.

Hence

$$\frac{dA}{dx} = 0 \qquad \frac{dA}{dy} = 0 \qquad \frac{dA}{dz} = 0$$

or A is a constant throughout the vessel. Also

$$\frac{dC}{dx} = 2\,A\,M\frac{d\psi}{dx} \qquad \frac{dC}{dy} = 2\,A\,M\frac{d\psi}{dy} \qquad \frac{dC}{dz} = 2\,A\,M\frac{d\psi}{dz}$$

or

$$C = 2\,A\,M(\psi + B)$$

a. This is a slip of the pen. Maxwell treats this function as $f_1(\xi_1, \eta_1, \zeta_1)$ through the rest of this draft.

13. [Draft of "On the Final State of a System of Molecules in Motion Subject to Forces of Any Kind"]

Cambridge University Library, Maxwell Collection, Scientific Papers, 6

We are now prepared to investigate the fundamental problem in the theory of molecules–the determination of the ultimate state of a multitude of moving molecules confined in a vessel in such a way that neither the molecules themselves, nor their energy, can escape.

For greater generality, we shall suppose the molecules to be of several different kinds and we shall distinguish the quantities <corresponding> belonging to the different sets of molecules by different suffixes.

We shall also suppose that the molecules are acted on by attractions or repulsions, towards fixed points. The effect of gravity on the motion of a set of molecules <is a familiar> will thus be included in our solution but for the sake of generality we shall only assume that the force acting on any set of molecules is such that it may be derived from a potential, and we <need not> may even suppose the form of this potential to be different for different sets of molecules.

Let x, y, z, be the coordinates of a molecule of mass M and let ξ, $\eta\,\zeta$ be the components of its velocity and let *Nbe* the number of such molecules in unit of volume and let ψ be the potential of the force which acts on this set of molecules.

<Let> The number of molecules of this kind which on an average have their coordinates between the limits x and $x + dx$, y and $y + dy$, and z and $z + dz$ and also the components of their velocity between ξ and $\xi + d\xi$, η

and $\eta + d\eta$, ζ and $\zeta + d\zeta$ must be a function of x, y, z; ξ, η, ζ; dx, dy, dz and $d\xi\, d\eta\, d\zeta$ of the form

$$\aleph = f(x, y, z, \xi, \eta, \zeta)\, dx\, dy\, dz\, d\xi\, d\eta\, d\zeta \tag{1}$$

We have to investigate the form of this function.

We shall begin be determining the manner in which this function depends on the components of velocity (ξ, η, ζ) before we proceed to investigate in what manner it depends on the coordinates (x, y, z).

For this purpose we shall consider the effect of an encounter between two molecules which we shall assume to be of the first and the second kind respectively, distinguished by the suffixes 1 and 2.

The whole number of molecules of the first kind in unit of volume at the given place, which have their component velocities within the given limits, may be written

$$f_1(\xi_1\, \eta_1\, \zeta_1)\, d\xi_1\, d\eta_1\, d\zeta_1 = n_1 \tag{2}$$

The number of molecules of the second kind selected in a similar manner according to their velocities may be written

$$f_2(\xi_2\, \eta_2\, \zeta_2)\, d\xi_2\, d\eta\, d\zeta_2 = n_2 \tag{3}$$

The number of pairs which can be formed by taking one molecule out of each of these groups is $n_1\, n_2$.

<Let us consider> The result of an encounter between the two molecules forming one of those pairs will depend on the elements of position of the molecules, that is to say on the arm of approach and on the plane of the encounter if the action <is central> between the molecules passes through their centre of mass. If the <law> mode of action depends on the angular position of each molecule about its centre of mass then this also must be taken into account. <We may>

Let us class all encounters as of the same kind when these elements of position <are either identical or be within certain limits> differ from certain specified values by less than certain given small quantities <from>

Thus the number of encounters between n_1 molecules of one kind and n_2 of the other the elements of position of which lie within [space]

Now it is shown in treatises on dynamics*[a] that if in any motion a certain variation δq_a in one of the <linear> elements of position

*Thomson and Tait's Natural Philosophy Vol. 1 p. 250

of the initial motion causes a variation dp_b [b] < δq_b > in one of the angular elements of <velocity> momentum in the final motion then < a variation

δq > if the direction of motion be reversed a variation δq_b equal in magnitude to δq_a in the initial element of position will cause a variation $-\delta p'_a$ equal in magnitude to δp_b in the final element of <velocity> momentum

$$\frac{dp'_a}{dq_b} = -\frac{dp_b}{dq'_a}$$

We have already shown that if the system be turned round 180° in the plane of the encounter and the direction of motion reversed we have an encounter in which the initial velocities agree in direction and magnitude with the final velocities in the original encounter and in which the final velocities agree with the original velocities in the first encounter.

We now see that equal variations in the initial elements of position lead to equal variations in the final elements of velocity in the two encounters. Let us now consider a set of encounters all the elements of which except the velocity of the centre of mass agree within very close limits <except the> but let the velocity of the centre of mass of the two molecules be anywhere between the limits $\overline{\xi}$ and $\overline{\xi} + d\overline{\xi}$, $\overline{\eta}$ and $\overline{\eta} + d\overline{\eta}$ and $\overline{\zeta}$ and $\overline{\zeta} + d\overline{\zeta}$. Since the velocities relative to the centre of mass are the same for all the encounters the limits of the velocity of the molecule A before the encounter will be ξ_1 and $\xi_1 + d\xi_1$, and η_1 and $\eta_1 + d\eta_1$, and ζ_1 and $\zeta_1 + d\zeta_1$ and those of B will be ξ_2 and $\xi_2 + d\xi_2$, and η_2 and $\eta_2 + d\eta_2$, and ζ_2 and $\zeta_2 + d\zeta_2$.

<Let> If the symbols of the velocities after the encounter are distinguished by accents their limits will be ξ'_1 and $\xi'_1 + d\xi_1$, and η'_1 and $\eta'_1 + d\eta_1$, and ζ'_1 and $\zeta'_1 + d\zeta_1$ for A and ξ'_2 and $\xi'_2 + d\xi$, and η'_2 and $\eta'_2 + d\eta$, and ζ'_2 and $\zeta'_2 + d\zeta$ for B.

We shall call these the final limits.

The number of pairs of molecules which have their original velocities lying within the given limits is

$$f_1(\xi_1\, \eta_1\, \zeta_1)\, f_2(\xi_2\, \eta_2\, \zeta_2)\, (d\xi\, d\eta\, d\zeta)^2 \tag{4}$$

The proportion of these whose elements of position lie within the assigned limits may be represented by the factor ρ.

Hence the number <which> of pairs which change their velocities from the given original to the given final velocities is

$$\rho f_1(\xi_1\, \eta_1\, \zeta_1)\, f_2(\xi_2\, \eta_2\, \zeta_2)\, (d\xi\, d\eta\, d\zeta)^2 \tag{5}$$

But we have seen that it is possible for two molecules to <change their velocities from> have an encounter in which the original and final velocities are the final and original velocities of the first encounter and that the factor which expresses the condition that the elements of the encounter shall be within given limits is the same in both cases. Hence the number of encounters of the second kind is

$$\rho f_1(\xi_1' \, \eta_1' \, \zeta_1') \, f_2(\xi_2' \, \eta_2' \, \zeta_2') \, (d \, d \, d)^2 \tag{6}$$

In these encounters the velocities change from the final to the original limits. Thus there is an exchange of molecules between the two groups whose velocities lie within the original and the final limits respectively. In the ultimate state of the system as many pairs of molecules must <exchange> pass from the final to the original as from the original to the final velocities. Hence in the ultimate state we may equate (5) and (6) and omitting common factors,

$$f_1(\xi_1 \, \eta_1 \, \zeta_1) \, f_2(\xi_2 \, \eta_2 \, \zeta_2) = f_1(\xi_1' \, \eta_1' \, \zeta_1') \, f_2(\xi_2' \, \eta_2' \, \zeta_2') \tag{7}$$

One and only one necessary relation exists between the variables before and after the encounter namely that which expresses the conservation of kinetic energy of the system

$$< M_1(\xi_1^2 + \eta_1^2 + \zeta_1^2) + >$$

$$M_1 V_1^2 + M_2 V_2^2 = M_1 V_1'^2 + M_2'^2 \tag{8}$$

where the symbol V denotes the velocity and

$$V^2 = \xi^2 + \eta^2 + \zeta^2 \tag{9}$$

Since $f(\xi, \eta, \zeta)$ cannot depend on the directions assumed for the axes of $\xi \, \eta \, \zeta$ we may write it < $f(V)$ or $e^{F(V)}$ Hence, taking the logarithm of equation> in the form

$$f(\xi, \eta, \zeta) = e^{F(V)} \tag{10}$$

Hence taking the logarithm of equation (7) it becomes

$$F_1(V_1) + F_2(V_2) = F_1(V_1') + F_2(V_2') \tag{11}$$

whatever be the values of $V_1 \, V_2 \, V_1' \, V_2'$ provided

$$M_1 V_1^2 + M_2 V_2^2 = M_1 V_1'^2 + M_2 V_2'^2 \tag{12}$$

From this we obtain

$$< F_1(V_1) = A \, M_1 V_1^2 + B_1, \qquad F_2(V_2) = A \, M_2 V_2 + B_2 >$$

$$F(V_1) = B_1 - A \, M_1 V_1^2, \qquad F_2 V_2 = B_2 - A \, M_2 V_2^2 \tag{13}$$

where A is a constant common to both functions and B_1, B_2 are independent. We may now write the expression (1) in the more definite form

$$\aleph = e^{B-AMV^2}\, d\xi\, d\eta\, d\zeta\, dx\, dy\, dz \tag{14}$$

where B is a function of x, y, z which may be different for different kinds of molecules while A, though it may, for aught we know be a function of x, y, z is the same for <all> the different kinds of molecules.

The relation between the velocities immediately before and immediately after a sudden encounter is not affected by the continuous action of finite forces in the molecules. Hence the above expression for the distribution of velocities at any given point is true whether external forces act or not. The effect of external forces is to alter the velocity of the molecules while they are describing their free paths.

Let us now confine our attention to molecules of the first kind and let us suppose that they are acted on by a force whose potential is ψ.

Consider the $\aleph$ molecules whose coordinates of position are x, y, z and whose components of velocity are $\xi\,\eta\,\zeta$.

During the time δt <these quantities will have the > the

Let us follow any one of those molecules in its motion. The rates of variation of the coordinates are

$$\frac{dx}{dt} = \xi \qquad \frac{dy}{dt} = \eta \qquad \frac{dz}{dt} = \zeta \tag{15}$$

The rates of variation of the component velocities are

$$\frac{d\xi}{dt} = -\frac{d\psi}{dx} \qquad \frac{d\eta}{dt} = -\frac{d\psi}{dy} \qquad \frac{d\zeta}{dt} = -\frac{d\psi}{dz} \tag{16}$$

Hence if we consider $\aleph$ as a function of x, y, $z\,\xi\,\eta\,\zeta$

$$\begin{aligned}\frac{d\log\aleph}{dt} = {} & \frac{dB}{dx}\xi + \frac{dB}{dy}\eta + \frac{dB}{dz}\zeta \\ & + 2\,A\,M\,(\xi\frac{d\psi}{dx} + \eta\frac{d\psi}{dy} + \zeta\frac{d\psi}{dz}) \\ & - M\,V^2\,(\frac{dA}{dx}\xi + \frac{dA}{dy}\eta + \frac{dA}{dz}\zeta)\end{aligned} \tag{17}$$

Now though the molecules have by their motion passed into this new condition <in which> their number must remain the same as before, so that the <quantity> right hand member of the above equation (17) must be identically zero, whatever be the values of $\xi\,\eta\,\zeta$

Hence

$$\frac{dA}{dx} = 0 \qquad \frac{dA}{dy} = 0 \qquad \frac{dA}{dz} = 0 \tag{18}$$

or A is <constant and> independent of x, y, z and is constant throughout the system. Also

$$\frac{dB}{dx} = -2\,A\,M\frac{d\psi}{dx} \qquad \frac{dB}{dy} = -2\,A\,M\frac{d\psi}{dy} \qquad \frac{dB}{dz} = -2\,A\,M\frac{d\psi}{dz} \tag{19}$$

whence

$$B = C - 2AM\psi \tag{20}$$

where C is independent of x, y, z.
Hence we obtain

$$< e^{A\,M(\xi^2+\eta^2\zeta^2+2\psi+2B)}dx\,dy\,dz\,d\xi\,d\eta\,d\zeta >$$

<To determine>

$$\aleph = e^{C_1-A M_1(2\psi_1+\xi^2+\eta^2+\zeta^2)}dx\,dy\,dz\,d\xi\,d\eta\,d\zeta \tag{21}$$

as the number of molecules of the first kind which at a given instant lie within the element of volume $dx\,dy\,dz$ and have their <component> velocities represented by points within the element $d\xi\,d\eta\,d\zeta$ of the diagram of velocities.

In this expression C_1 is a constant for each kind of molecule but may be different for different kinds of molecules.

A is a constant which is the same for all kinds of molecules in the ultimate state of the system.

ψ_1 is the potential of the force which acts on molecules of the first kind. The other kinds of molecules may be acted on by forces having potentials different from ψ_1.

To find the whole number of molecules within the element $dx\,dy\,dz$ we must integrate (21) between the limits $\pm\infty$ with respect to $\xi\,\eta$ and ζ so as to include all values of the velocity. <Calling N_1 the number of molecules in unit of volume we find>

Dividing the result by $dx\,dy\,dz$ we obtain N_1 the number of molecules of the first kind in unit of volume

$$N_1 = \frac{(\pi)^{3/2}}{A\,M_1}e^{C_1-A\,M_1 2\psi_1} \tag{22}$$

<The kinetic energy of the molecular system is obtained by integrating>
From this it appears that the density of the first medium, which is the <proportional to> the product of the number of molecules in unit of volume into the mass of a molecule, varies from one part of the vessel to another according to a law which is independent of the existence of molecules of

other kinds in the vessel. <For instance if a given quantity of oxygen and a given quantity of nitrogen are placed in a closed vessel under the action of gravity, the quantity of oxygen in unit of volume taken at any part of the vessel will be the same as if there had been no nitrogen in the vessel.>

For instance if a certain quantity of oxygen is placed in a closed vessel, it will be distributed according to the well known law of a gas under the action of gravity, being denser below and rarer above. If now a quantity of nitrogen be added to the oxygen in the vessel and time be allowed for a thorough diffusion to take place, the distribution of the oxygen will be exactly as before. That of the nitrogen will follow its own law diminishing in density in the upper part of the vessel but at a slower rate than in the case of oxygen because the mass of its molecule is less. This is Dalton's law of mixed gases according to which if several portions of gas whether of the same kind or of different kinds occupy the same region each portion behaves as a vacuum to the rest.

The kinetic energy of the medium in unit of volume is found by <integrating $\aleph$ > multiplying $\aleph$ by $1/2M_1V_1^2$ and integrating as before. By dividing this by N_1 we get the mean kinetic energy of a molecule

$$\frac{1}{2}M_1\overline{V_1^2} = \frac{3}{4}\frac{1}{A} \tag{23}$$

As the quantity A is in the ultimate state of the system the same for every kind of molecule and for every part of the vessel, it follows that the mean kinetic energy is the same for all molecules whatever be their masses.

Now when two bodies are placed in communication their temperatures tend to become equal and the temperature of each depends in some manner on the agitation of its parts. The relation between the kinetic energy of agitation and temperature is not known for all bodies but in the case of two gases mixed together we now see that what tends to become equal is the mean kinetic energy of a single molecule in each gas. Hence we may assert that in gases the temperature is a function of the kinetic energy of agitation of a single molecule.

It follows from this that in the ultimate state of the system of molecules acted on by gravity the temperature is the same in all parts of the vessel <a result which> The effect of gravity is to increase the density in the lower parts of the vessel but not to <alter> make any difference in the mean velocity of the molecules. This may appear in contradiction to the fact that in the free motion of any one molecule its velocity increases as it descends and diminishes as it ascends. To show that the contradiction is only apparent requires a little consideration which will form a good exercise for the student.

Another and still more important result is that equal volumes of different gases at equal temperatures and pressures contain equal numbers of molecules. This is Gay-Lussac's law of molecular volumes of gases.

For the pressure on unit of area is

$$p = \frac{1}{3} M N \overline{V^2} = \frac{1}{2} \frac{N}{A}$$

where A is a function of the temperature. Hence when the temperature and pressure are given, $N = 2Ap$ is given or the number of molecules in unit of volume is independent of the nature of the gas.

<Again–that part of the kinetic energy of a gas which depends on the motion of translation of the molecules is

$$\frac{1}{2} M N \overline{V^2} = \frac{3}{4} \frac{N}{A} \qquad \text{per unit of volume} >$$

Again, when heat is communicated to a gas that part of it which is spent in increasing the motion of agitation of the centres of the molecules is for each molecule the increment of $\frac{1}{2} M_1 \overline{V^2}$ or $\frac{3}{4} \frac{1}{A}$ which is the same for all gases for the same increment of temperature. Hence the specific heat per molecule is the same for all gases. <which is the> except in so far as their thermal energy may depend on the internal motions of each molecule. This is the law of molecular specific heats discovered by Dulong & Petit.

a. This refers to William Thomson and Peter Guthrie Tait, *Treatise on Natural Philosophy* (Oxford: Clarendon Press, 1867). The authors had intended further volumes on branches of physics beyond the published volume on mechanics. For an account of the intentions of Thomson and Tait and what was actually published see Crosbie Smith and Norton Wise, *Energy and Empire: A Biographical Study of Lord Kelvin* (Cambridge: Cambridge University Press, 1989), chap. 4.

b. dp_b, a slip of the pen. It is treated as δp through the rest of the notes.

14. "On the Final State of a System of Molecules in Motion Subject to Forces of Any Kind"

Rep. 42nd Meeting BAAS (1873): Trans. Sect. A, 29–32.

Since reading Principal Guthrie's first letter on this subject ('Nature,' May 22, 1873), I have thought of several ways of investigating the equilibrium of temperature in a gas acted on by gravity. One of these is to investigate the condition of the column as to density when the temperature is constant, and to show that when this is fulfilled the column also fulfills the condition that there shall be no upward or downward transmission of energy, or, in fact, of any other function of the masses and velocities of the molecules. But a far more direct and general method was suggested to me by the investigation of Dr. Ludwig Boltzmann [a] on the final distribution of energy in a finite system of elastic bodies; and the following is a sketch of this method as applied to the simpler case of a number of molecules so great that it may be treated as infinite.

Principal Guthrie's second letter is especially valuable as stating his case in the form of distinct propositions, every one of which, except the fifth, is incontrovertible. He has himself pointed out that it is here that we differ, and that this difference may ultimately be traced to a difference in our doctrines as to the distribution of velocity among the molecules of any given portion of the gas. He assumes, as Clausius (at least in his earlier investigations) did, that the velocities of all the molecules are equal, whereas I hold, as I first stated in the Philosophical Magazine for January 1860, that they are distributed according to the same law as errors of observation are distributed according to the received theory of such errors.[b] It is easy to show that if the velocities are all equal at any instant they will become unequal as soon as encounters of any kind, whether collisions or "perihelion passages" takes place.[c] The demonstration of the actual law of distribution was given by me in an improved form in my paper on the "Dynamical Theory of Gases," Phil. Trans. [R. Soc. London 157] 1866, and Phil. Mag. 1867; and the far more elaborate investigation of Boltzmann has led him to the same result. I am greatly indebted to Boltzmann for the method used in the latter part of the following sketch of the general investigation.

Let perfectly elastic molecules of different kinds be in motion within a vessel with perfectly elastic sides, and let each kind of molecules [sic] be acted on by forces which have a potential the form of which may be different for different kinds of molecules.

Let x, y, z be the coordinates of a molecule, M, and ξ, η, ζ the components of its velocity, and let it be required to determine the number of molecules of a given kind which, on an average, have their coordinates between x and $x + dx$, y and $y + dy$, z and $z + dz$, and also their component velocities between ξ and $\xi+d\xi$, η and $\eta+d\eta$, and ζ and $\zeta+d\zeta$. This number must depend on the coordinates and the components of velocities and on the differences of the limits of these quantities. We may therefore write it

$$dN = f(x, y, z, \xi, \eta, \zeta)\, dx\, dy\, dz\, d\xi\, d\eta\, d\zeta \qquad (1)$$

We shall begin by investigating the manner in which this quantity depends on the components of velocity, before we proceed to determine in what way it depends on the coordinates.

If we distinguish by suffixes the quantities corresponding to different kinds of molecules, the whole number of molecules of the first and second kind within a given space, which have velocities within given limits, may be written

$$f_1(\xi_1,\, \eta_1,\, \zeta_1)\, d\xi_1,\, d\eta_1,\, d\zeta_1 = n_1 \tag{2}$$

and

$$f_1(\xi_2,\, \eta_2,\, \zeta_2)\, d\xi_2,\, d\eta_2,\, d\zeta_2 = n_2 \tag{3}$$

The number of pairs which can be formed by taking one molecule of each kind is $n_1\, n_2$.

Let a pair of molecules encounter each other, and after the encounter let their component velocities be $\xi_1',\, \eta_1',\, \zeta_1'$ and $\xi_2',\, \eta_2',\, \zeta_2'$. The nature of the encounter is completely defined when we know $\xi_1 - \xi_2,\, \eta_1 - \eta_2,\, \zeta_1 - \zeta_2$ the velocity of the second molecule relative to the first before the encounter, and $x_1 - x_2,\, y_1 - y_2,\, z_1 - z_2$ the position of the centre of the second molecule relative to the first at the instant of the encounter. When these quantities are given, $\xi_2' - \xi_1',\, \eta_2' - \eta_1',\, \zeta_2' - \zeta_1'$, the components of the relative velocity after the encounter, are determinable.

Hence, putting α, β, γ for these relative velocities, and a, b, c for the relative positions, we find for the number of molecules of the first kind having velocities between the limits ξ_1 and $\xi_1 + d\xi_1$ &c., which encounter molecules of the second kind having velocities between the limits ξ_2 and $\xi_2 + d\xi_2$, &c., in such a way that the relative velocities lie between α and $\alpha + d\alpha$ &c., and the relative positions between a and $a + da$, &c.

$$\begin{aligned} &f_1(\xi_1,\, \eta_1,\, \zeta_1)\, d\xi\, d\eta\, d\zeta \,.\, f_2(\xi_2,\, \eta_2,\, \zeta_2)\, d\xi\, d\eta\, d\zeta \\ &\qquad\qquad .(a\, b\, c\, \alpha\, \beta\, \gamma)\, da\, db\, dc\, d\alpha\, d\beta\, d\gamma. \end{aligned} \tag{4}$$

and after the encounter the velocity of M_1 will be between the limits ξ_1' and $\xi_1' + d\xi$, &c., and that of M_2 between the limits ξ_2' and $\xi_2' + d\xi$ &c.

The differences of the limits of velocity are equal for both kinds of molecules, and that both before and after the encounter.

When the state of motion of the system is in its permanent condition, as many pairs of molecules must change their velocities $V_1,\, V_2$ to $V_1',\, V_2'$ as from $V_1',\, V_2'$ to $V_1,\, V_2$; and the circumstances of the encounter in the one case are precisely similar to those in the second. Hence, omitting for the sake of brevity the quantities $d\xi$ &c., and ϕ, which have the same values in the two cases, we find

$$f_1(\xi_1,\, \eta_1,\, \zeta_1)\, f_2(\xi_2,\, \eta_2,\, \zeta_2) = f_1(\xi_1',\, \eta_1',\, \zeta_1')\, f_2(\xi_2', \eta_2',\, \zeta_2') \tag{5}$$

If we now write

$$log f(\xi,\, \eta,\, \zeta) = F(MV^2,\, \ell,\, m,\, n) \tag{6}$$

where $\ell,\, m,\, n$ are the direction cosines of the velocity V of the molecule M, taking the logarithms of both sides of equation (5),

$$\begin{aligned} &F_1(M_1\, V_1^2\, \ell_1\, m_1\, n_1) + F_2(M_2\, V_2^2\, \ell_2\, m_2\, n_2) \\ &\qquad = F_1(M_1\, V_1^{'2}\, \ell_1'\, m_1'\, n_1') + F_2(M_2\, V_2^{'2}\, \ell_2'\, m_2'\, n_2') \end{aligned} \tag{7}$$

The only necessary relation between the variables before and after the encounter is

$$M_1 V_1^2 + M_2 V_2^2 = M_1 V_1^{'2} + M_2 V_2^{'2} \tag{8}$$

If the right-hand sides of the equations (7) and (8) are constant, the left-hand sides will also be constant; and since ℓ_1, m_1, n_1 are independent of ℓ_2, m_2, n_2, we must have

$$F_1 = -A\, M_1 V_1^2 \qquad \text{and } F_2 = -A\, M_2 V_2^2 \tag{9}$$

where A is a quantity independent of the components of velocity, or

$$f_1(\xi_1, \eta_1, \zeta_1) = C_1\, e^{-A\, M V_1^2}, \tag{10}$$

$$f_2(\xi_2, \eta_2, \zeta_2) = C_2\, e^{-A\, M V_2^2}, \tag{11}$$

This result as to the distribution of the velocities of the molecules at a given place is independent of the action of finite forces on the molecules during their encounter; for such forces do not affect the velocities during the infinitely short time of the encounter.

We may therefore write equation (1)

$$dN = C\, e^{-AM(\xi^2+\eta^2\zeta^2)}\, d\xi\, d\eta\, d\zeta\, dx\, dy\, dz, \tag{12}$$

where C is a function of $x\,y\,z$, which may be different for different kinds of molecules, while A is the same for every kind of molecule, though it may, for aught we know as yet, vary from one place to another.

Let us now suppose that the kind of molecules under consideration are acted on by a force whose potential is ψ. The variations of $x\,y\,z$ arising from the motion of the molecules during a time δt are

$$\delta x = \xi\, \delta t, \qquad \delta y = \eta\, \delta t, \qquad \delta z = \zeta\, \delta t; \tag{13}$$

and those of ξ, η, ζ in the same time due to the action of the force, are

$$\delta\xi = -\frac{d\psi}{dx}\, \delta t, \qquad \delta\eta = -\frac{d\psi}{dy}\, \delta t, \qquad \delta\zeta = -\frac{d\psi}{dz}\, \delta t, \tag{14}$$

If we put

$$c = log C, \tag{15}$$

$$log \frac{dN}{d\xi d\eta d\zeta dx dy dz} = c - AM(\xi^2 + \eta^2 + \zeta^2), \tag{16}$$

The variation of this quantity due to the variations δx_1, δy_1, δz_1, $\delta\xi_1$, $\delta\eta_1$, $\delta\zeta_1$, is

$$\left.\begin{aligned} \left(\xi\frac{dc}{dx}+\eta\frac{dc}{dy}+\zeta\frac{dc}{dz}\right)\delta t \\ +2\,A\,M\left(\xi\frac{d\psi}{dx}+\eta\frac{d\psi}{dy}+\zeta\frac{d\psi}{dz}\right)\delta t \\ -M(\xi^2+\eta^2+\zeta^2)\left(\xi\frac{dA}{dx}+\eta\frac{dA}{dy}+\zeta\frac{dA}{dz}\right)\delta t. \end{aligned}\right\} \tag{17}$$

Since the number of the molecules does not vary during their motion, this quantity is zero, whatever the values of ξ, η, ζ . Hence we have in virtue of the last term,

$$\frac{dA}{dx}=0, \qquad \frac{dA}{dy}=0, \qquad \frac{dA}{dz}=0, \tag{18}$$

or A is constant throughout the whole region traversed by the molecules.

Next, comparing the first and second terms, we find.

$$c=-2AM(\psi+B). \tag{19}$$

We thus obtain as the complete form of dN,

$$dN_1=e^{-AM(\xi_1^2+\eta_1^2+\zeta_1^2+2\psi_1+B_1)}\,dx\,dy\,dz\,d\xi\,d\eta\,d\zeta, \tag{20}$$

where A is an absolute constant, the same for every kind of molecule in the vessel, but B, belongs to the first kind only. To determine those constants, we must integrate this quantity with respect to the six variables, and equate the result to the number of molecules of the first kind. We must then by integrating

$$dN_1\frac{1}{2}\,M_1(\xi_1^2+\eta_1^2+\zeta_1^2+2\psi_1),$$

determine the whole energy of the system, and equate it to the original energy. We shall thus obtain a sufficient number of equations to determine the constant A, common to all the molecules, and B_1, B_2, &c., those belonging to each kind.

The value of A determines that of the mean kinetic energy of all the molecules in a given place, which is $\frac{3}{2}\frac{1}{A}$; and therefore, according to kinetic theory, it also determines the temperature of the medium at that place. Hence, since A, in the permanent state of the system, is the same for every part of the system, it follows that the temperature is everywhere the same, whatever forces act upon the molecules.

The number of molecules of the first kind in the element $dz\,dy\,dx$,

$$\left(\frac{\pi}{A}\right)^{3/2}e^{-AM_1(2\psi_1+B_1)}dx\,dy\,dz, \tag{21}$$

The effect of the force whose potential is ψ_1 is therefore to cause the molecules of the first kind to accumulate in greater numbers in those parts of the vessel towards which the force acts; and the distribution of each different kind of molecules [sic] in the vessel is determined by the forces which act on them in the same way as if no other molecules were present. This agrees with Dalton's doctrine of the distribution of mixed gases.

a Footnote at the bottom of the page. Studien über das Gleichgewicht der lebendigen Kraft [zwischen] bewegten materiellen Punkte, von Dr. Ludwig Boltzmann. Sitzb. d. Akad. Wissensch. [58] October 8, 1868 (Vienna) [517–560].

b. Maxwell,"Illustrations of the Dynamical Theory of Gases," *Phil. Mag.* 19 (1860): 19–32; 20 (1860): 21–37, reprinted in *Scientific Papers*, vol. 1, 377-409 and Garber, Brush, and Everitt, *Maxwell on Molecules and Gases*, 285–318.

c. This refers to the phrase in Document II–10 used by Guthrie to describe the paths of molecules during their interactions.

15. "On the Final State of a System of Molecules in Motion Subject to Forces of Any Kind"

Nature 8 (1873): 537–538, reprinted in *Scientific Papers*, vol. 2, 351–354.

This published paper is not reprinted here because its content is almost identical with that of Document II–14. This version of the paper is somewhat shorter as the paragraphs on Francis Guthrie are omitted in the *Nature* article. The only addition where Maxwell, discussing the characteristics of A adds "The quantity A is essentially negative."

By choosing only the final version of this paper published in *Nature* the editor of Maxwell's scientific papers W. D. Niven hides from the reader the essential history of the problem. This omission may well have been an editorial decision arising from space limitations and the editor's sense that his readers knew the context. The earlier exchange of correspondence with Guthrie and the latter's persistent questioning moved Maxwell to eventually consider the problem more closely. Without this context the paper itself stands in his *Scientific Papers* as an isolated production, unconnected in a very un-Maxwell like way except in very general terms, with his other work in this area.

16. Letter from Francis Guthrie, "Molecular Motion"

Nature 10 (1874): 123.[a]

In Prof. Maxwell's communication NATURE, vol.viii p. 537, on this subject, he assumes that if n_1 represent the number of molecules of a particular kind in a given element of space with a velocity given in direction and magnitude, which we will call v_1: and if n_2 represent the particles of another kind in the same element with the velocity v_2, then the number of encounters of these particles is proportional to $n_1 \times n_2$, and if out of these we select the particular encounters which give rise to a given set of resultant velocities v_1' and v_2', then we may assume that if the number of particles in the element which originally had the velocities v_1' and v_2' be called n_1' and n_2', then

$$n_1 n_2 = n_1' n_2'$$

This reasoning does not seem convincing. Assuming that in an element of space the average number of particles having a given velocity is the same, so that n_1 and n_2 are *not* functions of v_1 and v_2 then Mr. Maxwell's statement might be admitted; but if the number of particles in a given element is a function of its velocity in direction and magnitude, then although the average of the numbers in each direction is maintained, it does not follow that the average numbers of particles having the velocities v_1 and v_2 are directly restored from the particles having the velocities v_1' and v_2'. All that can be assumed is, that the average number of particles in a given element of space is maintained from the particles in that and the remaining elements. Just as in the case of an equilibrium of trade, the average course of exchange with respect to a given country is at par; but we cannot therefore safely assume that the same is the case relatively to any other individual country.

There are several other points in Mr. Maxwell's communication which seem to me to require fortification, but the subject has already assumed so technical a form that it would perhaps be uninteresting to your readers to point them out. My impression is that the whole subject is still somewhat beyond the grasp of strict mathematical reasoning, and is still open to experimental investigation.

Graaff Reinet College, Feb. 7 F. Guthrie

a. Francis Guthrie is criticizing Maxwell, "On the Final State of a System of Molecules in Motion Subject to Forces of Any Kind," *Nature* (1873): 537–538, and *Rep. 42nd Meeting BAAS* (1873): Trans. Sect. A, 29–32, and document II–14. Publication of this letter was evidently delayed until June 18th, 1874, after Maxwell's reply, Document II–17, was received.

17. Maxwell's reply to Guthrie, "Molecular Motion," *Nature* 1874[a]

Nature 10 (1874): 124.

This question is treated at length in my paper On the dynamical theory of gases (Phil. Trans. , 1866). It is there shown that if the average course of exchange is in a cycle from A to B, B to C, C to A, an equal reason may be given why it should be in the opposite cycle A to C, C to B, B to A, and thus it is shown that the exchange is at par between each pair of states separately. For a far more elaborate theoretical treatment of the subject Prof. Guthrie is referred to the papers of Prof. Ludwig Boltzmann in the Vienna Transactions since 1868. I fear we must delay the experimental investigation for some time, till we are able to count the molecules in a given space, to observe their velocities, and to repeat these operations millions of times in a second.—J. Clerk Maxwell.

a. This is Maxwell's reply to Guthrie, "Molecular Motion," *Nature* 10 (1874): 123, Document II–16.

18. [Draft of a review of *A Treatise on the Kinetic Theory of Gases*, by H. W. Watson]

Cambridge University Library, Maxwell Collection[a]

A Treatise on the Kinetic Theory of Gases by Henry William Watson, M.A.,[b] formerly fellow of Trinity College, Cambridge. Oxford, Clarendon Press, 1876.

This treatise is arranged in the form of thirteen propositions <in which> for the determination of the <final> ultimate average distribution of velocities among an immense number of elastic bodies in motion within a confined space.

In the first proposition, the bodies are supposed to be "rigid elastic" spheres all <of> equal <among each other> in mass. In the second, the spheres are supposed to belong to two different sets, the mass being different in each set. It appears from this proposition that the mean <The third> *vis viva* of a sphere is the same for both sets of spheres. This is a most significant proposition, as it is the dynamical basis of what is called in Chemistry the Law of volumes of gases. It is also the first example of a dynamical explanation of the equilibrium of heat between two bodies at the same temperature.

The third proposition investigates the distribution of the relative velocities of pairs of spheres, one from each set, and determines the number of collisions between such pairs in unit of time. The fourth proposition shows the effect on the system of a force <acting on the spheres and tending to fixed centres> such as gravity, having a potential which is a function of the coordinates. It appears that, in the final average distribution, the average *vis viva* of the spheres is the same in all parts of the region, but the average number of spheres in unit of volume <is greater> increases <as the

potential increases> according to a certain law as the potential decreases. Translated into physical language, this <statement may> is equivalent to the following statement–

If a mixture of any number of gases is confined in a vessel the sides of which are impervious to heat, the final distribution will be such that the temperature will be uniform throughout the vessel, and the density of each gas in any part of <processes which in the case of a mass> the vessel <is> will be the same as it would have been if <all> no other gas had been present.

If, therefore, the gases are acted on by gravity, the density of each gas will diminish from the bottom to the top of the vessel, but that of the denser gas will diminish most rapidly, so that the <mixture will contain the light> proportion of the lighter gases in the mixture will increase <as we> from the bottom to the top of the vessel.

In order that such a variation in the composition of the mixture should be large enough for <capable of> experimental verification the vessel in which the gases are confined must be a very tall one–not less than a hundred feet high; and in a vessel of such dimensions the time necessary for the mixture to attain even a rough approximation to its ultimate distribution by the slow processes of thermal conduction and gaseous diffusion <would be very great as it would require about a year before would be measured by> several years.

<At the same time>

It would be exceedingly difficult to maintain the temperature of the sides of such a vessel uniform; and the slightest inequality of temperature in the sides of the vessel would set up currents in the enclosed gas.

Of these currents, <that which ascends becomes colder as it expands,> and the descending one would <that which descends> becomes warmer as it reaches regions of greater pressure; and in this way the temperature of the upper part of the vessel would become lower than that of the under part according to the law known as that of the convective equilibrium of temperature.[c]

Another effect of the currents of convection is to mix up the whole mass of gas, and so to make its composition more uniform throughout <instead of> than it would be <in the final state of the gas at rest,> if it were left undisturbed for a very long time. <It is therefore very unlikely> It would be very difficult to devise an experiment by which the result of the Proposition <would> could be verified.

The clear demonstration given the proposition by Mr Watson is therefore <especially> of great scientific value <very necessary invaluable>, for almost every one of those who have <attempted the investigation> attacked the question <by less complete math> by insufficient methods of investigation have arrived at the conclusion that the temperature would diminish as the height increases.

Here insert pps 4 and 5[d]

By means of this proposition we may obtain a much clearer view of the dynamical significance of <the equat> thermal equilibrium than that permitted by Prop[osition] II and which has hitherto been used to illustrate what is meant by two gases being at the same temperature.

In Prop. [osition] II the two sets of spheres were completely mixed up together like two gases in the same vessel. We may say (as <I said in 1860> assumed in the original form of the theorem (Phil Mag Jan 1860)[e] that the two gases are at the same temperature because they are <in dynamical equilibrium> thoroughly mixed together. But we cannot test this <by> experimentally by putting a thermometer first into the one gas and then into the other.

But if we now call to our aid a system of forces acting on the molecules and tending to fixed centres and assume that these forces are such that the potential energy of a molecule of sphere of the set N is much greater <in the region B than> when the sphere is in one part of a vessel which we shall call the region B than when it is in the region A, and that the potential energy of a sphere of the set N' is much greater in A than in B then the N spheres will be kept almost entirely within the region A and the N' spheres within the region B. The two sets of spheres are thus kept in great measure separate while free to exchange their energy of motion.

<Now since by definition the temperatures of two bodies are equal if when placed in contact when neither communicates heat to the other the two sets of spheres>

Now by definition the temperature of two bodies are equal if when the bodies are placed in contact neither communicates heat to the other.

In Prop. [osition] I we could hardly speak of the N spheres and the N' spheres as two bodies because they were inextricably mixed together, but now we have got them almost <entir> completely separated from each other into two distinct regions. They are therefore practically two bodies and we can test their temperatures by placing a thermometer first in the region A and then in the region B.

Hence the statement that the temperature of two gases are equal when the kinetic energy of the centre of mass of a molecule is the same in each is true not only of gases mixed together—but also of two pure gases in different parts of the same vessel.

If we assume <that each molecules is acted on by a> that the force which acts on each molecule tends to a <centre> fixed centre belonging to that molecule, each molecule will always be very near to its own centre and the assemblage of molecules will behave like a solid body. But this case comes within Prop[osition] IV so that the relations between temperature and the kinetic energy of a single molecule must be extended even to solids <as well as gases>[f]

<Mr Watson then goes>

In the sixth proposition the mutual actions of two portions of the medium separated from each other by an imaginary plane is considered, as far as it arises from the motion of the spheres across the plane of separation.

The effect of this motion, <is to carry> in the case of any particular sphere, is to carry the sphere, with all its properties, from one portion of the medium, which we call A, into the other, which we may call B. By this transference, in the first place the mass of matter in the region A is diminished and that in the region B is increased by the mass of the sphere. In the ultimate state of the system, however, this effect will be balanced by the passage of as many spheres from B to A as from A to B.

But in the second place the sphere in passing from A to B, does not experience any change in its momentum, either <in> as regards direction or magnitude. Hence if we consider only that component of the momentum which is normal to the plane in the direction AB, the aggregate momentum of the medium in the region A is diminished and that in the region B is increased by that belonging to the sphere.

Now this transference of momentum cannot be balanced by means of spheres crossing from B to A, for these spheres have their velocity, and consequently their momentum, in the opposite direction, so that they transfer negative momentum from B to A; and <this effect will add to> thus the transfer of a sphere in either direction will diminish the positive momentum in region A, and increase that in region B.

<When the moment>

Now whatever produces change in the momentum of a body is called force, and <it> this force is measured by the change of momentum produced in unit of time.

In the case before us, the group of spheres which at any time belongs to the region A or the region B cannot be called in strictness a body, because it does not always consist of the same set of spheres, but to our senses it appears to be a body, because we cannot see the spheres passing into it or out of it. <We> The effect of the mutual bombardment between the two regions appears to us therefore to be a forces acting between the two different portions of gas.

Now when the mutual action between two bodies tends to separate them it is called a Repulsion, and when the two bodies are apparently in contact it is called a Pressure. Since this is the case with respect to the two regions A and B, their mutual action is called Pressure.

The value of this pressure on unit of surface is shown to be

$$p = \rho u^2$$

where ρ is the density of the medium and u^2 is the mean square of that <velocity> component of the velocity which is normal to the plane of separation. In the ultimate state of the system <this> u^2 will be one third of the mean square of the total velocity of agitation of the centres of the spheres.

In this investigation no notice has been taken of the mutual action between pairs of spheres the centre of one being in the region A and that of the other in the region B. It may be shown, however, that if the spheres act on each other only by impact, that part of the pressure between the two regions due to impact between spheres whose centres are in different regions will be

$$\frac{16}{3}\rho v^2 \varepsilon$$

where ε is the <ratio of the> total volume of the spheres contained in unit of volume of the medium. The total pressure is therefore

$$p = \frac{1}{3}\rho v^2(1 + 16\varepsilon)$$

In the seventh proposition the <method is> bodies whose motion is considered are no longer supposed to be smooth rigid elastic spheres but molecules, that is to say, material systems consisting of any number of parts acting on each other with forces of any kind consistent with the principle of the conservation of energy. All that is assumed about a molecule is that it is a material system having m degrees of freedom. This assumption <cannot be untrue for> is the most general that we can make.

It is also assumed that in the <that> in the enunciation that all the forces in the system are either forces <having a potential> tending to fixed centres, and functions of the distance from these centres or else forces acting between the parts of the same molecule, thus excluding forces acting between one molecule and another.

This restriction however does not appear necessary and indeed it is easy to remove it.

For the result of the proposition is to prove that if we define the group, (A), of molecules as consisting of those whose generalised coordinates are between certain limits and whose generalised momenta are within certain other limits then the number of molecules in the group is

$$Ae^{-h(\chi+T)}dp_1 \dots dp_m dq_1 \dots dq_m$$

where A is a constant <depending on the ma> which is the same for all the molecules of the same kind but is different for different gases in a mixture, but h is the same for all the gases χ is the potential energy, and T the kinetic energy of a molecule when in the state (A), and $dp_1 \ldots dp_m$ are the differentials of the components of momentum and $dq_1 \ldots dq_m$ those of the coordinates.

<If from> According to this result <we calculate> the mean <vis viva> kinetic energy of a molecule is $\frac{m}{2h}$, where m is the number of degrees of freedom of the action of the molecule.

If we also calculate the average number of molecules whose configuration is between certain limits we find this number is proportional to

$$Ae^{-h\chi}$$

where χ is the potential energy of the molecule arising from forces either internal to the molecule or tending to fixed centres but excluding inter-molecular forces.

But as our definition of a molecule is of the most general kind we can as easily take into account any inter-molecular forces by including within our "molecule" all those molecules between which inter-molecular forces are exerted.

For instance there is nothing to prevent us from defining as a molecule a material system consisting of any atom in Sirius and another in Arcturus and a third in Aldebaran. If the universe is supposed to have attained that condition of thermal equilibrium to which alone these propositions apply, the average kinetic energy of each of their atoms will be $\frac{3}{2h}$ because each has three degrees of freedom.

That of the system of three atoms will be evidently the sum of the kinetic energy of the three atoms, namely $\frac{9}{2h}$. We obtain the same result from the consideration that this system has nine degrees of freedom.

The centre of mass of the three atoms is a mathematical point at an immense distance from any of them. It has of course three degrees of freedom and the kinetic energy of a material point whose mass is the sum of the masses of the atoms and which moves as the centre of mass moves, is $\frac{3}{2h}$.

It appears therefore that <if> the mean kinetic energy of <the> every molecule is so far as it depends on the motion of its centre of mass is $\frac{3}{2h}$, and that this is true whether the parts of the molecule are close together and bound together by powerful forces or scattered about the distant regions so that we may define our molecule so as to take in as many atoms as we please providing we do not take in <every atom the whole matter> all the atoms which <set ea> encounter each other, in which case, of course the centre of mass is not agitated by the mutual actions of the atoms during their encounters.

Returning to the general expression for the number of molecules in a group, we may make it yield us information of various kinds. Thus if we wish to know the density of a <particular kind of molecule> gas at a given point of the vessel, we have only to make the limits of the group those of the element of volume, and we find the density proportional to $e^{-h\chi}$, where χ is the <average> potential energy of a molecule due to forces external to the gas such as gravity.

Again if we wish to know how many pairs of molecules are <at> in a given configuration with respect to each other we find that this number is proportional to $e^{-h\chi}$ where χ is the potential energy of the <two mole> pair of molecules in so far as it depends on their configuration.

For instance if the molecules <attract> act on each other, <then the> and if χ is the potential energy due to this <attr> action corresponding to the distance r, then < χ will be negative> the number of pairs of molecules whose distance is between r and $r + dr$ will be proportional to $r^2 e^{-h\chi}$.

In the case of attraction, χ is negative, so that there will be a greater number of pairs of molecules within <the sphere of each other at> these limits <a certain> distance than there would be if they did not attract each other.

In the case of repulsion, χ is positive, so that <a smaller number of molecules> the repulsion <makes> diminishes the number of pairs within these limits of distance. <smaller>

If the molecules are solid elastic spheres, the potential energy of a pair of molecules increases enormously when the distance of their centres becomes less than the sum of their radii, and therefore the number of pairs which at a given distance are much disfigured by each other is very small.

If the spheres are rigid the potential becomes infinite for all distances less then the sum of the radii, and consequently no pairs of rigid spheres are found within this distance.

It is easy to extend the interpretation of the results to the case of the simultaneous encounter of three or more molecules, so that these cases, which were formally excluded in the earlier proposition, do not in any way interfere with the absolute generality of the final result.

The clear way in which Mr Watson has demonstrated these propositions leaves us no escape from the terrible generality of the results. Some of these, no doubt are very satisfactory to us in our present state of opinion about the constitution of matter, but others are <calculated> so likely to startle us <from> out of our complacency and perhaps ultimately to drive us out of all the hypotheses in which we hitherto found refuge into that state of

<consciou fully> thoroughly conscious ignorance which is the prelude to <a more profound> any real advance in knowledge.

<In the first place enumerate some of the results.>

In the first place our notions of thermal phenomena have become more definite.

We distinguish that part of the motion of a system in virtue of which it is hot from <its> other kinds of motion by defining it as a motion of agitation of the molecule such that the motion motion of the centre of mass of all the molecules in any visible portion of the substance is insensible, and as irregular that the only way in which it can be transmitted from one part of a body to another is by <the> that slow process of diffusion called the conduction of heat.

We have also arrived at a dynamical definition of temperature which we <know to be> have already defined as the condition which determines the direction in which the conduction of heat takes place. The condition of the equilibrium of the motion of agitation among bodies of different kinds is, as we have seen, that the kinetic energy of each must be proportional to the number of its degrees of freedom. Hence temperature must be a function of the kinetic energy of a body divided by the number of its degrees of freedom. <and>

If we suppose bodies to be made up of atoms the number of degrees of freedom of each atom is three. The centre of mass of a body whether made up of atoms or not has three degrees of freedom. Hence we may adopt as our definition of <absolu> temperature a number proportional to the kinetic energy of <the centre of mass> of any body supposed to be concentrated at its centre of mass and to move as that point moves. It is manifest that the results cannot be confined to gases but must be equally applicable to liquids and to solids. For the existence of the forces which bind together the parts of liquids and solids does not interfere with the law of distribution of kinetic energy in the state of thermal equilibrium.

When heat passes into a body, the body receives an <certain> increment of energy equal to the dynamical value of the heat absorbed. The effects of the <heat> are to increase the kinetic energy of the body, to increase its intrinsic energy and to work against external pressure.

Of these three effects, the work done against external pressure can be measured experimentally and the increase of kinetic energy can be expressed in terms of the observed rise of temperature <if we know> and the number of degrees of freedom. Let us <take the case> now estimate the specific heat of a gas at constant volume, that is to say the quantity of heat which must be communicated to the gas to raise its temperature one degree.

We know that the part of the kinetic energy of the gas which depends on the motion of the centre of mass of each molecule is <two thirds> three halves of the product of the volume into the pressure. Hence as the volume is constant the increment of the energy is <two thirds> three halves of the product of the volume into the increment of pressure or < $\frac{2}{3}$ > $\frac{3}{2}\, v\, dp$. But if each molecule has m degrees of freedom, the whole kinetic energy is to that of the centre of mass as m to 3. Hence the <whole> increment of the whole kinetic energy is $\frac{m}{2}\, v\, dp$. It is probable that the mean potential energy is also increased <at the> at the same time with the kinetic energy, so that the total increment of energy is greater than $\frac{m}{2}\, v\, dp$. It appears therefore that the observed specific heat of a gas under constant volume furnishes us with a superior limit to the value of m for each molecule and we find in this way that for mercury gas m cannot exceed 3; for oxygen, hydrogen, nitrogen air carbonic oxide nitrous oxide and hydrochloric acid, it cannot exceed 5; and for chlorine, ammonia and sulphuretted hydrogen it cannot exceed 6.

It is true as <indeed> Boltzmann points out in his paper Über die Natur der Gasmolecüle Vienna Academy 14 Dec 1876[9] that if the molecules were perfectly <smooth> rigid elastic spheres their rotation would not be affected by their collisions and m would be <reduced to that if they were figures of revolution the velocity of rotation about the axis of figure would not be affected and m would be reduced to 5 w> that if the molecules were rigid elastic spheres of any form m would be 6, that if they were smooth figures of revolution the velocity of rotation about the axis of figure would not be affected by the collisions, so that m would be reduced to 5, and that if they were smooth spheres the three component velocities of rotation would each of them be unaffected by collision, so that m would be reduced to 3; and that their values are in striking agreement with the phenomena of the three groups of gases

But before we admit this somewhat promising view of the matter let us <consider> try to construct a rigid elastic body. We must not take an ordinary elastic solid of homogeneous matter and suppose its coefficients of elasticity increased without limit till the body becomes practically rigid. For however great the coefficients of elasticity may be, the body is still capable of internal vibrations and these of an infinite variety of types, so that the body has in reality an infinite number of degrees of freedom.

A truly rigid elastic body is one whose encounters <another> with other similar bodies take place as if they were elastic but which are not capable of being set into a state of vibration. We may construct one by taking a perfectly rigid body and <giving> conferring on it power to <act at a distance> repel all other bodies, but only when they come within a very short distance of its surface, the <law of> repulsion increasing as the distance diminishes so that under no circumstances whatever could the two bodies ever come into contact.

This appears to be the only way in which we can imagine a rigid-elastic body to be constructed. And now that we have got it the best thing we can do is get rid of the rigid nucleus altogether. For we have been obliged to admit forces acting at a distance in order to get the effect of a spring without admitting the vibration of the matter of which the spring is made, <and now there is no more need of spreading the matter of the nucleus over a finite region fo> and we may now suppose the matter of the nucleus to be concentrated into one or more mathematical points instead of spreading it in a continuous manner over a finite region and thus rendering it necessary to invoke a special system of forces to keep the parts of this nucleus together.

But Boltzmann's molecules are not absolutely rigid-elastic they are only approximately so. In fact as he tells us they do vibrate after severe collisions, and that in vibrations of several different types, as we learn from the spectroscope. But still he thinks these vibrations do not greatly disturb the figure of the molecules. He compares them to the billiard balls which when they strike each other vibrate for a short time but soon give up the energy of their vibration to the air which carries far and wide the sound of the click of the balls, <far and w> In like manner the light emitted by the molecules after each collision shows that their internal vibrations are <soon> quickly given up to the luminiferous aether.

<Of course> <In this case>

When the gas is enclosed in a vessel whose sides are at the same temperature as itself the energy lost by communication to the aether must be compensated by an equal amount of energy communicated by the aether to the molecules on account of the radiation from the sides of the vessel, but we have not the slightest evidence for believing that it is in this way that the energy of the internal motions of the molecules are prevented from attaining <their> its due proportion to the energy of the motion of agitation of their centres of mass.

For if a particular kind of motion of the aether is excited by particular <the> internal motions of molecules, then the kind of motion, <of the aether will> if already existing in the aether, <tend to excite> cannot fail to excite the corresponding kind of internal motion in the molecules. This is a consequence of a very general theorem of the reciprocity of dynamical actions which has been recently pointed out by Lord Rayleigh.[h]

The aether then cannot serve our purpose by taking <energy> from the molecules their energy of internal vibration and giving back to them energy of translation. It cannot in any way interfere with the natural ratio between these kinds of energy. All it can do is to take up its own due proportion of energy according to the number of its degrees of freedom. We leave to the authors of the "Unseen Universe" to follow out the consequences of this statement.[i]

We have thus the theorem that the specific heat of <a gas> unit of volume of a gas at constant <pressure> volume must be *at least*

$$< m\frac{v}{2}\frac{dp}{d\theta} > \qquad\qquad m\frac{v}{2}\frac{dp}{d\theta}$$

where m is the number of degrees of freedom of a molecule and this is combined with actual measurement tells us that in no case is m very large.

Hence molecules capable *at ordinary temperatures* of vibrating in several different ways are out of the question. But we know from *spectroscopic* observation that at high temperatures <most> the molecules of all gases vibrate in many different ways. We must therefore conclude that the parts of a molecule are bound together so tightly that they do not vibrate at all under the impacts of the other molecules at ordinary temperatures, but that they vibrate when struck with the velocity of the electric spark.

It is difficult, however, to reconcile this form of the theory with the fact that the <vibrations lines> bright lines of the gases are fixed, showing that the vibrations of the molecules are isochronous and therefore that the forces <under which they> which they call into play very directly as the relative displacements. If this be the character of the forces, <small impacts will produce vibrations> all impacts however small will produce vibrations. <If>

Besides this, we know that certain gases, as for instance vapour of <iolo> iodine, nitrous acid, and enchlorine, exhibit absorption bands when cold.

a. This is a draft of Maxwell's review of Watson's treatise that was published in *Nature* 16 (1877): 242–246. The review was not reprinted in Maxwell's collected papers.

b. Rev. Henry William Watson (1827–1903) was appointed Headmaster of Berkewell School, Coventry in 1865 and remained there until his retirement in 1902. In the 1880s his attention shifted from kinetic theory to the theory of electromagnetism and he coauthored a text with Samuel Hawksley Burbury. They also collaborated on the article on molecules in the *Encyclopedia Brittanica.*

c. William Thomson established the rate of cooling of an isolated column of gas. See Thomson, "On the convective Equilibrium of Temperature in the Atmosphere," *Mem. Manchester Lit. Phil. Soc.* 2 (1865): 125–131.

d. The editors followed Maxwell's instruction, to "insert pages 4 and 5 in here" in reconstructing the text. This was the obvious, logical development of Maxwell's argument.

e. Maxwell,"Illustrations of the Dynamical Theory of Gases. Part I On the Motions and Collisions of Perfectly Elastic Spheres," *Phil. Mag.* 19 (1860): 19–32, reprinted in *Scientific Papers,* vol. 1, 378–391 and Garber, Brush, and Everitt, *Maxwell on Molecules and Gases,* 285–318.

f. This is the end of page 5. We return to manuscript page 3.

g. Ludwig Boltzmann,"Über die Natur der Gasemolecüle," *Sitz. Math.-Naturwiss. Cl. Akad. Wien* 74 (1876): 553–560. The brackets [ ] are Maxwell's.

h. John William Strutt, 3rd Baron Rayleigh (1842–1919), "General Theorems Relating to Equilibrium and Initial and Steady Motions," *Phil. Mag.* 49 (1875): 218–224, reprinted in Rayleigh, *Scientific Papers,* (New York: Dover reprint of 1900 edition, 1964), vol. 1, 232–237.

i. This refers to Balfour Stewart and Peter Guthrie Tait, *The Unseen Universe, or Physical Speculation on a Future State* (London: 1875). The first edition appeared anonymously. See P. H. Heimann,"The *Unseen Universe*: Physics and the Philosophy of Nature in Victorian Britain," *Brit. J. Hist. Sci.* 6 (1972): 73–79, for an assessment of this work and its place in opposition to Tyndall's scientific naturalism. See also Silvan S. Schweber, "Demons, Angels and Probability: Some Aspects of British Science in the Nineteenth Century," in Abner Shimony and Herman Feshbach eds. *Physics as Natural Philosophy: Essays in Honor of Laszlo Tisza*, 319–363, 330–331.

19. "*A Treatise on the Kinetic Theory of Gases*, by Henry William Watson"

Nature, 18 (1877): 242–246.

A Treatise on the Kinetic Theory of Gases by Henry William Watson, M. A. formerly Fellow of Trinity College, Cambridge. (Oxford; Clarendon Press, 1876)

This book does not profess to treat of all that has been written about the kinetic theory of gases. It discusses the ultimate average condition of a material system, consisting of a very great number of parts in motion within a confined space, and it follows for the most part the methods of investigation given by Boltzmann. The discussion is arranged in the form of thirteen propositions, in which the different cases are considered in the order of their complexity. In the earliest propositions the moving bodies are supposed to be rigid-elastic spheres acting one each other only by impact, afterwards external forces are introduced, and finally the bodies are supposed to be material systems, the parts of which are held together by any system of forces consistent with the principle of the conservation of energy.

The ultimate average condition of such a system is investigated in a very satisfactory manner in this book. No part of mathematical science requires more careful handling than that which treats of probabilities and averages. Mathematicians, whose competence to deal with other questions is undoubted, have fallen into errors in treating of probabilities, and even the validity of certain methods is still apparently an open question.

Besides this, some of the consequences to which these theorems lead us are so startling that we are not prepared to admit them without an unanswerable proof, and of the investigations already given, some are so short and incomplete, and others so long and roundabout, that it requires no ordinary exercise both of penetration and of patience to find out whether they are proofs at all. Mr. Watson has conferred a great benefit on the students of the kinetic theory by placing before them in a series of distinct propositions, none of them too long for the mind to grasp, all the necessary steps leading to the result, and none of the superfluous evolutions in which the mental energy of the student is so often dissipated. The book, as we have said before, is confined to the investigation of the ultimate average condition of the system, and does not discuss the process of diffusion by which that ultimate condition is attained, such as the interdiffusion of gases, the diffusion of momentum by viscosity, and the diffusion of energy by thermal conduction. These have been recently treated in a larger work,[1] to which we may have occasion to refer.[a]

There are two very different methods of defining and investigating the state of a complex material system. According to the stricter dynamical method the particles of the system are defined in any sufficient manner, as, for instance, by their coordinates at a given epoch, and the position of any particle at any other time is then defined by its coordinates, expressed as functions of the time, the form of these functions being different from particle to particle, and not necessarily continuous in passing from one particle to another which was contiguous to it in the initial configuration.

According to this method our analysis enables us to trace every particle throughout its whole course, and therefore we can apply the laws of motion in all their strictness.

The application of this method to systems consisting of large numbers of bodies is out of the question. We therefore make use of another method which we may call the statistical method, on account of its analogy with the methods employed in dealing with the fluctuations of a large population.

We divide the bodies of the system into groups according to their position, their velocity, or any other property belonging to them, and we fix our attention not on the bodies themselves, but on the *number* belonging at any instant to one particular group. This number is, of course, subject to change on account of bodies entering or leaving the group, and we have therefore to study the conditions under which bodies enter or leave the group, and in so doing we must follow the course of the bodies according to the dynamical method. But as soon as the process is over, when the body has fairly entered the group or left it, we withdraw our attention from the body, and if it should come before us again we treat it as a new body, just as the turnstile at an exhibition counts the visitors who enter without respect to what they have done or are going to do, or whether they have passed through the turnstile before.

The first mode of grouping the bodies of the system is to class those together which, at a given time, are in a given region of space. This is called grouping according to configuration, and what we learn from it is the distribution of the positions or coordinates of the bodies in space.

The second mode of grouping is that according to velocity. The best way to understand this is to suppose a diagram of velocities constructed by drawing from a given point as origin a system of vectors representing in direction and magnitude the velocities of the different bodies. The extremities of these vectors are called the velocity-points of the bodies to which they correspond, and by grouping the bodies according to the regions of the diagram in which their velocity-points lie, we learn from the numbers in the groups the distribution of velocities among the bodies.

In like manner we may form groups defined in any other way, as, for instance, those pairs of bodies whose distance from one another lies between given limits, and by confining within sufficient narrow limits the values of all the properties of the bodies which form the group, we may consider all the bodies belonging to the group as practically in the same state. Whether at a given instant any body actually belongs to the group is, of course, another question.

The object of study in the statistical method is the probable number of bodies in each group. We may get rid of the idea of probability by supposing the system to continue under the same conditions for a very long time. During this time many bodies will enter the group, stay in it for a certain time, then leave it. If we add together the times of residence within the group of all these bodies, and divide the sum by the whole time of observation, we obtain a numerical quantity which we may call the average number of bodies in the group. The longer the time of observation, the nearer does this number approach to what we have called the probable number of bodies in the group.

The average number of bodies in a group depends on the limits which define the group, being, of course, greater when these limits are wide than when they are narrow. But it also depends on the character of the group, that is to say, the particular set of mean values of the conditions which entitle a body to be ranked in the group.

It appears from the investigation that if ϕ be any property of the body, such that if ϕ_1 and ϕ_2 are its values for two bodies before an encounter, and Φ_1 and Φ_2 its values after the encounter, and if under all circumstances $\phi_1 + \phi_2 = \Phi_1 + \Phi_2$, and if the number of bodies in each group varies as

$e^{-h\phi}$, then the distribution of the bodies in the groups will not be altered by the encounters between the bodies.

Now if we make ϕ equal to the sum of the kinetic and the potential energy of each body, the quantity $\phi_1 + \phi_2$ is not altered, either by an encounter between the two bodies or by external forces acting on them: so that a distribution according to the values of the function $e^{-h\phi}$ will satisfy the condition of permanence.

The most general case is that given in the seventh proposition. The bodies are no longer supposed to be smooth rigid-elastic spheres, but molecules, that is to say, material systems consisting of any number of parts acting on each other with forces of any kind consistent with the principle of the conservation of energy. The molecules of any one kind are supposed to have m degrees of freedom, this number being, in general, different in different kinds.

It is also assumed in the enunciation that all the forces in the system are either forces tending to fixed centres and functions of the distance from those centres, or else forces acting between the parts of the same molecule, thus excluding forces acting between one molecule and another except during the encounter of two molecules. This restriction, however, does not appear necessary, and indeed it is easy to remove it.

For the result of the proposition is to prove that if we define the group (A) of molecules as consisting of those whose generalised co-ordinates (q) are between certain limits (q and $q + dq$), and whose generalised momenta (p) are between certain other limits (p and $p+dp$), then the average number of molecules in the group is–

$$Ae^{-h(\chi+T)}\, dp_1 \dots dp_m\, dq_1 \dots q_m,$$

where A is a constant which is the same for all groups of molecules of the same kind, but is different for different kinds of molecules. χ is the potential energy, and T the kinetic energy of a molecule when in a state (A), and $dp_1 \dots dp_m$ are the differentials of the components of momentum, and $dq_1 \dots dq_m$ the differentials of the co-ordinates. The continued product of these differentials specifies the extent of the group.

By integrating this expression with respect to any one of the variables, we may ascertain the average number of molecules in a larger group, in which that variable does not form a ground of subdivision. For instance, if we integrate with respect to all the co-ordinates, we arrive at a group consisting of all the molecules whose momenta are between certain limits, or by integrating with respect to the momenta we form a group of molecules whose configurations lie within certain limits.

In this way we obtain two very important results:–

1. The average kinetic energy of a molecule is $\frac{m}{2h}$ where m is the number of degrees of freedom of the molecule. This is independent of the position of the molecules.

2. The average number of molecules whose configuration lies between certain limits is:–

$$Ae^{-h\chi}\, dp_1 \ldots dp_m\, dq_1 \ldots dq_m,$$

where χ is the potential energy of the molecule, arising from forces either internal to the molecule or tending to fixed centres, but (according to Mr. Watson) excluding intermolecular forces.

But as our definition of a molecule is of the most general kind, nothing is easier than to take into account any intermolecular forces by simply including within our "molecule" all those molecules between which intermolecular forces are exerted.

For instance, there is nothing to prevent us from defining as a molecule a material system consisting of one atom in Sirius, another in Arcturus, and a third in Aldebaran. If the universe is supposed to have attained the condition of thermal equilibrium to which alone these propositions apply, the average kinetic energy of each of these atoms will be $\frac{3}{2h}$, because each has three degrees of freedom.

That of the system of three atoms will be the sum of the kinetic energies of the three atoms, namely $\frac{9}{2h.}$ We might obtain the same result from the consideration that this system has nine degrees of freedom.

The centre of mass of the three atoms is a mathematical point at an immense distance from any of them. It has, of course, three degrees of freedom, and the kinetic energy of a material particle whose mass is the sum of the masses of the atoms and which moves as the centre of mass does is $\frac{3}{2h}$.

The value of the kinetic energy of the centre of mass will be the same for any system of atoms provided that every atom of the system is liable to encounters with atoms not belonging to the system. Of course if we take into our "molecule" all the atoms of a material system unconnected with any other system, its centre of mass will not be agitated at all by the mutual actions of the atoms during their encounters.

And here we must notice a point to which Mr. Watson has adverted in a note at the foot of p. 20—the definition of the motion of the medium distinguished from the motion of agitation of the molecules. For this is connected with the weak point of the demonstration and shows us a way to strengthen it.

The weak point of the demonstration is the tacit assumption that the sum of the potential and kinetic energies of a pair of molecules is the only function which does not change during their encounter. For there are other quantities which are not altered by the mutual action of the bodies such as their masses themselves, the sum of their momenta resolved in any given direction, and their angular momenta about a fixed axis.

Hence if instead of $T = \frac{1}{2} M(u^2 + v^2 + w^2)$ we write in the expression for the distribution of velocities—

$$E = T - MUu + Vv + Ww + p(wy - vx) + q(uz - wx) + r(vx - uy) + CM$$

the distribution of velocities will still be a permanent one.

In this expression the quantities U, V, W, p, q, r, may have any values provided they are the same for the whole system of bodies, but C may be different for different bodies, because it is multiplied by the mass of the body, which is invariable. But we arrive at the same expression by substituting in T for u, v and w the quantities–

$$u - U + qz - ry$$

$$v - V + rx - pz$$

$$w - W + py - qx$$

or, in other words, by substituting for the absolute velocities of the bodies their velocities relative to a system of axes moving in the most general manner possible; that is to say, the components of velocity of the origin being U, V, W and the components of the velocity of rotation being p, q, r, and at the same time adding to T the quantity–

$$\frac{1}{2} M(qz - ry)^2 + (rx - pz)^2 + (py - qx)^2 - C'$$

which depends on the co-ordinates only and not on the velocity of the body.

We now see that the most general case of permanent distribution is when the system of bodies is contained in a vessel of invariable form which moves with constant velocity along a screw, that is to say, in which one point is moving along a straight line with constant velocity, while the vessel rotates about an axis passing through this point with constant angular velocity.

When there is rotation, we must subtract from the potential energy a term depending on the co-ordinates, which shows that the rotation produces an effect similar to that of a centrifugal force at right angles to the axis of rotation.

Returning to the general expression for the number of molecules in a group, we may make it yield us information of other kinds. Thus, if we want to know the density of a particular gas at any given point in the mixture, we have only to make the limits of the group those of an element of volume, and we find the density proportional to $e^{-h\chi}$, where χ is that part of the energy of a single molecule which is due to external forces, such as gravity. In the case of gravity, χ is equal to $m\,g\,z$ where m is the mass of a molecule, g the intensity of gravity, and z the height. This leads to the ordinary expression for the density of a gas of uniform temperature in a vertical column, and it shows that in the ultimate distribution of a mixture of gases the density of each gas diminishes with the height according to its own law, that is to say that of the heaviest gases diminishes most rapidly, so that the proportion of the heaviest gases diminishes with height.

This law of the distribution of gases was asserted by Dalton as a consequence of his theory of gases,[b] and numerous experiments have been made on air collected at different heights in the atmosphere in order to detect a difference in their composition but we cannot say that such a difference has as yet been satisfactorily established.

The atmosphere, in fact, is eminently unfitted for testing the theory of the ultimate state of a mixture in equilibrium, for the inequalities of temperature in so large a body of gas produce powerful currents which continually carry masses of the mixture from one stratum into another. This tends to produce a uniformity of composition and a variation of temperature which are both of them contrary to our theory of the condition of equilibrium, and which seem to favour certain other theories.[c]

Nor is the case much improved if, instead of the open atmosphere, we substitute a mixture of gases contained in a vertical tube. For in order to obtain a difference of composition at the top and bottom of the tube large enough for experimental verification, the tube must be at least 100 metres high, and it would take more than a year for the contents of such a tube to approximate by one half to their final distribution. In the mean time the slightest difference of temperature in the sides of the tube would produce currents which would tend to equalise the composition of the mixture. To verify the other result of our theory—the uniformity of temperature in the ultimate state of the vertical column—would be attended with still greater difficulties.[d]

But it would be quite within the powers of experimental methods to verify the law of distribution of a mixture of gases in a rotating vessel. Let two bulbs be connected by a wide tube, say 10 cm, long, and let them be filled with equal volumes of hydrogen and carbonic acid, well mixed together. Let this apparatus be placed on a whirling machine, so that one of the bulbs shall be close to the axis, while the other is moving at the rate of fifty metres per second. The same degree of approximation to the final state, which would take years in a long tube, will be effected in minutes in this small apparatus, and the proportion of carbonic acid to hydrogen will be about $\frac{1}{140}$ greater in the bulb furthest from the axis.

The clear demonstration of this proposition given by Mr. Watson is of great scientific value, for almost every one of those who have attacked the question with insufficient methods of investigation have come to the conclusion that the temperature would diminish in a vertical column as the height increases; and those regard gaseous diffusion from a chemical rather than a dynamical point of view would probably expect the composition to be uniform at all heights.

But the profound scientific value of this proposition becomes more manifest when we make use of it in establishing the definition of temperature and the law of volumes of gases.

In Prop. II of this book, which corresponds to the original form of the theorem, as I gave it in the *Philosophical Magazine,* January 1860,[e] two sets of spheres are completely mixed up together in the same vessel, and it is proved that the average kinetic energy of a sphere is the same for either set. We may then assert, as I did, that the two gases are at the same temperature because they are thoroughly mixed together. But this assertion has no scientific meaning, because we cannot test its truth by putting a thermometer first into one gas and then into the other.

But if we now call to our aid a system of forces acting on the molecules and tending to fixed centres, we may obtain a result capable of experimental verification; for though we are not acquainted with natural forces acting exclusively on one kind of gas, we can calculate the effects of such forces.

Let us assume, then, that the forces are such that the potential energy of a sphere of the set N is much greater in one part of the vessel, which we shall call B, than when it is in part A, these two parts being separated by a stratum C, within which the potential varies continuously. The medium consisting of the spheres N will be dense in A, it will become rarer in the stratum C and there will be hardly any of these spheres in B.

Now let the potential energy of a sphere of the set N' be much greater when it is in A than when it is in B, and let it vary continuously from the one value to the other in the stratum C. Then the spheres of this set will be thickly scattered in B, will thin out in the stratum C, and will be very rare in A.

The two sets of spheres are thus kept in great measure separate in A and B, while free to exchange their kinetic energy within the stratum C.

Now by definition, the temperature of two bodies are equal if, when the two bodies are placed in contact, their thermal state remains the same. We cannot apply this definition to the two sets of spheres in Prop. II., for they were inextricably mixed up together, but we have now got them almost completely separated from each other into two distinct regions. They are therefore practically distinct bodies, and we can test their temperatures separately.

Hence the statement, that the temperatures of two gases are equal when the kinetic energy of the centre of mass of a molecule is the same in each, is true, not only of gases mixed together, but of two pure gases in different parts of the same vessel.

If we assume that a powerful external force acts on each molecule tending to a fixed centre belonging to that molecule, each molecule will always remain very near its own fixed centre of force, and the assemblage of molecules will behave like a solid body. But forces of this kind are included among those considered in Prop. IV., so that the relation between temperature and the kinetic energy of the centre of mass of a single molecule must be extended even to solids.

Returning once more to the general expression for the average number of molecules in a group, we may make it yield us information with respect to the average number of sets of molecules which, at a given instant, are in a given configuration with respect to each other.

For instance, if two molecules act on one each other, and if χ is the potential energy due to this action corresponding to a distance r, then the number of pairs of molecules whose distance is between r and $r + dr$ will be proportional to $r^2 e^{-h\chi}$. In the case of attraction χ is negative, so that there will be a greater number of pairs of molecules within these limits of distance than there would have been if they did not attract each other. In the case of repulsion, χ is positive, so that the repulsion diminishes the number of pairs within the distance of repulsion. If the potential energy of a pair of molecules rapidly increases to an enormous value when the distance between their centres becomes less than a given quantity, the number of pairs which are within the given distance will be practically zero, and the molecules will behave like smooth rigid-elastic spheres.

By making the "molecule" include three or more molecules, and making χ the potential energy of this system, we may extend the theorems to the simultaneous encounter of three or more molecules, so that these cases, which were formally excluded in the earlier propositions, do not in any way interfere with the absolute generality of the final results.

The clear way in which Mr. Watson has demonstrated these propositions leaves us no escape from the terrible generality of his results. Some of these, no doubt, are very satisfactory to us in our present state of opinion about the constitution of bodies, but there are others which are likely to startle us out of our complacency, and perhaps ultimately to drive us out of all the hypotheses in which we have hitherto found refuge into the state of thoroughly conscious ignorance which is the prelude to every real advance in knowledge.

If we know from observation either the specific heat of a gas at constant pressure, or the ratio of its specific heats at constant pressure and at constant volume, we can determine the ratio of the rate of increase of its total energy to the rate of increase of the energy of agitation of the centres of its molecules. Now if the molecule has m degrees of freedom, its total kinetic energy is to the energy of agitation of its centre of mass as m to 3. It is probable that the internal potential energy of the molecule increases as the temperature rises, and this would make the ratio of the whole energy to that of agitation of centres greater than that of m to 3, so that if we know this ratio by experiment, we can assert that m cannot exceed a certain value.

For chlorine, ammonia, and sulphuretted hydrogen, m cannot exceed 6; for hydrogen, oxygen, nitrogen, air, carbonic acid, nitrous oxide, and hydrochloric acid, it cannot exceed 5, and for mercury gas, according to the experiments of Kundt and Warburg,[f] it cannot exceed 3.

Now Boltzmann has pointed out in a paper: "Über die Natur der Gasmolecüle" (Vienna Acad, December 14, 1876),[g] that if the molecules were rigid-elastic bodies of any form m would be 6, that if they were smooth figures of revolution, the velocity of rotation about the axis of figure would not be affected by the collisions, so that m would be 5, and that if they were smooth spheres, the three component velocities of rotation would each of them be independent of collisions, so that m would be reduced to 3, and these values are in striking agreement with the phenomena of the three groups of gases.

But before we accept this somewhat promising hypothesis, let us try to construct a rigid-elastic body. It will not do to take a body formed of continuous matter endowed with elastic properties, and to increase the coefficients of elasticity without limit till the body becomes practically rigid. For such a body, though apparently rigid, is in reality capable of internal vibrations, and these of an infinite variety of types, so that the body has an infinite number of degrees of freedom.

The same objection applies to all atoms constructed of continuous, non-rigid matter, such as the vortex-atoms of Thomson.[h] Such atoms would soon convert all their energy of agitation into internal energy, and the specific heat of a substance composed of them would be infinite.

A truly rigid-elastic body is one whose encounters with similar bodies take place as if both were elastic, but which is not capable of being set into a state of internal vibration. We must take a perfectly rigid body and endow it with the power of repelling all other bodies, but only when they came within a very short distance from its surface, but then so strongly that under no circumstances whatever can any body come into actual contact with it.

This appears to be the only constitution we can imagine for a rigid-elastic body. And now we have got it, the best thing we can do is to get rid of the rigid nucleus altogether, and substitute for it an atom of Boscovich—a mathematical point endowed with powers of acting at a distance on other atoms.[i]

But Boltzmann's molecules are not absolutely rigid. He admits that they vibrate after collisions, and that their vibrations are of several different types, as the spectroscope tells us. But still he tries to make us believe that these vibrations are small importance as regards the principal part of the motions of the molecules. He compares the to billiard balls, which, when they strike each other, vibrate for a short time, but soon give up the energy of their vibrations to the air, which carries far and wide the sound of the click of the balls.

In like manner, the light emitted by the molecules show that their internal vibrations after each collision are quickly given up to the luminiferous ether.

If we were to suppose that at ordinary temperatures the collisions are not severe enough to produce any internal vibrations, and that these occur only at temperatures like that of the electric spark, at which we cannot make measurements of specific heats we might, perhaps, reconcile the spectroscopic results with what we know about specific heat.

But the fixed position of the bright lines of a gas shows that the vibrations are isochronous, and therefore that the forces which they call into play vary directly as the relative displacements, and if this be the character of the forces, all impacts, however slight, will produce vibrations.

Besides this, even at ordinary temperatures, in certain gases, such as iodine gas and nitrous acid, adsorption bands exist, which indicate that the molecules are set into internal vibration by the incident light.

The molecules, therefore, are capable, as Boltzmann points out, of exchanging energy with the ether.[j]

But we cannot force the ether into the service of our theory so as to take from the molecules their energy of internal vibrations and give it back to them as energy of translation. It cannot in any way interfere with the ratio between these two kinds of energy which Boltzmann himself has established.[k] All it can do is to take up its own due proportion of energy according to the number of its degrees of freedom.

We leave it to the authors of the "Unseen Universe" to follow out the consequences of this statement.[ℓ]

J. Clerk Maxwell

1 "Die kinetische Theorie der Gase in elementarer Darstellung, mit mathematischen Zusätzen." Von Dr. Oskar Emil Meyer, Professor der Physik an der Universitat Breslau. (Breslau, 1877).

a. However, in his preface Watson states that,

> To Professor Clerk Maxwell I am indebted for much kind assistance, and especially for access to some of his manuscript notes on this subject, from which I have taken many valuable suggestions.

Indeed, these notes are cited in the text. Maxwell's style and suggestions are visible throughout once you compare this text with Maxwell,"On the Final State of a System of Molecules in Motion Subject to Forces of Any Kind," *Rep. of the 42nd Meeting BAAS* (1873): Trans. of Section A, 29–32, reprinted as Document II–14. See also Garber, Brush, and Everitt, *Maxwell on Molecules and Gases*, Document II–33, 260–262.

b. John Dalton, "A New Theory of the Constitution of Mixed Aeriform Fluids, and Particularly of the Atmosphere," *J. Nat. Phil. Chem. Arts,* 51 (1801): 241–244, and, "On the Constitution of Mixed Gases," *Mem. Manchester Lit. Phil. Soc.* 5 (1802): 535–550. For a recent study of Dalton see Arnold Thackray, *John Dalton* (Cambridge MA; Harvard University Press, 1972).

c. Maxwell is referring to his recent exchange with Francis Guthrie on kinetic theory and the properties of the atmosphere. See op. cit. (note a), reprinted as document 14. The exchange between Guthrie and Maxwell appears as Documents II–7 and 8, 10 and 11.

d. Maxwell developed these ideas further, then published them, along with an extended discussion of the experiment in Maxwell, "On Boltzmann's Theorem on the Average Distribution of Energy in a System of Material Points," *Trans. Cambridge Phil. Soc.* 12 (1879): 547–570, reprinted in *Scientific Papers,* vol. 2, 713–741. See also Document V–7.

e. Maxwell, "Illustrations of the Dynamical Theory of Gases Part I On the Motions and Collisions of Perfectly Elastic Spheres," *Phil. Mag.* 19 (1860): 19–32, reprinted in *Scientific Papers* vol. 1, 377–401 and Garber, Brush, and Everitt, *Maxwell on Molecules and Gases,* 285–318.

f. August Kundt and Emil Warburg,"Über die specifische Wärme des Quecksilbergases, " *Berlin, Chem. Gesell. Sitz.* 8 (1875): 943–948; *Ann. Phys.* 157 (1876): 353–369.

g. Ludwig Boltzmann,"Über die Natur der Gasemolecüle," *Sitz. Math.-Naturwiss. Cl. Akad. Wiss. Wien* 74 (1876) [1879]: 553–560; abstract, *Phil. Mag.* 3 (1877): 320.

h. William Thomson,"On Vortex Atoms,"*Phil. Mag.* 34 (1867): 15–24; *Proc. Glasgow Phil. Soc.* 4 (1868): 197–206; *Proc. R. Soc. Edinburgh* 6 (1869): 94–105.

i. Roger Joseph Boscovich (1711-1787), first published his theory as *De viribus vivis* (Rome, 1745). See Boscovich *Theoria Philosophia Naturalis* (Venice, 1763) translated by J. M. Childs as *A Theory of Natural Philosophy* (Cambridge MA: MIT Press, 1966).

j. See Boltzmann, op. cit. (note g).

k. Ibid.

ℓ. Balfour Stewart and Peter Guthrie Tait, *The Unseen Universe: or Physical Speculations on a Future State* (London: Macmillan, 1875), first edition published anonymously. See P. M. Heimann, "*The Unseen Universe*: Physics and the Philosophy of Nature in Victorian Britain," *Brit. J. Hist. Sci.* 6 (1972): 73–79.

20. Postcard from Maxwell to Peter Guthrie Tait, 1878[a]

Cambridge University Library, Maxwell Collection, Maxwell-Tait Correspondence.

Return after translating

My better $\frac{1}{2}$, who did the real work of the kinetic theory[b] is at present engaged in other researches. When she is done I will let you know her answer to your enquiry.
Boltzmann would make the density at $\infty = \rho_0 e^{-g\rho_0 R/p_0}$
$[R =$ earths radius$]$ This supposes dynamical and thermal eq^{m}[c] established, our earth at rest.
The $3\theta\Delta\eta cs$[c] men, R. C & T[e] are getting it hot & strong. You have got an installment O. T.[f] already in Diffusion Enc. Britt.[g]
We have a Peleus instead of a Telephon.
Comparison of Herns_ Pos. Little squeaker. Comp. Vicar of Bray Sup. Aurora. Critical temperatures of water calculated from Regnault at 0°, 100° & 200° corresponds to 436°C
Cagniards[h] value 412° is certainly too low, by reason of his method
Compliments of the season with all their supplements

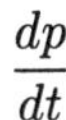

a. Undated, probably late 1878, early 1879. The equation given is for the density of a gas infinitely far from the earth. Maxwell was working at this time on the problem of the change in density with height. His remarks on this problem and an experiment to measure the change in density with distance is in Maxwell, "On Boltzmann's Theorem and on the Average Distribution of Energy in a System of Material Points," *Trans. Cambridge Phil. Soc.* 12 (1879): 547–570, reprinted in *Scientific Papers* vol. 2, 713–741, and Document V-7.

b. This refers to his wife's work on the experiment to measure the friction of air.

c. Equilibrium

d. Thermodynamics

e. Maxwell is referring to John Macquorn Rankine, Rudolph Julius Emmanuel Clausius and William Thomson.

f. O. T. , or O. T′ was Maxwell's salutation to Tait especially in postcards.

g. Maxwell is referring to his article, "Diffusion," *Encyclopedia Brittanica*, ninth edition, reprinted in *Scientific Papers*, vol. 2, 625–646 and Garber, Brush, and Everitt *Maxwell on Molecules and Gases*, 524–546.

h. Charles Cagniard de la Tour (1777–1859), "Sur les effets qu'on obtienne par l'application simultanée de la chaleur et de la compression á certain liquides," *Ann. Chim.* 22 (1823): 410–415; 26 (1823): 98–100.

III Documents: Thermodynamics

1. Letter from Maxwell to William Thomson, May 15, 1855[a]

Proc. Cambridge Phil. Soc. 32 (1936): 695–750, 705.

Trin. Coll. May 15/55

Dear Thomson,

Many thanks for your list of Electrical matter. I think I can get hold of all you mention. I am reading Weber's Elektrodynamische Maasbestimmungen[b] which I have heard you speak of. I have been examining his mode of connecting electrostatics with electrodynamics, induction etc., and I confess I like it not at first. He makes the attraction of two elements of electricity

$$= -\frac{ee'}{r^2}\left(1 - a^2\left(\frac{dr}{dt}\right)^2 + b\,\frac{d^2r}{dt^2}\right)$$

determining a and b from Ampère's laws.

But I suppose the rest of his views are founded on experiments which are trustworthy as well as elaborate.

I am trying to construct two theories, mathematically identical, in one of which the elementary conceptions shall be about fluid particles attracting at a distance which in the other nothing (mathematical) is considered but various states of polarization, tension etc., existing at various parts of space.[c] The result will resemble your analogy of the steady motion of heat. have you patented that notion with all its applications? for I intend to borrow it for a season, without mentioning anything about heat (except of course historically) but applying it in a somewhat different way to a more general case to which the laws of heat will not apply.[d]

By the way do you profess to account for what becomes of the vis viva of heat when it passes through a conductor from hot to cold? You must either modify Fourier's laws or give up your theory, at least so it seems to me....

a. Joseph Larmor, "Origins of Clerk Maxwell's Electrical Ideas, as described in Familiar Letters to William Thomson," Proc. Cambridge Phil. Soc. 32 (1936): 695–750, 705. See also *The Scientific Letters and Papers of James Clerk Maxwell* ed. P. M. Harman, (New York: Cambridge University Press, 1990), vol. 1, 305-313. This long letter is only reprinted in part. It contains much on color theory and the stability of fluid flow with examples. Enough has been included so that the context of Maxwell's remarks on irreversibility is clear.

b. Wilhelm Eduard Weber (1804–1881) introduced absolute measures for the electromagnetic units of electric current in 1846, then of electrical resistance in 1852. In the latter year he also introduced his own law of force, cited in Maxwell's letter, between the elements of electric current. Weber considered this force as a central force between moving charges. His work appeared in several massive papers: "Electrodynamische Maasbestimmungen," *Abh. bei Begrundung der König. Sachs. Gesells. der Wissenschaften am Tage der zweihundertjahigen Geburtstag für Leibnizen* (Leipzig: 1846): 211–378 abstracted in *Ann. Phys.* 73 (1846): 193–240 and "Electrodynamische Maasbestimmungen in besondere Widerstandsmessungen," *Abh. Math.-Phys. Cl. Gesell. Wissen. Leipzig* 1 (1852): 197–382, the paper Maxwell cited. Compare the comment on Weber's theory in Document III-16. For Maxwell's later views on Weber's theory see his *Treatise on Electricity and Magnetism* (Oxford: Clarendon, 1873), vol. 2, secs. 862–866.

c. The first attempt was published as Maxwell, "On Faraday's Lines of Force," *Trans. Cambridge Philos. Soc.* 10 (1857): 27–83, reprinted in *Scientific Papers,* vol. 1, 155–229.

d. William Thomson's papers on Fourier's mathematics appeared as a series of papers while he was still as student at Cambridge during the 1840s. Thomson explored the mathematical form of Fourier's equations to extend the mathematics of electrostatics. These are mathematics papers in the traditional of french physique-mathématique. He drew no physical conclusions from his mathematics until the 1850s. For the development of french mathematics see Ivor Grattan-Guinness *Convolutions in French Mathematics, 1800–1840: From the Calculus and Mechanics to Mathematical Analysis and Mathematical Physics* 3 vols. (Boston MA: Birkhäuser, 1990). For the content of Thomson's papers using Fourier see Garber,"Reading Mathematics, Constructing Physics: Fourier and his Readers, 1822–1850," in press. For a different interpretation of Thomson's work using Fourier see, Crosbie Smith and Norton Wise, *Energy and Empire: A Biographical Study of Lord Kelvin* (Cambridge: Cambridge University Press, 1989), 202–212.

2. Letter from Maxwell to C. J. Munro, May 20, 1857[a]

Lewis Campbell and William Garnett, *The Life of James Clerk Maxwell,* 267–268.

Glenlair, Springholm
Dumfries, 20th May 1857

I went to Old Aberdeen[b] for Fourier,... but I have forgotten what was to be discovered out of him.

The session went off smoothly enough. I had Sun all the beginning of Optics, and worked off all the experimental part up to Fraunhofer's lines, which were glorious to see with a water prism.[c] I have set up in the form of a cubical box, 5 inch side. The only things not generally done that I attempted last session were the undulatory medium made of bullets for the advanced class, and Plateau's experiments on a sphere of oil in a mixture of spirits and water of exactly its own density.[d]

I succeeded very well with heat.[e] The experiment in latent heat came out very accurate. That was my part, and the class could explain and work out the results better than I expected. Next year I intend to mix experimental physics with mechanics, devoting Tuesday and Thursday (what would Stokes say?) to the science of experimenting accurately.

I got a glorified top made at Aberdeen. I think you saw the wooden type at Cambridge. I have made it the occasion of a short screed on rotation coming out in the Roy. Soc., Edinburgh, presently.[f]

Last week I brewed chlorophyll (as the chemists word it), a green liquor, which turns the invisible light red. My pot of all the winter spinach that remained was portentous, so I exhibited the optical effects, which were allowed to be worth the potful.

My last grind was the reduction of equations of colour which I made last year. The result was eminently satisfactory.

a. Cecil J. Monro (1833–1882) was a college friend of Maxwell's. Later he contributed to the discussion on the individuality of molecules: see "Manufactured Articles" *Nature* 10 (1874): 481, and Maxwell's letter to Bishop Ellicott, November 22, 1876, in Campbell and Garnett, *Life,* (1882), 393. See Monro's obituary in *Proc. London Math. Soc.* 14 (1884): 323 for remarks on his correspondence with Maxwell and others.

b. At this time Maxwell was a professor at Marischal College, Aberdeen. Going to "Old Aberdeen" meant going to the library of the rival institution of King's College which is situated in Old Aberdeen: a dangerous journey in the minds of some Marischal College people. See, Campbell and Garnett, *Life,* 264.

c. Joseph Fraunhofer (1787–1826) had made observations on spectra and the sun beginning about 1814 in an effort to improve lens combinations to eliminate spherical aberration and coma. Fraunhofer found that, on comparing light from flames and the sun, the latter's spectrum was crossed by many fine dark lines some of which corresponded to the bright lines in flame spectra. His findings were reported in Fraunhofer, "Bestimmung des Brechungs- und Farbenzerstreuungsvermogens verscheidener Glasaerten, in Bezug auf die Vervolkommung achromatischer Fernrohre," *Denkschriften der Königl. Baierischen Akademie der Wissenschaften zu Munchen,* (1814–1815): 193-226, translated in *Edinburgh Philosophical Journal* 9 (1823): 288–299. The lines in the sun's spectrum had also been observed by William Hyde Wollaston in 1802.

d. Joseph Antoine Ferdinand Plateau (1801–1883) turned his attention from sense perception to the study of molecular forces about 1840. He performed a number of experiments with oil drops in mixtures of water and alcohol, the proportions of the latter being chosen to give the same density as that of the oil so as to effectively annul the force of gravity. See "Sur les phénomènes qui presente une masse liquide libre et soustraite a l'action de la pesanteur," *Mémoires de l'Academie des Sciences et Belles Lettres de Bruxelles,* 16 (1843), translated in Taylor's *Scientific Memoirs* 4 (1846): 16–43, and later papers on this subject.

e. Maxwell mentioned his course on heat at Marischal College in other letters, see Maxwell to Lewis Campbell, 6 February 1857, in Campbell and Garnett, *Life,* 264–266. He also mentioned his lectures on heat at Cambridge in the fall of 1855 in a letter to his father, 24 September 1855, in Campbell and Garnett, *Life,* 216. Maxwell wrote to William Thomson on the same subject, 17 December 1856 see, Larmor, "Origins of Maxwell's Electrical Ideas," *Proc. Cambridge Phil. Soc.* 32 (1936): 695–750, 722–725.

f. Maxwell, "On a Dynamical Top, for Exhibiting the Phenomena of the Motion of a System of Invariable Form About a Fixed Point, with Some Suggestions as to the Earth's Motion," *Trans. R. Soc. Edinburgh* 21 (1857): 559–570, reprinted in *Scientific Papers,* vol. 1, 248–262.

3. Letter from Peter Guthrie Tait to Maxwell, December 6, 1867

Cambridge University Library, Maxwell Collection, Maxwell-Tait Correspondence

Edinburgh
6/12/67

Dear Maxwell,

Many thanks for your letter, though you don't solve my difficulty about giving frictional or other resistance its true place by *geometrical* methods when too strong. I have tried several times, and got one (luckily the most important case) but the other is hopelessly complex.

Please to remember that you are a fellow of the R. S. E.,[a] and be good enough to send us a paper on Knots and their possible equations in 3 dimensions.[b] We devised all your figures (and many more) long ago—(Crum Brown[c] and I, working for Thomson[d])—but we never tried *equations.* Give us a paper on them like a good fellow—whether for the Trans.[e] or merely for the Proc.[f]

Give me also a reference as to your capillarity[g] investigations—for I am about to establish a working Laboratory for students, and will be delighted to get any hints as to keeping them to work & *useful* work.

Thomson's[h] letter was rather ambiguous, seeing that I dictated one half & he the other, and that it was not written by him but by his nephew,[i] whose hand is somewhat similar to his.

Nevertheless I hope your Treatise on Electricity will go on soon—whips and scorpions notwithstanding.[j]

Half the edition of our first vol. of Nat. Phil.[k] is already sold!!! It was published only 3 months ago.

You many understand my desire for a solution of the *cosh* question when I tell you that it would be extremely useful in our smaller volume now going through the press.[ℓ]

Are you sufficiently up with the history of Thermodynamics to critically examine & put right a little treatise I am about to print—and will you kindly apply your critical power to it?[m] You would greatly oblige me by doing so, as Clausius & others have cut up very rough about bits referring to them. I don't pretend to know the subject thoroughly and would be glad of your help. The fact seems to me to be that both Clausius[n] & Rankine[o] are bout [sic] as obscure in their writings as anyone can well be.

Shall we never see you in Edin^r?[p]

Yours Ever
P. G. Tait

P.S. Ponder this proposition. A man of your *originality,* and *fertility,* and *leisure,* is undoubtedly bound to furnish to the chief Society of his native land, numerous papers, however short.[q]

a. Royal Society of Edinburgh

b. For Tait's pioneering interest in the topology of knots, see C. G. Knott, *Life and Scientific Writings of Peter Guthrie Tait* (Cambridge: Cambridge University Press, 1911). The date of the letter (1867) and Tait's remark that he and Crum Brown had devised all Maxwell's figures long before, is interesting since Tait's first published paper on knots, "Applications of the Theorem that Two Plane Curves Intersect an Even Number of Times," *Proc. R. Soc. Edinburgh* 9 [1876] (1878): 237–246, was not published until nine years later. Tait's study of knots is sometimes said to originate with Thomson's vortex atom, and "working for Thomson" might be taken to imply as much, but Thomson's first published paper on vortex theory was written in 1867, which hardly seems "long ago," and the vortex atom did not emerge until 1870.

c. Alexander Crum Brown (1838–1922), Tait's brother-in-law.

d. William Thomson.

e. Transactions [of the Royal Society of Edinburgh]

f. Proceedings [of the Royal Society of Edinburgh]

g. This refers to Maxwell's apparatus to measure surface tension. Maxwell published no article on it but a description will be found in I. B. Hopley, "Clerk

Maxwell's Apparatus for the Measurement of Surface Tension," *Ann. Sci.* 13 (1957): 180–187.

h. William Thomson

i. The nephew was presumably James Thomson Bottomley who worked in Thomson's laboratory at Glasgow from about 1867 until the early 1900s.

j. The allusion is to 1 Kings 12:11 "My father hath chastised you with whips, but I will chastise with scorpions," but we do not know what Tait is referring to.

k. William Thomson and Peter Guthrie Tait, *A Treatise on Natural Philosophy* (Oxford: Clarendon Press, 1867). Of the volumes planned only this first one on mechanics was ever published. For details on what was planned and what actually published see, Crosbie Smith and Norton Wise, *Energy and Empire,* chap. 11. A second edition appeared in 1878 and a third in 1883.

ℓ. William Thomson and Peter Guthrie Tait, *Elementary Dynamics* (Oxford: Clarendon Press, 1867). For the cosh question see the next two letters.

m. Tait's historical sketch of thermodynamics formed the initial chapters of Tait, *Sketch of Thermodynamics* (Edinburgh: Edmonston and Douglas, 1868). Tait had previously published on the history of thermodynamics, "On the History of Thermodynamics," *Phil. Mag.* 28 (1864): 288–292 and managed to offend everyone involved except Joule and Thomson.

n. Rudolph Julius Emmanuel Clausius who protested Tait's version of the history of thermodynamics in Clausius, "A Contribution to the History of the Mechanical Theory of Heat," *Phil. Mag.* 43 (1872): 106–115.

o. William John Macquorn Rankine (1820–1872).

p. Edinburgh.

q. This request, reiterated by Tait in later letters did stimulate Maxwell to submit to the Royal Society of Edinburgh the long paper Maxwell, "On Reciprocal Figures, Frames and Diagrams of Forces," *Trans. R. Soc. Edinburgh* 26 (1872): 53–56, reprinted in *Scientific Papers,* vol. 2 161–207 for which he received the Keith Medal and the paper Maxwell, "On the Best Arrangement for Producing a Pure Spectrum on a Screen," *Proc. R. Soc. Edinburgh* 6 [1867-68] (1869): 238–242, reprinted in *Scientific Papers,* vol. 2, 96–100.

4. Letter from Maxwell to Peter Guthrie Tait, December 11, 1867[a]

Cambridge University Library, Maxwell Collection, Maxwell-Tait Correspondence.

I do not know in a controversial manner the history of thermodynamics, that is, I could make no assertions about the priority of authors without referring to their actual works. If I can help you in any way with your book I shall be glad, as any contributions I could make to that study are in the way of altering the point of view here and there for clearness or variety and picking holes here and there to ensure strength and stability.

As for instance I think you might make something of the theory of absolute <temp> scale of temperature by reasoning pretty loud about it and paying it due honour, at its entrance. To pick a hole—say in the 2nd law of θ^{cs} [thermodynamics] that if two things are <at the same temperature> in contact the hotter cannot take heat from the colder without external agency.

Now let A & B be two vessels divided by a diaphragm and let them contain elastic molecules in a state of agitation which strike each other and the sides.

Let the number of particles be equal in A and B but let those in A have greatest energy of motion. Then even if all the molecules have equal velocities, if oblique collisions occur between them their velocities will become unequal, and I have shown that there will be velocities of all magnitudes in A and the same in B, only the sum of the squares of the velocities is greater in A than in B.

When a molecule is reflected from the fixed diaphragm CD no work is lost or gained.

If the molecule instead of being reflected were allowed to go through a hole in CD no work would be lost or gained, only its energy would be transferred from one vessel to the other.

Now conceive a finite being who knows the paths and velocities of all the molecules by simple inspection but who can do no work except open and close a hole in the diaphragm by means of a slide without mass.

Let him first observe the molecules in A and when he sees one coming the square of whose vel.[ocity] is less than the mean sq. [square] velocity of the molecules in B let him open the hole and let it go into B. Next let him watch for a molecules of B, the square of whose velocity is greater than the mean sq. [square] vel. [velocity] in A, and when it comes to the hole let him draw the slide and let it go into A, keeping the slide shut for all other molecules.

Then the number of molecules in A and B are the same as at first, but the energy in A is increased and that in B diminished, that is, the hot system has got hotter and the cold colder and yet no work has been done, only the intelligence of a very observant and neat-fingered being has been employed.

Or in short if the heat is the motion of finite portions of matter and if we can apply tools to such portions of matter so as to deal with them separately, then we can take advantage of the different motion of different proportions to restore a uniform hot system to unequal temperatures or to motions of large masses.

Only we can't, not being clever enough.

a. Only the last part of this very long letter is reprinted here. The latter part of the letter is also quoted in Knott, *Life of Tait,* 213–214. We have restored a sentence omitted by Knott. In Knott a note is appended that there is a pencilled addition to the letter by William Thomson, "Very good. Another way is to reverse the motion of every particle of the Universe and to preside over the unstable motion this produces." The present editors consider the writing closer to Tait's than Thomson's. Tait added such marginal notes to other letters from Maxwell. The first half of this letter is on problems in mechanics, presumably in answer to enquiries from Tait in an earlier letter.

5. Letter from Peter Guthrie Tait to Maxwell, December 13, 1867[a]

Cambridge University Library, Tait Letters

6 G. G. E.
13/12/67

Dear Maxwell,

1) Can't you contrive, like your neighbours Dudgeon of Carsen, or like Fox Talbot,[b] to spend a winter now & then Edin[r]? In that case we should at once put you on the Council of the R. S. E. and get some good out of you. Also you should have the run of my laboratory (which is shortly to be considerably increased) as well as those of Playfair,[c] Crum Brown,[d] etc. Ponder the Point. Good.

2) Thanks for time horizontal, etc, & the dodge about drawing logarithmic curves, which may be very useful to me. I fear I can't avail myself of your system for an *elementary* book, for w[h] my spiral is well fitted.

3) I read the extract of y[r] note about a paper on "Theory of reciprocal rectil[r] figs. & diagrams of forces" to the Council R. S. E. this afternoon; and you are booked for the paper as soon as you can send it.[e]

4) I don't see why you shouldn't send us a "note" on "Nots" with a stereogram or two for our Proceedings. It would give you no trouble to do this—though it *might* bother you to get ready a paper for the Transactions.

5) I suppose your tension of films goes to the London R. S.—else we should be glad of a few scraps.[f]

6) I object to your infinitely sharp individual that he *lets his gases mix,* and so spoils the theorem. But let him wait long enough to catch a quick one from the colder medium & a slow one from the hotter w[h] are moving in the same line so as to impinge centrically when he moves the slide. How many Darwinian ages will that require? And, when he has caught these two, won't he have to wait longer for a repetition? Good.

7) The Quaternions have been out as long as *the* Book, but I have not yet heard how they sell, and whether they have sold anybody yet.[g]

8) Δ is required in 4^{ions} for its finite diff^{ce} meaning—so we do Δ or ∇ for the flux. I didn't know that Lamé had a Δ, though I knew he had a $\Delta^2 = \left(\frac{d}{dx}^2\right) + \ldots$, and a δ (I think) for Betti's $(dF/dx)^2 + \ldots$ But it is long since I looked at his papers. Still if he had a ∇, and not merely separate parts of it such as $\frac{dF}{dx}$, &c. *he must have anticipated Hamilton in discovering* 4^{ions}, so I wish you would give me the reference.[h]

9) If you read the last 20 or 30 pages of my book I think you will see that 4^{ions} are worth getting up, for there it is shown that they go into that ∇ business like greased lightning. Unfortunately I cannot find time to work steadily at them.

10) I'll send you a copy of my first two Chapters on Thermodynamics in a day or two, when I hear from Rankine, meanwhile let me know your mind on the graver subjects propounded at the threshold of this note

Yrs

P. G. Tait

J. C. Maxwell Esqr.

a. Tait is addressing systematically points raised in the previous letter from Maxwell, reprinted partially as Document III-3.

b. William Henry Fox Talbot (1800–1877) was a mathematician, philologist and pioneer photographer, inventor of the negative-positive process.

c. Lyon Playfair (1818–1898) a member of the famous Scottish scientific family, was then professor of chemistry at the University of Edinburgh. He had studied under Liebig at Giessen and had collaborated with Joule and Bunsen. His principle life-work was as an advisor to the British government on scientific questions.

d. Alexander Crum-Brown was a chemist and Tait's brother-in-law.

e. The paper Tait is referring to is probably Maxwell, "On Reciprocal Figures, Frames and Diagrams of Force," *Proc. R. Soc., Edinburgh* 7 (1872): 53–56 and *Trans. R. Soc. Edinburgh* 26 (1872): 1–40. He had previously published on the subject in Maxwell, "On Reciprocal Figures and Diagrams of Forces," *Phil. Mag.* 27 (1864): 250–261, and, "Reciprocal Diagrams in Space, and Their Relation to Airy's Function of Stress," *Proc. London Math. Soc.* 2 (1869): 58–60. R. S. E. here and elsewhere is the Royal Society of Edinburgh.

f. Maxwell's ideas on surface tension appeared in Maxwell, "Plateau on Soap Bubbles," *Nature* 10 (1874): 119–121, reprinted in *Scientific Papers,* vol. 2, 393–399 and Garber, Brush, and Everitt, *Maxwell on Molecules and Gases,* 163–169.

g. Presumably Maxwell had asked whether Tait, *Elementary Treatise on Quarternions* (Oxford: Clarendon Press, 1867) was out yet.

h. Gabriel Lamé (1795–1870) established the transformation of differential equations (such as those of Laplace and Fourier) into curvilinear coordinates so that a number of problems in mathematical physics were solved more easily. He is not, however, recognized as a discoverer of quarternions. See, Michael J. Crowe, *A History of Vector Analysis* (Notre Dame, Ind.: University of Notre Dame Press,

1967). Enrico Betti (1823–1892) was professor of mathematics at Pisa University from about 1865 until his death. He wrote articles on algebraic equations, elliptic functions and various branches of theoretical physics. See Maxwell, *Treatise on Electricity and Magnetism,* secs. 173, 862.

6. Letter from Maxwell to Peter Guthrie Tait (undated) "Catechism on Demons"

C. G. Knott *Life and Scientific Work of Peter Guthrie Tait* (Cambridge: Cambridge University Press, 1911), 214–215.

Concerning Demons

1. Who gave them this name? Thomson.
2. What were they by nature? Very small **BUT** lively beings incapable of doing work but able to open and shut valves which move without friction or inertia.
3. What was their chief end? To show that the 2nd Law of Thermodynamics has only statistical certainty.
4. Is the production of an inequality of temperature their only occupation? No, for less intelligent demons can produce a difference in pressure as well as temperature merely by allowing all particles going in one direction while stopping all those going the other way. This reduces the demon to a valve like that of the hydraulic ram, suppose.

7. Letter from Maxwell to Peter Guthrie Tait December 23, 1867.[a]

Cambridge University Library, Maxwell Collection, Maxwell-Tait Correspondence.

Glenlair
Dalbeattie
1867 Dec 23

Dear Tait,

I have received your histories of Thermodynamics and Energetics, and will examine them along with Robertson[b] on the Unconditioned who holds that our ultimate hope of sanity lies in sticking to metaphysics and letting physics go down the wind.[c]

I have read some metaphysics of various kinds and find it more or less ignorant discussion of mathematical and physical principles, jumbled with a little physiology of the senses. The value of the metaphysics is equal to the mathematical and physical knowledge of the author divided by his confidence in reasoning from the names of things.

You have also some remarks on the sensational system of philosophising (sensational in the American not the psychological sense). Beware also of the hierophantic or mystagogic style. The sensationalist says "I am now going to grapple with the Forces of the Universe and if I succeed in this extremely delicate experiment you will see for yourselves exactly how the world is kept going." The Hierophant says "I do not expect to make you or the like of you understand a word of what I say, but you may see for yourselves in what a mass of absurdity the subject is involved."

Your statement however seems tolerably complete considering the number of pages. One or two ideas should be brought in with greater pomp of entry, perhaps.

I do not understand how Verdet's[d] discovery that paramagnetic bodies produce rotation of the plane of polarization in the opposite direction to diamagnetic bodies *confirms* Faraday's doctrine that the diamagnetic body is only less paramagnetic than the field.[e] It is a pretty doctrine, but I do not think Faraday thought it certain and Verdet's phenomenon appears to me the strongest thing against it. I am myself sorry to part with it.[f]

Weber's doctrine[g] is that paramagnetic bodies have ready made electromagnets in them which are set in one direction by magnetic force, and therefore when they are all set parallel there is a limit to magnetization and that diamagnetics have currents set up in them by induction and that these are unopposed by resistance and these will give an opposite polarity. I do not say this theory is true but if it were Thomson's[h] revolving diamagnetic sphere would not be a prime mover any more than any other electromagnetic machine consisting of coils revolving in the presence of magnets.

There is a difference between a vortex theory ascribed to Maxwell at p. 57, and a dynamical theory of Electromagnetics by the same author in Phil Trans 1865.[i] The former is built up to show that the phenomena are such as can be explained by mechanism. The nature of this mechanism is to the true mechanism what an orrery is to the Solar System. The latter is built on Lagrange's Dynamical Equation and is not wise about vortices. Examine the first part which treats of the mutual actions of currents before you decide that Weber's is the only hypothesis on the subject.

I hope you will come to some result with your vulcanite magnet.

It will require great speed and you will require to guard the <mai> testing magnet from direct electromagnetic action of the revolving machinery.

You wrote me about experiments in the Laboratory. There is one which is of a high order but yet I think within the means and powers of students namely the determination of Joule's coefft. [coefficient] by means of mercury. Mercury is 13.57/.033 times better than water so that about 9 feet would give 10°*Fah.* [Fahrenheit]. You have a cistern which you keep full to a certain point with mercury by

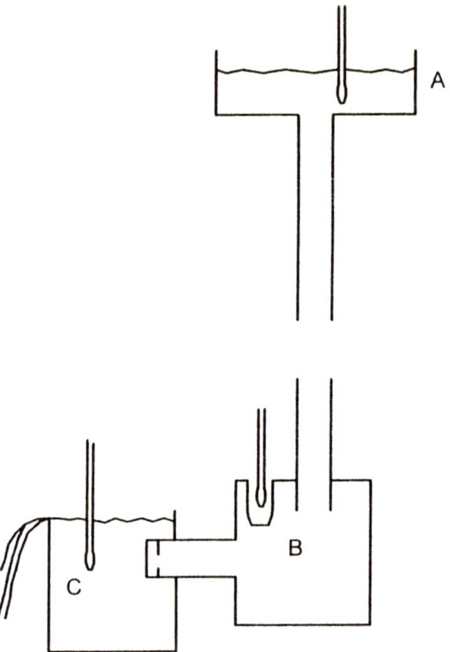

proper fillers and a tube of iron with a strong iron cistern at the bottom into which is let a place to put mercury and dip a thermometer without exposing it to pressure.

Out of this is a tube which enters an open cistern and out of this tube the mercury escapes either through a series of plates each having a fine hole in it or through a plug of compressed cotton or otherwise. The mercury and the tube is thus heated but the heat is all communicated (in time) to the mercury, for the part conducted back by the side of the tube must be small and can be estimated. The mercury rises and pours over a notch in the pen cistern below and students continually carry it up to the upper cistern mixing it with very cold mercury to keep the temperature even.

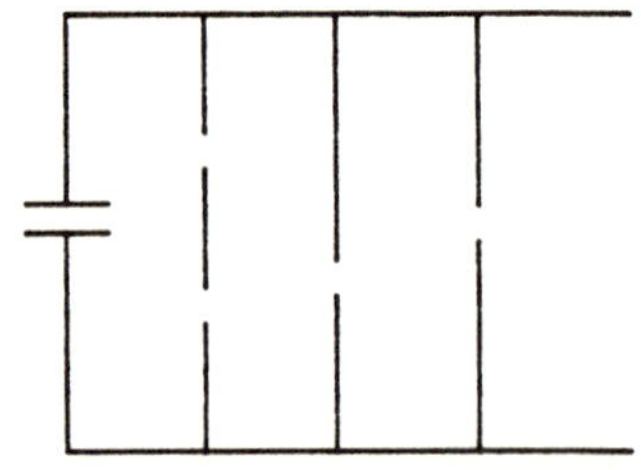

The experiments required are, 1. a comparison of the temperature at *A* and at *B* when the mercury flows pretty quick to determine the effect of pressure in altering temperature. 2. Rate of cooling (or heating) of *B* and

C with given difference of temperature from air. 3. Difference of level at A and C. This can be done with pointed screws dipping on the surface. 4. diff[c] [difference] of temperature at A and C can be got within $1/50°F$.

I have liberty from Joule to try this experiment but though I have considered the necessaries and do not think them unattainable I have no prospect of doing it. I think it a plan free from any mechanical difficulties and in a lofty room with plenty of mercury and strong iron works, and a cherub aloft to read the level and the thermometer and a monkey to carry up mercury to him (called Quicksilver Jack) the thing might go on for hours, the coefficient meanwhile converging to a value to be appreciated only by the naturalist.

Are you aware that if anything converges according to log, to fixed value v and if x, y, z are three equivalent values

$$v = \frac{xz - y^2}{x + y - zy}$$

I have sent the secretary of the R. S. E.[j] an article on the arrangement of a prism and a lens of the same material for spectrum observations, as for instance when you are restricted to quartz or rock salt and cannot use achromatic lenses.[k]

I may find something else but this is the first I could lay hands on for the Proceedings.

Yours truly,

J. Clerk Maxwell.

a. Quoted, partially in Knott, *Life of Tait,* 215. We have restored his omissions in the manuscript.

b. George Croon Robertson (1842–1892). As a philosopher his ideas developed from the work of J. S. Mill and Alexander Bain whose work he helped revise with Bain in 1867. Robertson did not publish very much especially at the beginning of his career due to ill health. His study of Thomas Hobbes was never completed as he had wished and was published very late in his career. He was the first editor of *Mind* from 1876 until his death in 1892.

c. There is more than meets the eye in Maxwell's remark that he will read Tait's *Sketch of Thermodynamics* along with Robertson on the Unconditional. Tait's hostilities extended to metaphysicians as a breed; he conducted a running battle for twenty-five years with his colleague Alexander Campbell Fraser, professor of Logic and Metaphysics at Edinburgh University and the successor to Sir William Hamilton. In connection with this see Maxwell, "Thomson and Tait's *Treatise on Natural Philosophy,*" *Nature* 20 (1878): reprinted in *Scientific Papers,* vol. 2, 778–779 for comments on the definition of "matter" given in Thomson and Tait's *Treatise.*

d. Marcel Émile Verdet (1824–1866). His work on the Faraday effect is in, Verdet, "Recherches sur les propertés optiques dévelopées dans les corps transparentes par l'action du magnétisme," *Ann. chim.* 41 (1854): 370–412; 42 (1858): 129–163; 43 (1858): 37–44.

e. Michael Faraday (1791–1867). His theory of diamagnetism appears in Faraday, "Experimental Researches in Electricity. Series 23. 29 On the Polar or Other Condition of Diamagnetic Bodies," *Phil. Trans.* 140 (1850): 171–188.

f. See Maxwell, *Treatise on Electricity and Magnetism* vol. 2 sec. 809 for his developed argument.

g. Wilhelm Weber, "Electrodynamische Maasbestimmungen," *Abh. Math.-Phys. Cl. Gesell. Wissen. Leipzig* 1 (1852): 197–382. For Maxwell's extended discussion of Weber's theory of magnetism see Maxwell, *Treatise,* vol. 2, secs. 838–845.

h. William Thomson, "A Mathematical Theory of Magnetism," *Phil. Trans. R. Soc. London* 141 (1851): 243–285, abstract *Phil. Mag.* 37 (1849): 540 and *Proc. R. Soc. London* 5 (1849): 975–978. See also Thomson, "On Electric Machines Founded on Induction and Convection," *Phil. Mag.* 35 (1868): 66–73.

i. Maxwell is contrasting his use of the vortex theory in Maxwell, "On Physical Lines of Force," *Phil. Mag.* 21 (1861): 161–175, reprinted in *Scientific Papers,* vol. 1, 451–513, and "A Dynamical Theory of the Electromagnetic Field," *Phil. Trans. R. Soc. London* 155 (1865): 459–512, reprinted in *Scientific Papers,* vol. 1, 526–597. The latter is based on Lagrange's equation.

j. Secretary of the Royal Society of Edinburgh.

k. The paper to which Maxwell refers appeared as Maxwell, "On the Best Arrangement for Producing a Pure Spectrum on a Screen," *Proc. R. Soc. Edinburgh* 6 (1869): 238–242, reprinted in *Scientific Papers,* vol. 2, 96–100.

8. Letter from Maxwell to Peter Guthrie Tait, March 5, 1868

Cambridge University Library, Maxwell Collection, Maxwell-Tait Correspondence.

$$\left.\begin{array}{l}\text{Fixed Point}\\ \text{of Reference}\end{array}\right\} = \left\{\begin{array}{l}\text{Glenlair}\\ \text{Dalbeattie}\end{array}\right.$$

I at 55° Long. 4 1868. 534

D[r] P. G. T. ,

Use no more the above triliteral expression on the outside of your communications as the expression Palace Garden Terrace will not avail in communicating with me.

Your hypothesis about the law of conductivity would be a sound one if the state of steady flow of heat were not a state involving the continual dissipation of energy. It may however be *true* though unsound. Does it agree with the *general* equations of steady flow as well as with those in one dimension? that is, Is the distribution of temperature in any case of steady flow such that if it were arrested at any instant the recoverable energy would be a min.[a]

I have improved my machine for finding the superficial tension of liquids. I find it decreases in water with the temperature very considerably and rapidly, say about 7.7grammes to the metre at 69° and 5.6 at 100°*F*. I am trying to get the temperature correctly by immersing my machine in hot water and I have succeeded in preventing the glass from being dimmed by the steam. I shall also try soap bubbles to test the effect of the soap.

I am doing figures of equipotential surfaces and lines of force. The figure for Thomson's Electrical Images is very pretty if well done.[b] I am also doing a few cases of conduction, etc. , in plane sheets (what Rankine calls plane water lines) and sections of the low spherical harmonics, and *map* of the surface harmonics of the 5th order as given in T. T'.[c]

I have a letter from Joule who says he is game to do the equivalent of heat by friction of mercury. For whom do you want my vote, Jupiter I know but who is Vulcan? Not he who fell on ice is it? and what the vote? Send me the vortices here

40227.-

J C M

a. minimum.

b. William Thomson,"Extrait d'une lettre sur l'application du principe des images à la solution de quelques problèmes relatifs à la distribution d'électricité," *J. math. pures appl.* 10 (1845): 364–367; "On Electrical Images," *Rep. 17th Meeting BAAS* (1847): 6–7.

c. This is a reference to William Thomson and Peter Guthrie Tait, *A Treatise on Natural Philosophy,* (Oxford: Clarendon Press, 1867).

9. Letter from Maxwell to the editor of *Saturday Review*, April 7, 1868[a]

Pattison Papers, Bodleian Library, Oxford

8 Palace Gardens Terrace
London W
1868 April 7

Sir

In the Saturday Review of April 4 is an article on Science and Positivism[b] the writer of which appears to take so much interest in metaphysics science and positivism that I should be obliged if you will communicate to him the following remarks on a portion of the article—

M. Caro, according to the article (for I have not yet seen his own work) uses in his argument the doctrine of the gradual conversion of all kinds of energy into the form of heat, and the ultimate uniform distribution of temperature over all matter.

As the speculation has important consequences I should like to point out to the writer of the article where he may find the data on which it is formed.

1 Fourier, in his great work on the conduction of heat, has given methods by which if we know the temperature of every part of a body at any time and the temperature of the surface at all times we can determine the temperatures of any point of the body at any *future* time as arising from the conduction of heat within it. (If the body be supposed to include the universe the condition about the surface is unnecessary.)

Now the formulae of Fourier for predicting the future temperatures are equally applicable to determine the *former* temperature of the body in all its parts <at any> by simply making the quantity denoting the time a negative quantity. If in this way we attempt to ascertain the state of the body previous to our observation of it, then (except in particular cases) we find that as we go back we arrive at an epoch at which the temperature varied in a discontinuous manner, and if we seek for the state of the body at any time still farther back we arrive [at] an impossible result. Hence the body could not have existed as a solid body and a conductor of heat before a certain epoch. At that time or after it something must have happened, e.g., two bodies at different temperatures may have joined together at a certain epoch and the present condition of the compound body will indicate when that was.

This is a purely mathematical result founded however on experimental data. It has been pointed out by Sir W. Thomson in several papers on the Secular cooling of the Earth & Sun.[c]

2 The general doctrine of the dissipation (not the destruction) of energy was first clearly stated by Sir W. Thomson in his papers on the Dynamical Theory of Heat (Trans. Royal Society of Edinburgh 1852, 3, 4, 5). It has also been treated at some length and with great labour by Prof. R. Clausius of Zurich under the name of "Entropy" which is an expression for the quantity of energy now rendered unavailable, a quantity always on the increase.

The data are

1st the fact that energy cannot be created or destroyed by physical agency.

2nd the fact that energy may change its form in two different ways

(α) in a certain class of conceivable cases the process by which the transfer takes place may be exactly reversed so that everything is at last in the same condition as at first.

(β) in another class of cases the process is not reversible by any physical agency e. g., the equalization of temperature by conduction of heat in a body & the production of heat by the electric current when it meets with "resistance."

By no contrivance can we arrange an example of class (α) without introducing processes belonging to class (β) so that in every action in nature part of the process is not capable of reversion. This part always tends in one direction to diminish the energy which is available for producing phenomena involving change, and to increase the energy that cannot be used. The ultimate condition is one of uniform temperature in which everything remains at the same distance from every other thing in so far as these are sensible objects.

I speak of sensible objects because according to a certain theory the phenomena of heat are due to the intestine motion of the smallest parts of hot bodies. A uniformly hot body apparently at rest is not on this theory devoid of motion for if we had the means of observing the very small<est> parts (I do not speak of atoms) of the body and of distinguishing the smallest intervals of time we should find at a given instant different parts moving in different ways, and if we could lay hold of these parts by machinery we might extract energy from this motion till the whole mass was reduced to stillness.

There is no evidence that this can be done, either by direct manipulation or by any physical process, and therefore in the present dispensation there remain a number of irreversible processes, all of which tend in the same direction, and therefore tend of themselves to an end and by reasoning backwards (if we knew enough) we should find an epoch before which the present order could not have existed.

The peculiar faith required of a positivist is in the universal validity of laws, the form of which he does not know, though he speculates to a certain extent on their results.

A strict materialist believes that everything depends on the motion of matter. He knows the form of the laws of motion though he does not know all their consequences when applied to systems of unknown complexity. Now one thing in which the materialist (fortified with dynamical knowledge) believes is that if every motion great & small were accurately reversed, and the world left to itself again, everything would happen backwards[.] [T]he fresh water would collect out of the sea and run up the rivers and finally fly up to the clouds in drops which would extract heat from the air and evaporate and afterwards in condensing would shoot out rays of light to the sun and so on. Of course all living things would regrede [sic] from the grave to the cradle and we should have memory of the future but not of the past.

The reason why we do not expect any thing of this kind to take place at any time is our experience of irreversible processes, all of one kind, and this leads to the doctrine of a beginning & an end instead of cyclical progression for ever.

The practical relation of metaphysics to physics is most intimate. Metaphysicians differ from age to age according to the physical doctrines of the age and their personal knowledge of them. Leibnitz is in advance of Descartes[.] Newton so far as he exposes himself is distinct. Locke Berkeley &c differ according to the degree in which they enjoyed the diffusion and dilution of the Galilean and Newtonian doctrines.

The Edinburgh & Dublin Hamilton differ in their metaphysical power in the direct ratio of their physical knowledge (not the inverse as most people suppose).[d]

On the other hand the effect of the absence of metaphysics may be traced in most physical treatises of the present century. I have been somewhat diffuse but I happen to be interested in speculations standing on experimental and mathematical data and reaching beyond the sphere of the senses without passing into that of words and nothing more.

I am Sir

yours truly

J. Clerk Maxwell

a. The editor was Mark Pattison. We thank Dr. Thomas Simpson for informing us of this and the following letter, Document III–10, and furnishing a copy of his transcriptions.

b. The article in question was "Science and Positivism," *Saturday Review* 25 (1868): 455–456, a review of E. Caro, *Le materialisme et la science* (1867). The reviewer notes that according to Caro the mechanical forces of nature are tending to convert themselves into atomic forces. The consequence is that all life and motion would eventually cease.

c. However, William Thomson drew more far reaching consequences than Maxwell from this mathematical result. See Crosbie Smith and Norton Wise, *Energy and Empire* (Cambridge: Cambridge University Press, 1989), and Joe D. Burchfield, *Lord Kelvin and the Age of the Earth* (New York: Science History Publications, 1975).

d. William Hamilton (1788–1856) was professor of Logic and Metaphysics at Edinburgh University when Maxwell was a student there in the late 1840s. William Rowan Hamilton (1805–1865) of Dublin was primarily known for his contributions to Optics and Dynamics. The harsh comment on the Edinburgh Hamilton is in contrast to Maxwell's biographers' claim of his influence. See Campbell and Garnett, *Life*, chap. v.

10. Letter from Maxwell to the editor of *Saturday Review*, April 13, 1868

Pattison Collection, Bodleian Library, Oxford University

8 Palace Gardens Terrace
London W
1868 April 13

Sir

I have received your letter and will do my best to answer your queries.[a] You must see that my acquaintance with positivism whether as stated by Comte, Littré, or Mill, is but slight. Comte certainly endeavoured to form a distinct picture of the shape, method and aim of a system of [the] sciences, some of which he considered as in a very immature state at present. This picture he considered as a sort of matrix in which these sciences would probably be developed by future labourers and one of its uses was to prevent them from wasting their efforts on attempts of a kind which would prove abortive. In this part of his work Comte was like other philosophers who try to make their "Principles of Human Knowledge" practically useful, and using every method—conjecture—imagination & himself for the purpose. But when he prescribes rules for the study of the sciences already formed, as Astronomy, all this part of the philosophic spirit is proscribed, and the astronomer is even forbidden to extend his views beyond the Solar System.[b]

I say therefore that though it is quite the part of a philosopher to try to form a conception of a state of science more developed than at present, and of the methods likely to be the most fertile, he who does so assumes that although he does not know *what* will be discovered he has some anticipation of the mould in which future discoveries will be cast. The statements of certain positivists therefore about human knowledge & science in general are apt to make one think that they are convinced that new truths will fall into the old moulds, even in sciences yet unknown.

I have no doubt however that in my last letter I went astray on this subject. I go on to your 3rd query, taking them in the reverse order.

I am not sure whether the "Matter" against which Berkeley argued has any existence now. I am little satisfied with most of the definitions of it. "That which is perceived by the senses" is utterly wrong both in excess and defect.

Lucretius says

"Facere et fungi sine corpora nulla potest res"

In Dynamics, the science of the motion of matter as affected by forces, matter is defined and measured solely with respect to the force required to move it in a certain manner and the force is likewise defined with respect to matter & motion.

Having defined equal times[,] equal distance and equal velocities.[sic] Force is defined to be that which produces change in a body's velocity and is *measured* by the change which it would produce in the velocity of a standard body (Imperial Pound) if it acted on it for a second.

Any other mass on which the same force would produce the same effect is called an equal mass whatever be its other properties. This is the definition of equal quantities of matter and it is found to lead to consistent statements.

The measurement of quantity of matter by weight is a secondary method founded on the fact, which Newton and others carefully verified, that the weight of all bodies known to us is proportional to the quantity of matter in them and independent of the *kind* of matter. When the word Inertia is used by the physicist since Newton, it generally means not metaphysical passivity but a measureable quantity namely the number of pounds in the body.

Suppose that a chemist asserted that the addition of a certain substance (say phlogiston) diminished the weight of a body, a physicist would say that there could not be very much of it in the Solar System or facts would be different. But if two bodies were found to be of equal mass (by giving them equal and oppos[ite] velocities and observing them come to rest after impact) and if phlogiston added to one of them diminished its mass *so measured* then either the mechanical properties of the body and the phlogiston are not simply added together, or the phlogiston by itself would have a negative mass, that is, a force acting on it would cause it to move against the force, which is pure nonsense.

Again on the undulatory theory of light, the medium is supposed to communicate motion from one part to another by the action of elastic force which gradually changes the motion of each portion. This implies that the medium is material and that the number of pounds of it in a cubic mile might be ascertained*[c] though as yet we have no evidence of gravitation acting on it.

These examples are meant to show that the true test of matter is in its relation to the force which alters its motion.

When matter is in motion it has two mathematical possessions

1 Momentum, a directed quantity equal to Mass × Velocity and in the direction of the velocity.

2 Kinetic Energy an absolute quantity, without direction measured by $\frac{1}{2}$ Mass $\times$ $(Velocity)^2$.

This was formerly called Vis Viva and the controversy between the Newtonians & Leibnitzians was about which (1) or (2) was the more excellent measure of the Motion of a Body.

The *facts* are that in *all* cases of the mutual action of a system of bodies
1 the *geometrical* sum of the momenta reckoned according to their directions remain constant.
2 In certain cases visibly, and in all cases if we could measure it the sum of the kinetic energies taken arithmetically together with the sum of certain other energies called potential energies remains constant.

Berkeley quotes (with disdain) a passage of Torricelli which seems appropriate,[d] "Matter is nothing but the enchanted vase of Circe, which serves for a receptacle of force and momenta of impulse. Power*[e] and impulse†[f] are such subtle abstracts, are quintessences so refined, that they cannot be enclosed in any other vessels but the inmost materiality of natural solids."

I so far agree with this, that I cannot admit any theory which considers matter as a system of points which are centres of force acting on similar points, and admits nothing but these forces. For this does not account for the perseverance of matter in its state of motion and for the measure of matter.

I am afraid I have been tiresome on this subject but I consider it of the first importance in physics to know what we mean by matter and how to measure it, and both natural and mental science writers often go astray at the beginning about these things.

You ask about the definition of Energy. Energy is of two kinds, Kinetic and Potential. Energy is the capacity which a body has of doing *work*. A moving body has energy due to its motion called kinetic energy and measured by $\frac{1}{2}$ mass $\times$ $(velocity)^2$. A body or system of bodies which are connected so that forces act between them tending to alter their relative position is capable of work and this capacity is called Potential Energy. Two heavenly bodies tending to approach (say = the earth and a weight) have potential energy due to gravitation. Electrified bodies have electrical potential. Elastic springs when bent &c &c are examples of potential energy.

Energy of both kinds is capable of exact measurement, and the progress of science at present is in the direction of measuring additional forms of energy.

Now the conception of Kinetic Energy is simply that of a moving mass which can do work till it is stopped. Potential Energy is force acting between bodies and capable of <acting> continuing to act between them while they yield to its action. Now the conception of a body in motion is more fundamental than that of the power of producing motion in other bodies (commonly called force of attraction, repulsion &c). Hence when any form of potential energy (such as the elasticity of air) can be explained by motion only, the explanation is a step in advance, and I suppose that gravitation itself will not be satisfactorily understood till it is explained in some way not involving action at a distance.

I have examined several attempts by various speculators, but the dynamics in all were faulty. I have also tried to apply to gravitation the same method which I found useful in electromagnetism, but I have found no opening for a theory of gravitation. I merely state this to show that there is a desire among men to explain action apparently at a distance by the intermediate action of a medium and <if possible> then to explain the action of the medium as much as possible by its motion, and so to reduce Potential Energy to a form of Kinetic Energy.

Energy is never destroyed, but some of its transformations are not reversible. I have investigated the case of a multitude of molecules moving in a confined space and occasionally deflecting each other from their paths. This is a question in pure dynamics but it gives results which apply with great exactness to the properties of gases.

In whatever way the motion is distributed among the particles at first, it is very quickly distributed according to a certain law according to which the number of particles having velocities between certain limits can be found. There remains always a great difference between the greatest and the least velocities <but> and the molecules are often gaining and losing velocity but a general law of distribution prevails like that of wealth in a nation by which the proportion having so much above or below the average is calculable. Now in a nation you can pick out the rich people as such, but in a gas you cannot pick out the swift molecules either by mechanical or chemical means. As a simple instance of an irreversible operation which (I think) depends on the same principle suppose so many black balls put at the bottom of a box and so many white above them. Then let them be jumbled together. If there is no physical difference between the white and black balls, it is exceedingly improbable that any amount of shaking will bring all the black balls to the bottom and all the white to the top again, so that the operation of mixing is irreversible unless either the black balls are heavier than the white or a person who knows white from black picks them and sorts them.

Thus if you put a drop of water into a vessel of water no chemist can take out that identical drop again, though he could take out a drop of any other liquid.

[Can you tell me without trouble what is meant by the expression "idem *numero*," in metaphysical or theological language as when James Bernoulli calls his spiral "carnis nostrae ... post varias alterationes—ejusdem *numero* resurrecturae symbolum."][g]

I think you are right in thinking that we are likely to arrive at physical indications of a beginning & an end. That end is not a destruction of matter or of energy but such a distribution of energy that no further change is possible without an intervention of an agent who need not create either matter or energy but only direct the energy into new channels.

I do not think that anyone can have a second-hand acquaintance with a physical *principle* such as the idea of matter, force, energy, motion. To every one it must be either a mere word or a true form of thought, whether he is a professional experimenter or a mathematician or a lover of wisdom. The power to understand and assimilate elementary physical ideas (such as those contained in the Definitions and Axioms in Newton's Principia) is not confined to professed mathematicians or experimenters. Experiments are often made to illustrate these principles but not to prove them, and these illustrations help to explain what is meant, say, by Action and Reaction but no one has proved them equal, or has required proof, after he knew the meaning of the terms.

All these things therefore, being the foundations of science are as much the property of one man as of another. The discovery of such principles is the work of eminent men like Archimedes, Galileo, Newton, Young, Faraday but when they are illustrated and explained they may be fully understood as a useful mental possession by any one whose mind is open to receive them. At the same time it is possible to become eminent both in mathematics and in experiment with a very faulty set of first principles provided they are not often appealed to and so brought to the test.

I think that Newton through himself, Desauguliers, Bentley, Locke, Gregory and others[h] exercised a very great influence on English thought among men who were neither mathematicians nor astronomers and that Voltaire's scientific writings produced a very great influence on French thought, though several most important ideas in Newton have been almost lost in the process of transmission so that I would recommend the reading of Newton's Definitions & Axioms to every scientific man who is not familiar with them <and others will find them the origin of several expressions which are now part of our mother tongue> If I have taken more time than I should have occupied yours less with trying to answer your queries.

I remain
Yours faithfully
J. Clerk Maxwell

a. This is presumably a reply to a letter from the person who received Document III-9, either Mark Pattison as editor of the *Saturday Review* or the author of the article, "Science and Positivism."

b. Auguste Comte, *Cours de Philosophie Positive* (Paris: Bachelier, 1835), tome II, 2. On the relation of positivism to scientific thought see Walter Simon, *European Positivism in the Nineteenth Century* (Ithaca NY: Cornell University Press, 1963); James Maurice Murphy, *Positivism in England: The Reception of Comte's Doctrines* unpublished PhD dissertation, Columbia University, 1968; Stephen G. Brush, *The Temperature of History* (New York: Franklin, 1978), chap. VI. Maxwell makes a similar remark in Document III-40.

c. Maxwell's footnote. "See Thomson on the Value of a Cubic Mile of Sunlight and the Density of the Aether Trans RSE 1854." That is William Thomson, "Note on the Possible Density of the Luminiferous Medium and on the Mechanical Value of a Cubic Mile of Sunlight," *Trans. R. Soc. Edinburgh* 21 (1854): 57–61 and *Phil. Mag.* 9 (1855): 36–40.

d. Maxwell used this passage from Berkeley quoting Torricelli several times when commenting on theories of matter. He paraphrased the passage in 1867 in his remarks to Tait, and his disagreements with the manuscript form of the first edition of Thomson and Tait's *Treatise on Natural Philosophy.* See Garber, Brush, and Everitt *Maxwell on Molecules and Gases*, 87-88. He returned to this criticism in his review of the second edition, Maxwell, "Thomson and Tait's *Natural Philosophy,*" *Nature* 20 (1879), reprinted in *Scientific Papers*, vol. 2, 776–785, and in the final sections of his treatise on electricity and magnetism where he critiques the notion of action at a distance and the theories of Neumann, Clausius, Reimann, and Weber. See Maxwell, *Treatise on Electricity and Magnetism* 2 vols (New York: Dover reprint of third edition), vol 2, chap. XXIII.

e. Maxwell's footnote "= Energy."

f. Maxwell's footnote "= Momentum in modern language."

g. The quotation is from James Bernoulli on the logarithmic spiral. See Carl B. Boyer, *A History of Mathematics* (Princeton NJ: Princeton University Press, 1985), 457-458.

h. For more details of Maxwell's ideas on dynamics and the history of mechanics see a series of letters to Lewis Campbell in 1877 in Campbell and Garnett, *Life*, 397–401.

11. Letter from Maxwell[a] to Peter Guthrie Tait, July 1868.

Cambridge University Library, Maxwell Collection, Maxwell-Tait Correspondence.

Pray direct and dispatch
enclosed to T.[b]

D^r T',[c]

I do not think it necessary to explore Bertrand[d] in the wake of H^2,[e] the latter seems to me tolerably unassailable. I have been transforming electromagnetic props.[f] into vortical ones e.g., "Two electric circuits act on one another with equal and opposite forces" becomes "Two ring vortices of any form affect each other's area so that the sum of the projections of the two areas on any plane remains constant." I notice what H^2 [e] says about conduction of heat and agree with him that something might be done from an investigation of irregularly distributed motion, but not with you that it could be got from a maximum question, not involving the connexion of the parts. But perhaps I do not understand you as you stand in print and probably you will yourself enlarge your standing ground.

Matthiessen[g] finds general similarity between the conductivities of metals for heat and for electricity. He also finds a wonderful equality in the effect of temperature on electric conductivity in all metals *except* **Iron** and Thallium.

But non-metallic bodies conduct heat proportionally much better than electricity and gutta percha, glass etc. conduct electricity the better the warmer. With respect to conductivity of irregular motions both Clausius and I make that of gases increase with temperature. I make it proportional to absolute T[h] (you make it in solids $\propto 1/T$) see dynamical theory of gases Phil Trans 1867.[i] I will write you about your treatise at earliest[j] but (1) I, personally, am satisfied with the book as a development of T' and as an account of a subject where the ideas are new and as I well know almost *unknown* to the most eminent scientific men. It is a great thing to get them expressed anyhow and I think you have done it intelligibly as well as accurately. *But* with respect to the bits of matter I sent you <should not> do you not think there are breaches of continuity between some, e.g., the statement about dynamical theories, and the context, if they do not actually contradict the context, at least the N.B.[k] Review part of it. If you disagree with anything of mine, out with it, for it is better to go into print having one opinion rather than with two opinions to throw the reader into perplexity.

(2) I shall see what case Clausius has.

(3) Who is Charles[ℓ] that I might believe in him. Is he B. Charles K and M, or is it he yclept[m] the Great of whom Ryan says of Almonte "His sight

He kept upon the standard, and the laurels
In fact and fairness are his earning, Charles."

I do not know where to find Charles, give reference.

JCM.

a. No date but from the subject matter of the letter and previous letters between Maxwell and Tait this was written in late July 1868.

b. T is William Thomson.

c. T' is Peter Guthrie Tait. Maxwell did not always use this designation consistently as T' became T in some of his later letters.

d. Joseph Louis Francois Bertrand (1822-1900).

e. H^2 always denoted Hermann von Helmholtz. Maxwell is referring to a dispute between Helmholtz and Bertrand on the flow of fluids. See Helmholtz, "Ueber discontinuirliche Flussigkeitsbewegungen," *Monat. Preuss. Akad., Berlin* (1868): 215–228, translated into *Phil. Mag.* 36 (1868): 337–346. Bertrand's theory was published as Bertrand, "Théorème relatif au mouvement le plus générale d'un fluide," *Comptes Rendus* 66 (1868): 1227–1230, and 67 (1868): 267–269. He commented on Helmholtz in Bertrand, "Observations nouvelles sur un mémoire de H. Helmholtz relatif à la théorie des fluides," *Comptes Rendus* 67 (1868): 469–472. Helmholtz had already commented upon Bertrand in Helmholtz, "Sur le mouvement plus générale d'un fluide," *Comptes Rendus* 67 (1868): 221–225, 754–757. Tait's interest in the dispute sprang from his work on vortices. Maxwell had read Helmholtz, "Ueber Integrale der hydrodynamischen Gleichungen welche den Wirbelbewegungen entsprechen," *J. reine angew. Math.*, 55 (1858): 25–55 and Tait's translation of it appeared in *Phil. Mag.* 33 (1867): 485–512. Tait subsequently published two papers on the subject of fluid motion, "On the Steady Motion of an Incompressible Perfect Fluid in two Dimensions," *Proc. R. Soc. Edinburgh* 7 (1870): 142–143, and "On the most General Motion of an Incompressible Perfect Fluid in two Dimensions," same journal, 7 (1870): 143–144.

f. Propositions.

g. August Matthiessen (1831–1870), who in the 1850's and 1860's studied the electrical, chemical and thermal properties of metals and their alloys. He wrote a long series of papers on the electrical and heat conducting powers of metals and alloys. The ones to which Maxwell refers are probably Matthiessen, "On the Thermoelectric Series," *Phil. Trans. R. Soc. London* 148 (1858): 369–382, and Matthiessen and Moritz von Bose, "On the Influence of Temperature on the Electrical Conducting Power of the Metals," same journal, 152 (1862): 1–27.

h. Maxwell's investigation into the conductivity of gases began with his first paper on the subject Maxwell, "Illustrations of the Dynamical Theory of Gases," *Phil. Mag.* 19 (1860): 19–32, and 20 (1860): 21–37; reprinted in *Scientific Papers*, vol. 1, 377–409, and Garber, Brush, and Everitt, *Maxwell on Molecules and Gases*, 285–318.

i. Tait's work on the thermal conductivity of solids was published over a decade. His preliminary results were presented to the British Association in 1869 and 1871. Tait's final report on this decade of work was published as Tait, "Thermal and Electric Conductivity," *Trans. R. Soc. Edinburgh* 28 (1878), reprinted in Tait, *Scientific Papers* (Cambridge: University Press, 1900), vol. 1, 363–392. The paper referred to is Maxwell, "The Dynamical Theory of Gases," *Phil. Trans. R.*

Soc. London 157 (1867): 49–88, reprinted in *Scientific Papers,* vol. 2, 26–78, and Garber, Brush, and Everitt, 419–472.

j. This paragraph refers to the proofs of Tait, *Sketch of Thermodynamics* (Edinburgh: Edmonston and Douglas, 1868) which Maxwell had undertaken to edit.

k. North British Review. Tait had published some of his remarks on the history of thermodynamics in this magazine. See, Tait, "The Dynamical Theory of Heat," *North British Review,* 1 (1864): 40–69, and "Energy," *North British Review,* 1 (1864): 337–368.

ℓ. Jacques-Alexandre-Cesar Charles (1746–1823), published nothing of significance in the theory of heat or gases and his work was unknown until Joseph Louis Gay-Lussac (1778–1850), published a critical account of Charles' method of measuring the thermal expansion of gases and his results. Charles had recognised that gases expanded equally when heated but Gay-Lussac did the systematic work necessary to establish ths quantitatively. See, Gay-Lussac, "Sur la dilatation des gaz et des vapeurs," *Ann. Chim.* 43 (1802): 137–175.

m. yclept means called.

12. Letter from Maxwell to Peter Guthrie Tait, August 3, 1868

Cambridge University Library, Tait Letters

Glenlair
1868 Aug. 3

Dear T'

I return the letter of T[a] and your prooves.[b] I have made a few marks on the letter which being in some measure repetitious of what I did before, you may class as remarks. But I shall remark also on what is to come. Have you given any evidence that when heat is communicated from one body to another by conduction, the one body loses as much as the other gains. This is the first axiom in measurement of a thing but especially when the thing is not a thing it requires proof.

I would put both bodies into a calorimeter and observe that the resultant effect was the same whether the one had had thermal intercourse with the other, or not.

e.g. a ball at 100°C will melt the same ultimate quantity of ice whether it be enclosed in a shell at 0°C or not. In the first case there is conduction which is eliminated in the second.

[2.] Have you a definition of temperature. *I* say "Temperature is the thermal state of a body considered with respect to its power of exchanging heat with other bodies."

[3.] In your §"What is Heat," you should eliminate the doctrine of Locke that Heat is a sensation or idea existing only in the mind and then only when it is felt. The heat in your book is only found in bodies, and is detected only by thermometers. I do not think that radiant heat is heat at all as long as it is radiant.[c]

With respect to my electrical treatise the Clarendon people have I believe accepted it.[d]

I am writing out the kinematic part (Ohms law and theory of Conduction).

For Electrolysis see (besides Faraday, Miller, etc.) Thomson Phil. Mag. 1851, expounded by Max[well] & Jenkin. Brit. Ass. Reports 1863 54 and followed up, very well for a Frenchman, by Georges Salet, *Laboratory* July 7, 1867. The subject looks temptingly simple but is not altogether so as yet.

My view of the energetics of magneto electric induction is to be found in the 1st part of my paper on the Field, and no where else.[f]

Rankine in a very short statement in the Phil. Mag. on Conservation has expressed several things very well about energy, force and effect.[g]

N. B. There are two kinds of Dissipation of E.[energy] one of which is possible in a strictly conservative universe provided it is ∞, namely the propagation of undulations to ∞ from vibrating body.

Stokes has just sent to the R. S. a paper on a sphere vibrating in S. H. [simple harmonics] in an elastic fluid. When the elasticity of the fluid is as great as in Hydrogen & the wave length therefore great as compared with the dimension of the vibrating regions of the sphere the motion is nearly that of an incompressible fluid and little sound is sent off to ∞.[h]

Thus a bell makes less noise in a mixture of air and H, than in the rarified air without the H. which is less dense than the mixture. The other dissipation is conversion to heat. Either kind causes a steady periodic driving power to produce a motion converging to a steady periodicity (without arbitrary functions). See a question about a vibrating disk in a tube Senate House 1867.[i]

Yours truly,

J. Clerk Maxwell

a. William Thomson.

b. Probably Tait's *Sketch of Thermodynamics* (Edinburgh: Edmonston & Douglas, 1868).

c. This remark might now seem unnecessary but in the first part of the nineteenth century the modern distinction between heat and "radiant heat" had not yet been accepted, see, Brush, *The Kind of Motion we call Heat* (1976), Chapter 9.

d. Maxwell, *Treatise on Electricity and Magnetism* (Oxford: Clarendon Press, 1873).

e. William Thomson, "On the mechanical theory of Electrolysis," *Phil. Mag.* 2 (1851): 429–444. J. C. Maxwell and Fleming Jenkin, "On the Elementary Relations of Electrical Quantities," *Rep. 33rd Meeting BAAS* (1863): 130–163. W. A. Miller, "On some Recent Research in Electrical Decomposition," *Chemist* 5 (1844): 348; "On Electro-Chemistry," *Rep. 27th Meeting BAAS* (pt. 2) (1857): 158–159. Georges Salet, "Affinity and Electricity," *Laboratory* 1 (1867): 248–250.

f. Maxwell, "A Dynamical Theory of the Electromagnetic Field," *Phil. Trans. R. Soc. London* 145 (1865): 459–512, reprinted in *Scientific Papers* vol. 1., 526–597.

g. Maxwell is probably referring to Rankine, "Conservation of Energy," *Phil. Mag.* 17 (1859): 250–253, where Rankine distinguishes force and energy explicitly.

h. George Gabriel Stokes, "On the Communication of Vibration from a Vibrating Body to a Surrounding Gas," *Phil. Trans. R. Soc. London* 158 (1868): 447–464.

i. This is a question set by Maxwell for the Cambridge Mathematical Tripos examination (held in the Senate House) for January 1867.

13. Letter from Maxwell to William Thomson, December 7, 1868

Kelvin papers, M30, Glasgow University Library

Glenlair
Dalbeattie
1868 Dec. 7

Dear Thomson

If you ever see White[a] would you ask him why he neither sends me a "wheel of life" nor answers my two letters asking him why he does not, during 7 weeks.

Can you give me a quite *elementary* Thermodynamics *Problem.* I have not the opportunity of setting bookwork so I am trying to get in a problem which shall be thermodynamically easy. If it is mathematically difficult it will be no fault in the eyes of the Cambridge men, though I would prefer it easy myself.

I think of asking a question about the deduction of the dilation per degree of temperature at constant pressure from the intrinsic energy expressed as a function of two variables either v & t or v and ϕ.

I think it is important to insert the wedge by the thin end and to "hold the eel of science by the tail."[b] Great mental inertia will be called into play if the new ideas are not fitted on to the old in a continuous manner.

If you could give me a hint of a problem I would thank you.

Yours truly
J. Clerk Maxwell

a. James White, the Glasgow instrument maker, with whom Thomson later formed a partnership (eventually Kelvin and White Limited) through which most of Thomson's scientific instruments were developed and marketed.

b. This refers to the process of introducing questions in physics, i.e., heat, electricity and magnetism, into the Cambridge Mathematical Tripos examination.

14. Letter from Maxwell to William Thomson November 16, 1869.

"Origins of Clerk Maxwell's Electrical Ideas, as described in Familiar Letters to William Thomson," ed. Joseph Larmor *Proc. Cambridge Phil. Soc.*, 32 (1936): 740–742.

Glenlair, Dalbeattie
16 Nov. 1869.

Dear Thomson,

Mr. Tatlock tells me you want to know about the conductivity of liquids for heat. The last thing on the subject is Prof. F. Guthrie[a] On the Thermal Resistance of Liquids (proc RS Jan 21, 1869). He states in his paper (I do not know if it is printed or to be printed) previous results. His experimental methods seem very good. His chief defect is that he never seems to know what he is going to measure. He works at the Royal Institution and has been so impregnated with radiant heat and otherwise Tyndallized[b] that he describes the specific resistance of a liquid to be the ratio of the quantity of heat *arrested* by the liquid to that arrested by an equal thickness of water.

He states his object to be "to determine the laws according to which heat travels by conduction through liquids" but he goes to work as if he wanted to find their absorption of radiant heat. The actual phenomena he observes are mainly phenomena of conduction.

He finds that heat gets through a millimeter of water in a minute very much better than any other liquid, their resistances being

Water	1	All solutions
Glycerine	3.84	of Salts
Acetic Acid glacial	8.38	increase
Alcohol	9.08	resistance
Oil of Turpentine	11.75	or diminish
Chloroform	12.10	conductivity

Of course mercury is not in this series. From my recollection of the paper the only results previously obtained were proofs that there is such a thing as true conduction of heat by liquids.

Perhaps Dr. Matthiessen[c] could give you information. He knows what conduction means and he collects the properties of bodies, and he will not tell you what he does not know.

For experiments on steady conduction an iron plate covered with copper on both sides might be useful as a measurer of the flow of heat. The two copper plates to be connected with a galvanometer. The galvanometer having great resistance compared with the plate, its indications may be trusted to give the mean electromotive force over the plate or the mean flow of heat per square inch. The value of the galvanometer readings must be found by regular calorimetric methods.

The compound plate being a much better conductor than most of the substances to be tried and also probably thinner the differences of temperature in the other bodies may be found by more direct methods of thermometry.

With respect to the stress in a medium arising from magnetism

$$\frac{d}{dx}\,p_{xx} + \frac{d}{dy}\,p_{yx} + \frac{d}{dz}\,p_{zx}$$

is the x-force on an element of the medium referred to unit of volume. If in that element there is neither electric current nor magnetization $X = 0$. If there is electric current the right expression comes out. If there is magnetization the case is more difficult because of the double def^n of force within a magnet.[d] If there is nothing but Ampère's currents and if these are recognized as currents then all is easy, and

$$X = \alpha\,\rho - \beta\,w + \gamma\,v$$

where α, β, γ are components of magnetic force u, v, w are components of currents, $\rho =$ density of imaginary magnetic matter

$$4\,\pi\,u = \frac{d\gamma}{dy} - \frac{d\beta}{dz}\,\text{etc} \qquad 4\,\pi\,\rho = \frac{d\alpha}{dx} + \frac{d\beta}{dy} + \frac{d\gamma}{dz}$$

Yours truly,

J. Clerk Maxwell.

a. Frederick Guthrie, "On the Thermal Resistance of Liquids," *Rep. 37th Meeting BAAS,* (1868): 15–17. An abstract of this work read to the Royal Society appeared in *Proc. R. Soc. London* 17 (1869): 234–236 is the paper to which Maxwell refers. The work was reported in full in Guthrie, *Phil. Trans. R. Soc. London* 159 (1869): 637–660. Frederick Guthrie must be distinguished from his brother Francis who interacts with Maxwell in Documents II-7, 8, 10 and 11. For details on Frederick see note a, Document II-7.

b. This refers to John Tyndall (1820–1893) who followed Faraday in 1863 as professor of natural philosophy at the Royal Institution. At the time Tyndall was working on radiant heat. His results were published as *Contributions to Molecular Physics in the Domain of Radiant Heat* (London, 1872) and was obviously affecting the conception of all heat phenomena in those who worked with him.

c. Augustus Matthiessen (1831–1870) who was working on the electrical and thermal properties of metals and their alloys.

d. definition. The "double" definition refers to the distinction between the vectors B and H, introduced by William Thomson in Thomson, "Mathematical Theory of Magnetism," *Phil. Trans. R. Soc. London* 141 (1851): 243–285, and developed further by Maxwell.

15. Letter from Maxwell to William Thomson, April 14, 1870

Cambridge University Library, Tait Letters

Glenlair
14 April 1870

[a]Dear Thomson,

The first table of dimensions that I know is in Fourier Theorie de Chaleur p. 157[b] and he makes frequent use of it. The dimensions are in length, time, and temperature. If you define the unit of heat as that which raises unit of volume of the substance in its actual state one degree then in the equation $\frac{dv}{dt} = -k\nabla v$ k is the conductivity with this unit of heat.

But to my question—In an infinite solid heated suddenly at the origin and then left to itself the temperature at a distance r after a time t is

$$v = A\,t^{-3/2}\,e^{r^2/4kt} \qquad p.\ 478$$

Hence if v_0 is the initial temperature at the point $a\,b\,c$ and v the actual temperature at time t at $x\,y\,z$, distant r from $a\,b\,c$

$$v = \iiint (da\,db\,dc\,v_0(kt)^{-3/2}\,e^{-r^2/4kt}$$

This tells us completely what the state of the infinite solid will be at any future time if we know its initial state. The temperature of every point is the mean of the original temperatures of all the points, the weight attributed to each point in taking the mean being $e^{-r^2/4kt}$. This all stands to reason.

Now suppose we have observed the actual state of the infinite solid and wish to deduce its previous state at a time $-t$, can we adapt Fourier's solution. In the formula there is the awkward quantity $(kt)^{3/2}$ which is objectionable when t is -ve. In the method of taking means $e^{r^2/4kt}$ is a very respectable quantity, only it <becom> gives the greatest weight to the most distant points and it does this *especially* when t is small,† [c] that is just before the time of observation, when if ever† [d] we ought to be able to deduce the previous state from the state of neighbouring points. I have not found any attempt at this inverse problem in Fourier. Has it been done?† [e] or shown to lead to insuperable difficulties? and if you do not know about it who does? I do not know any (Joseph) Fourierists.[f] He is principally known as having invented Fourier's Theorem, the ratio of which to a 3 day problem is an unknown quantity.[g]

Of course if you cut up the distribution of temperature into harmonics, you can work back till the harmonic series becomes divergent. If you invent <an arbitrary> a function to express the present temperature the divergence usually comes on as soon as you put t negative.[h] It is only when you write down an harmonic series of set purpose that you can avoid this.

It is because you have attended to the historical problem of the conduction of heat that I ask you. Everybody else inclines to the prophetical problem which is much easier.[i]

I am boiling all this down for my chapter on Conduction with pictures of the diffusion of heat and salts and gases.[j] The thermal view of an harmonic is a distribution of heat which cools down without altering the ratios of the temperatures of the parts.

To prove that every figure has a fundamental harmonic with a series of higher harmonics in order would be probably a stiff business.[k] These harmonics die away, the highest most rapidly, and at last, however the body is originally heat[ed] (provided the fundamental harmonic is not absent altogether) the distribution approximates to that of the fundamental harmonic.

I have arranged the paraffin between two prisms to determine its refractive index but have not got the angle measure till I Have a salt wick going.[ℓ]

I am greatly surprised that Joules combination of levers for magnifying magnetic disturbances does any good. I should have thought that if a microscope is to be used at all there should be no magnification except that done by levers of light which do not get shaken or affected by gravity or viscosity.

Yrs

$$\frac{dp}{dt}$$

a. Note written at upper left corner: "When do you go to Cambridge? Thanks for the Diffusion of Gases." This and the other notes are probably by Thomson.

b. Jean Baptiste Joseph Fourier (1768–1830), *Théorie Analytique de la Chaleur* (Paris: Chez Firmin Didot, 1822), trans. and ed. by A. Freeman as, *The Analytical Theory of Heat* (New York: Dover reprint of the Cambridge University Press 1878 edition, 1955).

c. Note written upside down at the top of the page: † "This is the analytical expression of the impossibility to find a physical antecedent of an arbitrary distribution. See an Essay " De Caloris Motu per Terrae Corpus" read in the Faculty room of the old College in 1846, and now to be met through all space (if there is no limit to the greatest velocity of an individual molecule of gas) combined with oxygen." If Thomson is referring to his early papers on Fourier they are on the mathematical aspects of Fourier's work only. See Garber, "Reading Mathematics, Constructing Physics: Fourier and his Readers, 1822-1850," in press. Another note is written along the right-hand margin of the page: † "Loc C M D Notes on Certain Points in the Theory of Heat or some other equally appropriate & suggestive title, about year 1864." Assuming that Thomson is referring to his own papers on Fourier locating this paper depends on deciphering C M D and unscrambling the date. C M D may well be the *Cambridge and Dublin Mathematical Journal.* However, it had become *The Quarterly Journal of Mathematics* in the 1850's. The paper is probably Thomson, "Note on Some Points in the Theory of Heat," *Cambridge Mathematical Journal* 4 (1844): 191-192. See Maxwell, *Theory of Heat* (London: Longmans Green, 1872), chap. XVIII.

d. Note at the bottom of the page: † "but <it never> this is just what cannot be done in a distribution (such as $v = e^{-r^2/t}t^{-1/2}$ when $t = 0$) which is essential in time."

e. The second remark quoted above as note c may be intended to apply here.

f. The term "Fourierist" was usually applied to followers of the French socialist Charles Fourier (1772–1837).

g. The remark on Fourier's theorem and a "3 day problem" refers again to the Mathematical Tripos and the strictly mathematical nature of that examination without reference to the physical deductions being discussed in this letter.

h. Note written above line: "Hear hear; see C M D"

i. Note added to this sentence: "because of the irreversibility of dissipation."

j. Maxwell is referring to his *Theory of Heat* chap. XVIII.

k. Note added after this sentence: "Not so very. It was a favorite ((unproved)) proposition of Liouville in 1846 Jan Feb March."

ℓ. The "salt wick" was probably used to produce the sodium spectral lines. See for example Stokes, *Mathematical and Scientific Papers* (Cambridge: Cambridge University Press, 1904), vol. 4, 134. Maxwell seems to be setting up an experiment to measure the refractive index of liquid paraffin to check his prediction from the electromagnetic theory of light that the refractive index is proportional to the square of the dielectric current.

16. Letter from Maxwell to John William Strutt, December 6, 1870

Robert John Strutt, fourth Baron Rayleigh, *Life of John William Strutt Third Baron Rayleigh* (Madison, Wisc.: University of Wisconsin Press, 1968) 47–48.

Glenlair, Dec. 6th, 1870.

Dear Strutt,—

If this world is a purely dynamical system, and if you accurately reverse the motion of every particle of it at the same instant, then all things will happen backwards to the beginning of things, the raindrops will collect themselves from the ground and fly up to the clouds, etc, etc, and men will see their friends passing from the grave to the cradle till we ourselves become the reverse of born, whatever that is. We shall then speak of the impossibility of knowing about the past except by analogies taken from the future and so on. The possibility of executing this experiment is doubtful, but I do not think it requires such a feat to upset the 2nd law of thermodynamics.

For if there is any truth in the dynamical theory of gases, the different molecules in a gas of uniform temperature are moving with very different velocities. Put such a gas into a vessel with two compartments and make a small hole in the AB wall about the right size to let one molecule through. Provide a lid or stopper for this hole and appoint a doorkeeper very intelligent and exceedingly quick, with microscopic eyes, but still an essentially finite being. Whenever he sees a molecule of great velocity coming against the door from A into B he is to let it through, but if the molecule happens to be going slow, he is to keep the door shut. He is also to let slow molecules pass from B to A but not fast ones. (This may be done if necessary by another doorkeeper and a second door.) Of course he must be quick, for the molecules are continually changing both their courses and their velocities.

In this way the temperature of B may be raised and that of A lowered without any expenditure of work, but only by the intelligent action of a mere guiding agent (like a pointsman on a railway with perfectly acting switches who should send the express along one line and the goods along another). I do not see why even intelligence might not be dispensed with and the thing made self-acting.

Moral. The 2nd law of thermodynamics has the same degree of truth as the statement that if you throw a tumblerful of water into the sea, you cannot get the same tumblerful of water out again.

Many thanks for your two papers; the electro-magnetic one has just come in time for me as I am at that part of the subject.[a] Have you seen Helmholtz on the Equations of Motion of Electricity in conductors at rest.[b] It is a very powerful paper.

I have been doing Weber's theories of magnetic and diamagnetic induction. There are some mistakes in integration but the theory of moveable magnetic molecules is of great use in explaining the phenomena, especially all about magnetization, demagnetization and remagnetization.

Yours truly,

J. Clerk Maxwell

I have improved my book[c] by means of three of your suggestions:–
1. Wrong sign in an equation about M.
2. Discussion of terms in kinetic energy involving products of currents and ordinary velocities.
3. Magnetization as a test of maximum current and as a cause of anomalies in galvanometry.
If therefore any more occur to you and you send them I shall be thankful.

a. The electromagnetic paper to which Maxwell refers is probably, Rayleigh, "On some Electromagnetic Phenomena considered in Connexion with the Dynamical Theory," *Phil. Mag.* 38 (1869): 1–15.

b. Hermann von Helmholtz, "Ueber die Bewegungsgleichungen der Electricität für ruhende leitende Körper." *J. reine angew. Math.* 72 (1870): 57–129.

c. Maxwell, *Treatise on Electricity and Magnetism* (Oxford: Clarendon Press, 1873) vol. 2, secs. 572–576 for the second of Strutt's suggestions.

17. Postcard from Maxwell to Peter Guthrie Tait, February 15, 1871[a]

Cambridge University Library, Maxwell Collection, Maxwell-Tait Correspondence.

D[r] T' You must explain the magnitude of the operators V and S. I supposed $VS = 0 = SV$, $VV = 1, SS = 1$.[b] ∇^c is a a mere operator, like d/dx, of dimensions -1 in length. What is $\int V.(V\,d\rho\,\nabla)\sigma$ compared with $\int V(d\rho.\nabla\sigma)$? Have you been introduced to Virial and Ergal? Ergal is an old Friend = Potential of a system in itself or $\sum m_1 m_2\,\phi_{12}$. Virial is $\sum m_1 m_2 r_{12}\,d\phi_{12}/dr_{12}$. He appears in dp/dt on Reciprocal Figures etc. p. 13 near bottom.

Clausius is now working along with these eminent artistes at the 2nd law of $\theta\Delta^{cs}$ [thermodynamics], but as far as I see they have not yet furnished him with the dynamical condition of the equilibrium of temperature. This is got by the celebrated principle of Assumption and Resumption. Pray read line 3 on opposite side after the Lion and the Unicorn.

Yrs $\dfrac{dp}{dt}$

a. Dated from the postmark.

b. This probably is part of Maxwell's attempts to use quarternions in his *Treatise on Electricity and Magnetism.* In quaternions if σ is a vector and ∇ is the operator $i\frac{d}{dx} + j\frac{d}{dy} + k\frac{d}{dz}$ where (i, j, k) are the unit vectors in the x, y, z directions, then $\nabla\sigma$ has two parts, a vector and a scalar part. In quarternion notation the vector part is denoted by $V\nabla\sigma$ which Maxwell called the Curl of σ and is equivalent to our cross product $\nabla \wedge \sigma$. The scalar part, denoted by $S\nabla\sigma$ which Maxwell called the convergence is equivalent to the dot product $\nabla \cdot \sigma$.

c. ∇ is the vector operator defined above. The two integrals are equivalent.

d. Virial and Ergal were introduced into gas theory by Rudolph Clausius in Clausius, "Über einen auf die Wärme anwendbaren mechanische Satz," *Sitz. Bonn* (1870): 114–119 and *Ann. Phys.* 141 (1870): 124–130, translated in *Phil. Mag.* 40 (1870): 122–127. Maxwell's paper is Maxwell, "On Reciprocal Figures, Frames and Diagrams of Force," *Proc. R. Soc. Edinburgh* 7 (1872): 53–56 and *Trans. R. Soc. Edinburgh* 26 (1872): 1–40 for which he was awarded the Keith Prize of the Royal Society of Edinburgh. The theorem on page 13 of this paper in the *Trans.* is on the equilibrium of a point under a system of attractions and repulsions. As Maxwell states the theorem it refers only to the static case in which $\sum Rr = 0$, where R is the force acting on the body.

e. This is a play on Maxwell's initials JCM in the form of the thermodynamic expression,

$$\frac{dp}{dt} = JCM$$

where p is the pressure and t the temperature, J Joule's constant, C Carnot's function and M the rate at which heat must be supplied per unit increase in volume when the temperature is constant. In this form it appeared in William Thomson's early papers and in Tait's *Sketch of Thermodynamics* as a form for the second law.

18. Postcard from Maxwell to Peter Guthrie Tait, May 2, 1871[a]

Cambridge University Library, Maxwell Collection, Maxwell-Tait Correspondence.

O T' Can you tell me the tale of the Florentine thermometers,[b] one of which is in your Apparatus room having glass beads for degrees? Who made them? of what date? Were any ancient observations made with them which have been translated into modern degrees since the discovery of the instrument. When were they lost? Who discovered them again and who wished they had not been discovered? Who gave one to the Edin^h Nat. Phil.[c] Is there anything in print about it? Information sent to dp/dt will receive attention.

Again Who is the author of the theorem $\iint s.\nabla\frac{1}{\rho}\, U\nu\, ds = 4\pi$ or 0 according as a closed surface encloses the origin or not. Is it Gauss or Stokes? I mean in its Cartesian form.

To conclude. Is $Xdx + Ydy + Zdz$ in certain cases, a *complete, exact, total* or what else? differential.
Which is the correct word.
Lastly. I thank you and praise you for turning me from the system of the hop to that of the vine.[d] I have perverted the whole of electromagnetics

to said. When you send me my proofs I will send you correct cards of the book. But tell me about thermometers at once if you can.

$$\frac{dp}{dt}$$

a. Dated by the postmark.

b. Galileo invented this thermometer in Florence in 1597.

c. The professor of Natural Philosophy at Edinburgh University. The holder of this office was Tait at the time of this postcard.

d. On changing from "the system of the hop to that of the vine" Maxwell refers to changing froma left-handed to a right-handed system of coordinates. See Maxwell, *Treatise on Electricity and Magnetism,* vol. 1, sec. 23, note: "Professor W. H. Miller has suggested to me that as the tendrils of the vine are right-handed screws and those of the hop left-handed, the two systems of relations in space might be called those of the vine and the hop respectively." William Hallowse Miller (1801–1880) was a crystallographer and should be distinguished from William Allen Miller (1817–1870) the chemist referred to in Document III-12 above.

19. Letter from Maxwell to James Thomson,[a] July 13, 1871.

Queens University Belfast.

Glenlair
Dalbeattie,
13 July 1871

Dear Sir,

In a book on heat which I am in the middle of I have given a short account of Dr Andrews'[b] researches, about which we had some conversation at Glasgow. I have since heard his lectures at the Royal Institution[c] and seen a little of the phenomena. My account of the facts and theories is therefore derived partly from Andrews and partly from you and is considerably modified in the process of boiling down. I should like to hear from you if the proof I sent gives a fair account of what Andrews and you have done and more particularly if you have told me anything in confidence that you have not yet published, mark it out. I hope however that you will publish some of what you told me for the speculation seemed of the fertile kind.

Sir William's relations between capillarity, curvature and pressure of vapour seem to me to have connection with the retardation of boiling and of condensation.[d]

The next difficulty is What determines the true boiling temperature of the steam which is found to be so constant?

If I do not hear from you in 10 days I shall suppose you are not at home and use my own discretion.

Remember me kindly to M[rs] Thomson and believe me,

Yours truly,

J. Clerk Maxwell.

Shall you be at Edinburgh at the B. A.?[e]

a. James Thomson (1822–1892), elder brother to William Thomson, was professor of Civil Engineering at Queens College Belfast, from 1857 to 1877, then became professor of Applied Mechanics at Glasgow University.

b. Maxwell is referring to *Theory of Heat* (London: Longman, 1872) in which he described Andrews' experiments on the liquefaction of carbon dioxide.

c. Thomas Andrews lectured at the Royal Institution on the continuity of the gaseous and liquid states, June 2nd 1871.

d. William Thomson, "On the Equilibrium of Vapour at a Curved Surface of a Liquid," *Phil. Mag.* 42 (1871): 448–452, read February 7th 1870 to the Royal Society of Edinburgh, see, *Proc. R. Soc. Edinburgh* 7 (1872): 63–68. William Thomson had been knighted in 1866 for his services in the successful laying of the first Atlantic telegraph cable.

e. The British Association for the Advancement of Science which met at Edinburgh in 1871. Maxwell mentions his text again in a letter to Tait, November 2nd, 1871. The letter is about quarternions but at the end Maxwell remarks "There appears to be a desire for thermodynamics in these regions [Cambridge] more than I expected, but there are some very good men to be found." Quoted from, Knott, *Life of Peter Guthrie Tait,* 101.

20. Letter from James Thomson to Maxwell, July 21 1871.[a]

Queen's University Belfast.

17, University Square,
Belfast, 21 July 1871

My dear Maxwell,

I have been a little longer writing to you than I hoped, having been away from home for two long days on a survey and much engaged at home on another day or two.

In the proof sheets which you sent me and which I now return,[b] there are a few sentences which I think would require a little amendment as I think a few of them scarcely attribute to D[r] Andrews the credit which is due to him for the revolution he has brought about in people's views in general on the relations of the gaseous and liquid states of matter.[c]

I think in justice to him as well as to make the historical allusions in your book correct in reference to this matter it ought to be stated that it was Dr Andrews who showed that the liquid and gaseous states are continuous: and that to him is due the true explanation of Cagnard de la Tour's experiment.[d] In your proofs I have indicated a few proposed alterations as suggestions for your consideration: but of course with the intention that you should do whatever you will think right and suitable yourself.

Then as to the passage relating to my suggestions (pages 122 and 123), I think you will find on consideration that what you have written will not hold good altogether. I think it is *not* possible for the substance at the pressure indicated by B to pass into the gaseous state and that if the liquid is in contact with its vapour at this pressure it is really *found that the liquid will not* begin to pass into the gaseous state. On the contrary I think under the circumstances stated it will all go down to the liquid state.[e]

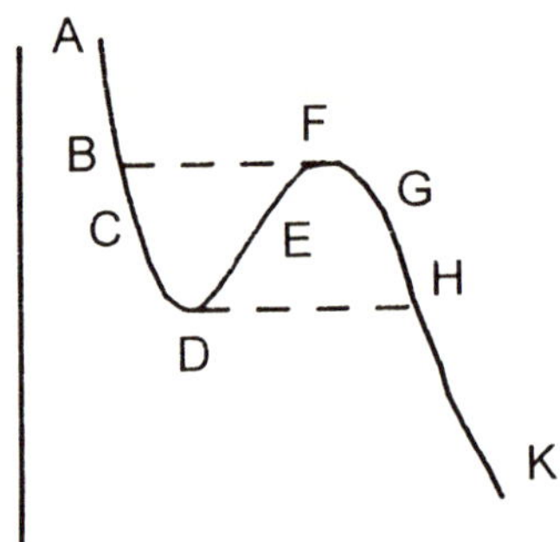

Again:- If there be any drops of liquid in the vessel when the pressure is that belonging to D and H, I think condensation will not *now* begin but on the contrary all the liquid will evaporate into the gaseous state. Then in respect to the question in your letter where you say "The next difficulty is:– *What determines the true boiling temperature of the steam which is found to be so constant?* I will answer rather a corresponding question which is quite to the same effect but suits better to the diagram before us in which the vertical ordinates represent pressures, and the horizontal axis presents volumes:– *What determines the true boiling pressure of the steam which is found to be as constant for any given temperature?*

Reply:- There is just one intermediate point of pressure between the pressure at F and the pressure at D, at which the liquid and its gas can be present together in contact with one another; and that is the boiling, or rather evaporating or condensing pressure, for the temperature to which the curve belongs. In using the name boiling pressure here we must understand not the very uncertain and variable pressure at which bubbles would form themselves in a continuous liquid when the boiling takes place with bumping; but the absolutely definite pressure at which the liquid and its gas can be present together in contact with each other when either evaporation or condensation may be going on, or no change either way may be taking place.

< *I* think there is no *"lingering"* in the liquid nor in the gaseous condi-

tion, when the two conditions of the same substances are present together, and the pressure is being altered while the temperature is fixed, (or equally when temperature is altered while pressure is fixed).>

<What determines the true boiling temperature of the steam which is found to be constant? Thus:–There is just one intermediate point of pressure, between the pressure at F and the pressure at D, at which the liquid and its gas can be present together in contact with one another and that is the boiling or rather evaporating *(not ebullition)* or condensing pressure for the temperature to which the curve belongs.>

I think there is no "lingering" in the liquid nor in the gaseous conditions when the two conditions of the same substance are present <together>.

I enclose to you a manuscript copy of my paper recently sent to the Royal Society on the subject and which I presume will very soon be published in the *Proceedings.*[f] It will explain my ideas I think quite clearly. I suppose it will be published before your book will come out; and I think if you would wish to refer to it as a published paper permission may easily be obtained to do so, or information as to the date or page at which it will appear in the Proceedings may easily be obtained by applying to the Secretaries of the Royal Society either of Prof. Stokes, Lensfield Cottage, Cambridge or to Walter White Esq., Royal Society, Burlington House, London. You might write or I might if you wish.

I send to you by this post a report of Dr. Andrews' lecture at the Royal Institution at which I think you say you were present; and also I send you a copy of "Nature" containing an article abridged from an essay by me giving an account of Dr Andrews' researches and conclusions.[g]

<We> Mrs Thomson and I hope to be at the Brit. Assoc^{n}.; and hope to see you there and with kind regards I am,

Yours truly,

James Thomson

a. This is a copy of a letter sent to Maxwell in James Thomson's hand and appears to be a rough draft rather than the copy of a finished letter. For details about James Thomson see note a, Document III-19.

b. These were proofs of Maxwell, *Theory of Heat* (London: Longman, 1872).

c. Thomas Andrews (1813–1885) whose experiments on carbon dioxide were first published in W. A. Miller, *Elements of Chemistry* (London, J. W. Parker and Sons, 1863). Andrews's energies went equally into teaching and administration as well as university politics and his research suffered. However, his experiments on the liquefaction of gases were presented as the Bakerian Lecture in 1869 and further work again in 1876. See Andrews, "On the Continuity of the Gaseous and Liquid States of Matter," *Phil. Trans. R. Soc. London* 159 (1869): 575–590; "On the Gaseous State of Matter," same journal 166 (1876): 421–449, and,

"On the Properties of Matter in the Gaseous and Liquid States under Various Conditions of Temperature and Pressure," same journal 178 (1888): 45–56, his most important papers on the subject.

d. Charles Cagnard de la Tour (1777–1859) attempted to vaporize liquids in hermetically sealed containers, maintaining a constant ratio between the volume of the liquid and that of the vessel. He found that above a certain temperature all liquids could be vaporized and measured the temperatures and pressure for this critical state. His results were published in a series of papers. See Cagnard de la Tour, "Exposé de quelques resultats obtenus par l'action conbinée de la chaleur et de la compression sur certains liquides tel que l'eau, l'alcool etc.," *Ann. Chim.* 21 (1822): 127–132; "Sur les effets qu'on obtene par l'application simultanée de la chaleur et de la compression à certaine liquides," same journal 23 (1823): 410–415.

e. The diagram is a P-V diagram with P the vertical axis, although the volume axis is not marked it clearly increases to the right, horizontally. AB represents the liquid phase and HK the gaseous phase of the substance.

f. The paper is James Thomson, "Considerations on the Abrupt Change at Boiling or Condensing in reference to the Continuity of the Fluid States of Matter," *Proc. R. Soc. London* 20 (1871): 1–8.

g. An account of Andrews' lecture at the Royal Institution, June 2nd, 1871 was published as, Andrews, "On the Gaseous and Liquid States of Matter," *Notice of the Proceedings of the Royal Institution of Great Britain,* (1871), reprinted in *The Scientific Papers of Thomas Andrews, together with a Memoir by P.G. Tait and Alexander Crum Brown* (London: Macmillan and Co., 1889). James Thomson's notice of the lecture appeared as Thomson, "On the Gaseous and Liquid States of Matter," *Nature* 4 (1871): 186–188.

h. British Association for the Advancement of Science which met in Edinburgh in 1871.

21. Letter from Maxwell to James Thomson, July 24 1871.

Queens University Belfast.

Glenlair,
Dalbeattie
24 July 1871

My Dear Thomson,

Many thanks for your letters, corrections and papers.[a] Pray give my thanks to Dr Andrews for his lecture. I have not had time to digest all you have sent. May I keep your copy of the R.S.[b] paper till I see you in Edinburgh? You are certainly right that there is a definite pressure at which evaporation takes place at a given temperature (putting aside capillary phenomena). But I never could see what that pressure must be with reference to your continuous curve. I think, however, that the following method determines it

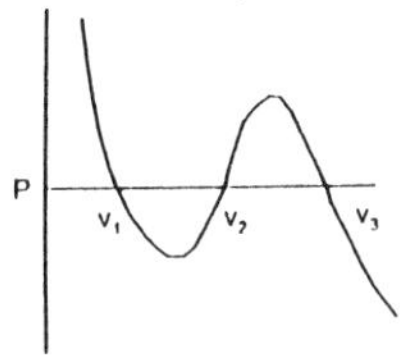

For given value of P the pressure and θ the absolute temperature. Let there be three values of V, the volume

V_1 the liquid volume
V_2 the unstable volume } for unit of mass
V_3 the gaseous volume

Let x be the mass of the liquid
Let y be the mass of the gas
$x + y = \text{const[ant]}$.
Let ϕ_1, ϕ_2, ϕ_3 be Rankine thermodynamic functions for the 3 states.
In general, if V and ϕ be made to vary the work done by the fluid is $P\,dV$.
The heat absorbed measured dynamically $\theta\,d\phi$.
Energy developed or emitted $P\,dV - \theta\,d\phi$
Hence if any variation takes place in the mixed mass of liquid and gas
Energy emitted $= x(P_1\,dV_1 - \theta_1\,d\phi_1) + y(P_3\,dV - \theta_3\,d\phi_3)$.
If the variation arises from a chance of pressure dP while the temperature θ remains the same then P and θ are the same throughout the expression and energy $= dP\,x\,(P\,\frac{dV_1}{dP} - \theta\,\frac{d\phi_1}{dP}) + y\,(P\,\frac{dV_3}{dP} - \theta\,\frac{d\phi_3}{dP})$, being constant in the differentiation. Next let x vary then $dx + dy = 0$, and the energy emitted is

$$dP\,dx\{P\,\frac{dV_1}{dP} - \theta\,\frac{d\phi_1}{dP} - (P\,\frac{dV_3}{dP} - \theta\,\frac{d\phi_3}{dP})\}$$

Now x will tend to increase or condensation will occur when this quantity is positive and to diminish, indicates evaporation, when it is negative. Hence the equilibrium of vapour and liquid occurs when

$$p\,\frac{dV}{dP} - \theta\,\frac{d\phi}{dP} \qquad (A)$$

is the same for the liquid and the gaseous states. But by thermodynamics $d\phi/dP = -dV/d\theta$ so that we may write the expression

$$P\frac{dV}{dP_{(\theta\,const)}} + \theta\frac{dV}{d\theta_{(P\,const)}} \qquad (B)$$

Here $-dV/dP_{(\theta\,const)}$ denotes the compressibility of unit mass at constant temperature and $dV/d\theta_{(P\,const)}$ the dilativity of unit mass at constant pressure. The expression (B) must be the same for the liquid and the gas at the point of equilibrium.

Take the case of steam. Suppose it obeys Boyle and Charles in the gaseous state then PV/θ is constant and $P\,\frac{dV}{dP} + \theta\,\frac{dV}{d\theta} = 0$. In the liquid state therefore $P\frac{dV}{dP}+\theta\frac{dV}{d\theta}=0$. In fact both the terms are very small and of opposite sign, but the second is the largest in most cases. Hence for the vapour $P\frac{dV}{dP}+\theta\frac{dV}{d\theta}$ is positive. We know that the compressibility is greater than that given by Boyle's law. Hence, a fortiori the dilation at constant pressure must be greater than that given by Charles' law. In fact the <presence of vapour> dilation of superheated vapours is much greater than that of permanent gases.

We may also write the expression (B)

$$\frac{dV}{dP_{(\theta\ const)}}\left(P-\frac{dP}{d\theta_{(V\ const)}}\right)$$

or still more simply $dE/dP_{(\theta\ const)}$.
This condition determines the value of P at which the liquid and its vapour can coexist, so that the condition is

$$\frac{d}{dP}(E_1-E_3)=0$$

or E_1-E_3 a maximum for the given value of the temperature, where E_1 and E_3 denote the intrinsic energy of unit of mass of the gas and the liquid at the same temperature and pressure.

I have made use of writing to you in order to get this matter into shape as I am very busy at present about electricity, so I was by no means clear about the matter when I began writing. I now see that E, the intrinsic energy is the best thing to look at. In a perfect gas it is constant for the same temperature whatever be the pressure, and is proportional to the temperature. In an incompressible liquid it is independent of the pressure and is proportional to the temperature and the specific heat.

In real liquids the part depending on temperature is still commonly greater than that depending on pressure. I think that this leads to a method of determining, from a complete knowledge of the continuous isothermal curve and the consecutive isothermal curves, the points of those curves which correspond to the state of equilibrium between the liquid and its vapour, and shows that I was wrong in supposing that there could be anything indefinite about it.

Have you stated anything about this in the paper you are going to print?

I remain

Yours very truly,

J. Clerk Maxwell

a. This refers to a lecture Andrews delivered at the Royal Institution, June 2nd, 1871. See note g, Document III-21.

b. Royal Society. This refers to a copy of a paper James Thomson and sent Maxwell previously that Thomson had submitted to the Royal Society for publication. See Thomson, "Considerations on the Abrupt Changes at Boiling or Condensing in reference to the Continuity of the Fluid State of Matter," *Proc. R. Soc. London* 20 (1872): 1–8, *Phil. Mag.* 43 (1872): 227–234.

22. Excerpts from Maxwell "Heat Engines"

Maxwell, *Theory of Heat* (London: Longman, 1871) chapter VIII, "Heat Engines," 152–153.[a]

Carnot's principle, then, is that the efficiency of a reversible engine is the greatest that can be obtained with a given range of temperature.

For suppose a certain engine, M, has a greater efficiency between the temperatures S and T than a reversible engine N, then if we connect the two engines, so that M by its direct action drives N in the reverse direction, at each stroke of the compound engine N will take from the cold body B of the heat h, and by the expenditure of work w give to the hot body A the heat H.

The engine M will receive this heat H, and by hypothesis will do more work while transferring it to B than is required to drive the engine N. Hence at every stroke there will be an excess of useful work done by the combined engine.

We must not suppose, however, that this is a violation of the principle of conservation of energy, for if M does more work than N would do, it converts more heat into work in every stroke, and therefore M restores to the cold body a smaller quantity of heat than N takes from it. Hence, the legitimate conclusion from the hypothesis is, that the combined engine will, by its unaided action, convert the heat of the cold body B into mechanical work, and that this process may go on till all the heat in the system is converted into work.

This is manifestly contrary to experience, and therefore we must admit that no engine can have an efficiency greater than that of a reversible engine working between the same temperatures. But before we consider the results of Carnot's principle we must endeavour to express clearly the law which lies at the bottom of the reasoning.

The principle of the conservation of energy, when applied to heat, is commonly called the First Law of Thermodynamics. It may be stated thus: When work is transformed into heat, or heat into work, the quantity of work is mechanically equivalent to the quantity of heat.

The application of the law involves the existence of the mechanical equivalent of heat.

Carnot's principle is not deduced from this law, and indeed Carnot's own statement involved a violation of it. The law from which Carnot's principle is deduced has been called the Second Law of Thermodynamics.

Admitting heat to be a form of energy, the second law asserts that it is impossible, by the unaided action of natural processes, to transform any part of the heat of a body into mechanical work, except by allowing heat to pass from that body into another at a lower tempera- ture. Clausius, who first stated the principle of Carnot in a manner consistent with the true theory of heat, expresses this law as follows:–

It is impossible for a self-acting machine, unaided by any external agency, to convey heat from one body to another at a higher temperature.

Thomson gives it a slightly different form:–

It is impossible, by means of inanimate material agency, to derive mechanical effect from any portion of matter by cooling it below the temperature of the coldest of the surrounding objects.[b]

By comparing together these statements, the student will be able to make himself master of the fact which they embody, an acquisition which will be of much greater importance to him than any form of words on which a demonstration may be more or less compactly constructed.

Suppose that a body contains energy in the form of heat, what are the conditions under which this energy or any part of it may be removed from the body? If heat in a body consists in a motion of its parts, and if we were able to distinguish these parts, and to guide and control their motions by any kind of mechanism, then by arranging our apparatus so as to lay hold of every moving part of the body, we could, by a suitable train of mechanism, transfer the energy of the moving parts of the heated body to any other body in the form of ordinary motion. The heated body would thus be rendered perfectly cold, and all its thermal energy would be converted into the visible motion or some other body.

a. These pages contain Maxwell's initial statements of the first and second laws of thermodynamics.

b. Both these statements of the second law of thermodynamics are actually taken from William Thomson, "On the Dynamical Theory of Heat," *Phil. Mag.* 4 (1852): 8–21. Clausius's statement is Thomson's version of Clausius and is on page 14. Thomson's own version is on page 13. Clausius's own statements were not yet put into the form of a theorem but his work in thermodynamics was based on the explicit understanding that whenever heat is used to produce work, then heat flows from the hotter body to a colder one. See Clausius, "On the Motive Force of Heat, and the Laws Regarding the Nature of Heat itself which are deducible therefrom," *Phil. Mag.* 2 (1851): 1–20, 102–119, 102–103.

[The following is Maxwell's attempt to incorporate Rankine's function into his account of the two laws of thermodynamics.]

Maxwell, *Theory of Heat,* (1871): 162–163.

It has also been proposed to define temperature so that equal increments of heat applied to a standard substance will produce equal increments of temperature. This method also fails to give results consistent for all substances, because the specific heats of different substances are not in the same ratio at different temperatures.

The only method which is certain to give consistent results, whatever be the substance employed, is that which is founded on Carnot's Function, and the most convenient form in which this method can be applied is that which defines the absolute temperature as the reciprocal of Carnot's Function. We shall see afterwards how a comparison can be made between the absolute temperature on the thermodynamic scale and the temperature as indicated by a thermometer of a particular kind of gas.

To draw an adiabatic line through any point requires only experiments on the substance. The series of adiabatic lines in our diagram is defined so that when the substancc cxpands at a certain temperature T, the same quantity of heat H causes it to pass from one adiabatic line to the next. If we make this quantity of heat measured dynamically (that is in units of work) equal numerically to T, the absolute temperature of the standard isothermal line, then, as we have already shown,the area of every quadrilateral contained between two consecutive isothermals and two consecutive adiabatics will be $CH = T_1/T_2 = I$. To measure any area on the diagram we have only to count the number of these quadrilateral contained in it. We must then mark the adiabatic lines, beginning with the line of no heat, with the indices $0, 1, 2, 3$, &c., up to ϕ, the index or number of the line. This quantity,ϕ, is called by Rankine the Thermodynamic Function.

It is probably impossible to deprive a body entirely of heat. If, however, we could do so, its temperature would be absolute zero, and the relation between its volume and its pressure at this temperature would be given by an isothermal line called the line of absolute cold. Of course we have no experimental data for determining the form of this line for any actual substance. We can only regard it as a limit beyond which no line in the diagram can extend.

[The following is Maxwell's initial understanding of Clausius's entropy and Thomson's dissipation of energy.]

Maxwell, *Theory of Heat,* "On the Available Energy of a System of Bodies," 185–188.

Let us suppose that a number of different substances are placed in a confined region, the volume of which is V, and that no heat is allowed to escape from this region, though it may pass from one body to another within it.

Let us also suppose that we are able to make use of all the work done by the expansion of the different substances. Since they are in a confined space, one substance can expand only by compressing others, and work will be done only as long as the expanding substance has a greater pressure than those which it compresses. Hence, when all the substances are at the same pressure, no more work can be done in this way.

Finally, let us suppose that we are able by means of a perfect heat engine to transfer heat from one of the substances to another. Work will be done only when the heat is transferred from a hotter to a colder substance. Hence, when all the substances are reduced to the same temperature, no more work can be obtained in this way.

If, therefore, a number of substances are contained in a vessel which allows neither matter nor heat to pass its walls, the energy which can be converted into mechanical work will be entirely exhausted when all the substances are at the same pressure and the same temperature. To obtain any more work from the system, we must allow it to communicate either mechanically or thermally with bodies outside the vessel. Hence only a part of the whole intrinsic energy of the system is capable of being converted into mechanical work by actions going on within the vessel, and without any communication with external space by the passage either of matter or of heat. This part is sometimes called the Available Energy of the system. Clausius has called the remainder of the energy, which cannot be converted into work, the Entropy of the system. We shall find it more convenient to adopt the suggestion of Professor Tait, and give the name of Entropy to the part which can be converted into mechanical work.

DEFINITION OF ENTROPY.–*The Entropy of a system is the mechanical work it can perform without communication of heat, or alteration of its total volume, all transference of heat being performed by reversible engines.*

When the pressure and temperature of the system have become uniform the entropy is exhausted.

The original energy of the system is equal to the sum of the entropy and the energy remaining in the state of uniform pressure and temperature.

The entropy of a system consisting of several component systems is the same in whatever order the entropy of the parts is exhausted. It is therefore equal to the sum of the entropy of each component system, together with the entropy of the system consisting of the component systems, each with its own entropy exhausted.

When the parts of a system are at different temperatures, and if there is thermal communication between them, heat will pass from the hotter to the colder parts by conduction and radiation. The result of conduction and radiation is invariably to diminish the difference of temperature between the parts of the system, and the final effect is to reduce the whole system to a uniform temperature.

During this process no external mechanical work is done, and when the process is completed, and the temperature of the system has become uniform, no work can be obtained from the thermal energy of the system.

Hence the result of the conduction and radiation of heat from one part of a system to another is to diminish the entropy of the system, or the energy, available as work, which can be obtained from the system.

The energy of the system, however, is indestructible, and as it has not been removed from the system, it must remain in it. Hence the intrinsic energy of the system, when the entropy is exhausted by thermal communication, conduction, and radiation, is equal to its original energy, and is of course greater than in the case in which the entropy is exhausted by means of the reversible engine.

Again, when the parts of the system are at different pressures, and there is material communication by open channels between them, there will be a tendency of the parts of the system to move, and the result of this motion will be to equalise the pressure in the different parts. If the energy developed in this way is not gathered up and used in working a machine it will be spent in giving velocity to the parts of the system.

But as soon as this motion is set up it begins to decay on account of the resistance which all substances, even in the gaseous state, offer to the relative motion of their parts. The resistance in the case of solid bodies sliding over each other is called Friction. In fluids it is called Internal Friction or Viscosity. In every case it tends to destroy the relative motion of the parts, and to convert the energy of this motion into heat.

If, therefore, the system contains portions of matter in which the pressure is different, and if there is a material communication (that is, an open passage of any kind) between these portions, the part of the entropy depending on difference of pressures will be converted first into visible motion of the parts, and, as this decays, into heat.

It is possible to prevent material communication between the parts of a system by enclosing the substances in vessels through which they cannot pass. But it is impossible to prevent thermal communication between the parts of the system, because no substance known to us is a non-conductor of heat. Hence the entropy of every system is in a state of decay unless it is supplied from without.

This is Thomson's doctrine of the Dissipation of Energy.[1] [a] Energy is said to be dissipated when it cannot be rendered available as mechanical energy. The energy which is not yet dissipated is what we have here, following Tait, called Entropy. The theory of entropy, in this sense, was given by Thomson in 1853.[2b] The name, however, was first employed by Clausius in 1854,[3c] and it was used by him to denote the energy already dissipated.

The law of communication of heat, on which we founded our first definition of temperature; the principle of Carnot, and the second law of thermodynamics; and the theory of Dissipation of Energy, may be considered as expressions of the same natural fact with increasing degrees of scientific completeness. We shall return to molecular theories.

a. Maxwell's footnote. 1. [William Thomson,] 'On a Universal Tendency in Nature to Dissipation of Energy.' -Phil. Mag. [4 (1852): 304–306] and Proc. R. S.[oc.] E[dinburgh]. 1852 [139–142].

b. Maxwell's footnote 2. [William Thomson,] 'On the Restoration of Mechanical Energy from an unequally heated Space.' -Phil. Mag. [4] Feb 1853 [102–105].

c. Maxwell's footnote 3. [Rudolph Clausius, "Ueber ein veränderte Form des zweiten Haupsatzes der mechanischen Wärmetheorie,"] Pogg. Ann. [93] Dec. 1854 [481–507].

23. Maxwell, "Limitation of the Second Law of Thermodynamics"

Theory of Heat first edition 1871, 308–309.

Before I conclude, I wish to direct attention to an aspect of the molecular theory which deserves consideration.

One of the best established facts in thermodynamics is that it is impossible in a system enclosed in an envelope which permits neither change of volume nor passage of heat, and in which both the temperature and the pressure are everywhere the same, to produce any inequality of temperature or of pressure without the expenditure of work. This is the second law of thermodynamics, and it is undoubtedly true as long as we can deal with bodies in one mass, and have no power of perceiving or handling the separate molecules of which they are made up. But if we conceive a being whose faculties are so sharpened that he can follow every molecule in its course, such a being, whose attributes are still as essentially finite as our own, would be able to do what is at present impossible to us. For we have seen that the molecules in a vessel full of air at uniform temperature are moving with velocities by no means uniform, though the mean velocity of any great number of them, arbitrarily selected, is almost exactly uniform. Now let us suppose that such a vessel is divided into two portions, A and B, by a division in which there is a small hole, and that a being, who can see the individual molecules, opens and closes this hole, so as to allow only the swifter molecules to pass from A to B, and only the slower ones to pass from B to A. He will thus, without expenditure of work, raise the temperature of B and lower that of A, in contradiction to the second law of thermodynamics.

This is only one of the instances in which conclusions which we have drawn from our experience of bodies consisting of an immense number of molecules may be found not to be applicable to the more delicate observations and experiments which we may suppose made by one who can perceive and handle the individual molecules which we deal with only in large masses.

In dealing with masses of matter, while we do not perceive the individual molecules, we are compelled to adopt what I have described as the statistical method of calculation, and to abandon the strict dynamical method, in which we follow every motion by the calculus.

It would be interesting to enquire how far those ideas about the nature and methods of science which have been derived from examples of scientific investigation in which the dynamical method is followed are applicable to our actual knowledge of concrete things, which, as we have seen, is of an essentially statistical nature, because no one has yet discovered any practical method of tracing the path of a molecule, or of identifying it at different times.

I do not think, however, that the perfect identity which we observe between different portions of the same kind of matter can be explained on the statistical principle of the stability of averages of large numbers of quantities each of which may differ from the mean. For if of the molecules of some substance such as hydrogen, some were of slighlty greater mass than others, we have the means of producing a separation between molecules of different masses, and in this way we should be able to produce two kinds of hydrogen, one of which would be somewhat denser than the other. As this cannot be done, we must admit that the equality which we assert to exist between the molecules of hydrogen applies to each individual molecule, and not merely to the average of groups of millions of molecules.

24. Postcard from Maxwell to Peter Guthrie Tait, February 3, 1872

Cambridge University Library, Maxwell Collection, Maxwell-Tait Correspondence.

p. 433ℓ^a 3 from bottom "infinitely mutual." What are the degrees of mutuality?

p. 448ℓ 5 "irrational" p. 462 14 for 1867 put 1847.

p. 462ℓ 17, 467 footnote emf p. 471 14 sensibility? or susceptibility.

p. 471ℓ 23 and 24 read from the substance in that direction.

478ℓ 16 AB and C (see equations), < $Eq^n(5)$ > bad arrangement of dashes. Should be

	A	B',	C''	when	$B' = A''$	
b	A''	B,	C'		$C' = B''$	line 28 made to turn
	A' [d]	B'	C		$A' = B''$ [c]	round a fixed axis [d]

Observe how my invincible ignorance of certain modes of thought has caused Clausius to disagree with me (in the digestive sense) so that I failed in my attempts to boil him down and he does not occupy the place in my book on heat to which his other virtues entitle him. If he can get himself assimilated now I shall appear in a state of disgregation. Ergal lusting against Virial, and Virial against Ergal. Any Prooves for dp/dt?

a. ℓ denotes line.

b. In Tait's handwriting is the note "I defend this."

c. In Tait's pen B' is scored through and replaced by C.

d. These may be corrections for Tait, "Thermoelectricity," *Nature,* 8 (1873): 122–124.

25. Letter from Maxwell to Peter Guthrie Tait, February 12, 1872[a]

Cambridge University Library, Maxwell Collection, Maxwell-Tait Correspondence.

---

11 Scroope Terrace,
Cambridge,
12 Feb. 1872

O T'

What makes you address to Glenlair? I have no time, strength, or fury to smash.

As for C.[b] though I imbibed my $\theta\Delta^{cs}$ [c] from other sources I know that he is a prime source and have in my work for Longman [d] been unconsciously acted on by the motive not to speak about what I don't know. In my spare moments, I mean to take such draughts of Clausiustical Ergon as to place me in that state of disgregation in which one becomes conscious of the increase of the general sum of Entropy. Meanwhile till

Ergal and Virial from their thrones be cast

And end their strife with suicidal yell,

I remain yrs $\frac{dp}{dt}$

Electromagnetic Trails are to be served up (on toast) by Stokes at R S[e] on Thursday. Note on Felici and Jochmann.[f]

a. This postcard is quoted in Knott, *Life of Tait,* 221, without the first two sentences or the postscript.

b. Rudolph Julius Emmanuel Clausius. The reference is to Clausius as founder of thermodynamics and an important source for understanding that science.

c. Thermodynamics.

d. This is a reference to Maxwell's text on the theory of heat published by Longman, Maxwell, *Theory of Heat,* (London: Longmans Green, 1871).

e. Royal Society. The paper Maxwell refers to is Maxwell, "On the Induction of Electric Currents in an Infinite Plane Sheet of Uniform Conductivity," read by Stokes as Secretary of the Royal Society on February 15th, 1872, and published in *Proc. R. Soc. London,* 20 (1872): 160–168.

f. This refers to papers on induction to which Maxwell refers in his Royal Society paper. Felici, "Saggio di una applicazione del calcolo alle correnti indotte dal magnetismo in movimento," *Tortoli Annali* 4 (1853): 173–183; "Sopra i fenomeni di induzione della bottiglia di Leid," same journal, 14 (1853): 237–238, and, "Sulla teoria matematica dell induzione electro-dinamica," same journal, 5 (1854): 35–58. Emil Jochmann, "Ueber die dur Magnetpole in rotirenden körperlichen Leitern inducirten electrischen Ströme," *Ann. Phys.* 122 (1864): 214–237, translated in *Phil. Mag.* 27 (1864): 506–528, and 28 (1864): 347–349.

26. Letter from Maxwell to Peter Guthrie Tait, December 1, 1873[a]

Cambridge University Library, Maxwell Collection, Maxwell-Tait Correspondence.

Natural Science Tripos 63 Dec 1873

O.T. For the flow of a liquid in a tube, axis z[b]

$$\mu\left(\frac{d^2\omega}{dx^2}+\frac{d^2\omega}{dy^2}\right)=\frac{dP}{dz} \tag{1}$$

$$\text{Surface condition } \mu\frac{d\omega}{d\nu}=\lambda\,\omega \tag{2}$$

where ν is the normal drawn towards the liquid. When the curvature is small, (2) is equivalent to supposing the walls removed back by μ/λ and then λ made ∞ or $W = 0$. For glass and water by Helmholtz and Pietrowski[c] $\mu/\lambda = 0$.

If so, and if the value for$W$ is $C\left(1 = \frac{x^2}{a^2} - \frac{y^2}{b^2}\right)$

$$2\mu\, C\left(\frac{1}{a^2} + \frac{1}{b^2}\right) + \frac{dP}{dz} = 0, \text{ which gives } C.$$

If not you may write

$$W = A + B\,r^2 + C_2\,r^2 cos\, 2\phi + C_4\,r^4 cos\, 4\phi + \text{etc.}$$

when $x = ar\,cos\,\theta$ and $y = br\,sin\theta$ and then

$$2\mu\, B(\frac{1}{a^2} + \frac{1}{b^2}) + \frac{dP}{dz} = 0,$$

and you satisfy (2) the best way you can when $r = 1$.

As to Ampère—of course you may lay on d_1(anything) where d_1 is with respect to the element of a circuit. Have you studied H^2 [d] on the potential of two elements? or Bertrand[e] who, with original bosh of his own rushes against the thicker bosches of H^2's buckler? and says that H^2 believes in a <const> force which does not diminish with the distance so that the reason why Ampère or H^2 or Bertrani[f] observe peculiar effects is because some philosopher in Centauri happens to be completing a circuit.

$XQqD$ [g] as I am surrounded by Naturals and cannot give references.

In introducing 4^{ions} [h] do so by blast of trumpet and tuck of drum. Why should $V.\alpha\beta\gamma$ come in sneaking without having his style and titles proclaimed by fugleman? Why even . should be treated with due respect and we should be informed whether he is attractive or repulsive. What do you think of "Space-variation" as the name of Nabla?[i]

It is only lately under the conduct of the professor Willard Gibbs that I have been led to recant an error which I had imbibed from your $\theta\Delta^{cs}$,[j] namely that the entropy of Clausius is unavailable energy while that of T'[k] is available energy. The entropy of Clausius is neither the one nor the other. It is only Rankine's[ℓ] Thermodynamic function and if we compare the vocabulary

Thermodynamic Function	Entropy (Clausius)
Entropy (Tait)	Available Energy

I think we shall prefer the 2nd column. Available energy, there is none in a system at uniform temperature and pressure.[m]

I have also great respect for the elder of those celebrated acrobats, Virial and Ergal, the Bounding Brothers of Bonn. Virial came out in my paper on Frames R. S. E., 1870, under the form $\sum Rr = 0$ when there is no motion.[n] When there is motion the <mean> time average of $\frac{1}{2}\sum Rr =$ time - average of $\frac{1}{2}Mv$ unless R is positive for attraction.

But it is rare sport to see those learned Germans contending for the priority of the discovery that the 2nd law of $\theta\Delta^{cs}$ is the Hamiltonische Princip,[o] when all the while they *assume* that the temperature of a body is but another name for the vis viva of one of its molecules, a thing which was suggested by labours of Gay-Lussac, Dulong etc., but first deduced from dynamical statistical considerations by dp/dt. The Hamiltonische Princip the while soars along in a region unvexed by statistical considerations while the German Icari flap their waxen wings in nephelococcygin, amid those cloudy forms which the ignorance and finitude of human science have invested with the incommunicable attributes of the invisible Queen of heaven.

Dictum of K and T'[p] concerning 3 points, p. 160. If their perps[q] intersect in G, the three points A, B, C will be in one plane. Inference (by a new logical formula invented by dp/dt). "All planes pass through 0."[r] General[s] exercise. Interpret every 4^{ion} expression in literary geometrical language, e.g., express in neat set terms the result of $\frac{\beta}{\alpha}.\gamma$.

$$\frac{dp}{dt}$$

a. This letter is quoted in large part in Knott, *Life of Tait,* 114–116. We note Knott's omissions, notes to this letter and our additions.

b. Note in Knott, "See Tait's Laboratory Notes [*Proc.*] *R. S.*[*oc.*] *E.*[*din burgh*]. viii, [1875] p. 208): On the Flow of Water through fine Tubes. The experiments were made by C. Michie Smith and myself with tubes of circular and elliptic bore. Tait had asked Maxwell to give him the theory of the phenomenon as a problem in viscosity."

c. Hermann von Helmholtz and G. von Piotrowski, "Ueber Reibung tropfbarer Flussigkeiten," *Sitz. Math.-Naturwiss. Cl. Akad. Wiss. Wien* 40 (1860): 607–658, where they deduced the phenomenon of "slip" between a moving viscous liquid and a solid. Kundt and Warburg discovered the analogous phenomenon in gases in 1875 that was crucial to Maxwell's investigation on stresses in rarefied gases. See the documents in chap. VI.

d. Hermann von Helmholtz. Note in Knott, "The reference is to H[ermann] H[elmholtz]'s electrodynamic investigation which supplied the true criterion in place of the hasty generalisation of §385 in the first edition of *Thomson and Tait.*" The specific paper of Helmholtz to which Maxwell refers is, Helmholtz, "Ueber der Theorie der Elektrodynamik," *Monats. Berlin* (1872): 247–256, and *J. reine und angew. Mathematik* 75 (1873): 35–66, translated in *Phil. Mag.* 64 (1872): 503–537. Maxwell reviewed the problem of the mutual action between two isolated current elements in his *Treatise on Electricity and Magnetism,* vol. 2, part 4, chap. 2. Innumerable writers have discussed it since. Leigh Page and N. I. Adams in *Amer. J. Phys.* 13 (1945): 141 have given a lucid solution in terms of Maxwell's theory.

e. Bertrand's objections were published in, "Observations sur la thèorie nouvelle des actions electro-dynamique proposée par M. Helmholtz," *Comptes Rendus* 75 (1872): 860–865.

f. Bertrand.

g. Excuse details. The squiggles representing the tails are omitted in Knott with the note that they exist in the letter.

h. Quarternions. Note in Knott, "See chapter on Quarternions for other remarks by Maxwell on Tait's quarternion work. Maxwell was reading Kelland and Tait, *Introduction to Quarternions* which he reviewed in *Nature* shortly after." The anonymous article, "Quarternions," appeared in *Nature,* 10 (1874). It was a review of P. Kelland and P. G. Tait, *Introduction to Quarternions with Numerous Examples* (London: Macmillan, 1873).

i. "Space Variation" refers to the operation of ∇ (Nabla) on a scalar quantity. Maxwell later coined the now standard term, gradient.

j. Thermodynamics. This is a reference to Tait, *A Sketch of Thermodynamics* (Edinburgh: Edmonston and Douglas, 1868).

k. "T"' is Tait. Note in Knott, "Tait suggested in the first edition of his *Thermodynamics* (contracted to $\theta\Delta$cs by Maxwell) that the work Entropy should be used in this sense. In the second edition he went back to the original meaning as given by Clausius."

ℓ. Rankine

m. Josiah Willard Gibbs (1839–1903). His first two papers were already in print and from Maxwell's remarks here and in the following letters he had already read them. Gibbs, "Graphical Methods in the Thermodynamics of Fluids," *Trans. Acad. Sci. Connecticut* 2 (1873): 309–342 and 382-404.

n. This is a reference to Maxwell, "On Reciprocal Figures, Frames and Diagrams of Force," *Trans. R. Soc. Edinburgh* 26 (1872): 1–40.

o. Hamilton's Principle of Least Action. This refers to Boltzmann's and Clausius's priority dispute over who first deduced the second law as a form of Hamilton's Principle.

p. This is a reference to Kelland and Tait,*Introduction to Quarternions,* (London: Macmillan, 1873).

q. perpendiculars.

r. The first sentence in this paragraph is omitted in Knott, *Life of Tait.* It is critical of Tait's work on Quarternions. Knott does not show anywhere in his book on Tait the very limited use Maxwell made of quarternions in his *Treatise.* His enthusiasm was not extensive.

s. This is an exercise in quarternion algebra.

27. Postcard from Maxwell to Peter Guthrie Tait, October 13, 1874

Cambridge University Library, Maxwell Collection, Maxwell-Tait Correspondence

$$\ell = \frac{4^\circ e^{15\,octnt} + \theta^\circ\, 10^l\, e^{t-15\,oct}}{e^{15\,oct-t} + e^{t-15\,oct}}$$

$$\lambda = \frac{(55^\circ e^{15\,oct-t} + 52\, e^{t-15\,oct})}{(e^{15\,oct-t} + e^{t-15\,oct})}$$

The good word shall return to you goodyer.
Before finishing up $\theta\Delta^{cs}$ [a] read Prof. J. Willard Gibbs on the surface whose coordinates are Volume, Entropy and Energy. See Trans. Acad. Connecticut Vol II. He has more sense than any German.[b]

a. Thermodynamics. This refers to the second edition of Tait, *A Sketch of Thermodynamics* which Tait was already planning.

b. Josiah Willard Gibbs, "Graphical Methods in the Thermodynamics of Fluids," *Trans. Connecticut Acad. Sci.* 3 (1873): 309–342, 382–404.

28. Letter from Maxwell to Thomas Andrews, November 1874[a]

P.G. Tait and Alexander Crum-Brown, "Memoir of Dr. Thomas Andrews," in *The Scientific Papers of the late Thomas Andrews with a Memoir* (London: Macmillan and Co., 1889): ix–lxii, liv–lv.

... What you told us at Belfast[b] about the properties of a mixture of carbonic acid and nitrogen has been often in my thoughts, and appears to me exceedingly important for the theory of gases and liquids. The most obvious way of considering the matter appears to be to compare the observed facts with Dalton's law of evaporation.

Conceive a vessel containing nitrogen at a given pressure and temperature. Now let a certain quantity of liquid carbonic acid be introduced. According to Dalton it will evaporate till the density of the gaseous carbonic acid has reached a value corresponding to the temperature, but if there is enough room it will all evaporate.

That is, it will all evaporate unless the total volume of the mixed matter is less than the volume of CO_2 gas at its maximum density.

Conversely, and even a fortiori, CO_2 gas will not liquefy till its density (independent of N) is at least the maximum density.

This is a hypothetical case founded on Dalton's law. If, in the real case, CO_2 gas can exist when mixed with nitrogen in a vessel of a certain volume, whereas, if the nitrogen were removed from the vessel, part of the CO_2 would liquefy, then the power of nitrogen to keep carbonic acid in the gaseous form is fully established.

I should be greatly obliged to you if you will let me know when you publish the experiments in full, as the numerical results will be of great value in discriminating between different theories of gases.

Are the numerical results of your former experiments on CO_2 published anywhere except in the Phil Trans?[c] ...

I have just finished a clay model of a fancy surface showing the solid, liquid and gaseous states and the continuity of liquid and gaseous states.

I am afraid that even CO_2 would not make a very compact model if worked truly to scale. But the data as to specific heat in the liquid and solid states are wanting as yet and also the latent heat of fusion and evaporation.

a. Thomas Andrews (1813–1885).

b. The British Association met in Belfast in 1874. Maxwell is probably referring to Andrews' address as president of the Chemical Section, see, *Rep. 43rd Meeting BAAS,* Trans. Section B (1874): 59–66.

c. Andrews' experiments were done over a protracted period of time and published as, Andrews, "On the Continuity of the Gaseous and Liquid States of Matter," *Phil. Trans. R. Soc. London* 159 (1869): 575–590.

29. Letter from Maxwell to James Thomson, March 27, 1875

Queens University Belfast.

Glenlair,
Dalbeattie,
27 March 1875

Dear Thomson,

I shall be very much obliged to you if you will let me know any suggestions, corrections, or amendments to my book on heat. A new edition is to come out soon and several things must be modified, e.g., all about Entropy and about the conditions of evaporation now solved by Gibbs.[a]

I dare say Poundians are better for the British Engineer than Dynes. The dynes are such a feeble folk that for engineering purposes the number of figures becomes difficult to remember. Of Free level I highly approve. I am glad to see you recognize[b] Bow's[c] method of drawing diagrams. He has made a decided advance in Graphical Statics, though I cannot yet say what Culmann has done.[d] Have you seen Levy (Maurice) La Statique Graphique?[e]

I have not seen him yet.

Have you considered "Pieaucellier's cell"[f] from a statical point of view. If $yu = yx$ and $uv = vx = xw = wu$ are the jointed pieces then the forces acting at y, w *and* v must pass all through one point z which must lie in a straight line with u and x.

The reciprocal diagrams is[sic] given below. In the upper diagram xy cuts wz in t. Hence in the reciprocal diagram xt and yt are equal and parallel and also wt and rt, and t is the space surrounding the diagram.

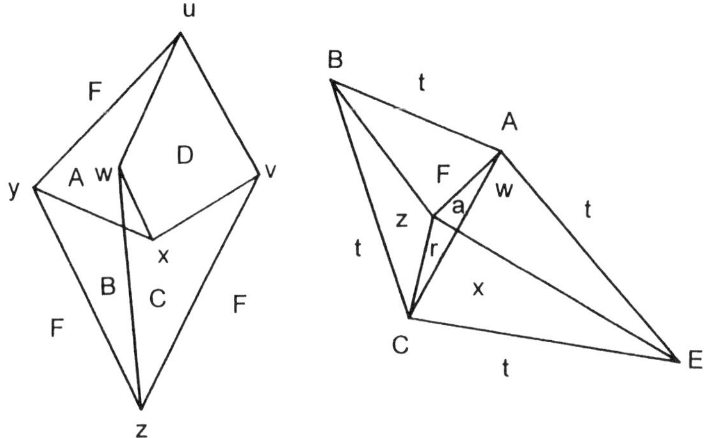

Yours truly, J. Clerk Maxwell.

a. The new edition to which Maxwell refers is the fourth edition of Maxwell, *Theory of Heat* (1875) in which extensive changes were made especially with respect to Maxwell's interpretation of the second law of thermodynamics. It also included a whole section on Gibb's thermodynamics. See Document III-32.

b. Rankine had investigated the method of drawing diagrams of forces and applied it to frames in Rankine, *Applied Mechanics.* Maxwell then treated the problem geometrically in Maxwell, "On Reciprocal Figures and Diagrams of Forces," *Phil. Mag.* 27 (1864): 250–261, and later generalised the method in Maxwell, "On Reciprocal Figures, Frames and Diagrams of Force," *Trans. R. Soc. Edinburgh* 26 (1872): 1–40.

c. Robert Henry Bow (1827-1909), was a civil engineer in Edinburgh and had invented a new notation to simplify the process of drawing diagrams of forces, Bow, *Economics of Construction* (London, 1873).

d. Karl Culmann (1821-1881), was professor of Engineering at the Polytechnic in Zurich from 1855 until his death. He developed the method of reciprocal diagrams, Culmann, "Ueber das Parallelagramm und uber die Zusammensetzung der Kräfte," *Vierteljahrsschrift Naturforschenden Gesellschaft in Zurich* 15 (1870): 1–24, and *Graphische Statik* (1875).

e. Maurice Levy (1838-1910), also developed the method of reciprocal diagrams, *La statique graphique et ses applications aux constructions* (Paris, 1874).

f. See Maxwell, "On Bow's Method of Drawing Diagrams in Graphical Statics with Illustrations from Pieaucellier's Linkage," *Proc. Cambridge Phil. Soc.* 2 (1876): 407–414.

30. Postcard from Maxwell to James Thomson, 1875[a]

Queens University, Belfast.

Corrections received and in great part adopted. More will be thankfully received up to 1st July.[b] I shall try and keep clear of the line of absolute cold but I have also to avoid breaking the ice, i.e., disarranging the stereotype plates too much.[c] Interface and volumined I approve. Poundal is good too but it is approximating to a final form and maybe better in a year.[d]

In German perspective a picture is the Schnitt of a pencil of rays to the eye, and the pencil of rays is the Schein of the picture. Of course Schnitt is section and Schein is shine but shine is not intelligible. Name it so.

Yours truly, $\frac{dp}{dt}$

Address Glenlair, Dalbeattie.

a. The postcard is dated from the content and other references in correspondence between James Thomson and Maxwell to the new edition of Maxwell's *Theory of Heat* in 1875 and Maxwell's request for corrections from Thomson.

b. The corrections are to proofs of Maxwell, *Theory of Heat,* 4th edition, (London, 1875).

c. Stereotype plates were permanent plates made by taking a mold from the composed type in papier mache and from this making a permanent mold cast in type metal. Obviously correcting such a permanent mold was an expensive and time-consuming job. Maxwell's book on heat had become so popular that the publisher thought it worth the trouble of making permanent type of its pages. While Maxwell's corrections to this edition were extensive because of his reading of Gibbs and his new understanding of the second law they were fitted into whole pages and minimised the expense of the changes. Most of the chapters and sections remained untouched, only pagination was affected. This was not mere profit motive as the audience for the text was working men as well as students and cost thus a very real factor.

d. Poundal is referred to as "poundian" in the previous document. Thus James Thomson was the originator and Maxwell the propagator of the term "poundal" used by British engineers for the force required to give a mass of one pound an acceleration of 1 foot/sec^2.

31. Letter from Maxwell to James Thomson, July 8, 1875.

Queens University Belfast.

Glenlair,
Dalbeattie,

8 July 1875

Dear Thomson,

Since I got your notes and correction I have been busy adapting them to the new edition of my book on heat, and have made about 5 pages of corrections. Have you any special contrivance for getting rid of <the> Rankine method of conceiving lines traced beyond the field of experiment?

I think the simplest way is to select for the field of operations a portion of the whole field explored by experiment with a convenient boundary, formed, say by two lines of equal temperature and two lines of equal pressure.

Then if the substance passes from the state A to the state B by any given path we may find the heat which enters the body or "the heat of the path," by drawing isentropics (adiabatics) ϕ_A and ϕ_B at A and B as far as the isothermal T.

The area between the path, ϕ_B, T, and ϕ_A together with $(\phi_B - \phi_A)T$ is the heat of the path. Similarly by drawing lines of equal volume V_A and V_B to the line of pressure P the work done by the body is the area between the path, V_B, P, and V_A together with $(V_B - V_A)P$.

It will give me great pleasure to receive on the part of the Cavendish Laboratory a cast of your thermodynamic model with the lines marked on it. We have now got an excellent case with glass front containing a thermometer by Il Gonfio (before 1660) Wollaston's optical and thermal apparatus etc., and we shall have a special place for models such as yours.

I enclose a rough sketch of the lines on Gibbs's surface,[a] coordinates

Volume Entropy Energy

in an imaginary substance in which the principal features of known substances can be represented on a convenient scale. The black lines are not in the same surface with the coloured lines. They represent mixed states and are straight lines. The points at which they cut the red and blue lines are their extremities only. Here they intersect the same pair of red and blue lines at both extremities the intermediate apparent inter-sections are only optical as the lines do not meet in the surface except at the extremities. At the critical point the red, blue, green, yellow and black lines all touch the black line being of course infinitely short.

There are some errors in the "solid" part of the blue line which I have not time to erase before post

Yours very truly,

J. Clerk Maxwell.

a. Maxwell's freehand diagram is not reproduced here as it is very confusing. The diagram is actually the mirror image of the diagram reproduced in Document III-32, from Maxwell, *Theory of Heat,* 4th edition, 207 along with Maxwell's description of it.

32. Excerpts from Maxwell, "Available Energy"

Maxwell, *Theory of Heat* (London; Longmans, Green and Co., 1875), 187-193, chapter VIII, fourth edition.

Available Energy

The sum of the work done by the body and the dynamical equivalent of the heat which it gives out during its passage from the state A to the state B is, as we have seen, the same whatever be the path by which the body passes from the state A to the state B. If, however, we suppose that the body is surrounded by a medium, the temperature of which is maintained constant, so that the body can give out heat only when its temperature is higher than that of the medium, and can take in heat only when its temperature is lower than that of the medium, then these conditions will confine the path within certain limits.

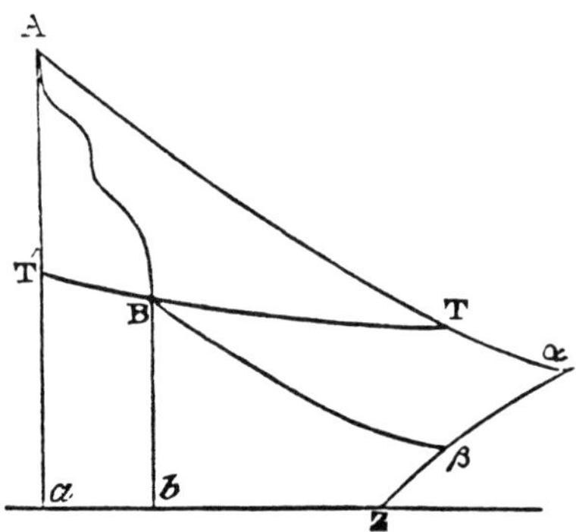

Draw the isothermal TT', representing the constsnt temperature of the surrounding medium. Then since the temperature of the body at A and at all points above the line TT' is higher than that of the medium, the body cannot receive heat from the medium. Hence its entropy cannot increase, and the path cannot rise above the adiabatic or isentropic A, drawn through A.

Again, when the body gives out heat to the medium, its temperature must be higher than that of the medium. Hence the path must be above the isothermal TT'.

The path formed by the isentropic AT and the isothermal TB is therefore the limiting form of the path, and is that wherein the work done by the body is a maximum, and the heat given out by it a minimum.

If we denote the energy of the body in the state A by ϵ, and its entropy by ϕ, and the energy and entropy of the body at the temperature and pressure of the surrounding medium (represented by B) by ϵ_0 and ϕ_0, then the total energy given out as work and heat during the passage from the state A to the state B is $\epsilon - \epsilon_0$.

The amount of heat which the body gives out during the process cannot be less than that corresponding to the path ATB, which is

$$(\phi - \phi_0)T$$

where T is the absolute temperature of the surrounding medium.

The amount of work done by the body during the process cannot, therefore, be greater than

$$\epsilon - \epsilon_0 - (\phi - \phi_0)T.$$

This, therefore, is the part of the energy which is available for mechanical purposes under the circumstances in which the body is placed, namely, when surrounded by a medium at temperature T and pressure P.

It appears, therefore, that the greater the original entropy, the smaller is the available energy of the body.[1 a]

If the system under consideration consists of a number of bodies at different pressures and temperatures contained within a vessel from which neither matter nor heat can escape, then the amount of energy converted into work will be greatest when the system is reduced to thermal and mechanical equilibrium by the following process.

1st. Let each of the bodies be brought to the same temperature by expansion or compression without communication of heat.

2nd. The bodies being now at the same temperature, let those which exert the greatest pressure be allowed to expand and to compress those which exert low pressure, till the pressures of all bodies in the vessel are equal, the process being conducted so slowly that the temperatures of all the bodies remain sensibly equal to each other throughout the process.

During the first part of this process, in which there is no communication of heat between the bodies, the entropy of each body remains constant. During the second part, the bodies are all at the same temperature, and therefore the communication of heat from one body to another diminishes the entropy of the one body as much as it increases that of the other, so that the sum of the entropy remains constant. Hence the total entropy of the system remains the same from the beginning to the end of the process. The work done against mechanical resistances during the establishment of thermal and mechanical equilibrium is greater when the process is conducted in this way than when conduction of heat is allowed to take place between bodies at sensibly different temperatures.

Hence the final state of the system is determined by the following conditions:

Let n be the number of bodies forming the system.

$m_1 \dots m_n$ be the masses of those bodies

$v_1 \dots v_n$ the volumes of unit of mass of each,

$\phi_1 \dots \phi_n$ the entropy of unit of mass of each,

$\epsilon_1 \dots \epsilon_n$ the energy of unit of mass of each,

$p_1 \dots p_2$ the pressure of each,

$\theta_1 \dots \theta_n$ the temperature of each.

The volume of the whole is

$$m_1 v_1 + \ldots + m_n v_n = \sum(mv),$$

and since the system is contained in a vessel of volume V,

$$\sum(mv) = V$$

during the whole process.

The entropy of the whole is

$$\sum m_1 \phi_1 + \ldots + m_n \phi_n = \sum(m\phi) = \Phi$$

When there is no communication of heat except between bodies of equal temperature, Φ remains constant. When there is communication of heat between bodies of different temperature, Φ increases.

In the final state of the system

$$p_1 = p_2 \ldots = p_n = P$$

$$\theta_1 = \theta_2 \ldots \theta_n = \Theta$$

There are therefore $n - 1$ conditions with respect to pressure, and $n - 1$ conditions with respect to temperature, together with one condition with respect to volume and one with repsect to entropy, or, in all, $2n$ conditions to be satisfied by the n bodies; and since the state of each body is a function of two variables, the conditions are necessary and sufficient to determine the final state of each of the n bodies.

The work done against resistances external to the system may be determined by comparing the total energy at the beginning of the process with the final energy; for, since no heat is allowed to escape, any diminution of energy must arise from work being done.

The total energy is

$$\sum(m\,\epsilon) = E.$$

If E be the original and E' the final value of this quantity, the energy available to produce mechanical work is

$$E - E'.$$

If during any part of the process by which the system reaches its final state of thermal and mechanical equilibrium there takes place a communication of a quantity H of heat from a body at temperature θ_1 to a body at temperature θ_2, the increase of the total entropy of the system arising from the communication is, as we have shown (at p. 163),

$$H\left(\frac{1}{\theta_2} - \frac{1}{\theta_1}\right),$$

and the final entropy, instead of being equal to the original entropy Φ, becomes

$$\Phi' = \Phi + H\left(\frac{1}{\theta_2} - \frac{1}{\theta_1}\right),$$

This increase of the final entropy involves a corresponding increase in the final temperature and the final energy.

If the rise of the final temperature is small, then, since the volume is constant, the increase of the final energy is

$$\Theta(\Phi' - \Phi) = H\Theta\left(\frac{1}{\theta_2} - \frac{1}{\theta_1}\right),$$

and the available energy is therefore diminished by this quantity on account of the passage of the quantity H of heat from a body at temperature θ_1 to a body at temperature θ_2.

Processes of this kind, by which, while the total energy remains the same, the available energy is diminished, are instances of what Sir W. Thomson has called the Dissipation of Energy. The doctrine of the dissipation of energy is closely connected with that of the growth of entropy, but is by no means identical with it.

The increment of the total entropy of a system arising from the communication of a given amount of heat, H, from a body at one given temperature, θ_1, to another given temperature, θ_2 is, as we have seen,

$$H\left(\frac{1}{\theta_2} - \frac{1}{\theta_1}\right)$$

a quantity completely determined by the state of the system when this communication takes place.

The energy dissipated or rendered unavailable as a source of mechanical work is

$$H\Theta\left(\frac{1}{\theta_2} - \frac{1}{\theta_1}\right)$$

into which a new factor,Θ, enters, and this factor denotes the final temperature of the system when it has reached the state of thermal and mechanical equilibrium. Θ, therefore, since it depends on the final state of the system, can only be calculated, when we know not only the relations between the thermodynamical variables for all the bodies, but the volume which they occupy in their final state.

The calculation of the amount of energy dissipated during any process is therefore much more difficult than that of the increase of total entropy.

If the system is allowed to reach its final state of thermal and mechanical equilibrium, in such a manner that no external work is done, and no heat is allowed to leave or enter the system, the condition is that the final energy is equal to the original energy.

Combining this with the other conditions, that the volume is unchanged, and that the final state with respect to pressure and temperature is common to all bodies, we may determine the final value of the temperature, pressure and total entropy.

The total entropy will now have the maximum value consistent with the original state of the system. The dissipation of the available energy will be complete.

a. 1 refers to a footnote at the bottom of the page. "In former editions of this book the meaning of the term Entropy, as introduced by Clausius, was erroneously stated to be that part of the energy which cannot be converted into work. The book then proceeded to use the term as equivalent to the available energy; thus introducing great confusion into the language of thermodynamics. In this edition I have endeavoured to use the work Entropy according to its original definition by Clausius."

Maxwell, *Theory of Heat,* 4th edition (1875), 193–208

Mechanical and Thermal Analogies

In studying thermodynamics we may find considerable assistance from a comparison between the thermal and the mechanical phenomena.

We have to do with energy in two forms, work and heat. When energy is being transformed from one body to another we can always tell whether the first body is doing mechanical work on the second or communicating heat to it. Work is done by motion against resistance. Heat is communicated from a hotter to a colder body.

But as soon as the energy has entered the second body, we can no longer distinguish by any legitimate process whether it is in the form of work or of heat. In fact we may remove it from the body under either of these forms.

If a fluid at a pressure p increases in volume from v to v' it performs work against external resistance, the amount of which work is

$$p(v' - v) = W.$$

If a body at temperature θ increases in entropy from ϕ to ϕ', an amount of heat must have entered it represented by

$$\theta(\phi' - \phi) = H.$$

If both of these processes take place, and if the energy of the body is thereby changed from E to E', then

$$E' - E = H - W = \theta(\phi' - \phi) - p(v' - v).$$

Here then we have two sets of quantities, one relating to work, the other to heat.

W	v	p
H	ϕ	θ

Of these quantities Work and Heat are simply two forms of Energy.

The volume is a quantity such that without a change of its value no work can be done. The amount of work done, however, is measured, not by the change of volume alone, but by that change multiplied by another quantity–the pressure.

In the same way the entropy is a quantity such that without a change in its value no heat can enter or leave the body. The amount of this heat, however, is not measured by the change of entropy, but by that change multiplied by another quantity–the absolute temperature.

Again, the pressure is a quantity such that its equality in two communicating vessels determines their mechanical equilibrium, while its excess in either determines a flow of fluid from that vessel to the other.

In like manner the temperature is a quantity such that its equality in two bodies in contact determines their thermal equilibrium, while its excess in either determines a flow of heat from that body to the other.

If we regard the energy of a body as determined by its volume and its entropy, then the pressure may be defined as the rate at which the energy diminishes with increase of volume, while the entropy remains constant.

The temperature may in like manner be defined as the rate at which the energy increases with increase of entropy, the volume remaining constant.

Representation of the Properties of a Substance by means of a Surface.

Professor J. Willard Gibbs, of Yale College, U. S., to whom we are indebted for a careful examination of the different methods of representing thermodynamic relations by plane diagrams, has introduced an exceedingly valuable method of studying the properties of a substance by means of a surface.[1 a]

According to this method, the volume, entropy, and energy of the body in a given state are represented by the three rectangular coordinates of a point in the surface, and this point on the surface is said to correspond to the given state of the body. We shall suppose the volume measured towards the east from the meridian plane corresponding to no volume, the entropy measured towards the north from a vertical plane perpendicular to the meridian, whose position is entirely arbitrary, and the energy measured downwards from the horizontal plane of no energy, the position of which may be considered as arbitary, because we cannot measure the whole energy existing in a body.

The section of this surface by a vertical plane perpendicular to the meridian represents the relation between volume and energy when the entropy is constant, that is when no heat enters or leaves the body.

If the pressure is positive, then the body, by expanding would do work against external resistance, and its intrinsic energy would diminish. The rate at which the energy diminishes as the volume increases is represented by the tangent of the angle which the curve of section makes with the horizon.

The pressure, is therefore, represented by the tangent of the angle of slope of the curve of section. The pressure is positive when the curve slopes downwards towards the west. When the slope of the curve is towards the east the corresponding pressure is negative.

A tension or negative pressure cannot exist in a gas. It may, however, exist in a liquid, such as mercury. Thus, if a barometer tube is well filled with clean mercury, and then placed in a vertical position, with its closed end uppermost, the mercury sometimes does not fall in the tube to the point corresponding to the atmospheric pressure, but remains suspended in the tube, so as to fill it completely.

The pressure in this case is negative in that part of the mercury which is above the level of the ordinary barometric column.

In solid bodies, as we know, tensions of considerable magnitude may exist.

Hence in our thermodynamic model the pressure of the substance is indicated by the tangent of the slope of the curve of constant entropy, and is reckoned positive when the energy diminishes as the volume increases.

The section of the surface by a vertical plane parallel to the meridian is a curve of constant volume. In this curve the temperature is represented by the rate at which the energy increases, as the entropy increases, that is to say, by the tangent of the slope of the curve.

Since the temperature, reckoned from absolute zero, is an essentially positive quantity, the curve of constant volume must be such that the entropy and energy always increase together.

To ascertain the pressure and temperature of the substance in a given state, we draw a tangent plane to the corresponding point of the surface. The normal to this plane through the origin will cut a horizontal plane at unit of distance above the origin at a point whose coordinates represent the pressure and temperature, the pressure being represented by the coordinate drawn towards the west, and the temperature by the coordinate drawn towards the north.

The pressure and temperature are thus represented by the direction of this normal, and if, at any two points of the surface, the directions of the normals are parallel, then in the two states of the substance corresponding to these two points the pressure and temperature must be the same.

If we wish to trace out on a model of the surface a series of lines of equal pressure, we have only to place it in the sunshine and turn it so that the sun's rays are parallel to the plane of volume and energy, and make an angle with the line of volume whose tangent is proportional to the pressure. Then, if we trace on the surface the boundary of light and shadow, the pressure at all points of this line will be the same.

In like manner, if we place the model so that the sun's rays are parallel to the plane of entropy and energy, the boundary of light and shadow will be a line such that the temperature is the same at every point, and proportional to the tangent of the angle which the sun's rays make with the line of entropy.

In this way we may trace out on the model two series of lines: lines of equal pressure, which Professor Gibbs calls Isopiestics; and lines of equal temperature, or Isothermals.

Besides these, we may trace the three systems of plane sections parallel to the coordinate planes, the isometrics or lines of equal volume, the isentropics or lines of equal entropy, which we formerly called, after Rankine, adiabatics, and isenergics or lines of equal energy.[b]

The network formed by these five systems of lines will form a complete representation of the relations between the five quantities, volume, entropy, energy, pressure and temperature, for all states of the body.

The body itself need not be homogeneous either in chemical nature or in physical state. All that is necessary is that the whole should be at the same pressure and the same temperature.

By means of this model Professor Gibbs has solved several important problems relating to the thermodynamic relations between two portions of a substance, in different physical states, but at the same pressure and temperature.

Let a substance be capable of existing in two different states, say liquid and gaseous, at the same temperature and pressure. We wish to determine whether the substance will tend of itself to pass from one of these states to the other.

Let the substance be placed in a cylinder, under a piston, and surrounded by a medium at the given temperature and pressure, the extent of this medium being so great that its pressure and temperature are not sensibly altered by the change of volume of the working substance, or by the heat which that body gives out or takes in

The two physical states which are to be compared are represented by two points on the surface of the model; and since the pressure and temperature are the same, the tangent planes at these points are either coincident or parallel.

The surface representing the thermodynamic properties of the surrounding medium must be supposed to be constructed on a scale proportional to the amount of this medium; and as we assume that there is a very great mass of this medium, the scale of the surface will be so great that we may regard the portion of the surface with which we have to do as sensibly plane; and since its pressure and temperature are those of the working substance in the given state, this plane surface is parallel to the tangent plane at the given point of the surface of the model. Let ABC be three points of the model at which the tangent planes are parallel, the energy being reckoned downwards.

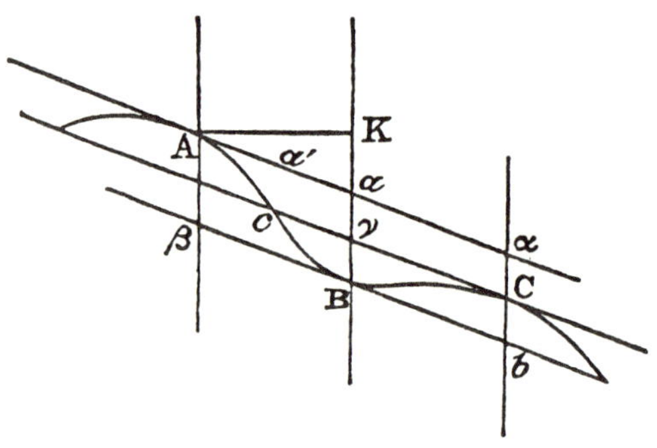

Let $Aa\alpha$ be the tangent plane at A, and let us consider it as part of the model representing the external medium, this model being so placed that volume, entropy, and energy are reckoned in the opposite directions from those in the model of the working substance.

Now let us suppose the substance to pass from the state A to the state B, passing through the series of states represented by the points of the isothermal line joining the points of equal temperature A and B.

Then since the working substance and the external medium are always at the same temperature, the entropy lost by the one is equal to that gained by the other.

Also the one gains in volume what is lost by the other.

Hence, during the passage of the working substance from the state A to the state B, the state of the external medium is always represented by a point in the tangent plane in the same vertical line as the point representing the state of the working substance.

For the same horizontal motion which represents a gain of volume or entropy of the one substance represents an equal loss of volume or entropy in the other.

Hence, when the state of the working substance is represented by the point B, that of the external medium will be represented by the point a, where the vertical line through B meets the tangent plane through A.

Now the energy is reckoned downwards for the working substance and upwards for the external medium. Hence, drawing AK horizontal, KB represents the gain in energy of the working substance and Ka the loss of energy of the external medium.

The line Ba, or the vertical height of the tangent plane above the point B, represents the gain of energy in the whole system, consisting of the working substance and the external medium, during the passage from the state A to the state B. But the energy of the system can be increased only by doing work on it.

But if the system can of itself pass from one state to another, the work required to produce the corresponding changes of configuration must be drawn from the energy of the system, and the energy must therefore diminish.

The fact, therefore, that in the case before us the energy increases, shows that the passage from the state A to the state B in the presence of a medium of constant temperature and pressure, cannot be effected without the expenditure of work by some external agent.

The working substance, therefore, cannot of itself pass from state A to the state B, if it lies *below* the plane which touches the surface at A.

We have supposed the substance to pass from A to B by a process during which it is always at the same temperature as the external medium. In this case the entropy of the system remains constant.

If, however, the communication of heat between the substances occurs when they are not at the same temperature, the entropy of the system will increase; and if in the figure the gain of entropy of the working substance is represented by the horizontal component of AB, the loss of entropy of the external medium will be represented by a smaller quantity, such as the horizontal component of Aa'. Hence a' will be to the left of a, and therefore higher. The gain of entropy of the system will therefore be represented by the horizontal part of aa'.

Now since temperature is essentially positive, a gain of entropy at a given volume always implies a gain of energy. Hence the gain of energy is greater when there is a gain of entropy than when the entropy remains constant.

There is, therefore, no method by which the change from A to B can be effected without a gain of energy, and this implies the expenditure of work by an external agent.

If, therefore, the tangent plane at A is everywhere above the thermodynamic surface, the condition of the working substance represented by the point A is essentially stable, and the substance cannot of itself pass into any other state while exposed to the same external influences of pressure and temperature.

This will be the case if the surface is convexo-convex upwards.

If, on the other hand, the surface, as at the point B, is either concave upwards in all directions, or concave in one direction and convex in another, it will be possible to draw on the surface a line from the point of contact lying entirely above the tangent plane, and therefore representing a series of states through which the substance can pass of itself.

In this case the point of contact represents a state of the substance which, if physically possible for an instant, is essentially unstable, and cannot be permanent.

There is a third case, however, in which the surface, as at the point C, is convexo-convex, so that the line drawn on the surface from the point of contact must lie below the tangent plane; but the tangent plane, if produced far enough, cuts the surface at C, so that the point A lies above the tangent plane. In this case the substance cannot pass through any continuous series of states from C to A, because any line drawn on the surface from C to A begins by dipping below the tangent plane. But if a quantity, however small, of the substance in the state A is in physical contact with the rest of the substance in state C, minute portions will pass at once from state C to state A without passing through the intermediate states.

The energy set at liberty by this transformation will accelerate the subsequent rate of transformation, so that the process will be of the nature of an explosion.

Instances of such a process occur when a liquid not in the presence of its vapour is heated above its boiling point, and also when a liquid is cooled below its freezing point, or when a solution of a salt, or of a gas, becomes supersaturated.

In the first of these cases the contact of the smallest quantity of vapour will produce explosive evaporation; in the second, the contact of ice will produce explosive freezing; in the third, a crystal of salt will produce explosive crystallization; and in the fourth, a bubble of any gas will produce explosive effervescence.

Finally, when the tangent plane touches the surface at two or more points, and is above the surface everywhere else, portions of the substance in states corresponding to the points of contact can exist in presence of each other, and the substance can pass freely from one state to another in either direction.

The state of the whole body when part is in one physical state and part in another is represented by a point in the straight line joining the centre of gravity of two masses equal respectively to the masses of the substances in the two states, and placed at the points of the model corresponding to these states.

Hence, in addition to the surface already considered, which we may call the primitive surface, and which represents the properties of the substance when homogeneous, all the points of the line joining the two points of contact of the same tangent plane belong to the secondary surface, which represents the properties of the substance when part is in one state and part in another.

To trace out this secondary surface we may suppose the doubly tangent plane to be made to roll upon the surface, always touching it at two points called the node-couple.

The two points of contact will thus trace out two curves such that a point in the one corresponds to a point in the other. These two curves are called in geometry the *node-couple* curves.

The secondary surface is generated by a line which moves so as always to join corresponding points of contact. It is a developable surface, being the envelope of the rolling tangent plane.

To construct it, spread a film of grease on a sheet of glass and cause the sheet of glass to roll without slipping on the model, always touching it in two points at least.

The grease will be partly transferred from the glass to the model at the points of contact, and there will be traces on the model of the node-couple curves, and on the glass of corresponding plane curves.

If we now copy on paper the curve traced out on the glass and cut it out, we may bend the paper so that the cut edges shall coincide with the two node-couple curves, and the paper between these curves will form the derived surface representing the state of the body when part is in one physical state and part in another.

There is one position of the tangent plane in which it touches the primitive surface in three points. These points represent the solid, liquid, and gaseous states of the substance when the temperature and the pressure are such that the three states can exist together in equilibrium.

The plane triangle, of which these points are the angles, represents all possible mixtures of these three states. For instance, if there are S grammes in the solid state, L grammes in the liquid state, and V grammes in the state of vapour, this condition of the substance will be represented by a point in the triangle which is the centre of gravity of masses S, L *and* V placed at the corresponding angular points.

From this position of the tangent plane it may roll on the primitive surface in three directions so as in each case to touch it at two points. We thus obtain three sheets of the derived surface, the first connecting the solid and liquid states, the second the liquid and gaseous states, and the third the gaseous and solid states. These three developable surfaces together with the plane triangle *SLV*, constitute what Professor Gibbs calls the Surface of Dissipated Energy.[c]

Of the three developable surfaces the first and third, those which connect the solid state with the liquid and gaseous, have been experimentally investigated only to a short distance from the triangle *SLV*; but the sheet which connects the liquid and gaseous states has been thoroughly explored.

The experiments of Cagniard de la Tour[d] and the numerical determinations of Andrews show that the curves traced out by the two points of contact of the double tangent plane unite in a point which represents what Andrews calls the critical state. At this point the two points of contact of the rolling tangent plane coalesce, and if the plane continues its roll on the surface it will touch it at one point only.

If the primitive surface forms a continuous sheet beneath the surface of dissipated energy, it cannot be at all points convexo-convex upwards. For let *AD* be the line joining two corresponding points of contact of the doubly tangent plane, and let *ABCD* be the section of the primitive surface by a vertical plane through *AD*, then it is manifest that the curve *ABCD* must in some part of its course be concave upwards.

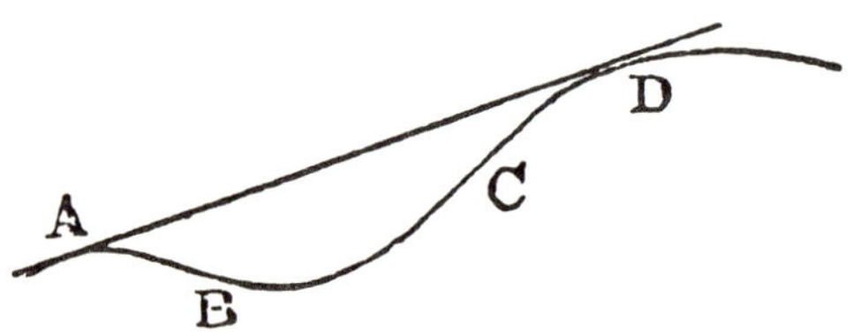

Now a point on the primitive surface at which either of its principal curvatures is concave upwards, represents a state of the body which is essentially unstable. Part of the primitive surface, therefore, if it is continuous, must represent states of the body essentially unstable. If, therefore, the primitive is continuous, there must be a region representing states essentially unstable, because one or both of the principal curvatures is concave upwards. This region is bounded by what is called in geometry the *spinode* curve. Beyond this curve the surface is convexo- convex, but the tangent plane still cuts the surface at some more or less distant point till we come to the curve of the node-couple, at which the tangent plane touches the surface at two points. Beyond this the tangent plane lies entirely above the surface, and the corresponding state of the body is essentially stable.

The region between the spinode curve and the node-couple curve represents states of the body which, though stable when the whole substance is

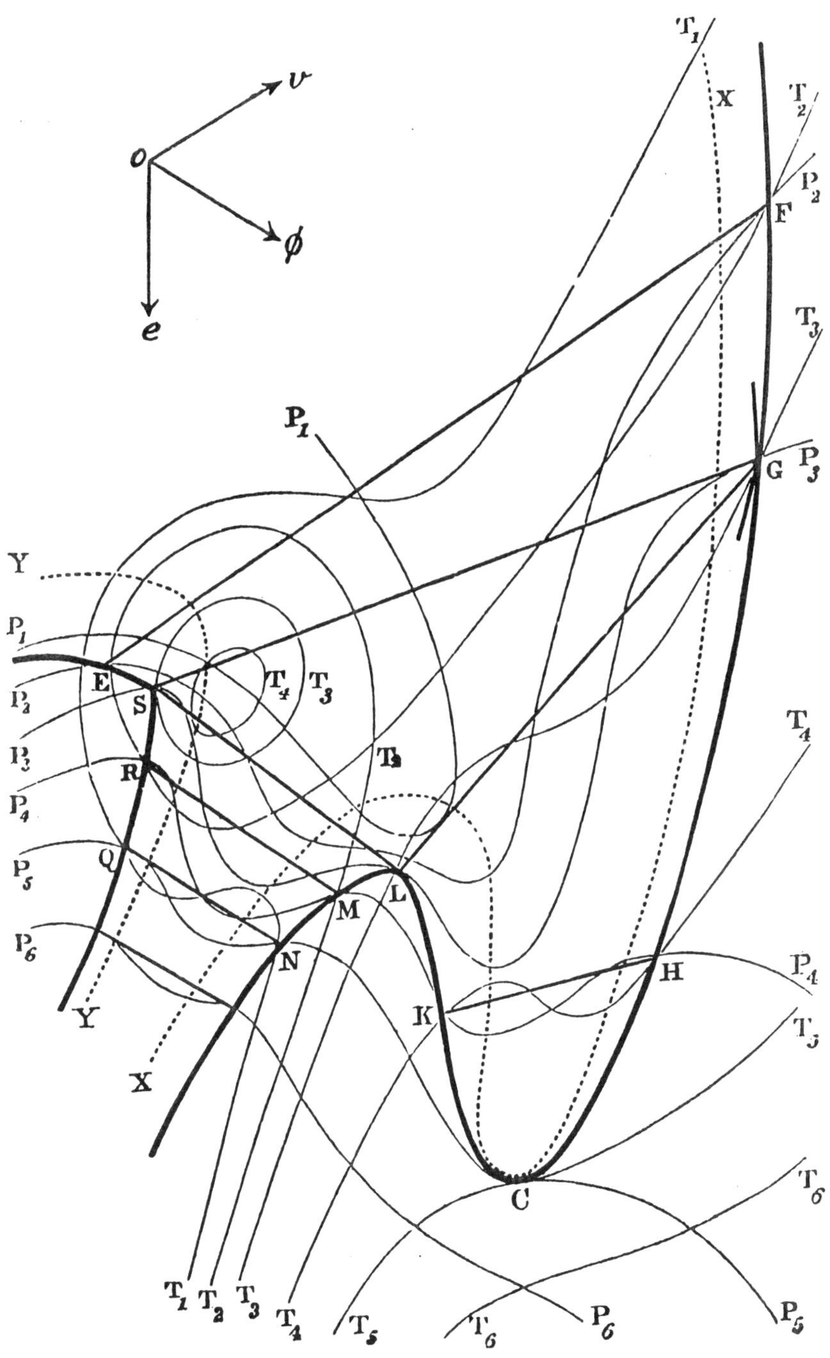

v
o
φ
e
T1
X
T2
P2
F
T3
P1
G
P3
Y
P1
E
S
T4
T3
P2
P3
R
T2
T4
P4
Q
P5
L
M
P6
N
H
P4
K
Y
T5
X
C
T6
T1
T2
T3
T4
T5
T6
P6
P5

homogeneous, are liable to sudden change if a portion of the same substance in another state is present.

Since every vertical section through two corresponding points of contact must cut the spinode curve at the points of inflexion B and C, the chord AD of the node-couple curve and the chord BC of the spinode curve must coincide at the critical point, so that at this point the spinode curve and the two branches of the node-couple curve coalesce and have a common tangent. This point is called in geometry the *tacnodal* point.

Note.–For these geometrical names I am indebted to Professor Cayley.

Thermal Lines on the Thermodynamic Surface

O	Origin
OV	Axis of volume
$O\phi$	Axis of entropy
Oe	Axis of energy
$P_1 \ldots P_6$	Isopiestics or lines of equal pressure

Of these P_1 represents a negative pressure, or, in other words, a tension, such as may exist in solids and in some liquids.

$T_1 ... T_6$, Isothermals, or lines of equal temperature. The curves T_3 and T_4 have branches in the form of closed loops.

$FGHC$. To the right of this line the substance is gaseous and absolutely stable. To the left of FG it may condense into the solid state, and to the left of GHC it may condense into the liquid state.

$CKLMN$. Below this line the substance is liquid and absolutely stable. To the right of LKC it may evaporate, to the left of LMN it may solidify.

$QRSE$. To the left of this line the substance is solid and absolutely stable. To the right of SRQ it may melt, and above SE it may evaporate.

C is the critical point of the liquid and gaseous states.

Below this line there is no discontinuity of states.

C is called in geometry the tacnodal point.

The curves FG, $GHCKL$, LMN, QRS and SE are branches of what is called in geometry the node-couple curve.

The curves XCX and YY are branches of the spinode curve. Above this curve the substance is absolutely unstable. Between it and the node-couple curve the substance is stable, but only if homogeneous.

The plane triangle SLG represents the state of uniform pressure and temperature at which the substance can be partly solid, partly liquid, and partly gaseous.

The straight lines represent states of uniform pressure and temperature in which two different states are in equilibrium.

SG and EF between solid and gaseous.

GL and KH between liquid and gaseous.

SL, RM and QN between solid and liquid.

The surface of dissipated energy consists of the plane triangle SLG and the three developable surfaces of which the generating lines are those above mentioned. This surface lies above the primitive thermodynamic surface and touches it along the node-couple curve.

a. 1. This refers to a footnote at the bottom of the page, *Transactions of the Academy of Sciences of Connecticut,* vol.ii This could refer to either, Gibbs, "Graphical Methods in the Thermodynamics of Fluids," (1873): 309–342, or, "A Method of Geometrically Representation of the Thermodynamic Properties of Substances by Means of Surfaces," 382–404, both are reprinted in Gibbs *Scientific Papers,* vol. 1: 1–32, 33-54. In view of Maxwell's discussion the second paper is the one Maxwell had in mind. Material from Maxwell's *Theory of Heat,* including some of the diagrams of Gibbs's thermodynamic surfaces was reprinted in the article "Heat" prepared after Maxwell's death by Thomas Box, which appeared in some printings of the 9th edition of the *Encyclopedia Brittanica* (in other printings the article on heat was by William Thomson).

b. Rankine introduced adiabatics in, "The Science of Energetics," *Edinburgh New Philosophical Journal* (1855): 120–121.

c. Gibbs, op cit (note a), in Gibbs *Scientific Papers,* vol. 1, 48.

d. Cagniard de la Tour, "Sur les effets qu'on obtienne par l'application simultanée de la chaleur et de la compression à la certaine liquides," *Ann. Chim.* 22 (1823): 410–415; same journal, 26 (1823): 98–100.

33. Letter from Maxwell to Thomas Andrews, July 15, 1875.

Cambridge University Library, Maxwell Collection. Copy of a letter at Department of Physics, Queens University Belfast

Glenlair,
Dalbeattie,
15 July 1875

Dear Dr. Andrews,

Your letter has been forwarded to me I am sorry that I cannot give a lecture to the Working Man's Institute this winter, and that I must decline your hospitable invitation to stay with you, but I have avoided all engagements other than my regular work at Cambridge, and must continue to do so for some time.

I shall be greatly interested in the account of your new experiments, I am busy with a new edition of my book on heat, and I hope to be able to bring it up to recent dates as regards things pertaining to an elementary treatise; and though your experiments are not elementary as regards manipulation, they are so fundamental that the sooner we learn something about the results the better.

I think the plane glass apparatus would be best for our work. I am getting a cathetometer with an additional object glass to the telescope, so as to act as a microscope at short distances and such an instrument cannot be used to observe what is inside a cylinder.

I think you know Prof. J. Willard Gibbs's (Yale College Connecticut) graphical methods in thermodynamics. Last winter I made several attempts to model the surface which he suggests, in which the three coordinates are volume, entropy and energy.

The numerical data about entropy can only be obtained by integration from data which are for most bodies very insufficient, and besides it would require a very unwieldy model to get all the features, say of CO_2, well represented, so I made no attempt at accuracy, but modelled a fictitious substance, in which the volume is greater when solid than when liquid; and in which, as in water, the saturated vapour becomes superheated by compression.

When I had at last got a plaster cast I drew on it lines of equal pressure and temperature, so as to get a rough motion of their forms. This I did by placing the model in sunlight, and tracing the curve when the rays just grazed the surface.

For if V(olume), ϕ (entropy), and E(nergy) are the coordinates

$$P(ressure) = \frac{dE}{dV} \qquad \theta\,(temperature) = \frac{dE}{d\phi}$$

I have not the model here, but I have been trying to trace the lines of temperature and pressure as far as we can conjecture their forms, and I think such graphical methods are better fitted for purely conjectural applications of the principle of continuity (beyond the range of experiment) than any empirical formulae.

I send you a sketch of these lines,[a] in which the isothermals are red, the isopiestics (equal pressure) blue, the node couple curve (limit of <stab> equilibrium of two different states) green. The spinode curve (extreme limit of stability in a homogeneous state) yellow. The black lines represent mixed states of the body. They do not lie in the thermodynamic surface but form a developable surface which is the envelope of a tangent plane touching the thermodynamic surface always at two points. These 2 points are corresponding points on the pair of branches of the node-couple curve.

The yellow spinode curve is where the curvature of the surface changes from convexo-convex to convexo-concave. <It> represents the state of water when superheated or cooled below freezing point.

At the critical point the spinode curve *touches* the node couple curve and also the isothermal and isopiestic curves.

Prof. Cayley who has furnished me with these geometrical names calls this point the tacnodal point. It is the plane where the valley of instability of the thermodynamic surfaces works itself out. Gibbs shows that at this critical point the equilibrium is stable enough though it is on the verge of instability.

Yours very truly,

J. Clerk Maxwell.

a. This freehand diagram is not reproduced here. The curves in the diagram sent to Andrews are those reproduced in Document III-32, from Maxwell, *Theory of Heat* fourth edition, 207.

34. Letter from Thomas Andrews to Maxwell, July 25, 1875.

Cambridge University Library, Maxwell Collection, Miscellaneous Correspondence.

20, Bedford Square
London. July 25,1875

Dear Professor Maxwell,

Your very interesting letter has been forwarded to me here, and I have endeavoured in the hurried intervals at my command to master the elaborate sketch you have been so kind as to send me of the lines of temperature, volume, etc. The application of the sunlight to the tracing of the isothermals and isopiestics on your model was a singularly happy thought, and a similar method will not be found useful in other cases. To an investigator like myself who can only keep my thoughts steady when the higher mathematics is used for constantly reducing the calculations to numerical values, the use of models is suggestive of the best lines of research would be very great, and I shall not fail to take advantage in this way of your suggestive note. I enclose a second copy of my recent note to the R. S.[a] Since it was published or rather communicated, I have made a few additional experiments on the value of α at different temperatures and pressures under a constant volume and with more CO_2 (p. 7). The general results are entire accordance with those given in the note, i.e., the coefficient of expansion increases with the pressure and diminishes with the temperature; but the implication stated at the foot of p. 7 does not hold good under more varied conditions of pressure. I head for the continent on Monday. I shall be away for some two months, but on my return I intend to complete this part of the enquiry and then to send the whole of my results to the R. S. If your new edition is not issued before these experiments are finished, I shall be very glad to give you the general results.

The working classes will sustain a loss from having not the advantage of hearing [you]: but I fully appreciate and concur in the propriety of your excuse. For myself I am running a race with time, and I know not how I can hope to complete the work I have in view in the few working years which remain,

Believe me

Yours very truly,

Thomas Andrews.

P.S. I have written to [?] to supply you with the plate glass [?] which are the infinitely fitted for research.

a. Royal Society. The paper Andrews refers to is, Andrews, "Preliminary Notice of Further Researches on the Physical Properties of Matter in the Liquid and Gaseous States under varied Conditions of Pressure and Temperature," *Proc. R. Soc. London,* 23 (1875): 514–521.

35. Letter from Maxwell to George Gabriel Stokes, August 3, 1875.

Memoir and Scientific Correspondence of the Late Sir George Gabriel Stokes, ed. J. Larmor (Cambridge: Cambridge University Press, 1907), vol. 2, 33–35.

Glenlair, Dalbeattie,
3 August 1875

Dear Stokes,

I have done nothing about the reflective power of gold, silver, or platinum, but am willing to continue my labours if so required.

Browning is making a divided circle to carry Jellett's prism, a 1/4 undulation plate or a Bibinet's compensator, any or all of them, and it is to be capable of being fastened to an inclined arm so as to examine the ray reflected from a liquid.

Yours very truly,

J. Clerk Maxwell

I have been thinking about Dr. Andrews's experiments on mixtures of N and CO_2.

I find the conditions of equilibrium between two mixtures in different states to be:

Let ϵ = energy of unit of mass of mixture ($\epsilon = f(v, \phi, q_1 \cdots q_{n-1})$

v = volume of do.[a]

ϕ = entropy of do.[b]

$q_1, q_2, \cdots q_{n-1}$ the masses of the substances $1, 2 \ldots n-1$ in unit of mass of the mixture (there being n constituents).
Then

$$\text{pressure} = p = -d\epsilon/dv$$

$$\text{temperature} = \theta = d\epsilon/d\phi$$

$$\text{“reaction”} = r_1 = d\epsilon/dq_1$$

$$r_2 = d\epsilon/dq_2 \text{ etc.}$$

and the conditions of equilibrium between two mixtures, one of which is distinguished by an accent, are

$$p = p'.$$

$$\theta = \theta'.$$

$$r_1 = r_1';$$

$$r_2 = r_2';$$

etc., and

$$\epsilon + p\,v - \theta\,\phi - r_1\,q_1 - r_2\,q_2 - \text{etc.} = e' + p\,v' - \theta'\,\phi' - r_1'\,q_1' - r_2'\,q_2' \text{ etc.}$$

These conditions seem to me to be all right. The difficulty is to conceive clearly the energy as a function of volume, entropy, and composition, say in liquid CO_2 saturated with absorbed N in contact with a gaseous mixture of CO_2 and N.

But it seems plain that if the thermodynamic surface (coordinates, volume, entropy and energy) has a valley or hollow place which works itself out at the critical point, after which the surface is convexo-convex, and if the surface for N is everywhere convex, then the effect of mixing N with CO_2 will be to make the head of the valley more convex, that is it will work itself out sooner, or the critical point will be lowered.

But the difficulty is to see how to form the surface for a mixture when those of the constituents are given.

However, I must interpret my conditions in this case, and in that of two liquid mixtures, as of benzol and (alcohol and water), and in solutions of salts (supersaturated etc.), and in solutions of gases (supersaturated etc.).[c]

a. Volume of unit mass of the mixture.

b. Entropy of unit mass of the mixture.

c. These notes appear to be written before Maxwell read Gibbs third paper on thermodynamics in which this problem is discussed. See Gibbs, "On the Equilibrium of Heterogeneous Substances," *Trans. Acad. Sci. Connecticut,* 3 (1876): 108–248, 343–524.

36. "On the Thermodynamics of Solutions of Variable Strength"

Cambridge University Library, Maxwell Collection Scientific Papers-7.

The properties of a substance of constant composition may be expressed as functions of two variables. If however, the substance is a mixture of any number of constituent substances its properties will depend on the proportions of these substances. If the mixture is homogeneous, the properties of each unit of mass of the mixture will be independent of the presence of the rest of the mixture and the properties depending on the variable composition of the mixture will be a function of the ratios of the components, that is, if there are n components of $n-1$ variables.

The properties of <a substance> unit of mass of a mixture of n constituents may therefore be expressed as functions of $n+1$ variables.

The most convenient variables are the volume, the entropy and the $n-1$ ratios of the masses of the $n-1$ of the constituents to the whole mass.

Then let $m_1, m_2, \ldots m_n$ be the masses of each of the constituents in the mixture, M being the whole mass then the composition of the mixture may be expressed in terms of the $n-1$ ratios

$$q_1 = \frac{m_1}{M} \qquad q_2 = \frac{m_2}{M} \cdots q_{n-1} = \frac{m_{n-1}}{M}$$

We may call q_1 the strength of the first constituent and so on. The strength of the last constituent need not be denoted by a separate symbol, for it is evidently $1 - q_1 - q_2 \cdots - q_{n-1}$.
The energy of unit of mass of the mixture will be denoted by ϵ, which is a function of the volume of unit of mass, v, of the entropy of unit of mass ϕ and of the strengths $q_1, q_2 \ldots q_{n-1}$, of all the constituents but the last.

The whole energy, volume and entropy of the mixture will be denoted by E, V, Φ where

$$E = M\epsilon \qquad V = Mv \qquad \Phi = M\phi.$$

Let us consider the variation of the whole energy E due to the variation of one of the constituents m_1,

$$\frac{dE}{dm_1} = \frac{dM}{dm_1}\epsilon + M\frac{d\epsilon}{dm_1}$$

$$= \epsilon + M\left[\frac{d\epsilon}{dq_1}\frac{dq_1}{dm_1} + \frac{d\epsilon}{dq_2}\frac{dq_2}{dm_2} + \ldots \frac{d\epsilon}{dq_{n-1}}\frac{dq_{n-1}}{dm_{n-1}}\right]$$

$$\left\langle = \epsilon + \frac{d\epsilon}{dq_1} - q_1\left(\frac{d\epsilon}{dq_1} + \frac{d\epsilon}{dq_2} + \quad + \frac{d\epsilon}{dq_{n-1}}\right)\right\rangle$$

$$= \epsilon + \frac{d\epsilon}{dq_1} - \left(q_1\frac{d\epsilon}{dq_1} + q_2\frac{d\epsilon}{dq_2} + \ldots q_{n-1}\frac{d\epsilon}{dq_{n-1}}\right)$$

The differential coefficient $d\epsilon/dq_1$ <which> denotes the rate of variation of the energy of unit mass of the mixture on account of a variation of the strength q_1 of the first constituent, the volume and entropy of unit of mass being constant. We shall denote $d\epsilon/dq_1$ by the single symbol r_1 and call it the reaction of the first constituent. It is manifest that

$$q_1 + q_2 \ldots + q_{n-1} + q_n = 1$$

$$r_1 + r_2 \ldots + r_{n-1} + r_n = 0.$$

Let us suppose two mixtures to be in contact and let us ascertain the conditions under which any constituent will pass from the first <portion> mixture to the second. We shall distinguish quantities belonging to the second <portion of the> mixture by an accent. We shall make use of the method given by Prof. J. Willard Gibbs.

We shall suppose that the total volume remains constant or

$$dV + dV' = 0$$

By the principle of Clausius the total entropy cannot diminish or

$$d\Phi + d\Phi' \geq 0$$

but since no <change can spontaneous change> motion can arise without the transformation of part of the energy into kinetic energy the quasi potential energy cannot increase, or

$$dE + dE' \leq 0$$

Hence if θ is the temperature and P the pressure of the whole

$$dE + P\,dV - \theta\,d\Phi + dE' + P\,dV' - \theta\,d\Phi' \leq 0$$

Now $< dE = M >$ $\qquad \frac{d\epsilon}{dV} = -P \qquad \frac{d\epsilon}{d\phi} = \theta$

Hence $dE =$

$$-M\,PdV + M\,\theta d\phi + \left\{\epsilon + \frac{d\epsilon}{dq_1} - (q_1 r_1 + q_2 r_2 + \ \ldots\ q_{n-1} r_{n-1}\right\} dm_1 + \text{etc.}$$

$$P\,dV = P\,M dv + P\,v(dm_1 + dm_2 + \text{etc})$$

$$\theta\,d\Phi = \theta\,M d\phi + \theta\phi\,(dm_1 + dm_2 + \text{etc})$$

Hence $dE + PdV - \theta\,d\Phi =$

$$[\epsilon + P\,v - \theta\,\phi + r_1 - (q_1\,r_1 + q_2\,r_2 + \text{etc}]\,dm_1$$
$$+[\epsilon + P\,v - \theta\,\phi + r_2 - (q_1\,r_1 + q_2\,r_2 + \text{etc}]\,dm_2$$

Adding, remembering that $dm_1 + dm_1' = 0$ equation ( ) becomes

$$[\epsilon + P\,v - \theta\,\phi + r_1 - (q_1\,r_1 + q_2\,r_2 + \text{etc})$$
$$-(\epsilon' + P\,v' + \theta\,\phi' + r_1') + q_1'\,r_1' + q_2'\,r_2' + \text{etc}]\,dm_1$$
$$+[\epsilon + P\,v - \theta\,\phi + r_2 - (q_1\,r_1 + q_2\,r_2 + \text{etc})$$
$$-(\epsilon' + P\,v' + \theta\,\phi' + r_2') + q_1'\,r_1' + q_2'\,r_2' + \text{etc}]\,dm_2 + \text{etc} \leq 0$$

Since m_1, m_2 etc. are independent variables each of the terms on the left hand side of the equation must be separately zero or negative. If we put

$$\epsilon + P\,v - \theta\,\phi - (q_1\,r_1 + q_2\,r_2 + \text{etc} + q_{n-1}\,r_{n-1}) = k$$

then for equilibrium we have

$$(k - k' + r_1 - r_1') < dm_1 \leq 0 > \equiv 0$$

$$(k - k' + r_2 - r_2') < dm_2 \leq 0 > = 0 \qquad \text{etc.}$$

Adding and remembering that $r_1 + r_2 + \ldots + r_n = 0$ we find as the condition for equilibrium.

$$k = k' \qquad r_1 = r_1' \qquad r_2 = r_2' \qquad \text{etc.}$$

that is to say that not only the pressure and temperature but the reactions of each of the constituents must be equal in the two mixtures and beside this we must have $k = k'$. These are the conditions of equilibrium of two mixtures in contact.

If the quantity

$$k = \epsilon + P\,v - \theta\,\phi - (q_1\,r_1 + q_2\,r_2 + \text{etc})$$

in which P, θ, r_1, r_2, etc. are to be regarded as constant, is an absolute maximum with respect to all values of v, ϕ, q_1, q_2, etc., that is to say if no other values of these quantities can make k greater than its original value then the mixture is absolutely stable.

If k is a maximum with respect to <all values of the variables> to all small variations of the variables though not with respect to certain large variations, then the mixture may be stable as long as it is not in contact with any other mixture in which a larger value of k occurs. If however it is in contact with such a mixture exchange will take place and the composition will be altered.

37. [Draft of "On the Equilibrium of Heterogeneous Substances"]

Cambridge University Library, Maxwell Collection, Scientific Papers-5.

thermodynamical.

The problem of the equilibrium of heterogeneous substances was first attacked by Kirchhoff in 1855.[a]

The study of science, however, is greatly facilitated when those who recognise the significance of an idea, even though that idea may be <expressed> imaged only in the form of a complex mathematical formula, give the idea a name and so render it an individual object of thought.

Thus the idea of the intrinsic energy of a material system was developed under various early and imperfect forms, such as those indicated by the terms "vis viva," "virtual velocities," "force-function" etc., till Thomson[b] introduced the phrase into Thermodynamics where it <immediately> was immediately adopted and assimilated by the principle promoters of that science.

Another thermodynamical idea, <that of Entropy> was independently discovered by Rankine[c] and by Clausius[d] and what Rankine called the Thermodynamic Function Clausius called Entropy, a much more convenient name. Entropy, however, did not make its way into scientific language as rapidly as Energy had done, <and for its original> for Clausius who invented it has had other still more abstract ideas to expound <some of which> .

Carl Neumann in his recent book on heat has used methods similar to those of Kirchhoff.[e]

Willard Gibbs of Yale College U.S. has, however, <developed the> paid greater attention to the idea of entropy and has expounded it in a very clear way, and his statement of the general conditions of equilibrium of heterogeneous substances is I think well worth the attention of chemists as well as of pure physicists.

The energy of a material system may be expressed in terms of two sets of variables.

Of these variable one set consists of quantities which can be combined by addition, such as volumes, masses, etc. The other set represents the intensities of properties of the system rather than the magnitude of the components of the system. They are pressures, temperatures, etc. <In different portions of a homogeneous substance the volumes, pressures and entropies are proportion vary with the amount of each portion and the>

The volume, the mass and the entropy of a material system is the sum of the volumes, masses and entropies of the different <components> parts of the system whatever these may be, but if the whole system can be said to have a certain pressure, temperature or potential it can only be in virtue of all the parts of the system having that pressure, temperature or potential. <It is not necessary to define the volume.>

We have to consider a fluid material consisting of one or more <connected> parts and occupying a connected region of space. Each of these parts is supposed to be homogeneous within itself, so that <as we cannot divide it into smaller parts of sensible magnitude having different ph> the smallest parts into which we are able to divide it have the same physical properties as the whole, that is to say, the same pressure, temperature, density and composition.

It is not necessary to define the volume of one of those parts. If the fluid does work it can only be by increasing its volume and the amount of work so done is measured not by the increase alone but by this increase multiplied by a variable quantity of the second kind–the pressure.

In like manner if heat enters, it can only do so by increasing the entropy, but the amount of heat which enters is increased not by the increase of entropy alone but by this increase multiplied by a variable of the second kind–the temperature.

This gives us a method of defining the entropy of a material system in a given state.

First cause each part of the system to expand or contract without communication of heat till it arrives at the standard temperature and then <let it be kept at this t> keeping it at the standard temperature bring it into the standard state as defined by <temperature and pr> its volume and pressure. The quantity of heat which must be removed from the system during the latter process, divided by the standard temperature on the thermodynamic scale gives the entropy of the system.

If the homogeneous <substance> mass is a compound one we must take account of the mass of each of its components. Distinguishing these components by suffixes we shall have

$$m = m_1 + m_2 + \text{etc.} + m_n$$

where m denotes the mass of the whole and m_1 < m_2 etc. > the mass of the first component etc.

It is not necessary to regard these components as chemical elements. They may be chemical compounds, or, solutions of any kind provided the substance is made up of them.

Thus in an ordinary mixture if oxygen, hydrogen and vapour of water we must regard all these three gases as components because at ordinary temperature oxygen and hydrogen neither combine nor become dissociated.

In other words these substances oppose a passive resistance <both to> to change of state whether by combination or dissociation which cannot be overcome at ordinary temperatures except by the condensing action of such substances as platinum. We have next to consider the principal addition which Prof. Willard Gibbs has made to the language of the subject. He defines the Potential of a particular substance with respect to a homogeneous compound as follows.

Def. If to any homogeneous mass we suppose a infinitesimal quantity of any substance to be added, the mass remaining homogeneous and its entropy and volume remaining unchanged, the increase of the energy of the mass divided by the mass of the substance added is the potential of that substance in the mass considered.[f]

a. Gustav Robert Kirchhoff (1814-1887),"Ueber einen Satz der mechanischen Wärmetheorie und einige Anwendung deselben," *Ann. Phys.* 103 (1855): 177–206.

b. William Thomson, "On the Universal Tendency in Nature to the Dissipation of Mechanical Energy," *Proc. R. Soc. Edinburgh* 3 (1852): 139–142 and *Phil. Mag.* 4 (1852): 304–306.

c. William John Macquorne Rankine, "On the Mechanical Action of Heat," *Trans. R. Soc. Edinburgh* 20 (1853): 1–21.

d. R. J. E. Clausius, "Ueber verscheiden für die Anwendung bequem Formen der Hauptgleichungen der mechanischen Wärmetheorie," *Ann. Phys.* 125, (1865): 353–400.

e. Carl Gottfried Neumann (1832-1925), *Vorlesungen über die mechanischen Wärmetheorie* (Leipzig: B. G. Teubner, 1875).

f. From these remarks it is clear that Maxwell had now read Gibbs, "On the Equilibrium of Heterogeneous Substances," *Trans. Acad. Sci. Connecticut,* 3 (1876): 108–248, 343–524; reprinted in *Scientific Papers of J. Willard Gibbs,* vol. 1, 55–353. The definition occurs on p. 93.

38. "On the Equilibrium of Heterogeneous Substances"

Phil. Mag. 16 (1908): 819–824, introduction by Joseph Larmor, 818–819.[a]

The warning which Comte[b] addressed to his disciples, not to apply dynamical or physical ideas to chemical phenomena, may be taken, like several other warnings of his, as an indication of the direction in which science was threatening to advance.

We can already distinguish two lines along which dynamical science is working its way to undermine at least the outworks of Chemistry, and the chemists of the present day, instead of upholding the mystery of their craft, are doing all they can to open their gates to the enemy.

Of these two lines of advance one is conducted by the help of the hypothesis that bodies consist of molecules in motion, and it seeks to determine the structure of the molecules and the nature of their motion from the phenomena of portions of matter of sensible size.

The other lines of advance, that of Thermodynamics, makes no hypothesis about the ultimate structure of bodies, but deduces relations among observed phenomena by means of two general principles–the conservation of energy and its tendency towards diffusion. The thermodynamical problem of the equilibrium of heterogeneous substances was attacked by Kirchhoff in 1855,[c] when the science was yet in its infancy, and his method has been lately followed by C. Neumann.[d] But the methods introduced by Professor J. Willard Gibbs, of Yale College, Connecticut,[e] seem to me to be more likely than any others to enable us, without any lengthy calculations, to comprehend the relations between the different physical and chemical states of bodies, and it is to these that I now wish to direct your attention.

In studying the properties of a homogeneous mass of fluid, consisting of n component substances, Professor Gibbs takes as his principal function the energy of the fluid, as depending on its volume and entropy together with the masses, m_1, m_2 ... m_n of its n components, these $n+2$ variables being regarded as independent. Each of these variables is such that its value for any material system is the sum of its values for the different parts of the system.

By differentiating the energy with respect to each of these variables we obtain $n+2$ other quantities, each of which has a physical significance which is related to that of the variable to which it corresponds.

Thus, by differentiating with respect to the volume, we obtain the pressure of the fluid with its sign reversed; by differentiating with respect to the entropy, we obtain the temperature on the thermodynamic scale; and by differentiating with respect to the mass of any one of the component substances, we obtain what Professor Gibbs calls the potential of that substance in the mass considered.

As this conception of the potential of a substance in a given homogeneous mass in a new one, and likely to become important in the theory of chemistry, I shall give Professor Gibbs' definition of it.

"If to any homogeneous mass we suppose an infinitesimal quantity of any substance added, the mass remaining homogeneous and its entropy and volume remaining unchanged, the increase of the energy of the mass, divided by the mass of the substance added, is the potential of that substance in the mass considered."

These $n + 2$ new quantities, the pressure, the temperature, and the n potentials of the component substances, form a class differing in kind from the first set of variables. They are not quantities capable of combination by addition, but denote the intensity of certain physical properties of the substance. Thus the pressure is the intensity of the tendency of the body to expand, the temperature is the intensity of its tendency to part with heat; and the potential of any component substance is the intensity with which it tends to expel that substance from its mass.

We may therefore distinguish between these two classes of variables by calling the volume, the entropy and the component masses the *magnitudes,* and the pressure, the temperature, and the potentials the *intensities* of the system.

The problem before us may be stated thus:—Given a homogeneous mass in a certain phase, will it remain in that phase, or will the whole or part of it pass into some other phase?

The criterion of stability may be expressed thus in Professor Gibbs's words:–"For the equilibrium of any isolated system it is necessary and sufficient that in all possible variations of the state of the system which do not alter its energy, the variation of its entropy shall either vanish or be negative.

"The condition may also be expressed by saying that for all possible variations of the state of the system which do not alter its entropy, the variation of its energy shall either vanish or be negative."

Professor Gibbs has made a most important contribution to science by giving us a mathematical expression for the stability of any given phase (A) of matter with respect to any other phase (B).

If this expression for the stability (which we may denote by the letter K) is positive, the phase A will not of itself pass into the phase B, but if it is negative the phase A will of itself pass into phase B, unless prevented by passive resistances.

The stability (K) of any given phase (A) with respect to any other phase (B), is expressed in the following form:–

$$K = \epsilon - V\,P + \eta T - m_1\,\mu_1 - \text{etc.} - m_n\,\mu_n$$

where ϵ is the energy, V the volume, μ the entropy, and m_1, m_2, etc., the components corresponding to the second phase (B) while P is the pressure, T the temperature, and μ_1, μ_2, etc., the potentials corresponding to the given phase (A). The intensities therefore were those belonging to the given phase (A), while the magnitudes are those corresponding to the other phase (B).

We may interpret this expression for the stability by saying that it is measured by the excess of the energy in the phase (B), above what it would have been if the magnitudes had increased from zero to the values corresponding to the phase B, while the values of the intensities were those belonging to phase (A).

If the phase (B) is in all respects except that of absolute quantity of matter the same as phase (A), K is zero; but when the phase (B) differs from phase (A), a portion of the matter in the phase (A) will tend to pass into the phase (B) if K is negative, but not if it is zero or positive.

If the given phase (A) of the mass is such that the value of K is positive or zero with respect to every other phase (B), then the phase (A) is absolutely stable, and will not of itself pass into any other phase.

If, however, K is positive with respect to all phases which differ from phase (A) only by infinitesimal variations of the magnitudes, while for a certain other phase, (B), in which the magnitudes differ by finite quantities from those of the phase (A) K is negative, then the question whether the mass will pass from phase (A) to phase (B) will depend on whether it can do so without any transportation of matter through a finite distance, or, in other words, on whether matter in the phase B is or is not in contact with the mass.

In this case the phase (A) is stable in itself, but is liable to have its stability destroyed by contact with the smallest portion of matter in certain other phases.

Finally, if K can be made negative by any infinitesimal variations of the magnitudes of the system (A), the mass will be in unstable equilibrium, and will of itself pass into some other phase.

As no such unstable phase can continue in any finite mass for any finite time, it can never become the subject of experiment; but it is of great importance in the theory of chemistry to know how these unstable phases are related to those which are relatively or absolutely stable.

The absolutely stable phases are divided from the relatively stable phases by a series of pairs of coexistent phases, for which the intensities P, T, μ etc. are equal and K is zero. Thus water and steam at the same temperature and pressure are coexistent phases.

As one of the two coexistent phases is made to vary in a continuous manner, the other may approach it and ultimately coincide with it. The phase in which this coincidence takes place is called the Critical Phase.

The region of absolutely unstable phases is in contact with that of absolutely stable phases at the critical point. Hence, though it may be possible by preventing the body from coming in contact with certain substances to bring it into a phase far beyond the limits of absolute stability, this process cannot be indefinitely continued, for before the substance can enter a new region of stability it must pass out of the region of relative stability into one of absolute instability, when it will at once break up into a system of stable phases.

Thus in water for any given pressure there is a corresponding temperature at which it is in equilibrium with its vapour, and beyond which it cannot be raised when in contact with any gas. But if, as in the experiment of Dufour, a drop of water is carefully freed from air and entirely surrounded by liquid which has a high boiling-point, it may remain in the liquid state at a temperature far above the boiling-point corresponding to the pressure, though if it comes in contact with the smallest portion of any gas it instantly explodes.

But it is certain that if the temperature were raised high enough the water would enter a phase of absolutely unstable equilibrium, and that it would then explode without requiring the contact of any other substance.

Water may also be cooled below the temperature at which it generally freezes, and if the water is surrounded by another liquid of the same density the pressure may also be reduced below that of the vapour of water at that temperature. If the water when in this phase is brought in contact with ice it will freeze, but if brought in contact with a gas it will evaporate.

Professor Guthrie[h] has recently discovered a very remarkable case of equilibrium of a liquid which may be solidified in three different ways by contact with three different substances. This is a solution of chloride of calcium in water containing 37 percent of the salt. This solution is capable of solidification at $-37°C.$, when it forms the solid cryohydrate having the same composition as itself. But it may be cooled somewhat below this temperature, and then if it is touched with a bit of ice it throws up ice, if it is touched with the anhydrous salt it throws down anhydrous salt, and if it is touched with the cryohydrate it solidifies into cryohydrate.

a. Joseph Larmor notes that this is the address Maxwell delivered at the South Kensington Conference, 24th May 1876, and while it was printed in the conference report it was no longer available. The paper was not included in Maxwell's collected papers and was therefore being reprinted. The paper under the above title in Maxwell, *Scientific Papers,* 2, 498–500 is only an abstract of the paper reproduced above.

b. Isidore Auguste Marie Francois Xavier Comte (1798–1857) systematically developed his positive philosophy for science in *Cours de philosophie positive,* 6 vols., translated into English and reduced to 2 vols. (London: 1853).

c. G. R. Kirchhoff, "Ueber einen Satz der mechanischen Wärmetheorie und einige Anwendung deselben," *Ann. Phys.* 103 (1855): 177–206.

d. Carl Gottfried Neumann (1832–1925), "Ueber die mechanische Energie der Schwefelsaure," *Leipzig Berichte Math. Phys.* 21 (1869): 221–256, and *Math. Ann.* 3 (1871): 581 610; and *Vorlesungen über die mechanischen Wärmetheorie,* (1875).

e. Josiah Willard Gibbs, "On the Equilibrium of Heterogeneous Substances," *Trans. Acad. Sci. Connecticut* 3 (1876): 108–248, 343–524. In this paper Maxwell refers to material in the first part of Gibbs's paper only, reprinted in *Scientific Papers of J. Willard Gibbs,* (New York: Dover, 1961), vol. 1, 55–353.

f. Gibbs, *Scientific Papers,* vol. 1, 93.

g. Gibbs, *Scientific Papers,* vol. 1, 56.

h. Frederick Guthrie, "On Salt Solutions and attached Water, Parts I-VI," *Proc. R. Soc. London,* 1 (1876): 53–101; 2, 1–14; 53–79; 87–108, reprinted in *Phil. Mag.* 49 (1875): 1–20, 206–218, 266–276.

39. [Abstract of "On the Equilibrium of Heterogeneous Substances"]

Proc. Cambridge Phil. Soc. 11 (1876): 427–30; *Scientific Papers,* vol. 2, 498–500. This is an abstract of Document III-38.

[From the *Proceedings of the Cambridge Philosophical Society*, Vol. II., 1876.]

LXXVI. *On the Equilibrium of Heterogeneous Substances.*

THE thermodynamical problem of the equilibrium of heterogeneous substances was first attacked by Kirchoff in 1855, who studied the properties of mixtures of sulphuric acid with water, and the density of the vapour in equilibrium with the mixture. His method has recently been adopted by C. Neumann in his *Vorlesungen über die mechanische Theorie der Wärme* (Leipzig, 1875). Neither of these writers, however, makes use of two of the most valuable concepts in Thermodynamics, namely, the intrinsic energy and the entropy of the substance.

It is probably for this reason that their methods do not readily give an explanation of those states of equilibrium which are stable in themselves, but which the contact of certain substances may render unstable.

I therefore wish to point out to the Society the methods adopted by Professor J. Willard Gibbs of Yale College, published in the *Transactions of the Academy of Sciences of Connecticut*, which seem to me to throw a new light on Thermodynamics.

He considers the intrinsic energy (ϵ) of a homogeneous mass consisting of n kinds of component matter to be a function of $n+2$ variables, namely, the volume of the mass v, its entropy η, and the n masses, m_1, $m_2 \ldots m_n$, of its component substances.

Each of these variables represents a physical quantity, the value of which, for a material system, is the sum of its values for the parts of the system.

By differentiating the energy with respect to each of these variables (considered as independent), we obtain a set of $n+2$ differential coefficients which represent the intensity of various properties of the substance. Thus,

$\frac{d\epsilon}{dv} = -p$, where p is the pressure of the substance;

$\frac{d\epsilon}{d\eta} = \theta$, where θ is the temperature on the thermodynamic scale;

$\frac{d\epsilon}{dm_1} = \mu_1$, where μ_1 is the potential of the component (m_1) with respect to the compound mass.

Each of the component substances has therefore a potential with respect to the whole mass.

The idea of the potential of a substance is, I believe, due to Prof. Gibbs. His definition is as follows:—

If to any homogeneous mass we suppose an infinitesimal quantity of any substance to be added, the mass remaining homogeneous, and its entropy and volume remaining unchanged, the increase of the energy of the mass, divided by the mass of the substance added, is the *potential* of that substance in the mass considered.

The condition of the stable equilibrium of the mass is expressed by Prof. Gibbs in either of the two following ways:

I. *For the equilibrium of any isolated system it is necessary and sufficient that in all possible variations of the state of the system which do not alter its energy, the variation of its entropy shall either vanish or be negative.*

II. *For the equilibrium of any isolated system it is necessary and sufficient that in all possible variations of the state of the system which do not alter its entropy, the variation of the energy shall either vanish or be positive.*

The variations here spoken of must not involve the transportation of any matter through any finite distance.

It follows from this that the quantities θ, p, $\mu_1 \ldots \mu_n$ must have the same values in all parts of the mass. For if not, heat will flow from places of higher to places of lower temperature, the mass as a whole will move from places of higher to places of lower pressure, and each of the several component substances will pass from places where its potential is higher to places where it is lower, if it can do so continuously.

Hence Prof. Gibbs shews that if Θ, P, $M_1 \ldots M_n$ are the values of θ, p, $\mu_1 \ldots \mu_n$ for a given phase of the compound, and if the quantity

$$K = \epsilon - \Theta\eta + Pv - M_1 m_1 - \&c. - M_n m_n,$$

is zero for the given fluid, and is positive for every other phase of the same components, the condition of the given fluid will be stable.

If this condition holds for all variations of the variables the fluid will be absolutely stable, but if it holds only for *small* variations but not for certain finite variations, then the fluid will be stable when not in contact with matter in any of those phases for which K is positive, but if matter in any one of these phases is in contact with it, its equilibrium will be destroyed, and a portion will pass into the phase of the substance with which it is in contact.

Thus in Professor F. Guthrie's experiments, a solution of chloride of calcium of 37 per cent. was cooled to a temperature somewhat below $-37°$ C. without solidification.

In this state, however, the contact of three different solids determines three different kinds of solidification. A piece of ice causes ice to separate from the fluid. A piece of the cryohydrate of chloride of calcium determines the formation of cryohydrate from the fluid, and the anhydrous salt causes a precipitation of anhydrous salt.

The phase of the fluid is such that K is positive for all phases differing slightly from its own phase, and its equilibrium is therefore stable, but for certain widely different phases, namely, ice, cryohydrate and anhydrous salt, K is negative.

If none of these substances are in contact with the fluid, the fluid cannot alter in phase without a transport of matter through a finite distance, and is therefore stable; but if any one of them is in contact with the fluid, part of the fluid is enabled to pass into a phase in which K is negative. The conditions of consistent phases are that the values of θ, p, $\mu_1 \ldots \mu_n$, and K are equal for all phases which can coexist in equilibrium, the surface of contact being plane.

This was illustrated by Mr Main's experiments on co-existent phases of mixtures of chloroform, alcohol and water.

40. Postcard from Maxwell to Peter Guthrie Tait, July 29, 1876

Cambridge University Library, Maxwell Collection, Maxwell-Tait Correspondence.

Glenlair, 29 July 1876.
Mohr sent on.[a] Still game to lynξ proofs. Busy with Cavendish on potentials, capacities, dielectric constants and effect of heat on them spreading of electricity etc.[b]
Have you read Willard Gibbs on Equilibrium of Heterogeneous substances,[c] Refreshing after H. Spencer on the Instability of the Homogeneous.[d]

a. This is a reference to Carl Friederich Mohr (1806–1879), "Ansichten über die Natur der Wärme," *Z. Phys.* 5 (1837): 419–432, 433–445, which was brought to Tait's attention by Alexander Crum Brown. Tait's translation of Mohr was published in *Phil. Mag.* 2 (1876): 110. Tait used Mohr's work to undermine J. Robert Mayer's claims for the discovery of the conservation of energy.

b. This is a reference to Maxwell's study of Henry Cavendish's work in electricity during which he repeated many of Cavendish's experiments. The study was published as Maxwell, *The Unpublished Electrical Writings of Hon. Henry Cavendish,* (Cambridge, 1879).

c. Josiah Willard Gibbs, "On the Equilibrium of Heterogeneous Substances," *Trans. Acad. Sci. Connecticut,* 3 (1876): 108–248, 343–524. Maxwell is here referring to the first part which was published the earlier in 1876. He had already received a reprint of it.

d. Herbert Spencer (1820–1903) developed a theory of social development based on the idea of evolution, Lamarck's not Darwin's, and the principles of thermodynamics as he interpreted them. Spencer translates the first law of thermodynamics into the "Persistence of Force." This principle stated that the homogeneous was unstable and all change produced multiple effects that led to more heterogenity and complexity. Increasing complexity was the end result of all change, physical or social. Thus Spencer could account for change in social systems, social evolution, and its necessity from basic first principles. He assumed, as did most social theorists of the late nineteenth century, that the laws of nature applied directly to society. Maxwell had argued with Spencer on his interpretation of the first law of thermodynamics. . This law appeared in Spencer, *First Principles,* (1862), his attempt to reconcile science and theology. See Garber, Brush, and Everitt, *Maxwell on Molecules and Gases,* 156–161.

41. Letter from Maxwell to Peter Guthrie Tait, October 13, 1876[a]

Cambridge University Library, Maxwell Collection, Maxwell-Tait Correspondence.

Glenlair,

Dalbeattie,
13th Oct. 1876.

O T',

I return the last page of Clausius. I have got the whole vollum from the author.

When you wrote the Sketch[b] your knowledge of Clausius was somewhat defective. Mine is still, though I have spent much labour upon him and have occasionally been rewarded, e.g., *earlier* papers on molecular stotting,[c] electrolysis, entropy and concentration of rays. N. B., In the latter paper (reprinted in the vollum) the name of Hamilton[d] does not occur. When you are atrouncing, trounce him for that. Only perhaps Kirchhoff ignored Hamilton first and Clausius followed him unwittingly not being a constant reader of the R. I. A.[e] transactions and knowing nothing of H[f] except (lately) his Princip.[g] which he and others try to degrade into the $2^n d$ law of $\theta\Delta$ [thermodynamics] as if any pure dynamical statement would submit to such indignity.

With respect to your citation of Thomson[h] it would need to be more explicit.

The likiest thing I find to what you give is in the 1st paper on D. T. of H[i] (17 March 1851, p. 272 and 273) but I do not find dq divided by anything like t.[j]

I think Rankine by introducing his Thermodynamical function ϕ which is $\int dq/t$ made a great hit, because ϕ is a real quantity whereas q is not, only $dq = t\,d\phi$.

There are many things in T[k] which are equivalent to this because T has worked at the subject and worked correctly, and all mathematical truth is one, but you cannot expect C.[ℓ] to see this unless it stands very plain in print. In short Rankine's statements on are *identical* with those of C but T's are only *equivalent.*

It is quite possible Challis may think you have an overweening estimate of foreigners and that a particular foreigner may think that your estimate of him is underweening as compared with his own.[m]

With respect to our knowledge of the condition of energy inside a body, both Rankine and Clausius pretend to know something about it.

We certainly know how much goes in and comes out and we know whether at entrance or exit it is in the form of heat or work, but what disguise it assumes when in the privacy of bodies or as Torricelli says "nell' intima corpalenza de' solidi naturali" is known only to R, C, and Co..[n]

To Cambridge on the 19th.

Cox,[o] a new fellow of Trinity has written a very grand dissertation "an account of Hamilton's papers on Systems of Rays." It will help in the boiling down of these most important papers,

Yours $\frac{dp}{dt}$

1st May 1854 Part V
Art 101, p. 126

$$\frac{H_t}{t} + \frac{H_t'}{t'} \text{ etc } = 0$$

Is this the true tepr?[p]

a. Some of this letter is quoted in Knott, *Life of Tait,* 222.

b. Tait, *A Sketch of Thermodynamics,* (Edinburgh, 1867).

c. "Stotting" is a Scottish and North of England dialect word meaning "bumping against one another." The noun Stot means a young ox or steer, so the verb to stot is closely parallel to the verbs butt or jostle. Maxwell is here referring to Clausius's papers on molecular collisions, Clausius, "Ueber die Art der Bewegung welche wir Wärme nennen," *Ann. Phys.* 100 (1857): 353–380, translated in *Phil. Mag.* 14 (1857): 108–127, and, "Ueber die mittlere Länge der Wege, welche bei der Molekülarbewegung gasförmiger Körper von den einzelnen Molekülen zurückgelegt werden nebst einigen anderen Bemerkungen über die mechanischen Wärmetheorie," *Ann. Phys.* 105 (1858): 239–258, translated in *Phil. Mag.* 17 (1859): 81–91. Clausius's early work on electrolysis was Clausius, "Ueber die Elektric-Leitung in Elektrolyten," *Ann. Phys.* 101 (1857): 338–360, translated in *Phil. Mag.* 15 (1858): 94–109. Clausius's paper on entropy refered to here was Clausius, "Über verscheiden für die Anwendung bequem Formen der Hauptgleichungen der mechanischen Wärmetheorie," *Ann. Phys.* 125 (1865): 353–401. The final paper mentioned is Clausius, "Über die Concentration von Wärme- und Lichtstrahlen und die Grenzen ihrer Wirkung," *Ann. Phys.* 121 (1864): 1–44.

d. William Rowan Hamilton (1805–1865).

e. Royal Irish Academy Transactions in which most of Hamilton's papers were published.

f. Hamilton.

g. Hamilton's Principle of Least Action which Clausius had used in his efforts to reduce the second law of thermodynamics to mechanical form. See Clausius, "Über die Zurückfuhrung des zweiten Haupsatzes der mechanischen Wärmetheorie auf allgemeinen mechanische Principien," *Ann. Phys.* 142 (1871): 433–461.

h. William Thomson.

i. Dynamical Theory of Heat. William Thomson read two papers to the Royal Society of Edinburgh on this subject in 1851. The first, read March 17th 1851 is the one to which Maxwell refers. See Thomson, "On the Dynamical Theory of Heat," *Trans. R. Soc. Edinburgh* 20 (1853): 261–288, *Phil. Mag.* 4 (1852): 8–21. Maxwell is correct in that no form of the second law appears here.

j. Tait still claimed that the expression for entropy was already in Thomson's paper on energy dissipation of 1852. See Thomson, "On a Universal Tendency in Nature to the Dissipation of mechanical Energy," *Proc. R. Soc. Edinburgh* 3

(1852): 139–142, *Phil. Mag.* 4 (1852): 304–306. However, no explicit form of the second law appears here either. Tait's assertion remained in his text despite Maxwell's observation. See Tait, *Sketch of Thermodynamics* second edition (1877), chap. 3, where Tait deduces the second law from Carnot's principle concluding with the remark that $\int \frac{dQ}{T}$ is the heat dissipated during a cycle, a result due to Thomson.

k. William Thomson.

ℓ. Clausius.

m. James Challis (1803–1882).

n. Rankine, Clausius and company.

o. This appears to be Homersham Cox, a Cambridge mathematician. For Maxwell's work on Hamilton's Characteristic function see Maxwell, "On Hamilton's Characteristic Function for a Narrow Beam of Light," *Proc. London Math. Soc.* 6 (1874), "On the Relation of Geometrical Optics to Other Parts of Mathematics and Physics," *Proc. Cambridge Philos. Soc.* 2 (1874), "On the Application of Hamilton's Characteristic Function to the Theory of an Optical Instrument symmetrical about its Axis," *Proc. London Math. Soc.* 6 (1875), all reprinted in Maxwell, *Scientific Papers*, vol. 2, 381–390, 391–392, 439–444, respectively. See also C. W. F. Everitt, *James Clerk Maxwell* (New York: Scribners, 1975), 167–169. Maxwell is referring back to the problem of note i. This other reference is to William Thomson, "On the Dynamical Theory of Heat Part IV," (read May 1st, 1854), *Trans. R. Soc. Edinburgh* 21 (1857): 123–171. This is the first analytical expression of the second law given by Thomson.

p. Temperature.

42. Letter from Maxwell to Peter Guthrie Tait, December 28, 1876[a]

Cambridge University Library, Tait Letters

Glenlair, Dalbeattie 28 Dec.

O T' I have not seen any prooves about Rumford & Davy. Wither did you send them? Precession.—Show how to construct the earth so as to be spherical outside and get to preced and nutate all the same.

Fig ⊕[b] If the ocean were spherical and in spite of the ⊕'s rotation show that precession would be recession.

If, in addition apparent gravity were uniform as well as towards the ⊕['s] centre calculate the recession.

General remark. In Tripos examination credit should not be assigned to an extensive acquaintance with the errors of popular text books. It is enough for the Senior Wrangler if he knows <facts> correct methods and established facts.

Capillarity—If the rim is circular you may go to work by considering equilibrium of forces resolved along the axis. But to have a figure not spherical you must have two circular rims. Then you get the meridian circle, that traced by the focus of the conic which rolls on the axis.

Pendulum which *just* makes *a* complete revolution (to make revolutions requires an eternity +)

Heat

In James Thomson's figure of the continuous isothermal show that the horizontal line representing mixed liquid and vapour cuts off *equal areas* above and below the curve. Do this by Carnots cycle. That I did not do it in my book shows my invincible stupidity.[c]

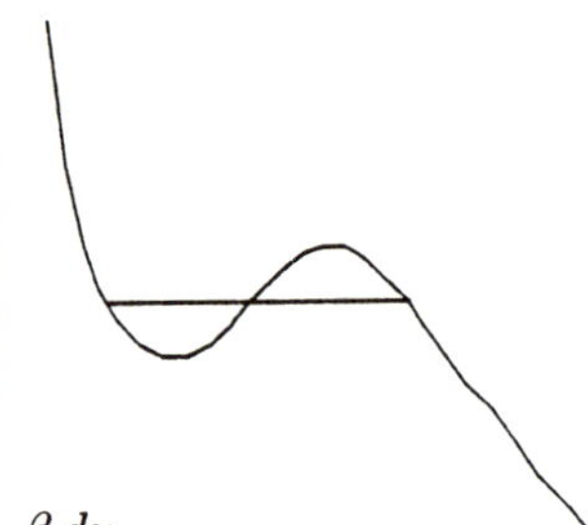

Physical meaning of complete difls [differentials]

$$d\phi = \alpha\, dx + \beta\, dy$$

depends on what you are studying. The general problem–given a mathematical expression, determine what physical meaning it may have–has not yet been solved in such a way as to be adapted to Senate House.

Yours $\dfrac{dp}{dt}$

N.B. I burnt the proof of the 2nd Day as I was loaded with Natural Tripos papers & Math Tripos was in danger.

Observed the Girton College new spelling: air in a barometer is most endurious as being most expansible.

This law wholds good.

a. The year on this letter is assumed from from its contents. Maxwell is referring to passages in the second edition of Tait's text on thermodynamics, published in 1877, that Maxwell proofread. 1876 would seem the most reasonable date.

b. Figure of the earth.

c. The following is on the reverse. "The proof is given at the end of Chapter VI in later editions of Maxwell's *Theory of Heat.*"

43. Letter from R. J. E. Clausius to Maxwell, November 8, 1877.

Cambridge University Library, Maxwell Collection, Miscellaneous Correspondence.

Bonn, 8. Nov. 1877

Hochgeehrter Herr,

Ich beeile mich, Ihre Anfrage wegen des Namens Entropie zu beantworten, indem ich das grösste Gewicht darauf lege, mit Ihnen nicht nur in Bezug auf die theoretischen Ansichten, sondern auch in Bezug auf die Benennungen in Uebereinstimmung zu sein.

Den Namen Entropie habe ich eingefuhrt in einer Abhandlung von 1865, welche ich Ihnen beifolgend unter Kreuzband übersende.[a] Die Stelle befindet sich auf Seite 390; es ist aber zweckmässig, zum besseren Verständnisse auch die vorhergehenden Seiten, etwa von S. 387 an, zu lesen. Die frühere Abhandlung von 1862, welche dort erwähnt ist, füge ich ebenfalls bei.[b]

Beide Abhandlungen sind auch in meiner Abhandlungen–Sammlung als Abhandlung VI and IX abgedruckt und befinden sich auch in der von Hirst herausgegebenen englischen Uebersetzung, wo die Stelle über den Namen Entropie sich auf S. 357 befindet.[c]

Mit der Versicherung der aus gezeichnetsten Hochachtung verbleibe ich

ganz der Ihrige

R. Clausius.

a. Clausius, "Über verschiedene für die Anwendung bequeme Formen der Hauptgleichungen der mechanischen Wärmetheorie," *Ann. Phys.* 125 (1865): 353–401.

b. Clausius, "Über die Anwendung des Satzes von der Aequivalenz der verwandlungen auf die innere Arbeit," *Ann. Phys.* 116 (1862): 73–112, translated in *Phil. Mag.* 24 (1862): 81–97, 201–213.

c. Clausius, *Abhandlungen über die mechanischen Wärmetheorie,* (Braunschweig, 1864–1867), 2 vols., translated by W. Hirst (London, 1876).

44. Letter from Maxwell to Peter Guthrie Tait, December 12, 1877

Cambridge University Library, Maxwell Collection, Maxwell-Tait Correspondence.

11, Scroope Terrace,
Cambridge,
12 Dec., 1877.

D^r T,

Can you tell me what sort of opinion the R. S. E.[a] wish about a Thesis for the degree of S. D.[b] Have I been made a University Examiner as well as a F. R. S. E.[c] or has the Pharisee as such the right of nominating or vetoing Doctors, or should I report that the Thesis ought or ought not to be printed in the Transactions (by leave of University) who may have their own opinion on its merits.

I have received and read $\theta\Delta^d$ which I have also in the scales. Can you tell me anything about the cubic compressibility of ice. Also its <dilation as it> change of volume by temperature. I have made casts of Gibbs Thermodynamic surface (fancy medium). Would you like one with isopiestics, isothermals etc., etc., drawn on it? But this is a pure fancy medium with its features distorted so as [to] get them all into the model. So I am now brewing a model for water-substance: showing ice, water and steam.
Qu. How to show, say, the volumes of ice, water at 0° at 4° at 100°, at critical point, steam at do., at 100° at 0° together with their entropies and energies all in the same model under 20 cm in the side?

Note received with thanx 2 naughty except for vacation reading: picture greatly admired and selected for Xmas decorations.

When the man of Bonn receives from his <bookseller> publisher the profits (±) of his 2nd edition I hope he will be able to distinguish how much is due to the "internal work" expended in concocting the same, and how much to the aequivalenz werth of the product.

Some of the materialists suppose that some definite proportion exists between these two quantities but the whole of experience, both ours and other men's goes to establish the contrary.

Tolver Preston[f] was once with T and F and fortune smiled propitious on his lot but since cables are slack he has been living on his funds and earning a precarious subsistence by publishing books on the aether at his own great expence. I have asked him what his forte is as regards gross matter as I do not know but what his powers if directed thereto might lead to butter as well as bread. He is by no means a paradoxer though a fierce speculator, and what is rare among such folk he improves and amends his errors. I told him he could do mathematics if he tried (not of course for anything but to assist speculation).

Have you seen Naegli in Nature?[g] He is of the school which supposes that all we think as well as all we do is the result of the deliberations of n plastidule souls in every one of the α molecules of us. Now if m Masters of Arts, sleeping within r of Great St. Mary constitute the Electoral Roll, the wisdom of the soul of that Roll must be $mn\alpha$ times that of a plastidule soul. The acts of the Roll are on record. Calculate the wisdom 1st of an M. A. 2nd of a plastidule soul,

$$\frac{dp}{dt}$$

a. Royal Society of Edinburgh.

b. Doctor of Science.

c. Fellow the the Royal Society of Edinburgh.

d. Thermodynamics. This refers to the proofs of the second edition of Tait's text on thermodynamics, Tait, *A Sketch of Thermodynamics* (Edinburgh: David Douglas, 1877), which Maxwell proofread and also reviewed, see Document III-47.

e. Rudolph Clausius, "the man of Bonn" and the first volume of the second edition of Clausius, *Abhandlungen über die mechanischen Wärmetheorie* volume 1 (Braunschweig: F. Viewig and Sohn, 1876). The second volume of this edition was published in 1887 and the third posthumously, edited from Clausius's notes by Max Planck and Carl Pulfrich in 1890. "Internal work" and "aequivalenz werth" refer to aspects of Clausius's thermodynamics. The latter appears in Clausius, "Ueber eine verwandte Form der zweite Hauptsatzes der mechanischen Wärmetheorie," *Ann. Phys.* 93 (1854): 481–507, translated in Phil. Mag. 12 (1856): 81–98. When heat was transformed into either heat at the same or a lower temperature or into work at the same temperature it generated an equivalence value of Q/T or $\frac{Q_1}{T_1} - \frac{Q_2}{T_2}$ with the change in temperature. This led to Clausius's first expression for the second law. Internal work was the work done by a gas against the forces between the molecules as it expanded. The concept appeared in Clausius's papers on disgregation. See Clausius, "Über die Anwendung des Satzes von der Aequivalenz der verwandlungen auf die innere Arbeit," *Ann. Phys.* 116 (1862): 73–112, translated in *Phil. Mag.* 24 (1862): 81–97, 201–213.

f. Samuel Tolver Preston was born in 1844. The date of his death is uncertain. Poggendorff lists him as still active in 1908, although no death date appears in the next edition. Preston acknowledged Maxwell's help in his paper, "On the View of the Propagation of Sound demanded by the Kinetic Theory of Gases," *Nature* 18 (1878): 253–255. He afterwards wrote many papers on heat, light, electromagentism and gravitation, not all of them speculative. "T and F" means Thomson and H. C. Fleming Jenkin, and "now that cables are slack" means that little work was available in the cable laying business.

g. Carl Wilhelm von Naegeli (1817–1891), botanist. Naegeli accepted the mechanism of evolution but not its implications about the operation of nature. He still accepted spontaneous generation of cells and remained a vitalist. The article referred to Naegeli, "The Limits of Natural Knowledge," *Nature* 16 (1877): 530–535, 559–563, explored some points of his vitalist philosophy. For Maxwell on Naegeli see Campbell and Garnett, *Life,* 456–458.

45. Postcard from Maxwell to Peter Guthrie Tait, February 28, 1878

Cambridge University Library, Maxwell Collection, Maxwell-Tait Correspondence

$$160 \text{ Eq } 10 \text{ should be } = \frac{e}{f}\left\{E - \frac{1}{2}\frac{e\,a^3}{f(f^2-a^2)}\right\}^a$$

as in MS but the 2 went wrongly to the upper 'wom. New edition in hand so corrections will be valued.[b] I will not do the Ranking[c] of rankine $(-t)^{1/2}$. Let the irreprintable be selected by editors who can appreciate what they are rejecting. Aether[d] is in Blacks Balaam Box. When ains come aboil it will go to you. As long as Nigellus is in the E I will judge his Cause. Is he shipping Westward yet? When $M^c F$ on $\sum \pi \alpha \rho \xi$ [e] is out ask him to send a copy to Cavendish Lab: Your thermodynamic surface will be sent this week.

Cavendish electrical researches is going to press here. Would you allow a traducer to overset you into fonetik Deutsch, leaving out the hau's as tun, tat teil and erteilen.
28/2/78

$$\frac{dp}{dt}$$

a. The equation is from Maxwell, *Treatise on Electricity and Magnetism,* (New York: Dover, 1954) vol. 1, sec. 160, 251. The equation expresses the energy due to the mutual action of an electrified point, charge e, at a distance f from the center of a sphere of radius a with charge E on the sphere.

b. The new edition refers to Maxwell *Treatise.*

c. The "Ranking of rankine" means to help Tait and W. J. Millar decide which papers of Rankine's should be reprinted in the *Miscellaneous Scientific Papers of W. J. M. Rankine,* ed. W. J. Millar, with a "Memoir" of the author by P. G. Tait (London, 1881).

d. Maxwell's article on "Ether" for the *Encyclopaedia Britannica,* published by A. and C. Black. "In Black's Balaam Box" may mean in Black's box for blessing.

e. $M^c F$ on $\sum \alpha \rho \xi$ refers to Alexander Macfarlane's doctorate of science dissertation on electrical discharges through air and other dielectrics. See, Knott *Life of Tait,* 96.

46. Postcard from Maxwell to Peter Guthrie Tait, 1878[a]

Cambridge University Library Maxwell Collection, Maxwell-Tait Correspondence.

Sparx[b] may be sent to Cavendish Laboratory till Easter.
To study them under diver's conditions requires a knowledge of the conductivity of salt water. Cavendish and Kohlrausch[c] agree in this most wonderfully. "Nature" wants my views on $\theta\Delta^{cs}$.[d] What $\theta\Delta^{cs}$? Is a new edition out yet of which I saw a preface?[e] But beware lest having hired me to curse him of Bonn I may be constrained to bless him partially.[f]
Can T[g] do anything for Tolver.[h] He was ambitious a grievous fault which he hath grievously answered, but he has really a good head if it were only trained a little and is no paradoxer. Only he finds he cannot live on air, still less on aether, and he requires some paying work to keep him from going to the U. U.[i] as a gasiform vertebrate.

a. There is no postmark or date on this postcard. It can be dated approximately from Maxwell's reference to *Nature's* request for Maxwell's views on thermodynamics. This refers to the second edition of Tait, *Sketch of Thermodynamics* (Edinburgh: David Douglas, 1877) which Maxwell was asked to review. See Maxwell, "Tait's *Thermodynamics,*" *Nature* 17 (1878): 257, reprinted in *Scientific Papers,* vol. 2, 660–671 and Document III-47. The postcard therefore was written during the first half of 1878.

b. Sparks.

c. This is a reference to Maxwell's work on Henry Cavendish's unpublished papers in electricity. The papers of Friedrich Wilhelm George Kohlrausch (1840–1910) on electrolysis are, Kohlrausch, "Ueber das elektrische Leitungsvermogen des Wassers und der Sauren," *Munchen Ber.* 5 (1875): 284–305; "Ueber das Leitungsvermogen der Wassergeloten Elektrolyte im Zusammenhang mit der Wanderung ihrer Bestandtheile," *Göttingen Nachrichten* (1876): 213–224.

d. Thermodynamics.

e. Maxwell is referring to the second edition of Tait, *A Sketch of Thermodynamics* (Edinburgh, 1877).

f. "Hire me to curse him" refers to the Book of Numbers and the story where Balak hired the soothsayer Balaam to curse the children of Israel, and instead he blessed them on three different occasions (Numbers 24: 10). "Him of Bonn" refers to Rudolph Clausius with whom Tait had a running battle over the history of thermodynamics. Tait never conceded Clausius's role in the development of the second law of thermodynamics.

g. William Thomson.

h. Samuel Tolver Preston. See note f, Document III-44.

i. This refers to Balfour Stewart and P.G. Tait, *The Unseen Universe* (London: Macmillan, 1875) which proposed to defeat materialism through scientific argument. The book caused a sensation and was popular enough to go through four editions within a year. See P. Heimann, "The *Unseen Universe*: Physics and the Philosophy of Nature in Victorian Britain," *British J. Hist. Sci.* 6 (1972): 73–79.

47. "Tait's *Thermodynamics,*"

Nature 17 (1878), 257–258 and 278–280, reprinted in *Scientific Papers,* vol. 2, 660–671.

[From *Nature*, Vol. XVII.]

XCI. *Tait's "Thermodynamics."*

THIS book, as we are told in the preface, has grown out of two articles contributed in 1864 by Prof. Tait to the *North British Review*. This journal, about that time, inserted a good many articles in which scientific subjects were discussed in scientific language, and in which, instead of the usual attempts to conciliate the unscientific reader by a series of relapses into irrelevant and incoherent writing, his attention was maintained by awakening a genuine interest in the subject.

The attempt was so far successful that the publishers of the *Review* were urged by men of science, especially engineers, to reprint these essays of Prof. Tait, but the *Review* itself soon afterwards became extinct.

Prof. Tait added to the two essays a mathematical sketch of the fundamental principles of thermodynamics, and in this form the book was published in 1868. In the present edition, though there are many additions and improvements, the form of the book is essentially the same.

Whether on account of these external circumstances, or from internal causes, it is impossible to compare this book either with so-called popular treatises or with those of a more technical kind.

In the popular treatise, whatever shreds of the science are allowed to appear, are exhibited in an exceedingly diffuse and attenuated form, apparently with the hope that the mental faculties of the reader, though they would reject any stronger food, may insensibly become saturated with scientific phraseology, provided it is diluted with a sufficient quantity of more familiar language. In this way, by simple reading, the student may become possessed of the phrases of the science without having been put to the trouble of thinking a single thought about it. The loss implied in such an acquisition can be estimated only by those who have been compelled to unlearn a science that they might at length begin to learn it.

The technical treatises do less harm, for no one ever reads them except under compulsion. From the establishment of the general equations to the end of the book, every page is full of symbols with indices and suffixes, so that there is not a paragraph of plain English on which the eye may rest.

Prof. Tait has not adopted either of these methods. He serves up his strong meat for grown men at the beginning of the book, without thinking it necessary to employ the language either of the nursery or of the school; while for younger students he has carefully boiled down the mathematical elements into the most concentrated form, and has placed the result at the end as a *bonne bouche*, so that the beginner may take it in all at once, and ruminate upon it at his leisure.

A considerable part of the book is devoted to the history of thermodynamics, and here it is evident that with Prof. Tait the names of the founders of his science call up the ideas, not so much of the scientific documents they have left behind them in our libraries, as of the men themselves, whether he recommends them to our reverence as masters in science, or bids us beware of them as tainted with error. There is no need of a garnish of anecdotes to enliven the dryness of science, for science has enough to do to restrain the strong human nature of the author, who is at no pains to conceal his own idiosyncrasies, or to smooth down the obtrusive antinomies of a vigorous mind into the featureless consistency of a conventional philosopher.

Thus, in the very first page of the book, he denounces all metaphysical methods of constructing physical science, and especially any *à priori* decisions as to what may have been or ought to have been. In the second page he does not indeed give us Aristotle's ten categories, but he lays down four of his own:—matter, force, position, and motion, to one of which he tells us, "it is evident that every distinct physical conception must be referred," and then before we have finished the page we are assured that heat does not belong to any of these four categories, but to a fifth, called energy.

This sort of writing, however unlike what we might expect from the conventional man of science, is the very thing to rouse the placid reader, and startle his thinking powers into action.

Prof. Tait next handles the caloric theory, but instead of merely shewing up its weak points and then dismissing it with contempt, he puts fresh life into it by giving (in the new edition) a characteristic extract from Dr Black's lectures, and proceeds to help the calorists out of some of their difficulties, by generously making over to them some excellent hints of his own.

The history of thermodynamics has an especial interest as the development of a science, within a short time and by a small number of men, from the condition of a vague anticipation of nature to that of a science with secure foundations, clear definitions, and distinct boundaries.

The earlier part of the history has already provoked a sufficient amount of discussion. We shall therefore confine our remarks to the methods employed for the advancement of the science by the three men who brought the theory to maturity.

Of the three founders of theoretical thermodynamics, Rankine availed himself to the greatest extent of the scientific use of the imagination. His imagination, however, though amply luxuriant, was strictly scientific. Whatever he imagined about molecular vortices, with their nuclei and atmospheres, was so clearly imaged in his mind's eye, that he, as a practical engineer, could see how it would work.

However intricate, therefore, the machinery might be which he imagined to exist in the minute parts of bodies, there was no danger of his going on to explain natural phenomena by any mode of action of this machinery which was not consistent with the general laws of mechanism. Hence, though the construction and distribution of his vortices may seem to us as complicated and arbitrary as the Cartesian system, his final deductions are simple, necessary, and consistent with facts.

Certain phenomena were to be explained. Rankine set himself to imagine the mechanism by which they might be produced. Being an accomplished engineer, he succeeded in specifying a particular arrangement of mechanism competent to do the work, and also in predicting other properties of the mechanism which were afterwards found to be consistent with observed facts.

As long as the training of the naturalist enables him to trace the action only of particular material systems without giving him the power of dealing with the general properties of all such systems, he must proceed by the method so often described in histories of science—he must imagine model after model of hypothetical apparatus till he finds one which will do the required work. If this apparatus should afterwards be found capable of accounting for many of the known phenomena, and not demonstrably inconsistent with any of them, he is strongly tempted to conclude that his hypothesis is a fact, at least until an equally good rival hypothesis has been invented. Thus Rankine*, long after an explanation of the properties of gases had been founded on the theory of the collisions of molecules, published what he supposed to be a proof that the phenomena of heat were invariably due to steady closed streams of continuous fluid matter.

The scientific career of Rankine was marked by the gradual development of a singular power of bringing the most difficult investigations within the range of elementary methods. In his earlier papers, indeed, he appears as if battling with chaos, as he swims, or sinks, or wades, or creeps, or flies,

> "And through the palpable obscure finds out
> His uncouth way;"

but he soon begins to pave a broad and beaten way over the dark abyss, and his latest writings shew such a power of bridging over the difficulties of science, that his premature death must have been almost as great a loss to the diffusion of science as it was to its advancement.

The chapter on thermodynamics in his book on the steam-engine was the first published treatise on the subject, and is the only expression of his views addressed directly to students.

In this book he has disencumbered himself to a great extent of the hypothesis of molecular vortices, and builds principally on observed facts, though he, in common with Clausius, makes several assumptions, some expressed as axioms, others implied in definitions, which seem to us anything but self-evident. As an example of Rankine's best style we may take the following definition :—

> "A PERFECT GAS is a substance in such a condition that the total pressure exerted by any number of portions of it, at a given temperature, against the sides of a vessel in which they are enclosed, is the sum of the pressures which each portion would exert if enclosed in the vessel separately at the same temperature."

Here we can form a distinct conception of every clause of the definition, but when we come to Rankine's Second Law of Thermodynamics we find that though, as to literary form, it seems cast in the same mould, its actual meaning is inscrutable.

* "On the Second Law of Thermodynamics," *Phil. Mag.* Oct. 1865, § 12, p. 244; but in his paper on the Thermal Energy of Molecular Vortices, *Trans. R. S. Edin.* xxv. p. 557 [1869], he admits that the explanation of gaseous pressure by the impacts of molecules has been proved + be possible.

"*The Second Law of Thermodynamics.*—If the total actual heat of a homogeneous and uniformly hot substance be conceived to be divided into any number of equal parts, the effects of those parts in causing work to be performed are equal."

We find it difficult enough, even in 1878, to attach any distinct meaning to the total actual heat of a body, and still more to conceive this heat divided into equal parts, and to study the action of each of these parts; but as if our powers of deglutition were not yet sufficiently strained, Rankine follows this up with another statement of the same law, in which we have to assert our intuitive belief that—

"If the absolute temperature of any uniformly hot substance be divided into any number of equal parts, the effects of those parts in causing work to be performed are equal."

The student who thinks that he can form any idea of the meaning of this sentence is quite capable of explaining on thermodynamical principles what Mr Tennyson says of the great Duke :—

"Whose eighty winters freeze with one rebuke
All great self-seekers trampling on the right."

Prof. Clausius does not ask us to believe quite so much about the heat in hot bodies. In his first memoir, indeed, he boldly dismisses one supposed variety of heat from the science. Latent heat, he tells us, "is not only, as its name imports, hidden from our perceptions, but has actually no existence;" "it has been converted into work."

But though Clausius thus gets rid of all the heat which, after entering a body, is expended in doing work, either exterior or interior, he allows a certain quantity to remain in the body as heat, and this remnant of what should have been utterly destroyed lives on in a sort of smouldering existence, breaking out now and then with just enough vigour to mar the scientific coherence of what might have been a well compacted system of thermodynamics.

Prof. Tait tells us :—

"The source of all this sort of speculation, which is as old as the time of Crawford and Irvine—and which was countenanced to a certain extent even by Rankine—is the assumption that bodies must contain a certain quantity of actual, or thermometric, heat. We are quite ignorant of the condition of energy in bodies generally. We know how much goes in, and how much comes out, and we know whether at entrance or exit it is in the form of heat or of work. But that is all."

If we define thermodynamics, as I think we may now do, as the investigation of the dynamical and thermal properties of bodies, deduced entirely from what are called the First and Second laws of Thermodynamics, without any hypotheses as to the molecular constitution of bodies, all speculations as to how much of the energy in a body is in the form of heat are quite out of place.

Prof. Tait, however, does not seem to have noticed that Prof. Clausius, in a footnote to his sixth memoir *, tells us what he means by the heat in a body. In the middle of a sentence we read :—

"......the heat actually present in a unit weight of the substance in question—in other words, the *vis viva* of its molecular motions"......

Thus the doctrine that heat consists of the *vis viva* of molecular motions, and that it does not include the potential energy of molecular configuration—the most important doctrine, if true, in molecular science—is introduced in a footnote under cover of the unpretending German abbreviation "d.h."

Prof. Clausius is himself the principal founder of the kinetic theory of gases. The theory of the exchanges of the energy of collections of molecules was afterwards developed by Boltzmann to a much greater extent than had been done by Clausius, and it appears from his investigations that whether we suppose the molecules to be acted on by forces towards fixed centres or not, the condition of equilibrium of exchange of energy, or in other words the condition of equality of temperature of two bodies, is that the average kinetic energy of translation of a single molecule is the same in both bodies.

We may therefore define the temperature of a body as the average kinetic energy of translation of one of its molecules multiplied into a constant which is the same for all bodies. If we also define the total heat of the body as the sum of the whole kinetic energy of its molecules, then the total heat must be equal to the temperature multiplied into the number of molecules, and by the ratio of the whole kinetic energy to the energy of translation, and divided by the above constant.

The kinetic theory of gases has therefore a great deal to say about what Rankine and Clausius call the actual heat of a body, and if we suppose that molecules never coalesce or split up, but remain constant in number, then we may also assert, all experiments notwithstanding, that the real capacity for heat (as defined by Clausius) is constant for the same substance in all conditions.

* Hirst's translation, p. 230, German edition, 1864, p. 258, "wirklich vorhandene Wärme, d.h. die lebendige Kraft seiner Molecularbewegungen."

Rankine, indeed, probably biassed by the results of experiments, allowed that the real specific heat of a substance might be different in different states of aggregation, but Clausius has clearly shewn that this admission is illogical, and that if we admit any such changes, we had better give up real specific heat altogether.

Statements of this kind have their legitimate place in molecular science, where it is essential to specify the dynamical condition of the system, and to distinguish the kinetic energy of the molecules from the potential energy of their configuration; but they have no place in thermodynamics proper, in which we deal only with sensible masses and their sensible motions.

Both Rankine and Clausius have pointed out the importance of a certain function, the increase or diminution of which indicates whether heat is entering or leaving the body. Rankine calls it the thermodynamic function, and Clausius the entropy. Clausius, however, besides inventing the most convenient name for this function, has made the most valuable developments of the idea of entropy, and in particular has established the most important theorem in the whole science,—that when heat passes from one body to another at a lower temperature, there is always an increase of the sum of the entropy of the two bodies, from which it follows that the entropy of the universe must always be increasing.

He has also shewn that if the energy of a body is expressed as a function of the volume and the entropy, then its pressure (with sign reversed) and its temperature are the differential coefficients of the energy with respect to the volume and the entropy respectively, thus indicating the symmetrical relations of the five principal quantities in thermodynamics.

But Clausius, having begun by breaking up the energy of the body into its thermal and ergonal content, has gone on to break up its entropy into the transformational value of its thermal content and the disgregation.

Thus both the energy and the entropy, two quantities capable of direct measurement, are broken up into four quantities, all of them quite beyond the reach of experiment, and all this is owing to the actual heat which Clausius, after getting rid of the latent heat, suffered to remain in the body.

Sir William Thomson, the last but not the least of the three great founders, does not even consecrate a symbol to denote the entropy, but he was the first to clearly define the intrinsic energy of a body, and to him alone are due the ideas and the definitions of the available energy and the dissipation of energy. He has always been most careful to point out the exact extent of the assumptions and experimental observations on which each of his statements is based, and he avoids the introduction of quantities which are not capable of experimental measurement. It is therefore greatly to be regretted that his memoirs on the dynamical theory of heat have not been collected and reprinted in an accessible form, and completed by a formal treatise, in which his method of building up the science should be exhibited in the light of his present knowledge.

The touchstone of a treatise on thermodynamics is what is called the second law.

Rankine, as we have seen, founds it on statements which may or may not be true, but which cannot be considered as established in the present state of science.

The second law is introduced by Clausius and Thomson as an axiom on which to found Carnot's theorem that the efficiency of a reversible engine is at least as great as that of any other engine working between the same limits of temperature.

If an engine of greater efficiency exists, then, by coupling this engine with Carnot's engine reversed, it is possible to restore to the hot body as much heat as is taken from it, and at the same time to do a certain amount of work.

If with Carnot we suppose heat to be a substance, then this work would be performed in direct violation of the first law—the principle of the conservation of energy. But if we regard heat as a form of energy, we cannot apply this method of *reductio ad absurdum*, for the work may be derived from the heat taken from the colder body.

Clausius supposes all the work gained by the first engine to be expended in driving the second. There is then no loss or gain of heat on the whole, but heat is taken from the cold body, and an equal quantity communicated to the hot body, and this process might be carried on to an indefinite extent.

In order to assert the impossibility of such a process in a form of words having sufficient verisimilitude to be received as an axiom, Clausius, in his first memoir, simply says that this process "contradicts the general deportment of heat, which everywhere exhibits the tendency to equalize differences of temperature, and therefore to pass from the warmer to the colder body *."

* Und das widerspricht dem sonstigen Verhalten der Wärme, indem sie überall das Bestreben zeigt, vorkommende Temperaturdifferenzen auszugleichen und also aus den wärmeren Körpern in die kälteren überzugehen.

In its obvious and strict sense no axiom can be more irrefragable. Even in the hypothetical process, the impossibility of which it was intended to assert, every communication of heat is from a warmer to a colder body. When the heat is taken from the cold body it flows into the working substance which is at that time still colder. The working substance afterwards becomes hot, not by communication of heat to it, but by change of volume, and when it communicates heat to the hot body it is itself still hotter.

It is therefore hardly correct to assert that heat has been transmitted or transferred from the colder to the hotter body. There is undoubtedly a transfer of energy, but in what form this energy existed during its middle passage is a question for molecular science, not for pure thermodynamics.

In a note added in 1864 Clausius states the principle in a modified form, "that heat cannot of itself pass from a colder to a warmer body" * and finally, in the new edition of his *Theory of Heat* (1876) he substitutes for the words "of itself" the expression "without compensation†."

With respect to the first of these emendations we must remember that the words "of itself" are not intended to exclude the intervention of any kind of self-acting machinery, and it is easy, by means of an engine which takes in heat from a body at 200° C., and gives it out at 100° to drive a freezing machine so as to take heat from water at 0°, and so freeze it, and also a friction break so as to generate heat in a body at 500°. It would therefore be necessary to exclude all bodies except the hot body, the cold body, and the working substance, in order to exclude exceptions to the principle.

By the introduction of the second expression, "without compensation," combined with a full interpretation of this phrase, the statement of the principle becomes complete and exact; but in order to understand it we must have a previous knowledge of the theory of transformation-equivalents, or in other words of entropy, and it is to be feared that we shall have to be taught thermodynamics for several generations before we can expect beginners to receive as axiomatic the theory of entropy.

Thomson, in his "Third Paper on the Dynamical Theory of Heat" (*Trans. R. S. Edin.* xx. p. 265, read March 17, 1851), has stated the axiom as follows :—

* Dass die Wärme nicht von selbst aus einem kälteren in einem wärmeren Körper übergehen kann.

† Ein Wärmeübergang aus einem kälteren in einem wärmeren Körper kann nicht ohne Compensation Statt finden.

"It is impossible, by means of inanimate material agency, to derive mechanical effect from any portion of matter by cooling it below the temperature of the coldest of surrounding objects."

Without some further restriction this axiom cannot be considered as true, for by allowing air to expand we may derive mechanical effect from it by cooling it below the temperature of the coldest of surrounding objects.

If we make it a condition that the material agency is to be left in the same state at the end of the process as it was at first, and also that the mechanical effect is not to be derived from the pressure of the hot or of the cold body, the axiom will be rendered strictly true, but this brings us back to a simple re-assertion of Carnot's principle, except that it is extended from heat engines to all other kinds of inanimate material agency.

It is probably impossible to reduce the second law of thermodynamics to a form as axiomatic as that of the first law, for we have reason to believe that though true, its truth is not of the same order as that of the first law.

The first law is an extension to the theory of heat of the principle of conservation of energy, which can be proved mathematically true if real bodies consist of matter "as per definition," acted on by forces having potentials.

The second law relates to that kind of communication of energy which we call the transfer of heat as distinguished from another kind of communication of energy which we call work. According to the molecular theory the only difference between these two kinds of communication of energy is that the motions and displacements which are concerned in the communication of heat are those of molecules, and are so numerous, so small individually, and so irregular in their distribution, that they quite escape all our methods of observation; whereas when the motions and displacements are those of visible bodies consisting of great numbers of molecules moving altogether, the communication of energy is called work.

Hence we have only to suppose our senses sharpened to such a degree that we could trace the motions of molecules as easily as we now trace those of large bodies, and the distinction between work and heat would vanish, for the communication of heat would be seen to be a communication of energy of the same kind as that which we call work.

The second law must either be founded on our actual experience in dealing with real bodies of sensible magnitude, or else deduced from the molecular theory of these bodies, on the hypothesis that the behaviour of bodies consisting of millions of molecules may be deduced from the theory of the encounters of pairs of molecules, by supposing the relative frequency of different kinds of encounters to be distributed according to the laws of probability.

The truth of the second law is therefore a statistical, not a mathematical, truth, for it depends on the fact that the bodies we deal with consist of millions of molecules, and that we never can get hold of single molecules.

Sir William Thomson * has shewn how to calculate the probability of the occurrence within a given time of a given amount of deviation from the most probable distribution of a finite number of molecules of two different kinds in a vessel, and has given a numerical example of a particular case of the diffusion of gases.

The same method might be extended to the diffusion of heat by conduction, and the diffusion of motion by internal friction, which are also processes by which energy is dissipated in consequence of the motions and encounters of the molecules of the system.

The tendency of these motions and encounters is in general towards a definite state, in which there is an equilibrium of exchanges of the molecules and their momenta and energies between the different parts of the system.

If we restrict our attention to any one molecule of the system, we shall find its motion changing at every encounter in a most irregular manner.

If we go on to consider a finite number of molecules, even if the system to which they belong contains an infinite number, the average properties of this group, though subject to smaller variations than those of a single molecule, are still every now and then deviating very considerably from the theoretical mean of the whole system, because the molecules which form the group do not submit their procedure as individuals to the laws which prescribe the behaviour of the average or mean molecule.

Hence the second law of thermodynamics is continually being violated, and that to a considerable extent, in any sufficiently small group of molecules belonging to a real body. As the number of molecules in the group is increased, the deviations from the mean of the whole become smaller and less frequent; and when the number is increased till the group includes a sensible portion of the body, the probability of a measurable variation from the mean occurring in a finite number of years becomes so small that it may be regarded as practically an impossibility.

* "On the Kinetic Theory of the Dissipation of Energy," *Proc. R. S. Edin.*, February 16, 1874, Vol. VIII. p. 323, also in *Nature*, Vol. IX. p. 441.

This calculation belongs of course to molecular theory and not to pure thermodynamics, but it shews that we have reason for believing the truth of the second law to be of the nature of a strong probability, which, though it falls short of certainty by less than any assignable quantity, is not an absolute certainty.

Several attempts have been made to deduce the second law from purely dynamical principles, such as Hamilton's principle, and without the introduction of any element of probability. If we are right in what has been said above, no deduction of this kind, however apparently satisfactory, can be a sufficient explanation of the second law. Indeed some of them have already indicated their unsoundness by leading to determinations of physical quantities which have no existence, such as the periodic time of the alternations of the volume of particular gases*.

* Szily, *Phil. Mag.*, October, 1876; Clausius, *Pogg. Ann.* CXLII. p. 433; *Pogg. Ann.* CXLVI. p. 585, May, 1872; J. J. Müller, *Pogg. Ann.* CLII. p. 105.

IV Documents: Virial Theorem & Equation of State

1. Letter from Maxwell to Peter Guthrie Tait, July 4, 1874

Cambridge University Library, Maxwell Collection, Maxwell-Tait Correspondence

New statement of (2)

<Particles are flying>

<The universe is>

In every part of infinite space except within the sphere of action of a fixed centre of force particles are <flying about> uniformly scattered and are flying about with velocities equal to a given velocity <The par> and in all directions <without> indifferently without interfering with each other. Shew that at a finite distance from the centre of force the number of particles in unit volume will be directly proportional to the velocity of the particles at that distance and that their <directions> of motion will be in all direction indifferently.

(4)

If the molecules of gas are elastic spheres <such that the volume of unit of> of density σ and if $\overline{v^2}$ is the mean square of their velocity, show that the pressure of the gas, taking account of the size of the molecules is

$$p = \frac{1}{3}\rho\overline{v^2}\left(1 + 16\frac{\rho}{\sigma}\right)$$

where ρ is the density of the gas and σ the density of the molecules which act only by impact. Hence show that this is not the constitution of CO_2. Van der Waals makes the only effect of the extension of the molecules a diminution of the effective volume wherein the gas moves by 4 times the volume of the molecules. I have not lynxed his proof enough. The above increment of pressure is due to the impacts between molecules whose centres are on opposite sides of the plane area over which p is measured the other part being due to the passage of molecules through the plane carrying their momentum with them.

The form of the result is quite different from the Dutchman's

$$\frac{dp}{dt}$$

2. Letter from Maxwell to Peter Guthrie Tait, September 2, 1874

Cambridge University Library, Maxwell Collection, Maxwell-Tait Letters

Particles are <moving> flying <through infinite space> in all directions with <equal> velocities equal to a given velocity and at any instant are uniformly scattered through infinite space except near a fixed centre of force Show that <if a fixed centre of force is introduced> the number of particles in unit volume at a given distance from this centre will be directly proportional to the velocity of the particles at that distance.

- - - - -

Show that if the stress in <an> a homogeneous (not isotropic) elastic solid arises from the direct action between different parts of the body then if A and B are two such parts and if the coordinates of B exceed those of A by x, y, z and if the components of the force with which A acts on B be $X\,Y\,Z$ then the <stresses> components of stress are as follows

$$P = \sum(Xx) \qquad Q = \sum(Yy) \qquad R = x\sum(Zz)$$

$$S = \sum(Yz) = \sum(Zy), \qquad T = \sum(Zx) = \sum(Xz)$$

$$U = \sum(Xy) = \sum(Yx)$$

where the summation is extended to all such pairs of parts in unit of volume and P, Q, R, S, T, U have the same meaning as in the Archiepiscopal Treatise.[a]

Hence prove the connexion between pressure and Virial.

Here is a Four-year problem (not to be set in the 3 days)[b]

A gas consists of an immense number of molecules which strike against each other irregularly. Each molecule when struck executes vibrations to the tune of $\cos(nt+\alpha)$ but after a time τ it is struck by another molecule and begins a new series of vibrations having the same amplitude and periodic time but an entirely different phase.

Shew that the light emitted by the gas will not be homogeneous but will consist of light of all kinds, the intensity of the kind of light whose periodic time is $\frac{2\pi}{p}$ being proportional to

$$\frac{sin^2(p-n)\,\frac{\tau}{2}}{(p-n)^2} + \frac{sin^2(p+n)\,\frac{\tau}{2}}{(p+n)^2}$$

the last term being insensible when $n\tau$ is a large number. Why does the spectrum of a gas consist of bright lines or of a continuous luminosity, acc^{g} [according] as the gas is at low or at high pressure

a. The Archiepiscopal Treatise refers to William Thomson and Peter Guthrie Tait, *Treatise on Natural Philosophy* (Oxford: Clarendon Press, 1867), second edition 1879. The discussion on Stress and its definition are in secs. 658–662 of the second edition. Archibald Tait was the archbishop of Canterbury and with William Thomson were creating as much stir in the Church of England as their namesakes in physics.

b. "A four year problem (not to be set in the 3 days)" refers to the division of the Mathematical Tripos examination into two parts. Easier questions were set in the first three days, the examination for those working towards the general (or Poll) degree. The next four days were required only of the serious candidates. "Four year problem" means one suitable for four year men taking the combined Mathematics and Natural Tripos.

3. "Van der Waals on the Continuity of the Gaseous and Liquid States"

Nature, 10 (October 15, 1874): 477–480, reprinted in *Scientific Papers,* vol. 2: 407–415.

O ver de continuiteit van den gasen vloeistofloestand. Academisch proefschrift. Door Johannes Diderik van der Waals. (Leiden: A.W. Sijthoff, 1873.)

That the same substance at the same temperature and pressure can exist in two very different states, as a liquid and as a gas, is a fact of the highest scientific importance, for it is only by the careful study of the difference between these two states, the conditions of the substance passing from one to the other, and the phenomena which occur at the surface which separates a liquid from its vapour, that we can expect to obtain a dynamical theory of liquids. A dynamical theory of "perfect" gases is already in existence; that is to say, we can explain many of the physical properties of bodies when in an extremely rarefied state by supposing their molecules to be in rapid motion, and that they act on one another only when they come very near one another. A molecule of a gas, according to this theory, exists in two very different states during alternate intervals of time. During its encounter with another molecule, an intense force is acting between the two molecules, and producing changes in the motion of both. During the time of describing its free path, the molecule is at such a distance from other molecules that no sensible force acts between them, and the centre of mass of the molecule is therefore moving with constant velocity and in a straight line.

If we define as a perfect gas a system of molecules so sparsely scattered that the aggregate of the time which a molecule spends in its encounters with other molecules is exceedingly small compared with the aggregate of the time which it spends in describing its free paths, it is not difficult to work out the dynamical theory of such a system. For in this case the vast majority of the molecules at any given instant are describing their free paths, and only a small fraction of them are in the act of encountering each other. We know that during an encounter action and reaction are equal and opposite, and we assume, with Clausius, that on an average of a large number of encounters the proportion in which the kinetic energy of a molecule is divided between motion of translation of its centre of mass and motions of its parts relative to this point approaches some definite value. This amount of knowledge is by no means sufficient as a foundation for a complete dynamical theory of what takes place during each encounter, but it enables us to establish certain relations between the changes of velocity of two molecules before and after their encounter.

While a molecule is describing its free path, its centre of mass is moving with constant velocity in a straight line. The motions of parts of the molecule relative to the centre of mass depend, when it is describing its free path, only on the forces acting between these parts, and not on the forces acting between them and other molecules which come into play during an encounter. Hence the theory of the motion of a system of molecules is very much simplified if we suppose the space within which the molecules are free to move to be so large that the number of molecules which at any instant are in the act of encountering other molecules is exceedingly small compared with the number of molecules which are describing their free paths. The dynamical theory of such a system is in complete agreement with the observed properties of gases when in an extremely rare condition.

But if the space occupied by a given quantity of gas is diminished more and more, the lengths of the free paths of its molecules will also be diminished, and the number of molecules which are in the act of encounter will bear a larger proportion to the number of those which are describing free paths, till at length the properties of the substance will be determined far more by the nature of the mutual action between the encountering molecules than by the nature of the motion of a molecule when describing its free path. And we actually find that the properties of the substance become very different after it has reached a certain degree of condensation. In the rarefied state its properties may be defined with considerable accuracy in terms of the laws of Boyle, Charles, Gay-Lussac, Dulong and Petit, &c., commonly called the "gaseous laws." In the condensed state the properties of the substance are entirely different, and no mode of stating these properties has yet been discovered having a simplicity and a generality at all approaching to that of the "gaseous laws." According to the dynamical theory this is to be expected, because in the condensed state the properties of the substance depend on the mutual action of molecules when engaged in close encounter, and this is determined by the particular constitution of the encountering molecules. We cannot therefore extend the dynamical theory from the rarer to the denser state of substances without at the same time obtaining some definite conception of the nature of the action between molecules when they are so closely packed that each molecule is at every instant so near to several others that forces of great intensity are acting between them.

The experimental data for the study of the mutual action of molecules are principally of two kinds. In the first place we have the experiments of Regnault and others on the relation between the density, temperature, and pressure of a various gases. The field of research has been recently greatly enlarged by Dr. Andrews in his exploration of the properties of carbonic acid at very high pressures. Experiments of this kind, combined with experiments on specific heat, on the latent heat of expansion, or on the thermometric effect on gases passing through porous plugs, furnish us with the complete theory of the substance, so far as pure thermodynamics can carry us.

For the further study of molecular action we require experiments on the rate of diffusion. There are three kinds of diffusion–that of matter, that of visible motion, and that of heat. The inter-diffusion of gases of different kinds, and the viscosity and thermal conductivity of a gaseous medium, pure or mixed, enable us to estimate the amount of deviation which each molecule experiences on account of its encounter with other molecules.

M. Van der Waals, in entering on this very difficult inquiry, has shown his appreciation of its importance in the present state of science; many of his investigations are conducted in an extremely original and clear manner; and he is continually throwing out new and suggestive ideas; so that there can be no doubt that his name will soon be among the foremost in molecular science.

He does not, however, seem to be equally familiar, as yet, with all parts of the subject, so that in some places, where he has borrowed results from Clausius and others, he has applied them in a manner which appears to me erroneous.

He begins with the very remarkable theorem of Clausius, that in stationary motion the mean kinetic energy of the system is equal to the mean virial. As in this country the importance of this theorem seems hardly to be appreciated, it may be as well to explain it a little more fully.

When the motion of a material system is such that the sum of the moments of inertia of the system about three axes at right angles to each other through its centre of mass does not vary by more than small quantities from a constant value, the system is said to be in a state of stationary motion. The motion of the solar system satisfies this condition, and so does the motion of the molecules of a gas contained in a vessel.

The kinetic energy of a particle is half the product of its mass into the square of its velocity, and the kinetic energy of a system is the sum of the kinetic energy of its parts.

When an attraction or repulsion exists between two points, half the product of this stress into the distance between the two points is called the Virial of the stress, and is reckoned positive when the stress is an attraction, and negative when it is a repulsion. The virial of a system is the sum of the virial of the stresses which exist in it.

If the system is subjected to the external stress of the pressure of the sides of a vessel in which it is contained, the amount of virial due to this external stress is three halves of the product of the pressure into the volume of the vessel.

The virial due to internal stresses must be added to this.

The theorem of Clausius may now be written–

$$\frac{1}{2}\sum(m\overline{v^2}) = \frac{1}{2}pV + \frac{3}{2}\sum\sum(Rr)$$

The left-hand member denotes the kinetic energy.

On the right hand, in the first term, p is the external pressure on unit of area, and V is the volume of the vessel.

The second term represents the virial arising from the action between every pair of particles, whether belonging to different molecules or to the same molecule. R is the attraction between the particles, and r is the distance between them. The double symbol of summation is used because every pair of points must be taken into account, those between which there is no stress contributing, of course, nothing to the virial.

As an example of the generality of this theorem, we may mention that in any framed structure consisting of struts and ties, the sum of the products of the pressure in each strut into its length, exceeds the sum of the products of the tension of each tie into its length, by the product of the weight of the whole structure into the height of its centre of gravity above the foundations. (See a paper on “Reciprocal Figures, &c.” Trans., S., Edin., vol. xxvi. p. 14. 1870.)

In gases the virial is very small compared with the kinetic energy. Hence, if the kinetic energy is constant, the product of the pressure and the volume remains constant. This is the case for a gas at constant temperature. Hence we might be justified in conjecturing that the temperature of any one gas is determined by the kinetic energy of unit of mass.

The theory of the exchange of the energy of agitation from one body to another is one of the most difficult parts of molecular science. If it were fully understood, the physical theory of temperature would be perfect. At present we know the conditions of thermal equilibrium only in the case of gases in which encounters take place between only a pair of molecules at once. In this case the condition of thermal equilibrium is that the mean kinetic energy due to the agitation of the centre of mass of a molecule is the same, whatever be the mass of the molecule, the mean velocity being consequently less for the more massive molecules.

With respect to substances of more complicated constitution, we know, as yet, nothing of the physical condition on which their temperature depends, though the researches of Boltzmann on this subject are likely to result in some valuable discoveries.

M. Van der Waals seems, therefore, to be somewhat too hasty in assuming that the temperature of a substance is in every case measured by the energy of agitation of its individual molecules, though this is undoubtedly the case with substances in the gaseous state.

Assuming, however, for the present that the temperature is measured by the mean kinetic energy of a molecule, we obtain the means of determining the virial by observing the deviation of the product of the pressure and volume from the constant value given by Boyle's law.

It appears by Dr. Andrews' experiments that when the volume of carbonic acid is diminished, the temperature remaining constant, the product of the volume and pressure at first diminishes, the rate of diminution becoming more and more rapid as the density increases. Now, the virial depends on the number of pairs of molecules which are at a given instant acting on one another, and this number in unit of volume is proportional to the square of the density. Hence the part of the pressure depending on the virial increases as the square of the density, and since, in the case of carbonic acid, it diminishes the pressure, it must be of the positive sign, that is, it must arise from *attraction* between the molecules.

But if the volume is still further diminished, at a certain point liquefaction begins, and from this point till the gas is all liquefied no increase of pressure takes place. As soon, however, as the whole substance is in the liquid condition, any further diminution of volume produces a great rise of pressure, so that the product of pressure and volume increases rapidly. This indicates negative virial, and shows that the molecules are now acting on each other by *repulsion*.

This is what takes place in carbonic acid below the temperature of

$30.92°C$. Above that temperature there is first a positive and then a negative virial, but no sudden liquefaction.

Similar phenomena occur in all the liquefiable gases. In other gases we are able to trace the existence of attractive force at ordinary pressures, though the compression has not yet been carried so far as to show any repulsive force. In hydrogen the repulsive force seems to prevail even at ordinary pressures. This gas has never been liquefied, and it is probable that it never will be liquefied, as the attractive force is so weak.

We have thus evidence that the molecules of gases attract each other at a certain small distance, but when they are brought still nearer they repel each other. This is quite in accordance with Boscovich's theory of atoms as massive centres of force, the force being a function of the distance, and changing from attractive to repulsive, and back again several times, as the distance diminishes. If we suppose that when the force begins to be repulsive it increases very rapidly as the distance diminishes, so as to become enormous if the distance is less by a very small quantity than that at which the force first begins to be repulsive, the phenomena will be precisely the same as those of smooth elastic spheres.

M. Van der Waals makes his molecules elastic spheres, which, when not in contact, attract each other. His treatment of the "molecular pressure" arising from their attraction seems ingenious, and on the whole satisfactory, though he has not attempted a complete calculation of the attractive virial in terms of the law of force.

His treatment of the repulsive virial, however, shows a departure from the principles on which his investigation is founded. He considers the effect of the size of the molecules in diminishing the length of their "free paths," and he shows that this effect, in the case of very rare gases, is the same as if the volume of the space in which the molecules are free to move had been diminished by four times the sum of the volumes of the molecules themselves. He then substitutes for V, the volume of the vessel in Clausius' formula, this volume diminished by four times the molecular volume, and thus obtains the equation–

$$\left(p+\frac{a}{V^2}\right)(V-b)=R(1+\alpha t)$$

where p is the externally applied pressure, a/V^2 is the molecular pressure arising from the attraction between the molecules, which varies as the square of the density, or inversely as the square of the volume. The first factor is thus what he considers the total effective pressure. V is the volume of the vessel, and b is four times the volume of the molecules. The second factor is therefore the "effective volume" within which the molecules are free to move.

The right hand member expresses the kinetic energy, represented by the absolute temperature, multiplied by a quantity, R, constant for each gas.

The results obtained by M. Van der Waals by a comparison of this equation with the determinations of Regnault and Andrews are very striking, and would almost persuade us that the equation represents the true state of the case. But though this agreement would be strong evidence in favour of the accuracy of an empirical formula devised to represent the experimental results, the equation of M. Van der Waals, professing as it does to be derived from the dynamical theory, must be subjected to a much more severe criticism.

It appears to me that the equation does not agree with the theorem of Clausius on which it is founded.

In that theorem p is the pressure of the sides of the vessel, and V is the volume of the vessel. Neither of these quantities is subject to correction.

The assumption that the kinetic energy is determined by the temperature is true for perfect gases, and we have no evidence that any other law holds for gases, even near their liquefying point.

The only source of deviation from Boyle's law is therefore to be looked for in the term $\frac{1}{2}\sum\sum(Rr)$, which expresses the virial. The effect of the repulsion of the molecules, causing them to act like elastic spheres, is therefore to be found by calculating the virial of this repulsion.

Neglecting the effect of attraction, I find that the effect of the impulsive repulsion reduces the equation of Clausius of the form–

$$pV = \frac{1}{3}\sum(m\overline{v^2})\left\{1 - 2log.(1 - 8\frac{\rho}{\sigma} + 17\frac{\rho^2}{\sigma^2} - \&c.)\right\}$$

where σ is the density of the molecules and ρ the mean density of the medium.

The form of this equation is quite different from that of M. Van der Waals, though it indicates the effect of the impulsive force in increasing the pressure. It takes no account of the attractive force, a full discussion of which would carry us into considerable difficulties.

At a constant temperature the effect of the attractive virial is to diminish the pressure by a quantity varying as the square of the density, as long as the encounters of the molecules are, on the whole, between two at a time, and not between three or more.

The effect of the attraction in deflecting the paths of the molecules is to make the number of molecules which at any given instant are at distances between r and $r+dr$ of each other greater than the number in an equal volume at a greater distance in the proportion of the velocities corresponding to these distances. As the temperature rises, the volume being constant, the ratio of these velocities approaches to unity, so that the distribution of molecules according to distance becomes more uniform, and the virial is thus diminished.

If there is a virial arising from repulsive forces acting through a finite distance, a rise of temperature will increase the amount of this kind of virial.

Hence a rise of temperature at constant volume will produce a greater increase of pressure than that given by the law of Charles.

The isothermal lines at higher temperatures will exhibit less of the diminution of pressure due to attraction, and as the density increases will show more of the increase of pressure due to repulsion.

I must not, however, while taking exception to part of the work of M. Van der Waals, forget to add that to him alone are due the suggestions which led me to examine the theory of virial more carefully in order to explore the continuity of the liquid and the gaseous states.

I cannot now enter into the comparison of his theoretical results with the experiments of Andrews, but I would call attention to the able manner in which he expounds the theory of capillarity, and to the remarkable phenomenon of the surface tension of gases which he tells (p. 38) has been observed by Bosscha in tobacco smoke. As tobacco smoke is simply warm air with a slight excess of carbonic acid, carrying solid particles along with it, the change of properties at the surface of the cloud must be very slight compared with that at the surface where two really different gases first come together. If, therefore, the phenomenon observed by Bosscha is a true instance of surface-tension, we may expect to discover much more striking phenomena at the meeting-place of different gases, if we can make our observations before the surface of discontinuity has been obliterated by the inter-diffusion of the gases.

J. CLERK-MAXWELL

4. "Report on Dr Andrews' paper 'On the Gaseous State of Matter'."[a]

Royal Society Archives

This paper contains the result of further investigations into the properties of Carbonic acid gas made with the apparatus described in the Bakerian Lecture for 1869.[b]

Further details are given of the method of preparing and filling the apparatus, and of deducing the true volume of the gases from the measured length of the column of gas in the tube, where the section of the tube is <being> altered by change of pressure and temperature. With respect to the correction for pressure, I would suggest that a specimen of the tube should be tested by Dr Everett of Queens' College Belfast, whose method of determining the coefficients of compressibility ànd rigidity is more satisfactory, than that adopted by Regnault.

The question as to whether air or carbonic acid are absorbed by mercury as they are by water, alcohol and other liquids is here, I think, suggested for the first time. The experiments to test this question had the advantage of employing the gases at very high pressures, and the disadvantage of a very small surface of contact between the gas and the mercury. The result, however, was a negative one, showing that the principal phenomena investigated in the paper were not sensibly disturbed by this cause.

The main object of the paper, however, is to ascertain the relations between pressure, temperature and volume in carbonic acid in the gaseous form, including <at leas> the cases in which the substance is partly liquid and partly gaseous. The importance of accurate determinations of this kind at high pressures depends on the value of such evidence <in omo> as a basis for a true theory of gases. The only additional data which we require are measurements of the heat absorbed or emitted during the different changes described, but these measurements, which are always difficult for gases, are rendered almost impossible at these high pressures on account of the large mass of any vessel strong enough to hold the gas. The only method of making such measurements is to operate on a large quantity of the gas passing continuously through a small calorimetric apparatus.

Dr Andrews' experiments were conducted at three well defined temperatures that of the water-supply about $6.5°C$, that of the vapour of methyl-alcohol boiling at atmospheric pressure, which may be taken at $64.°C$, and that of steam, about $100°C$. The actual temperatures, however, of each experiment were carefully observed.

The critical temperature for carbonic acid is $31°C$, so that the two higher temperatures are both above the critical temperature. Dr Andrews (p. 37) observes that empirical formulae are of little use for scientific purposes as they are apt to mislead the investigator and rarely aid research. He has therefore refrained from expressing the results of his experiments or of any portion of them in such formulae. But nevertheless he has arranged his results in such a way that they may be compared with those of the dynamical theories of gases <as> at present existing.

An empirical formula may be defined as one so framed as to give results consistent with experiments within the range of those experiments, but which we have no reason, founded on the general principles of physics, for believing to be applicable beyond that range.

The degree of accuracy of an empirical formula is a test of its merit, but not of whether it is empirical or not.

Thus Ohm's law of the resistance of conductors appears to be accurate to a very high degree, and through a very extensive range, both of substance and strengths of current. But it is a purely empirical law. The only attempt to deduce it from dynamical principles is that of Weber, who arrives at the result that Ohm's law should not be strictly true for strong currents. Experiments on strong currents seem, however, to show that Ohm's law is accurate, and that Weber's hypothesis is not correct.

Boyle's law of gases, on the other hand, was an empirical law till it was shown to be a result of a particular hypothesis about the molecules of gases, namely, that they are in motion, and that they do not act on each other except by collision. If this hypothesis were correct Boyle's law would be strictly true, and it would not be an empirical formula but a law of nature.

But Boyle's law is not strictly true. What, then is its proper designation?

The dynamical hypothesis as above stated may be modified by admitting that the action between the molecules is not like a collision, confined to a definite distance between the encountering molecules, but extends through a certain range, and experiments like those of this paper and others already published are sufficient to indicate what this action is, whatever may be the physical cause of it.

<It consists of>

As two molecules approach each other, the action between them is < attractive. It is> insensible at all sensible distances. At some exceedingly small distance it begins as an attractive force, reaches a maximum at a still smaller distance, and then becomes repulsive. <For> In certain cases such as that of two kinds of molecules which can enter into chemical combination but which do not so combine when simply mixed, we must admit that within the <distance> region of repulsion there is a second region of attraction, and if we continue to believe that two bodies cannot be in the same place, we must also admit that the force becomes repulsive, and that in a very high degree, when the atoms are as near together as is possible.

These attractive and repulsive forces may be regarded as facts established by experiment, like the fact of gravitation, without assuming either that they are ultimate facts or that they are to be explained in a particular way.

<Now it is easily shown that as long as the volume>

Now if we <call the sphere> described a sphere about the centre of a molecule with a radius such that if the centre of any other molecule is outside this sphere the action between them will be insensible, and if we call this sphere the sphere of action of the molecule, then we may assert that if the sum of the volumes of the spheres of molecular action is small compared with the whole volume occupied by the gas, the pressure will exceed or fall short of that given by Boyle's law by a quantity proportional to the square of the density. This will be true as long as the temperature is the same. For different temperatures the average velocity of collision will be different so that the effects of the attractive and repulsive forces will <come in> enter in different proportions.

If, however, the density be so great that more than two molecules are within the sphere of action at once, and still more if a large proportion of pairs are driven within the sphere of repulsive action, the deviation from Boyle's law may assume a different form.

Now what Dr Andrews has observed is expressed in <the> his formula

$$(\ell - pv)\, v = c$$

where c is a constant for a constant temperature, and ℓ is the value of pv for that temperature when p is very small.

We may therefore write the expression

$$(\ell - \frac{pv}{R\theta})\, v = c$$

where θ is the temperature from the zero of the gas-thermometer and R is a constant equal to $p_0 v_0/\theta_0$, where p_0 is small.

This gives $$p = \frac{R\theta}{v} - R\theta\,\frac{c}{v^2}$$

a result in exact accordance with the molecular theory, only we do not know what function c may be of θ.

Dr Van der Waals, of Leyden, expresses the relation between the pressure and volume of gas by the formula

$$(p + \frac{a}{v^2})\,(v - b) = R\theta$$

where a is a quantity depending on molecular attraction and b is a quantity proportional to the volume occupied by the solid molecules. The dynamical reasoning in support of this formula is by no means sound, but it expresses with considerable felicity the "march" of the phenomena of the compression of carbonic acid. If we make $a = R\theta\, c$ and $b = 0$ it becomes identical with Dr Andrews formula.

Prof. Athanese Dupré of Rennes in his very ingenious memoirs on the mechanical theory of heat (Paris, Gauthier Villars 1869 p. 61)[c] uses <the> an equation of the form

$$p = \frac{R\theta}{v + k}$$

where k is the "covolume" as he calls it < (I give it a negative sign to correspond to>

No physical interpretation of the covolume is given but it is easy to see that according to M. Dupré's formula

$$p = \frac{R\theta}{v} - k\frac{R\theta}{v^2} + k^2\frac{R\theta}{v^3} + \&c$$

so that the deviation of the pressure from that given by Boyle's law is nearly as the square of the density as we find it to be

Thomson and Joule, in their experiments on the thermal effects of gases in motion, found the deviation from the ideal properties of a "perfect gas" to vary as the square of the density.[d]

<The quantity c in Dr Andrews' equation>

It appears from Dr Andrews' experiments that the quantity c which is approximately constant for constant temperature, decreases rapidly as the temperature rises. The range of temperature is not sufficient to determine the law of variation, but it is at least as rapid as the inverse fourth power of the absolute temperature (from $-274°C$.)

The equation of Clausius in the dynamical theory of gases is

$$pv = \frac{1}{3}\sum(mc^2) - \frac{1}{2}\sum\sum(Rr)$$

where p is the pressure, and v the volume of the gas
m the mass of a molecule and c its velocity
r the distance between a pair of molecules, and
R the attraction between them at that distance

The expression $\frac{1}{2}\sum\sum(Rr)$ is termed by Clausius the Virial of the system. It is formed by summing Rr for every pair of molecules.

We have strong reasons for believing that the average value of mc^2 is the same for all gases, and that it varies as the absolute temperature on the thermodynamic scale. We may therefore write

$$\frac{1}{3}\sum(mc^2) = \frac{P_0V_0}{\Theta_0}\theta$$

where P_0 is the *very small* pressure corresponding to the *very great* volume V_0 and the standard temperature Θ.

We have next to determine $\sum\sum(Rr)$.

Let Q be the potential energy of a molecule depending on its position in space, then by Boltzmann's theorem the number of molecules in unit of volume is

$$C\varepsilon^{\frac{-Q}{a\theta}}$$

where a is an absolute constant, the same for all gases, and C is constant within the vessel considered.

Now consider the effect of a single molecule in altering the number of molecules within the sphere of action. Let Q be a function of r, the distance from the given molecule. If this molecule where[?] the number of other molecules in unit of volume would be C, but on account of the presence of the molecule it becomes $C\varepsilon^{-Q/a\theta}$. Hence if we write

$$4\pi \int_0^r (\varepsilon^{-Q/a\theta} - 1)\, r^2 dr = A_r$$

the whole increase of the number of molecules within the sphere of action (whose radius is r) is CA_r.

If the volume occupied by the gas is large compared with the sum of the spheres of action of the molecules, the regions common to two or more spheres of motion will count for very little and the whole increase in unit of volume will be

$$NCA_r = N - C$$

where N is the actual number of molecules in unit of volume.
Hence

$$C = \frac{N}{1 + NA_r}$$

(If <two or more> the spheres of action of two or more molecules have a common region then within this region the density is

$$c\varepsilon^{\frac{-Q_1+Q_2+\&c}{a\theta}}$$

which is different from $C(e^{Q_1/a\theta} + e^{-Q_2/a\theta} + \&c)$ and therefore our reasoning will not apply.)

We have next to find $\sum\sum(Rr)$ on the hypothesis that the regions common to the spheres of action of two or more molecules are negligible.

The force between two particles at distance r is $R = \frac{dQ}{dr}$.

The number of molecules whose distances from a given molecule are between r and $r + dr$ is $4\pi r^2\, dr\, Ce^{-Q/a\theta}$. Hence if we write

$$\int_0^r 4\pi r^3 \frac{dQ}{dr} e^{-Q/a\theta}\, dr = B_r$$

the value of $\sum(Rr)$ for a single molecule is $C\,B_r$.

If we multiply this by Nv we shall obtain $2\sum\sum(Rr)$ since each pair will enter twice. Hence <for unit of volume>

$$\sum\sum(Rr) = \frac{1}{2} NCB_r v$$

$$= \frac{1}{2}\frac{N^2\, B_r\, v}{1 + NA_r}$$

But we find by differentiation

$$3A_r - \frac{1}{a\,\theta}\,B_r = r^3\,(e^{-Q/a\theta} - 1)$$

When r is the radius of the sphere of action $Q = 0$ and the second member of the equation disappears, so that $B = 3\,a\theta\,A$ and in volume v

$$\sum\sum\theta\,(Rr) = \sum\sum\frac{3}{2}\frac{N^2\,a\theta\,Av}{1 + NA}$$

Hence the quantity which Dr Andrews calls c is

$$c = \left(1 - \frac{pv}{P_0 V_0}\,\frac{\Theta}{\theta}\right) v = \frac{1}{2} P_0 V_0 \theta_0\, N^2\, v^2\, \frac{aA}{1 + NA}$$

Here the only variables are N and v and Nv is the total number of molecules so that we may write

$$\begin{aligned} c &= \left\langle \frac{k}{\frac{1}{A} + b\rho} \right\rangle \\ &= \frac{k}{\frac{1}{A} + \frac{\rho}{m}} \end{aligned}$$

where $k = \frac{1}{2} P_0 V_0 \Theta_0 N^2\, v^2\, a$ and $< b$ is the$>$ m is the mass of a molecule. Hence when the density is small, c is constant $= Ak$.

As the density increases, c diminishes as Dr Andrews finds it does. But the formula ceases to be applicable when the substance is liquefied for c ought then to be negative and very large. But in this case the spheres of action must intersect, and there can be no doubt that in the liquid state, the interaction of the molecules is of a far too complicated kind for our present mathematical methods to grapple with. It is probable that a careful study of such measurements as those of Dr Andrews, under the guidance of sound principles in Thermodynamics, may throw great light on those molecular hypotheses by which we endeavour to feel our way into the minute structure of bodies.

I therefore think that Dr Andrews' paper should be printed in the Transactions.

J. Clerk Maxwell

a. The paper Maxwell referred was probably Thomas Andrews, "On the gaseous State of Matter," *Proc. R. Soc. London* 24 (1876): 455–459; *Phil. Mag.* 3 (1877): 63–66, published in full *Phil. Trans. R. Soc. London* 166 (1876): 421–449.

b. The Bakerian lecture was Andrews,"On the Continuity of the Gaseous and Liquid States of Matter," *Proc. R. Soc. London* 18 (1870): 42–45; *Phil. Mag.* 39 (1870): 150–153, printed in full in *Phil. Trans. R. Soc. London* 159 (1869): 575–590.

c. Athanese Dupré, *Théorie mécanique de la chaleur* (Paris: Gauthier Villars, 1869).

d. William Thomson and James Prescott Joule, "On the thermal Effects of Fluids in Motion," *Proc. R. Soc. London* 8 (1856): 41–42; *Phil. Mag.* 7 (1856): 466.

5. [Theory of pressure in a molecular medium]

Cambridge University Library, Maxwell Collection, Scientific Papers-7

Let us <suppose> consider the exchange of molecules through a plane surface which for convenience we may suppose horizontal. If the gas is apparently at rest and the surface also at rest the number of molecules which pass upwards will be equal to the number which pass downwards through the surface. If the vessel which contains the gas is made to move with uniform velocity upwards or downwards, while the horizontal surface moves with the same velocity as the vessel, there will still be an <equality> equilibrium of exchange. But in this case the apparent velocity of the gas is equal to that of the vessel. Hence the <definition of the> velocity of a molecular medium resolved in any direction may be defined to be the velocity <of> with which a surface normal to that direction must move that there <exchang> may be an equilibrium of exchange of molecules through that surface.

Theory of pressure in a molecular medium

Consider two portions of the medium separated by a plane surface which moves with the same velocity as the gas. In this case as we have seen the number of molecules which pass through the plane in opposite directions (upward and downward) is the same. Let a molecule be considered as above or below the plane when its centre of mass is above or below the plane.

Each molecule in crossing the plane from below to above enters the upper region in precisely the same state as it left the lower region. It therefore carries over into the upper region not only its mass but its momentum and its kinetic energy.

Let there be a group of n molecules in every unit of volume, the mass of each of which is m and having <the velocity u upwards> a velocity the components of which are u upwards and v and w in horizontal directions at right angles to each other.

The number of these molecules which pass upwards through unit of area of the horizontal plane is nu.

The momentum of each of these molecules is mu upwards and mv and mw in horizontal directions.

The kinetic energy of translation of each molecule is $\frac{1}{2}\, m(u^2 + v^2 + w^2)$.

Hence the momentum communicated to the upper part of the medium in unit of time on account of the nu molecules which enter from below is mnu^2 in an upward direction, and $mnuv$ and $mnuw$ in the horizontal directions.

Now a force of any kind acting on a body is measured by the momentum which is communicated to <the body on> it in unit of time, and the force with which one portion of a medium acts upon another contiguous portion is called pressure.

It appears therefore that the part of the pressure of the medium below the <surface> plane on the medium above which is due to the motion of the group of molecules under consideration has for its components mnu^2 upwards and $mnuv$ and $mnuw$ in the horizontal directions. Now let us take into account all the other molecules in unit of volume of the medium. These will have different values for the components of velocity but if we express by the symbol $\sum(nu^2)$ the sum of all products of the form nu^2, we may write the components of the pressure

$$X = m\sum(nu^2), \qquad Y = m\sum(nuv), \qquad Z = m\sum(nuw)$$

When the medium has attained its final state of equilibrium of exchange of the molecules which have a given upward velocity u as many will have positive as negative values of v or of w. Hence the quantities $\sum(nuv)$ and $\sum(nuw)$ will in this case vanish. If however the medium is in a state of apparent motion these quantities may have finite values <and>

Hence in the first place when the medium is apparently at rest the components Y and Z disappear and the pressure is equal to X in a direction normal to the plane <according to> which agrees with the theory of the pressure of fluids in equilibrium.

The value of this pressure estimated per unit of surface may be written

$$< X = m\sum nu^2 \text{ which may be } >$$

$p = \sum(mu^2)$ <and is> where p, the pressure, is expressed as the sum of the values of mu^2 for all the molecules in unit of volume, where m is the mass of each molecule, and u is the vertical <part> component of its velocity.

<When> In a gas at rest the pressure is equal in all directions so that

$$p = \sum(mu^2) = \sum(mv^2) = \sum(mw^2)$$

but the kinetic ener

[space]

If we express the mean value of any quantity for all the molecules by putting a bar over the symbol of that quantity then $\overline{u^2}$ will denote the mean value of the square of the vertical component of the velocity, and we may write the value of the <pressure components> pressure

$$p = m\,N\,\overline{u^2}$$

where m denotes the mass of each molecule, N the number of molecules in unit of volume, and $\overline{u^2}$ the mean value of the square of the vertical velocity.

In a gas at rest <if $\overline{V^2}$ is the mean square of the actual velocity since>

$$\overline{u^2} = \overline{v^2} = \overline{w^2} = \frac{1}{3}\,\overline{V^2}$$

where $\overline{V}^2$ denotes the mean square of the actual velocity of a molecule.

Also if ρ denotes the density of the medium, that is to say the mass contained in unit of volume

$$\rho = mN$$

We may therefore write equation ( ) in the form

$$p = \frac{1}{3}\rho\,\overline{V^2}$$

If we suppose that in a gas the pressure is entirely due to the molecular bombardment <of mol> which goes on <across> between one portion of the medium and a contiguous portion, it is easy to calculate the value of $\overline{V^2}$ for $\overline{V^2} = 3\frac{p}{\rho}$

For hydrogen at the temperature of melting ice we find $\overline{V^2} = 36593916$ in feet and seconds

The square root of this quantity is

$$\sqrt{\overline{V^2}} = 6097\,\text{feet per second.}$$

But though in the case of gases the main part of the action between one portion of the medium and a contiguous portion arises from the exchange of momentum between those portions, another kind of action is to be considered, the amount of which increases as the medium becomes more condensed. This is the direct action between molecules whose centres are on opposite sides of the plane of separation.

To estimate the amount of this let r be the distance between the centres of two molecules and let the components of r be x, y, z. Let R be the mutual action between the molecules which, when positive, we shall understand to be an attraction and let the components of R be X, Y, Z.

Then if n is the number of pairs molecules in unit of volume which are related to each as above, the number of such pairs which are <so placed

that> on opposite sides of the plane of separation is nx and the attractive force between the upper and lower portions of the medium arising from their mutual actions is nxX. Summing the actions of all the molecules we obtain an attraction

$$\sum (nxX)$$

or if we denote by N the whole number of molecules in unit of volume and by $\overline{xX}$ the mean value of all products such as Xx we may write the result $\overline{NxX}$.

This <forms> constitutes a force of attraction which diminishes the value of the pressure. We obtain for the corrected value of the pressure

$$p = mN\overline{u^2} - N\,\overline{xX}$$

But $xX + yY + zZ = rR$ and in a gas at rest $xX = yY = zZ = \frac{1}{3}rR$ so that we may write

$$p = \frac{1}{3}\,mN\,\overline{V^2} = \frac{1}{3}N\,\overline{rR}$$

The importance of the quantity $\sum(rR)$ was first pointed out by Clausius who obtained it however as the result of a different calculation[a] from that which we have just given. He has given to this quantity the name of *Virial* and this name we shall use when we have occasion to refer to it.

When the force between two molecules is an attraction the virial is positive, when it is a repulsion, the virial is negative.

< An important The simplest case of repulsion is that which takes place between two hard elastic spheres at the moment of impact >

If we suppose the molecules to be <hard> elastic bodies of any form acting on one another by impact, the part of the virial which depends on the force of impact may be expressed in a more convenient manner by <writing> taking <the mean value> instead of the virial at a given instant the mean value of the virial during a given time. That may be done by integrating

or,

$$\frac{1}{T}\int_0^T \sum \overline{rR}\,dt$$

< In the case of impact > During impact the variation of r is insensible and $\int R\,dt$ is measured by the change of momentum of either body so that if v is the relative velocity of the bodies M_1 & M_2, and if θ is the angle between v and r

$$\int rR\,dt = -\frac{2\,M_1M_2}{M_1 + M_2}\,v\cos\theta$$

and the virial is the sum of all the values of this quantity for all the impacts which take place in the time T divided by that time or <for > more briefly for all the impacts which take place in unit of time.

Since it is only during encounters between the molecules that the virial has a sensible value and as when the velocity of agitation is the same the number of encounters in unit of time is as the square of the number of molecules the <value of the> virial of unit of volume of a gas <will when the> at constant temperature will vary as the square of the density as long as the gas is so rare that a molecule always describes a free path between successive encounters. When <this ceases to be the case> the substance becomes so dense that one encounter begins before another ends the virial is no longer proportional to the square of the density.

In <most of> the gases which have been subjected to experiment <there is> the pressure is nearly but by no means exactly proportional to the density but as the density increases the deviation from proportionality increases and in most cases the pressure <is less than> increases in a smaller ratio than the <pressure> density. This shows that the virial is positive or in other words the action between the molecules is on the whole attractive. Hydrogen is the only gas in which the deviation from proportionality is in the opposite direction showing that repulsive action, whether at a distance or by impact, takes place between the molecules.

Another mode of investigating the nature of the virial is by allowing the substance to expand in such a way that no external work is done as when a gas <passes compressed in> passes through a porous plug from a place of high pressure to a place of low pressure. If there is attraction between the molecules their velocity will diminish as the gas expands. If there is repulsion <the velocity> it will increase. It is found that in most gases a cooling effect takes place showing that [?] the velocity of agitation has diminished and that the action between the molecules is on the whole attractive.

a. Inserted in the text is the reference *Berichte der Niederrhein. Gesellsch für Natur- und Heilkunde, Juni 1870. Pogg. Ann. Bd. 141 p. 124. The paper referred to is Clausius, "Über einen auf die Wärme anwendbaren mechanischen Satz," *Sitz. Bonn* (1870): 114–119; *Ann. Phys.* 141 (1870): 124–130, translated in *Phil. Mag.* 40 (1870): 122–127.

6. To find the mean value of the potential energy, the kinetic energy and the virial in a system of moving molecules

Cambridge University Library, Maxwell Collection, Scientific Papers-5.

Let us first consider two sets of molecules, one set having velocity components between the limits ξ_1 & $\xi_1 + d\xi_1$ η_1 & $\eta_1 + d\eta_1$ & ζ_1 & $\zeta_1 + d\zeta_1$

Let the mass of each molecule be M_1 and let n_1 molecules of this set exist in unit of volume.

Let the corresponding quantities for the second set be distinguished by the suffix $_2$.

Let V be the relative velocity of a molecule of the one set with respect to a molecules of the other set.

Then if each molecule moved in a straight line the number of pairs of molecules, one of each set which would pass each other at a minimum distance between b and $b + db$ in unit of time is

$$< 2\pi V\, b\, db > \qquad 2\pi\, n_1 n_2\, V b\, db \tag{1}$$

Let us now suppose that each of these pairs of molecules act on one another during their encounter, the action not being disturbed by the action of any other molecule. When the distance between the two molecules is r, let their relative velocity be v and let the angle between the directions of r and v be β.

The rate of description of areas by the one molecule about the other is, by our hypothesis, constant, for no other molecule <acts on> disturbs them during their encounter. Hence equating the area before the encounter with the area when the distance is r

$$Vb = v\, r\ \sin\beta \tag{2}$$

If Q is the work done by the attraction of two molecules while they are brought from an infinite distance to a distance r from one another. Q is a function of r and <since the kinetic energy> is equal to the kinetic energy developed by the attraction, or

$$Q = \frac{1}{2}\,(v^2 - V^2)\,\frac{M_1 M_2}{M_1 + M_2} \tag{3}$$

Hence v is a function of V and r and is independent of b and β.

We may therefore suppose V, v & r constant, and consider b as a function of β in which case the expression (1) becomes

$$2\pi\, n_1\, n_2\, \frac{v^2 r^2}{V}\ \sin\beta\, \cos\beta\, d\beta \tag{4}$$

This is the number of pairs of molecules which in unit of time pass through the configuration of being at distance r, and having their relative velocity in a direction making with v an angle between β and $\beta + d\beta$.

Since $\quad \dfrac{dr}{dt} = v\,\cos\beta$ the time during which one of these molecules is between the distances r and $r + dr$ is $dt = \dfrac{dr}{v\,\cos\beta}$

Let us next find the number of <molecules> pairs of molecules in unit of volume whose distance at a given instant is between r and $r + dr$, the value of β lying between the limits β and $\beta + d\beta$.

If N is the required number then the number which pass through the configuration mentioned above in unit of time is

$$N \frac{dr}{dt} = N v \cos\beta \tag{5}$$

Equating this to (4) we obtain

$$N = 2\pi\, n_1 n_2 \frac{v}{V} r^2 \sin\beta\, dr\, d\beta \tag{6}$$

Integrating with respect to β from $\beta = 0$ to $\beta = \pi$ we find for the number of pairs of molecules in unit of volume which at a given instant are at a distance between r and $r + dr$ of each other.

$$4\,\pi\, n_1 n_2 \frac{v}{V} r^2\, dr \tag{7}$$

If the two systems of molecules are in a state of kinetic equilibrium I have shown in my paper on the Dynamical Theory of Gases that

$$\langle n_1 = N_1/(T/M_1^{3/2})\rangle$$

$$n_1 = N_1 \left(\frac{M_1}{T}\right)^{3/2} \frac{1}{\sqrt{\pi}}\, e^{-M_1(\xi_1^2+\eta_1^2+\zeta_1^2)/T}\, d\xi_1\, d\eta_1\, d\zeta_1 \tag{8}$$

where N_1 is the number of molecules of the first kind in unit of volume and T is a quantity <proportional to the abs> depending only on the temperature. There is a similar expression for n_2.

The number of pairs of molecules whose relative velocity when undisturbed is between V and $V + dV$ is

$$4\pi^{-1/2}\, N_1\, N_2 \left(\frac{M_1 M_2}{T(M_1 + M_2)}\right)^{3/2} e^{-V^2 M_1 M_2/T(M_1+M_2)}\, V^2\, dV \tag{9}$$

The number of these which are at a distance between r and $r + dr$ is found by substituting the quantity for $n_1 n_2$ in (7) which gives

$$16\,\pi^{1/2}\, N_1 N_2 \left(\frac{M_1 M_2}{T(M_1 + M_2)}\right)^{3/2} e^{-V^2 M_1 M_2/T(M_1+M_2)}\, vV\, r^2\, dV dr \tag{10}$$

for the number of pairs in unit of volume whose original relative velocity is between V and $V + dV$ and whose actual distance is between r and $r + dr$.

We may express V in terms of v by equation (3) which <gives> makes the expression become

$$16\pi^{1/2}\, N_1 N_2 \left(\frac{M_1 M_2}{T(M_1+M_2)}\right)^{3/2} e^{\frac{2Q}{T}}\, e^{-v^2 M_1 M_2/T(M_1+M_2)}\, v^2\, r^2\, dv dr \quad (11)$$

where the limits of integration for v corresponding to $V=0$ and $V=\infty$ are $v^2 = 2Q\,\dfrac{M_1 M_2}{M_1+M_2}$ and $v^2=\infty$.

If the molecules, when their centres are at distance r rebound from each other like elastic spheres the part of the virial depending on the impacts will be negative.

The number of pairs which during the time dt <are at distance r and have> pass through the configuration $(r,\, \beta)$ is by (4)

$$2\pi\, n_1 n_2\, \frac{v^2 r^2}{V}\, \sin\beta\, \cos\beta\, d\beta\, dt$$

Multiplying this expression by Rr and remembering that during the impulse

$$\int R\, dt = -\frac{2M_1 M_2}{M_1+M_2}\, v\, \cos\beta$$

we obtain

$$-4\pi\, n_1 n_2\, \frac{M_1 M_2}{M_1+M_2}\, \frac{v^3 r^3}{V}\, \sin\beta\, \cos^2\beta\, d\beta$$

Integrating with respect to β from $\beta=0$ to $\beta=\pi$ we get

$$-\frac{8}{3}\,\pi\,,\, n_1 n_2\, \frac{M_1 M_2}{M_1+M_2}\, \frac{v^3 r^3}{V}$$

When there is no attraction $v = V$ and if we also < make > suppose the molecules all of one kind we have < v^2 the square of the > the mean square of the relative velocity equal to twice the mean square of the absolute velocity we find for the virial in unit of volume

$$-\frac{8}{3}\pi\, N^2\, M v^2\, r^3$$

Let b be the volume of unit of mass of the spherical molecules, then the volume of one molecule is $Mb = -\frac{4}{3}\,\pi\left(\frac{r}{2}\right)^3$.

Remembering also that $NM=\rho$ the virial becomes

$$-16\, b\, \rho^2\, \overline{v^2}$$

where $\overline{v^2}$ is the mean square of the velocity of agitation. Hence in this case the pressure is

$$p = \frac{1}{3}\,\rho\,\overline{v^2}\,(1 + 16\,b\rho)$$

If σ is the molecular density $b\,\sigma = 1$ and taking into account the occupation of space by the molecules we get

$$\begin{aligned} p &= \frac{1}{3}\,\rho\,\overline{v^2}\,\{1 + 16\frac{\rho}{\sigma}(\frac{\sigma}{8\rho}\,\log{[1 - 8\frac{\rho}{\sigma}]})\} \\ &= \frac{1}{3}\,\rho\,\overline{v^2}\,\{1 - 2\,\log{(1 - 8\,\frac{\rho}{\sigma})}\} \end{aligned}$$

[The following appears to be a supplementary calculation, not intended to be included in the text]

Number of molecules whose modulus is α which have a relative velocity with respect to v between r and $r + dr$

$$N\,\frac{1}{\alpha\,\sqrt{\pi}}\,\frac{r}{v}\left\{e^{-(r-v)^2/\alpha^2} - e^{-(r+v)^2/\alpha^2}\right\}$$

Mean velocity of the system (α) with respect to v, $\phi\dfrac{v}{\alpha} = \displaystyle\int_0^v e^{-\frac{v^2}{\alpha^2}}\,dv$

$$V = \frac{1}{\sqrt{\pi}}\left\{e^{-v^2/\alpha^2} + (2\,\frac{v}{\alpha} + \frac{\alpha}{v})\,\phi(\frac{v}{\alpha})\right\}$$

Integral of $\displaystyle\int_0^v V\frac{4}{\beta^3\sqrt{\pi}}\,v^2\,e^{-v^2/\beta^2}\,dv$ is

$$\begin{aligned} &4\,\frac{\alpha^2 + \beta^2}{\alpha\beta\pi}\,\phi(\frac{\alpha^2 + \beta^2}{\alpha^2\beta^2}\,v^2) \\ &\qquad - \frac{2}{\alpha\beta\pi}\,(2v^2 + 2\beta^2 + \alpha^2)e^{-v^2/\alpha^2}\;\phi(v^2/\alpha^2)\,\frac{2}{\alpha\beta\pi}\,v\,e^{-\frac{\alpha^2+\beta^2}{\alpha^2\beta^2}\,v^2} \end{aligned}$$

When $v = \infty$ this is $\dfrac{2\sqrt{\alpha^2 + \beta^2}}{\sqrt{\pi}}$ as it ought to be

7. On the probability of certain distributions of points in space by J. Clerk Maxwell

---

Cambridge University Library, Maxwell Collection

The investigation of the motion of systems of molecules has led to the consideration of several problems in pure geometry.

One of these is as follows.

In a very large vessel the volume of which is V are placed NV spheres of radius r_1 the only condition being that the spheres do not intersect each other. What is the probability that the centres of none of the spheres lie within a given small region of volume v.

This is equivalent to a distribution of points such that no two lie within a distance of 2r of each other.

<The introduction of one such point at a sufficient distance from the bounding surface of the vessel is equivalent to diminishing the region available for other points by the volume of a sphere of radius $2r$ or $6\pi r^3$. if there are a number of such points the diminution of the available room in the region outside a>

If there are already n points none of which lie within the region v, the probability that the $n+1^{th}$ point will not lie within v is

$$\frac{\text{available room not including} v}{\text{available room including} v}$$

We have therefore to calculate the available room after n points have been introduced into the region no two of which are within $2r$ of each other.

The available room is that exterior to a system of n spheres of radius $2r$ which may intersect each other provided no two <have their centres in> include each others centres.

The available room is therefore

The whole <room> volume
- Volume of n spheres of radius $2r$
+ Volume common to two spheres or more
- 2 x Volume common to three or more spheres
+ 3 x Volume common to four or more spheres
- &c

The greatest number of equal spheres which can be placed in contact with each other and with another equal sphere is 12.

Hence the above series cannot have more than 13 terms.

The probably available room must next be deduced by considering all the possible positions of the points.

The volume actually filled by N spheres of radius r is $\frac{4}{3}\pi r^3 N$. Calling this b, we may define b as the actual volume of the spheres in unit of volume.

The volume common to two spheres of radius $2r$ having their centres at distance ρ is $\frac{2\pi}{3}(16r^3 - 6r^2\rho + \frac{1}{8}\rho^3)$

The probable number of spheres <whose> the distance of whose centres from that of a given sphere is between ρ and $\rho + d\rho$ is $N\,4\pi^2\,d\rho$. Hence we find the probable volume of a given sphere which is common to it and to at least one other sphere to be $\frac{32\times 17}{9}N\,\pi^2\,r^6$ and the probable volume

<occupied by at least> common to two spheres at least in unit of volume is $\frac{16}{9} 17 N^2 \pi^2 r^2 = 17 b^2$.

To find the volume common to three spheres is a much more difficult problem and I have not attempted to find the probable volume common to three spheres at least in unit of volume. It will however be of the form $B_3 b^3$ where B_3 is a numerical quantity.

We should thus obtain as the probable value of the available room wherein to place the centre of a new sphere in a region of volume V containing a number of equal spheres whose united volume is Vb.

If therefore in a region of volume V there are already a number of equal spheres whose united volume is Vb, the probable value of the available room wherein to place the centre of another equal sphere [is]

$$V(1 - b + 17 b^2 - B_3 b^3 - B_4 b^4 - \&c)$$

In this expression there cannot be more than 13 terms.

It must become zero when the spheres are placed so that every sphere is in contact with twelve other spheres, in which case $b =$

For this is the closest possible arrangement of a number of equal spheres.

If points (A) are distributed on a plane so that there are N_a points in unit of area, and if another set (B) of points are N_b in unit area, then the number of points (b) which are at a distance between c and $c + dc$ of a given point $N_b 2\pi\, r dr$,[a] and the number of pairs of points whose distance lies between those limits, the first of them being within unit of area, is $N_a N_b\, 2\pi\, cdc$.

If a third set of points (C) has a surface density N_c, the number of these whose distance from A is between b and $b + db$ and from B between a and $a + da$ is $N_c \frac{da\, db}{\sin C}$

Hence the number of sets of three points (ABC) which have these limits of distance, and of which A lies in a given unit of area is

$$N_a N_b N_c\, 2\pi \frac{c\, da\, db\, dc}{\sin C} = N_a N_b N_c \frac{4\pi\, abc\, da\, db\, dc}{2\nabla}$$

where ∇ denotes the area of the triangle ABC.

If the volume densities of points (A) and (B) are N_a and N_b then the number of pairs at distances between (ab) and $(ab) + d(ab)$ in volume V is

$$V N_a N_b\, 4\pi\, (ab)^2\, d(ab)$$

The number of points (C) at distances between (ac) and $(ac) + d(ac)$ from A and (bc) and $(bc) + d(bc)$ from B is

$$N_c\, 2\pi\, (ac) \frac{\sin\, CAB}{\sin\, ACB}\, d(ac)\, d(bc) = N_c\, 2\pi \frac{(ac)\,(cb)}{ab}\, d(ac)\, d(bc)$$

Hence the number of sets of three within the volume V is

$$V\, N_a N_b N_c\, 8\pi^2\, (ac)\, (ca)\, (ab)\, d(bc)\, d(ca)\, d(ab)$$

The number of points E at distances between (ae) and $(ae) + d(ae)$ from A, (bc) and $(bc) + d(bc)$ from B and (ce) and $(ce) + d(ce)$ from C is

$$\frac{d(ae)\, d(be)}{\sin AEB} \left\langle \frac{d(ce)}{\sin CE.AEB} \right\rangle \frac{d(ce)}{\sin E}$$

where $< \sin ABE >$ E is the angle between CE and the plane AEB.

If T be the volume of the tetrahedron $ABCE$,

$$T = \frac{1}{3}\, (ce)\, \sin E\, \frac{1}{2}\, (ae)\, (bc)\, \sin AEB$$

The number of points (e) may therefore be written

$$\frac{\sin AEB\; d(bc)\, d(ca)\, (ac)\, (bc)\, (ca)}{6T}$$

and the whole number of sets of four within these limits will be

$$V\, N_a N_b N_c\, 8\pi^2\, \frac{(ab)(ac)(ae)(be)(bc)(ca)\; d(ac)\, d(ae)\, d(ac)\, d(ba)\, d(ce)}{6T}$$

a. Evidently r should have been changed to c in this expression,as it is elsewhere in the paragraph.

8. [On the virial theorem of Clausius]

Cambridge University Library, Maxwell Collection, Scientific Papers-5

Clausius*[a] has pointed out the importance of the quantity formed by taking the sum of the products of the tension between every pair of points in a material system into the distance between these points. He distinguishes this quantity by the name of Virial and expresses it by the sum $\sum Rr$ where R is the attraction between two points of the system and r is the distance between them. He has shown from the equations of motion that the mean kinetic energy of the system is equal to the mean virial provided the motion is not such as to produce a <permanent change in the mean density of the system> continual increase or diminution of the moments of inertia of the system.

The most important applications of this quantity however are to the establishment of general theorems relating to the internal stresses in a molecular medium and for this purpose the following method of obtaining fundamental equations seem to me to be worth attending to on account of the ease with which it adapts itself to the detailed explanation of the mutual action between two parts of a molecular medium.

Consider the action which takes place between two portions of the medium across unit of area of the surface which separates them. We shall suppose that the exchange of molecules across this surface is equal in both directions. We shall reckon a molecule to belong to one portion or to the other of the medium <acro> as its centre of mass is on the one side or the other of the bounding surface. The <pressure> stress between the first portion of the medium and the second is measured in magnitude and direction by the momentum communicated by the first portion to the second in unit of time. This momentum is communicated in two different ways. Molecules may pass from the first portion to the second carrying their momentum with them, or there may be direct action between pairs of molecules one of which belongs to the first portion and the other to the second. The <pressures> stresses due to these two causes may be called the kinetic and the static <pressures> stresses respectively.

Let us suppose the surface of separation, across which the stress takes place to be normal to the axis of x.

Let the number of molecules whose components of velocity lie between $\dot{x}$ and $\dot{x} + d\dot{x}$ $\dot{y}$ and $\dot{y} + d\dot{y}$ $\dot{z}$ and $\dot{z} + d\dot{z}$ be $f\, d\dot{x}\, d\dot{y}\, d\dot{z}$ in unit of volume, where f is a function of $\dot{x}$, $\dot{y}$, $\dot{z}$ then the number which cross the unit of areas of the surface in unit of time is $< xdr >$ $f\, \dot{x}\, d\dot{x}\, d\dot{y}\, d\dot{z}$

The momentum of each of these has for its components

$$m\dot{x} \qquad m\dot{y} \qquad m\dot{z}$$

where m is the mass of the molecule.

Hence the whole momentum communicated by the first portion of the medium to the second by the passage of these molecules has for its components

$$f\, m\dot{x}^2\, d\dot{x}\, d\dot{y}\, d\dot{z} \qquad f\, m\dot{x}\dot{y}\, d\dot{x}\, d\dot{y}\, d\dot{z} \qquad f\, m\dot{x}\dot{z}\, d\dot{x}\, d\dot{y}\, d\dot{z}$$

Integrating with respect to $< d\dot{x}\, d\dot{y}\, d\dot{z} >$ $\dot{x}$ $\dot{y}$ and $\dot{z}$ for all values of these quantities and putting N for the whole number of molecules in unit of volume and denoting the mean value of any quantity by a bar placed over the symbol or combination of symbols which expresses that quantity we have for the components of <stress> the force exerted by the <first> portion on the negative side of the plane yz on the portion on the positive side through unit of area of that plane

$$P < X > = N m \overline{\dot{x}^2} \qquad U < Y > = N m \overline{\dot{x}\dot{y}} \qquad T < Z > = N m \overline{\dot{x}\dot{z}}$$

This is the kinetic part of the stress.

The static part arises from the direct action <between> of molecules on the negative side of the plane on molecules on the positive side of the plane, by attraction repulsion or <direct> pressure of contact. Let r be the distance between two molecules and let the components of r be x, y, z. Let the number of pairs of molecules in unit of volume for which these components lie between x and $x + dx$, y and $y + dy$, and z and $z + dz$ be $F\,dx\,dy\,dz$ where F is a function of x, y, z. Also let R be the attraction between the molecules <that> then $R\frac{x}{r}$, $R\frac{y}{r}$ and $R\frac{z}{r}$ will be its components.

The number of pairs of molecules <in unit of volume> satisfying these conditions and also lying on opposite sides of the plane of separation is

$$x\;F\;dx\,dy\,dz$$

Hence the action of the first portion of the medium on the second has for its components

$$P \;<X> = -\iiint \frac{x^2}{r} R\,F dx\,dy\,dz,$$

$$U \;<Y> = -\iiint \frac{xy}{r} R\,F dx\,dy\,dz,$$

$$T \;<Z> = -\iiint \frac{xz}{r} R\,F dx\,dy\,dz$$

This is the static part of the stress.

In an isotropic fluid medium the stress-components S, T, U disappear and $P\,Q\,R$ become equal and we have for the whole stress which is now a hydrostatic pressure

$$p = \frac{1}{3} N\,M\,\overline{v^2} - \frac{1}{3}\overline{rR}$$

or the pressure per unit of surface is equal to one third of the product of the density into the mean square of the velocity of agitation, minus <together with> one third of the virial in unit of volume.

a. This presumably would have been a footnote had Maxwell completed these notes for a paper. It could refer to any one of several papers Clausius published on the virial in the early 1870s.

9. "Virial"

Cambridge University Library, Maxwell Collection

Virial

No of pairs in unit of volume whose relative velocity *at a distance* is between V & $V + dV$

$$N = N_1 N_2 \frac{1}{(\alpha_1^2+\alpha_2^2)^{3/2}\sqrt{\pi}} V^2\, e^{V^2/\alpha_1^2+\alpha_2^2}\, dV$$

Actual velocity at distance $r = u$ making angle β with v

Work of attraction from ∞ to $r = Q$ $\quad v^2 = V^2 + 2Q \quad vdv = VdV$

Number of encounters between b and $b + db$ in <a second> time dt $\quad n = N\,V\,2\pi b\, db\, dt$

$$\text{Area} \qquad bV = r\,v\,\sin\beta \qquad b\,db = \frac{v^2 r^2}{V^2}\sin\beta\,\cos\beta\, d\beta$$

$$\frac{dr}{dt} = v\,\cos\beta$$

$$n = N\,2\pi\,\frac{v\,r^2}{V}\,\sin\beta\, d\beta\, dr$$

Integrate from $\beta = 0$ to $\beta = \pi$ $\qquad 4\pi\,N\frac{v\,r^2}{V}\,dr$

For Virial $\displaystyle\iint 4\pi\,N_1N_2\frac{1}{(\alpha_1^2+\alpha_2^2)^3/2}\,\frac{1}{\sqrt{\pi}}\,Ve^{-v^2/\alpha_1^2+\alpha_2^2}v^2r^3\,R\,dV dr$

For <Ergal> vis viva

$$\iint 4\pi\,N_1N_2\frac{1}{(\alpha_1^2+\alpha_2^2)^3/2}\,\frac{1}{\sqrt{\pi}}\,Ve^{-v^2/\alpha_1^2+\alpha_2^2}\frac{1}{2}v^3r\,dV\,dr$$

$$\int_0^\alpha V\,v\,e^{-V^2/\alpha_1^2+\alpha_2^2}\,dV = \int_{2Q}^\infty v^2e^{-v^2-2Q/(\alpha_1^2+\alpha_2^2)}dV$$

$$= \frac{\alpha_1^2+\alpha_2^2}{2}\sqrt{2Q} + e^{2Q/\alpha_1^2+\alpha_2^2}\int_{2Q}^\infty e^{-v^2/\alpha_1^2+\alpha_2^2}$$

Let $\displaystyle\int_x^\infty e^{-x^2}dx = F(x^2)$ $\qquad$ then $\qquad \displaystyle\int_v^\infty e^{v^2/\alpha^2}dv = \alpha F(\frac{v^2}{\alpha^2})$

$$\int_0^\infty V\,v\,e^{-V^2/\alpha_1^2+\alpha_2^2}dV$$

$$= \frac{\alpha_1^2+\alpha_2^2}{2}\left(\sqrt{2Q} + e^{2Q/\alpha_1^2+\alpha_2^2}\sqrt{\alpha_1^2+\alpha_2^2}F\left(\frac{2Q}{\alpha_1^2+\alpha_2^2}\right)\right)$$

$$\int_0^\infty V v^3 e^{-V^2/\alpha_1^2+\alpha_2^2}dV$$

$$= (\alpha_1^2+\alpha_2^2)\left(Q + \frac{3}{4}(\alpha_1^2+\alpha_2^2) + \frac{3}{4}(\alpha_1^2+\alpha_2^2)^{5/2}F\left(\frac{2Q}{alpha_1^2+\alpha_2^2}\right)\right)$$

For Force function Ergal

$$\iint 4\pi N_1 N_2 \frac{1}{(\alpha_1^2+\alpha_2^2)^{3/2}} \frac{1}{\sqrt{\pi}} V e^{-V^2/\alpha_1^2+\alpha_2^2} v\, r^2\, Q\, dV\, dr$$

10. [Virial Theorem]

Cambridge University Library, Maxwell Papers, Scientific Papers-5.

It is only a small part of the investigation into the constitution of bodies which has as yet been put in the form of accurate deductions from known facts. <Science cannot advance without If the pioneers of science were forbidden to make any advances on unknown ground before they had established a secure base of a complete system of communications with a secure base of operations, we might be spared many erroneous statements, but we should obtain few new suggestions. The more regular operations of science>
To conduct the operations of science in a perfectly regular manner by means of methodical experiments and strict reasoning require a strategic skill which we must not look for even amóng those <who furnish us> to whom science is most indebted for new facts and suggestions. It does not detract from the merit of the pioneers of science that their advances, being made on unknown ground, are often cut off for a time from that system of communications with a secure base of operations which is essential to the permanent extension of science.

I shall begin therefore by directing your attention to what may be considered as established by strict dynamical methods.

We shall first consider that property of bodies which belongs most evidently to dynamics. that namely by which they resist the action of external forces which tend to alter their volume or their form.

When a body, acted on by external forces is in equilibrium, these external forces form a system which if applied to any other body of the same form would <be in equilib> keep it in equilibrium. It is customary to express this by saying that the external forces themselves form a system in equilibrium. <If we now conceive the body divided in imagination into two parts without altering in any way the physical connections of its parts the external forces which act on one of these parts will not, in general, be in equilibrium. To account for the actual equilibrium of this part of the body we must admit that besides the external forces, there is an action the other part of the body acts on it across the imaginary surface of which separates the two parts of the body. The resultant of this action must be exactly sufficient to balance the external forces acting on the part.>

If we now confine our attention to one part only of the body the equilibrium of this part must be maintained by the external forces which act on it <and by> together with the internal forces arising from its connexion with the other part of the body. The resultant of these internal forces is therefore equal and opposite to that of the external forces acting on the part of the body under consideration and is therefore known when the external forces are known.

The action and reaction between the parts of a body is called Stress. This word stress was borrowed from ordinary language and invested with a precise scientific meaning by the late Professor Rankine to whom we are indebted for several other valuable scientific words.

When the stress between two parts of a material system tends to draw those parts together or to keep them from separating it is called a Tension or an Attraction. When it tends to make them separate or to keep them from approaching it is called a Pressure or a Repulsion.

The words Tension and Pressure are commonly used when we conceive the stress exerted through the medium of a continuous body like a rope or a pole. The words Attraction and Repulsion are used when we <do not> are describing the mutual action of distant bodies without reference to the manner in which they are enabled to act on each other.

<The stresses in a solid body are sometimes neither tensions nor>

The stress between one part <and an> of a solid body and the contiguous part is sometimes neither a tension nor a pressure but an oblique stress.

Let us now consider the stress between to portions of a material system on opposite sides of an imaginary plane surface. If the particles composing the system are at rest, this stress can arise only from the action between pairs of particles one of which is on one side and the other on the other of this plane.

When a particle is partly on the one side of this plane and partly on the other we shall reckon it to belong to that side on which its centre of mass lies.

We shall suppose the system to be made up of particles, that is to say of small parts. This supposition is equally true whether <one> these small parts are physical molecules <or whether they are> each of which has a natural right to be treated as an individual or whether they are small portions <arbitrarily selected from the mass> into which we suppose the mass to be divided in any arbitrary manner merely in order to be able to think of them separately.

We use the word particle therefore without introducing any molecular or atomic hypothesis about the constitution of matter.

The particles may be either at rest or in motion. To suppose them in motion is the more general hypothesis. If we form a theory of particles on motion we have only to suppose the motion to be zero in order to obtain a theory of particles at rest.

Let us now consider a system of particles in motion and let us imagine a fixed plane surface dividing the system into two parts. We shall consider any particle to belong to that side of the plane on which its centre of mass happens to be. As we are not able to observe the individual particles our results must apply to the two systems made up of such particles, and lying on opposite sides of the plane of separation.

Two kinds of action may take place between these two systems. A particle of the one system may act directly on a particle od the other system, or a particle may pass from the one system into the other across the plane of separation. The first of these kinds of motion produces what we call the static part of the stress as it depends in the forces exerted between the particles and not on their motion. The second kind produces what may be called the kinetic stress as it depends on the motion of the particles.

We shall begin with the static stress in a uniform medium.

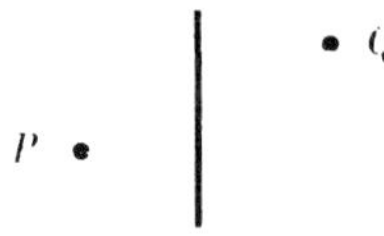

Let us consider the stress across unit of area of the plane, yz, normal to the axis of x arising from the mutual actionbetween pairs of particles on opposite sidesof the plane.

Let the coordinates of the first particle, P on the negative side of the plane, be x, y, z and those of the second, Q, $x+\xi, y+\eta, z+\zeta$.
Let the components of the force with which P acts on Q be X, Y, Z, then the component of the stress <between the two parts> of the negative side on the positive side of the medium will be

$$\sum(X) \qquad \sum(Y) \qquad \sum(Z)$$

where all pairs of molecules are included in the summation provided that the two particles lie on opposite sides of the separating plane. That this may be the case we must have x negative and $x+\xi$ positive, that is, the particle P must lie within a stratum of thickness ξ and area unity. <If V is the volume of the medium and N the number of particles in it the number each corresponding to P each of these having a particle in the relative position corresponding to Q at the number of particles corresponding to P within the required stratum will be

$$\frac{N\xi}{V_1}$$

Each of these contributes to the stress in the direction of x the force X so that the whole stress due to this set of pairs of particles in this configuration

is $\frac{1}{V} N X \xi$ where N is the whole number of such pairs in the medium. Hence $\frac{1}{V} \sum \sum X \xi$ will be the whole normal component.>
If V is the volume occupied by the medium the probability of any <partic> given particle lying in the stratum is $\frac{\xi}{V.}$ Hence the probable value of the <stress> x-component of stress contributed by any pair of particles is $\frac{X \xi}{V}$ and the actual value of this component is

$$\frac{1}{V} \sum \sum (X \xi)$$

where the double sign of summation indicates that the value of $X \xi$ is to be found for every pair of particles in the system and the sum of all such products is to be taken.

This then is the normal component of the stress between the two portions of the medium. We need not at present investigate the other components as they do not exist in fluids.

The normal stress in fluids is called pressure. We thus find that the statical part of fluid pressure–that which depends repulsion between the particles of the medium is

$$p = \frac{1}{V} \sum X \xi$$

In fluids the pressure is equal in all directions so that

$$\sum X \xi = \sum Y \eta = \sum Z \zeta$$

but if R is the <repulsion> attraction and r the distance of the particles

$$X\xi + Y\eta + Z\zeta = -Rr$$

Hence we may write the statical part of the pressure

$$pV = -\frac{1}{3} \sum \sum (Rr)$$

The importance of this quantity $\sum \sum (Rr)$ was first pointed out by Professor Clausius of Bonn who has thus added to <the> his valuable to whom the recent development of molecular science is mainly due and who has thus <added to> given us <not only> in addition to the results of <long and> his elaborate calculations <but> a new dynamical idea by the use of which important conclusions may be established without any calculation at all.

To assist us in remembering this quantity Professor Clausius has given it a name.

The product of an attraction into the distance between the attracting points he calls the Virial of that stress. When the stress <attraction> is <a> repulsive the Virial becomes negative.

The sum of the virials of all the <forces> stress is [that] which exist[s] between pairs of particles in the system he calls the Virial of the system.

We may therefore express our result in words as follows

<The Virial of a fluid mass, co>

If a fluid consists of particles at rest, its virial together with three times the product of the pressure into the volume, is zero.

Let us see how this result bears on the theory that <air consists of> the elasticity of air depends on repulsive forces between its particles.

It is well known that for a given quantity of air at constant temperature the product of the pressure into the volume is approximately constant. This is Boyles Law of the Spring of Air. Hence <if> on the theory that the particles of air are at rest the virial also must be negative and constant. In other words Rr must be constant, or the repulsion between two particles must be inversely as their distance. This law of force will account for the relation between volume and pressure in a given mass of air.

But it leads to inadmissable consequences. For if Rr is constant for all values of r it <is> must be the same for every pair of particles in the whole mass of air and therefore the whole virial is proportional to the number of pairs of particles. Now the number of pairs of particles varies as the square of the number of particles so that according to this law the product of pressure and volume should vary as the square of the mass of air whereas we know that it varies as the mass simply. The pressure, in fact would not depend on the density only but on the <absolute> product of the density into the whole mass of air considered <so that the law would be different in a large vessel and in a small one> so that for the same density the pressure would be greater in a large vessel that in a small one and greater in the open air than in any ordinary vessel.

The pressure of air cannot therefore be <mainly due to the repulsion of its particles these particles> explained by supposing its particles at rest and endowed with repulsive forces.

Let us therefore investigate the effect of the motion of particles in producing pressure.

A Force in general is measured by the momentum which it communicates in a unit of time to the body on which it acts.

Now a particle which passes from the negative to the positive side of the plane of separation carries its momentum with it. This momentum which originally was part of the momentum belonging to the medium on the negative side of the plane is now transferred to the medium on the

positive side of the plane.
If m is the mass of the particle and u is the component of velocity of the particle in the direction of x <and if m is the mass of the part> then the momentum transferred by the passage of the particle is mu. <That this passa> If V is the volume occupied by the medium then the probability that a given particle having the velocity u will pass through a given unit of area of the separating plane in a determinate unit of time is, $\frac{u}{V}$.
That this passage may take place during a given interval of time whose duration is t, the particle must be within a distance ut of the plane and if the passage takes place through a definite area of unit surface the particle must originally be in a region whose volume is ut. If V is the whole volume of the medium <and N the number of particles in it> the probability of a particle passing through the given unit of surface in time t is

$$\frac{u}{V}t$$

The probable value of the momentum communicated in this way in unit of time by a particle having u for one component of its velocity is

$$\frac{mu^2}{V}$$

and if we add the values of this expression for every particle in the medium we obtain for the whole momentum in the direction of u transferred across unit of area in unit of time

$$\frac{1}{V}\sum mu^2$$

Now this transfer of momentum is equivalent to a pressure and is in fact that part of the pressure which depends on the motion of the particles. Adding the statical and the kinetic parts of the pressure we find

$$Vp = \sum(mu^2) + \sum\sum(X\xi)$$

<In a fluid medium> if c denotes the velocity of a particle without regard to direction and $u\,v\,w$ its c[omponents]

In a fluid medium the velocities of the particles are distributed in one direction as much as in another, so that

$$\sum(mu^2) = \frac{1}{3}\sum mc^2$$

where c is the velocity without regard to direction and since the kinetic energy is $\frac{1}{2}\sum mc^2$ we may write the equation

$$Vp = \frac{2}{3}\sum(\frac{1}{2}mc^2) - \frac{1}{3}\sum\sum Rr$$

11. [Virial of a system of molecules]

Cambridge University Library, Maxwell Collection

If the mass of every portion of a material system is multiplied by the square of its distance from <one of the> a given <coordinate planes> plane, the sum of all such products is called the Moment of Inertia of the system, with respect to that plane. Thus if

$$A = \sum (mx^2) \tag{1}$$

A is the moment of inertia of the system with respect to the plane of yz. We may write B and C for the moments of inertia with respect to the plane of zx and xy and H for the sum of $A + B + C$.

Now $$\frac{1}{2}\frac{d}{dt}(mx^2) = mx\frac{dx}{dt}$$

and $$\frac{1}{2}\frac{d^2}{dt^2}(mx^2) = m\left(\frac{dx}{dt}\right)^2 + xm\frac{d^2x}{dt^2} \tag{2}$$

If we write the components of velocity

$$\frac{dx}{dt} = u \qquad \frac{dy}{dt} = v \qquad \frac{dz}{dt} = w \tag{3}$$

and the components of the force acting on m

$$m\frac{d^2x}{dt^2} = X \qquad m\frac{d^2y}{dt^2} = Y \qquad m\frac{d^2z}{dt^2} = Z \tag{4}$$

we may write [the] equation

$$\frac{1}{2}\frac{d^2A}{dt^2} = \sum(mu^2) + \sum(Xx) \tag{5}$$

Similarly

$$\frac{1}{2}\frac{d^2B}{dt^2} = \sum(mv^2) + \sum(Yy) \tag{6}$$

$$\frac{1}{2}\frac{d^2C}{dt^2} = \sum(mw^2) + \sum(Zz) \tag{7}$$

Adding these equations and dividing by two we find

$$\frac{1}{2}\frac{d^2H}{dt^2} = \left\langle \frac{1}{2}\frac{d^2}{dt^2}(A + B + C)\right\rangle$$
$$= \frac{1}{2}\sum m(u^2 + v^2 + w^2) + \frac{1}{2}\sum(Xx + Yy + Zz) \tag{8}$$

The left-hand member of this equation is the second differential coefficient with respect to the time of <half> one quarter of the moment of inertia of the system *with respect to the origin.*

The first term of the right hand member is half the Vis Viva of the system according to the old method of reckoning or <twice> the kinetic energy, and may be written $2T$.

The last term is twice the quantity defined by Clausius as the "Virial" of the system, so that the equation of Clausius may be written

$$\frac{1}{4}\frac{d^2H}{dt^2} = \left\langle \frac{1}{4}\frac{d^2}{dt^2}(A+B+C) \right\rangle$$
$$= \text{Kinetic Energy} + \text{Virial} \tag{9}$$

Now in a system of which the average distribution of mass is not in process of change, <we may w> the average value of the first member does not differ sensibly from zero, so that in such a system, which is defined by Clausius as in a "stationary" condition,

$$\text{Kinetic energy} + \text{Virial} = C \tag{10}$$

This is the form in which Clausius has expressed the result of his investigation. We shall find it convenient however to take up the equations in the more general form as given in (5), (6), (7), and to add to them the three following obtained by differentiating the products of inertia

$$D = \sum(myz) \qquad E = \sum(mzx) \qquad F = \sum(mzy) \tag{11}$$

thus

$$\frac{dD}{dt} = \sum m\,(y\frac{dz}{dt} + z\frac{dy}{dt})$$
$$\frac{d^2D}{dt^2} = 2\sum(m\frac{dy}{dt}\frac{dz}{dt}) + \sum m\,(y\frac{d^2z}{dt^2} + z\frac{d^2y}{dt^2})$$

or

$$\frac{1}{2}\frac{d^2D}{dt^2} = \sum(mvx) + \frac{1}{2}\sum(Yz+Zy) \tag{12}$$

Similarly

$$\frac{1}{2}\frac{d^2E}{dt^2} = \sum(mwu) + \frac{1}{2}\sum(Zx+Xz) \tag{13}$$
$$\frac{1}{2}\frac{d^2F}{dt^2} = \sum(muv) + \frac{1}{2}\sum(Xy+Yx) \tag{14}$$

Let us now consider the mutual action of two particles m_p and m_q. Let us suppose that they <attract> are at a distance r where

$$r_{pq}^2 = (x_p - x_q)^2 + (y_p - y_q)^2 + (z_p - z_q)^2 \tag{15}$$

Let λ, μ, ν be the direction cosines of r_{pq} <so that> Let R_{pq} be the stress between m_p and m_q and let it be reckoned positive when it is an *attraction.* <Its components will be> Let X_p be the whole x-component of the force on m_p

$$X_p = \sum_q X_{pq} + X'_p \tag{16}$$

where X'_p is the force acting on m_p arising from bodies external to the system and X_{pq} is the force on m_p <in the direct> arising from m_q. <Hence $\sum_p X_p x_p = \sum p(\sum)$ >

$$\left\langle X_p = \sum_q R_{pq} r_{pq} \lambda^2 + X'_p \right\rangle$$

$$X_p = \sum_q \left(R_{pq} \frac{x_q - x_p}{r_{pq}} x_p \right) + X'_p x_p \tag{17}$$

$$\sum_p (X_p x_p) = \sum_p \left[\sum_q R_{pq} \frac{x_p - x_q}{r_{pq}} x_p + X'_p x_p \right]$$

$$= - \sum_p \sum_q \frac{R_{pq}}{r} (x_q - X_p)^2 + \sum_p (X'_p x_p) \tag{18}$$

where in the double summation every pair of particles is to be taken only once. It will not be necessary to retain the suffixes in order to distinguish the particles, so that we may write

$$\sum Xx = - \sum \sum Rr\, \lambda^2 + \sum X'x \tag{19}$$

Similarly

$$\sum Yy = - \sum \sum Rr\, \mu^2 + \sum Y'y \tag{20}$$

$$\sum Zz = - \sum \sum Rr\, \nu^2 + \sum Z'z \tag{21}$$

Adding <and dividing by> we find

$$\sum (Xx + Yy + Zz) = - \sum \sum Rr + \sum (X'x + Y'y + Z'z) \tag{22}$$

one of the expressions for the virial given by Clausius.

Similarly we find

$$\sum (Yz + Zy) = -2 \sum \sum (Rr\mu\nu) + \sum (Y'z + Z'y) \tag{23}$$

$$\sum(Zx + Xz) = -2\sum\sum(Rr\nu\lambda) + \sum(Z'x + X'z) \tag{24}$$

$$\sum(Xy + Yx) = -2\sum\sum(Rr\lambda\mu) + \sum(X'y + Y'x) \tag{25}$$

Let us now suppose that the material system is contained in a vessel in the form of a parallelepiped bounded by the six planes $x = 0$, $y = 0$, $z = 0$, $x = a$, $y = b$, $z = c$. Let the components of the force arising from the action of bodies external to the system on the plane face be normal to x for which $x = a$ be

$$X' = bcP_{xx} \qquad Y' = bcP_{xy} \qquad Z' = bcP_{xz} \tag{26}$$

where the first suffix indicates the direction of the normal to the face on which the force acts, and the second suffix indicates the direction of the force itself.

Let the components of the force acting on the opposite face be equal and opposite to these and let the forces on the other faces be denoted by a notation of the same kind. Then the external forces acting on the system are such that if they acted on a homogeneous elastic body <of the> in the form of the parallelepiped *abc* they would produce in it a homogeneous stress whose components are P_{xx}, P_{xy} &c. We have supposed, on the contrary that the material system on which these forces act consists of molecules in motion and acting on each other by attractions and repulsion.

But if this system is kept in a stationary state by these forces, neither expanding nor contracting on the whole and since we can test the nature of a material system only by the action of external forces on it, then if the material system is so small that we cannot divide it but must deal with it as a whole, we shall not be able to distinguish it from a homogeneous body in which a state of stress exists which is specified by the six components $P_{xx}, P_{yy}, P_{zz}, P_{yz}, P_{zx}, P_{zy}$.

We must therefore <substitute> write the equations (19), (20), (21)

$$\sum(X'x) = bc\,P_{xx}\,a = abc\,P_{xx} \tag{27}$$

$$\sum(Y'y) = abc\,P_{yy} \tag{28}$$

$$\sum(Z'z) = abc\,P_{zz} \tag{29}$$

and in equations (23), (24), (25)

$$\sum(Y'z + Z'y) = ab\,P_{zy}\,c + ca\,P_{yz}\,b$$

or since $P_{zy} = P_{yz}$

$$\sum(Y'z + Z'y) = 2\,abc\,P_{yz} \tag{30}$$

similarly

$$\sum (Z'x + X'z) = 2\,abc\,P_{zx} \tag{31}$$

$$\sum (X'y + Y'x) = 2\,abc\,P_{xy} \tag{32}$$

Substituting in (5), (6), (7), the values of the last terms, we find

$$\sum (mu^2) - \sum\sum (Rr\,\lambda^2) + abc\,P_{xx} = 0 \tag{33}$$

$$\sum (mv^2) - \sum\sum (Rr\,\mu^2) + abc\,P_{yy} = 0 \tag{34}$$

$$\sum (mw^2) - \sum\sum (Rr\,\nu^2) + abc\,P_{zz} = 0 \tag{35}$$

Similarly <in> equations (12), (13), (14) become

$$\sum (mvw) - \sum\sum (Rr\,\mu\nu) + abc\,P_{yz} = 0 \tag{36}$$

$$\sum (mwu) - \sum\sum (Rr\,\nu\lambda) + abc\,P_{zx} = 0 \tag{37}$$

$$\sum (muv) - \sum\sum (Rr\,\lambda\mu) + abc\,P_{xy} = 0 \tag{38}$$

If we suppose that Rr, the product of the attraction between two molecules into the distance between them, becomes insensible for all distances greater than ε, a small distance which we may call the limit of molecular force, then if we suppose a sphere described with the radius ε concentric with a molecule we may call this the sphere of action of that molecule, and only those other molecules which happen to have their centres within this sphere will contribute to the sum $\sum Rr$ for that molecule.

<Hence though if the sides of the vessel are many times longer than ε >

Hence if the sphere of action of a molecule is entirely within the vessel, the contribution of that molecule to the components of virial will depend only on the <state of the sys> average density of the distribution of molecules within its sphere of action and if this density is proportional to the average density of the whole system the total virial $\sum\sum Rr$ will be proportional to the <number> product of the whole number of molecules into the number per unit of volume, and if we suppose the volume of the vessel to be unity, or $abc = 1$, the virial $\sum\sum Rr$ will be proportional to the square of the density.

Hence, if the elasticity of a gas<es> were due entirely to repulsive forces between <their> its molecules acting only at insensible distances, the pressure of the gas would vary as the square of the density.

[rest of the page is blank]

It may be instructive to consider these equations from a different point of view.

Let us consider an imaginary plane dividing the material system into two parts and let us study the action of one of these parts on the other. The action-and-reaction constitutes the internal stress of the system with respect to this plane. Let us suppose the plane parallel to that of yz, then the components of the force with which the part of the system on the positive side of the plane acts on the part of the negative side are denoted by P_{xx}, P_{xy}, P_{xz}, in the directions of xyz, respectively.

This stress arises from two distinct kinds of action,
1st from <particles whose> the direct action between pairs of particles whose centres of mass are on opposite sides of the plane.
2nd from particles <crossin> passing from one side of the plane to the other, carrying their momentum with them.

To calculate the effect of the direct action between pairs of molecules let us suppose that in unit of volume there are N pairs of particles A, B, whose distance lies between r and $r + dr$, <and> the direction cosines of r lying between λ and $\lambda + d\lambda$ and μ and $\mu + d\mu$ (since ν is a function of λ and μ its limits are not specified). If the first of these particles, A, is between the plane $x = 0$ and the plane $x = -r\lambda$, the other particle will be on the other side of the plane $x = 0$ and the stress between the particles will contribute to the stress across the plane. The volume of unit of area of this stratum is $r\lambda$ and the number of <first> particles A in it is $Nr\lambda$.

The contributions of each pair of particles to P_{xx}, P_{xy}, P_{xz} are $R\lambda$, $R\mu$ and $R\nu$ respectively, so that the $N\nu\lambda$ particles contribute $< N\nu R >$ $NRr\lambda^2$, $NRr\lambda\mu$ and $NRr\lambda\nu$, to the three components of the stress.

This is the contribution of the group of pairs of particles just specified, and the contributions of all pairs of molecules will be expressed by

$$\sum\sum Rr\lambda^2, \qquad \sum\sum Rr\lambda\mu \text{ and } \sum\sum Rr\lambda\nu$$

as in the equations ( ) ( ) ( )

To estimate the <second> effect of the transference of particles from one side of the plane to the other let us suppose that there are N particles in unit of volume whose velocity components are between u and $u + du$, v and $v + dv$, and w and $w + dw$.

If one of those particles is between the plane $x = 0$ and the plane $x = -u$ it will cross the plane $x = 0$ during the succeeding unit of time. Hence the number of such particles which cross unit of area of the plane in unit of time will be Nu. Now each particle carries with it a momentum mu parallel to x, mv parallel to y and mw parallel to z, so that the group of molecules just specified transfer a quantity of momentum whose components are

$$Nu^2 m, \qquad Nuvm \qquad \text{and} \qquad Nuwm$$

across unit of area in unit of time, and the components of the whole momentum transferred by all such groups that is to say by the whole system, are

$$\sum mu^2 \qquad \sum muv \qquad \sum muw$$

Now all forces are measured by the momentum which they communicate to a body in unit of time. In the case before us the "body" which we consider is the collection of particles which happen to be on the negative side of the plane and this "body" in unit time parts with the momentum just estimated. So that is to be reckoned as a negative stress.

It is true that this takes place by an exchange of matter so that <each> neither of the interacting "bodies" continues to consist of the same individual particles, but the effect as measured by the external forces exerted by the sides of the vessel in order to equilibrate the internal actions is the same whether this action <is of the> arises from stress between pairs of particles or from <transference> motion of particles.

V Documents: Statistical Mechanics

1. "On the Motions and Encounters of Molecules"[a]

Cambridge University Library, Maxwell Collection

Application of the Hamilton Method of the Hodograph to represent velocities of Molecules

Method of representing the velocities of molecules

The velocity of a body is a vector, that is a quantity which is determinate in direction and magnitude. It may therefore be conveniently represented by the finite straight line which would be traced by the body in unit of time if the velocity of the body remained the same in magnitude and direction during that time.

The position of the first point of the straight line is a matter of indifference so long as the line is considered only with respect in its magnitude and direction; so that although it might appear most natural to <draw take> draw the line from the <point> actual position of the body at the given time it is more convenient to draw all lines representing velocities from one point, called the Origin.

This method has been long known in the construction of what is called the "parallelogram of velocities" but its value was first clearly shown by Sir W. R. Hamilton who by drawing from one point a series of lines representing the successive velocities of a body, determined a series of points in a curve which he called the Hodograph.[b] The Hodograph is a curve each point of which corresponds to a point in the path of the body, and by studying the correspondence of these curves the force acting on the body and the whole circumstances of the motion may be ascertained.

In our present investigation we use the same method to compare the simultaneous velocities of different bodies as well as the successive velocities of each.

We may regard this method as an example of one of the most <fertile> powerful instruments of mathematical research–the simultaneous contemplation of two systems so related to each other that every element in the one has its corresponding element in the other. <Besides the> In pure geometry this study of corresponding elements in two figures has led to the establishment of a Geometry of Position by which results are obtained by pure reasoning without calculation the verification of which by Cartesian analysis would <occup> fill many pages with symbols.

In Statics the same method has enabled us to construct diagrams of stress by which <indicate> without calculation the stresses of the pieces of a frame are all represented and in Geometrical Optics the study of the correspondence between the object and the image <between the> has been of almost equal service to the theory of optical instruments and to pure geometry.

In all these instances we have to construct a figure the relative positions of whose elements indicates not the relative *position* of the corresponding elements of the original system but their relative velocity, the force acting between them or some other physical quantity not apparent to the eye in the original system.

In <the study of> molecular science this method is especially valuable when we wish to form a mental representation of the motion of an immense number of molecules at a given instant. Instead of <attempting first thinking> confusing the space we have already formed of the configurations of the system molecules by <and their> trying to attach to each molecule <some> an arrow or some other symbol to indicate its velocity, we form our image of the velocities on an entirely new field in which their positions are not represented at all <and in which each molecule is repres> . In this figure, to every molecule corresponds a point, <the dis> and the velocity of the molecule is represented in magnitude and direction by a line drawn from the origin to this point. Since in all the motions of real bodies the phenomena depend not on their absolute but on their relative velocities, and as these are indicated in the diagram of velocities by the distances between the points of the figure, the position of the origin itself may be transferred from one part of the figure to another without altering the relative position of the points just as the motion system of molecule <may move> as a whole may be varied while the relative motion of the molecules may be unaltered.

<If there are a great number of molecules having velocities different from each other, the diagram of velocities will contain as many points distributed over the diagram. If we take the a small element of volume of the diagram and consider these points which lie within it, these points correspond to

molecules whose velocities differ little from each other, either in magnitude or direction.>

In studying the motion of the system it is found convenient to <break> these <molecu> divide the molecules into groups <that> according to their velocities, those molecules whose velocities lie within certain limits with respect to magnitude and direction being placed in the same group.

In the diagram of velocities, these molecules are indicated at once by the points which correspond to them being included within a certain small region of the diagram, the boundaries of this region <being> corresponding to the given limits of velocity. We shall also find it convenient to use the term velocity-density to indicate the <ratio number of molecules> result of dividing the number of molecules whose velocities lie between the given limits by the volume of the corresponding region in the diagram of velocities.

a. Written lightly at the top of the page is "Rough notes which you may make any use of, but return before May" in Maxwell's hand. This note was probably directed to William Henry Watson. Maxwell himself published a discussion on the hodograph in Maxwell, *Matter and Motion* (London: Dover reprint of the 1876 edition of Society for the Propagation of Christian Knowledge, n.;), 107, and Maxwell, "Diagrams," *Encyclopedia Brittanica* reprinted in *Scientific Papers*, vol. 2, 647–659, 652–653.

b. Sir William Rowan Hamilton, "The Hodograph, or a New Method of expressing in Symbolical Language the Newtonian Law of Attraction," *Proc. R. Irish Acad.* 3 (1845–47): 308–309, reprinted in *Mathematical Papers* (Cambridge: At the University Press, 1931) , vol. 2 *Dynamics,* 287–292. Thomas L. Hankins discusses Hamilton's Hodograph in Hankins, *Sir William Rowan Hamilton* (Baltimore: Johns Hopkins Press, 1980). According to Hankins of all those informed of Hamilton's hodograph only William Thomson was enthusiastic. However, Tait published on it in, Tait, "Note on the Hodograph," *Proc. R. Soc. Edinburgh* 6 (1869): 221–226, and a discussion of it was included in Thomson and Tait, *Treatise on Natural Philosophy* (Oxford: Clarendon Press, 1867) vol. 1 § 37.

2. To determine the average distribution as to position and velocity of a finite number of material particles forming a conservative system[a]

Cambridge University Library, Maxwell Collection

Dynamical Specification.

We shall begin by supposing the system to be of the most general type, having its configuration determined by the n variables $q_1 \ldots q_n$, and its velocity determined by the n variables $\dot{q}_1 \ldots \dot{q}_n$. We shall suppose the forces which act between the parts of the system to be of the most general kind consistent with the conservation of energy. We shall therefore express the potential energy of the system by the symbol V which is a function of $q_1 \dots q_n$.

The kinetic energy of the system is denoted by T and the total energy by $E = V + T$. <and> The components of momentum <by> are $p_1 \dots p_n$.

To determine the configuration and velocity of the system requires $2n$ independent variables. These may be the n coordinates and the n velocities or the n coordinates and the n momenta, or we may take the < $n - 1$ > coordinates and $n - 1$ of the velocities or of the momenta, together with E the total energy of the system. This last method has the advantage that during <any given> the actual motion of the system E remains constant.

<The> Any particular state of the system, specified by its configuration and velocity may be called a *phase,* and the series of phases through which the system passes during its actual motion is called the *path* of the system.[b]

2 Relation between the initial and final variations

If the kinetic energy, <is> expressed as a function of the coordinates and momenta, is denoted by T_p, and if

$$E = V + T_p \tag{1}$$

then the equations of motion are

$$\frac{\partial q_r}{\partial t} = \frac{dE}{dp_r} \tag{2}$$

$$\frac{\partial p_r}{\partial t} = -\frac{dE}{dq_r} \tag{3}$$

where q_r and p_r are the corresponding coordinate and momentum.

If the initial coordinates are $q'_1 \dots q'_n$ and the final coordinates $q_1 \dots q_n$ then these $2n$ quantities together with E are sufficient to specify the whole motion. If

$$A = \int 2T dt \tag{4}$$

or <be> twice the time-integral of the kinetic energy, be called the Action of the system during the motion, and be expressed in terms of the $2n + 1$ quantities $q'_1 \dots q'_n; q_1 \dots q_n; E$ then

$$\frac{dA}{dq'_r} = -p'_r \tag{5}$$

$$\frac{dA}{dq_r} = p_r \tag{6}$$

$$\frac{dA}{dE} = t$$

Hence

$$\frac{dp'_r}{dq_s} = -\frac{d^2 A}{dq'_r dq_s} = \frac{dp_s}{dq_r} \tag{7}$$

where the indices r and s may be the same or different.

3 Relation between the initial and final products of variations

Now consider the product of the differentials

$$dq_1' \dots dq_n' \quad dp_1' \dots dp_n'$$

being those of the coordinates and velocities in the initial phase and let us express this in terms of the differentials of the initial and final coordinates. It becomes

$$dq_1' \dots dq_n' \quad dp_1' \dots dp_n' \, \Delta\binom{p'}{q}$$

where $\Delta\binom{p'}{q}$ is the functional determinant

$$\sum \pm \left(\frac{dp_1'}{dq_1} \cdots \frac{dp_n'}{dq_n} \right)$$

or

$$\begin{vmatrix} \frac{dp_1'}{dq_1} & \frac{dp_1'}{dq_2} & \cdots & \frac{dp_1'}{dq_n} \\ \frac{dp_2'}{dq_1} & \frac{dp_2'}{dq_2} & \cdots & \frac{dp_2'}{dq_n} \\ \cdots & & & \\ \frac{dp_n'}{dq_1} & \frac{dp_n'}{dq_2} & \cdots & \frac{dp_n'}{dq_n} \end{vmatrix}$$

But by equation (7) it appears that the terms in $\Delta\binom{p'}{q}$ are <identical> equal each to each <with> to those in $\Delta\binom{p}{q'}$ with their signs changed so that $\Delta\binom{p'}{q} = -\Delta\binom{p}{q'}$.
Hence

$$\begin{aligned} dq_1' \dots dq_n' \, dp_1' \dots dp_n' &= dq_1' \dots dq_n' \, dq_1 \dots dq_n \, \Delta\binom{p'}{q} \\ &= -dq_1' \dots dq_n' \, dq_1 \dots dq_n \, \Delta\binom{p}{q'} \\ &= -dp_1 \dots dp_n \, dq_1 \dots dq_n \\ < &= dq_1 \dots dq_n \, dp_1 \dots dp_n > \\ dq_1' \dots dq_n' \, dp_1' \dots dp_n' &= dq_1 \dots dq_n \, dp_1 \dots dp_n \end{aligned} \tag{8}$$

In this expression $q_1 \dots p_1$, the final coordinates and velocities, are to be considered as functions of $q_1' \dots p_n'$, the initial coordinates and velocities, <and> the time of the motion being regarded as constant.

4 Modification of the relation when E is constant

Let us now make another change of the independent variables which define the initial phase from

$$q_1' \ldots q_n' \, p_1' \ldots p_{n-1}' p_n' \text{ to } q_1' \ldots q_n' \, p_1' \ldots p_{n-1}' E$$

where instead of one of the momenta, p_n', we have E the total energy of the system, the other coordinates and momenta being as before.
Then since

$$\frac{dE}{dp_n'} = \frac{\partial q_n'}{\partial t} \tag{9}$$

$$dq_1' \ldots q_n' \, dp_1' \ldots dp_n' = dq_1' \ldots dq_n' \, dp_1' \ldots dp_{n-1}' \, dE \, \frac{1}{\frac{\partial q_n'}{\partial t}} \tag{10}$$

Similarly

$$dq_1 \ldots dq_n \quad dp_1 \ldots dp_n = dq_1 \ldots dq_n \quad dp_1 \ldots dp_{n-1} \quad dE \, \frac{1}{\frac{\partial q_n}{\partial t}} \tag{11}$$

Since the quantities on the left hand side of these equations have been proved to be equal, and since E is constant during the motion so that dE is the same in both equations

$$dq_1' \ldots dq_n' \, dp_1' \ldots dp_{n-1}' \frac{1}{\dot{q}_n'} = dq_1 \ldots dq_n \, dp_1 \ldots dp_{n-1} \frac{1}{\dot{q}_n} \tag{12}$$

This equation is applicable to the case in which the total energy is supposed not to vary from one particular case of motion to another and when therefore the $2n$ initial variables are no longer independent but, being subject to the equation of energy are reduced to $2n - 1$.

We shall now simplify our equations by supposing that the system consists of ν material particles whose masses are m, m', &c, whose coordinates are x, y, z : x', y', z' &c and whose velocity components are ξ, η, ζ : ξ', η', ζ'&c. This will be a particular case of the most general form if we make $n = 3\nu$; $q_1 = x$, $q_2 = y$, $q_3 = z$: $q_5 = x'$, $q_6 = y'$, $q_7 = z'$ &c $m_1 = m_2 = m_3 = m$; $m_4 = m_5 = m_6 = m'$ &c. The relation between any momentum p_r and the corresponding velocity $\dot{q}$ is

$$p_r = m_r \, \dot{q}_r \tag{13}$$

and

$$2T = \sum m_r \, \dot{q}_r^2 = \sum \frac{1}{m_r} \, p_r^2 \tag{14}$$

Statistical specification

Let us now widen the definition of a phase by saying that the system is in the phase A whenever the values of the coordinates are such that $q_1$is between b_1 and $b_1 + db_1$ and so on and q_n between b_n and the momenta such that p_1 is betwen a_1 and $a_1 + da_n \dots$ and p_{n-1} between a_{n-1} and $a_{n-1} + da_{n-1}$ the value of p_n being determined by the equation

$$E = V + 1/2 \sum_{r=1}^{n-1} m_r^{-1} p_r^2 + 1/2\, m_n^{-1} p_n^2 \tag{15}$$

Let us next suppose that there are a great many systems, each consisting of a set of particles the properties of which are the same as those of the first system, the only difference being that the initial values of the $2n - 1$ variables are different in each system, the value of E being the same for all. The motion of one system is not affected by that of the others. Let N be the whole number of systems, and let N_{At} be the number of these which at the time t are in the state A, then the statistical law of distribution is expressed by the equation

$$\begin{gathered} N_{At} = F(N, b_1, \dots a_{n-1},\, db_1, \dots da_{n-1},\, t) \\ < N_{At} = F(N, p_1, \dots q_{n-1},\, dp_1, \dots dp_{n-1},\, t) > \end{gathered} \tag{16}$$

It is manifest that N can only enter the function as a factor, for the different systems do not act on each other.

Also any differential, as db_1 can only enter as a factor, for the number of systems within any phase must vary in the ratio of the breadth within which the phase is limited. We may therefore write

$$\begin{gathered} < N_{At} = N f(b_1, \dots a_{n-1},\, t)\, db > \\ N_{At} = N f(b_1, \dots a_{n-1},\, t)\, db_1 \dots da_{n-1} \end{gathered} \tag{17}$$

We shall now follow the motion of these N_{At} particles from the time t to the time t'. We shall <indicate> distinguish the <value> symbols of quantities at the time t' by accents.

Then $$N_{A't'} = N f(b_1' \dots a_{n-1}',\, t')\, db_1' \dots da_{n_1}' \tag{18}$$

Since the number of particles does not change during their motion

$$N_{A't'} = N_{At} \tag{19}$$

Also by equation (12)

$$db_1' \dots da_{n-1}' (\dot{b}_n')^{-1} = db_1 \dots da_{n-1} (\dot{b}_n)^{-1} \tag{20}$$

Hence by (17), (18), (19), (20),

$$\dot{b}'_n\, f(b'_1, \ldots a'_{n-1}, t) = \dot{b}_n\, f(b_n, \ldots a_{n-1}, t) =< 1/m_n > C_1 \tag{21}$$

where C_1 is a constant for all phases of the same motion <at the time> as the group of particles passes through them in succession.
Since $a_n = m_n \dot{b}_n$.

$$f(b_1 \ldots a_{n-1}, t) = C_1 \, \frac{m_n}{a_n} \tag{22}$$

where $b_1 \ldots a_n$ are the values of these variables belonging to the group of particles at the time t.

If the distribution is such that the form of f does not <a function> vary with the time the motion of the set of systems is called steady. In this case

$$f(b_1 \ldots a_{n-1}) = C_1 \, \frac{m_n}{a_n} \tag{23}$$

where C_1 is constant for every phase belonging to the same path.

If therefore the initial distribution of coordinates and momenta is such that C_1 in equation (23) is constant for all phases consistent with the equation of energy

$$E = V + \frac{1}{2}\frac{1}{m_1}\, a_1^2 + \&c + \frac{1}{2}\frac{1}{m_n}\, a_n^2 \tag{24}$$

and zero for all phases which that equation shows to be impossible then the law of distribution will not change with the time and C_1 will remain an absolute constant and this will be a solution of the problem of finding a permanent law of distribution.

If the motion is such that it will of itself pass through every possible phase this will be the only solution. But if certain phases, though consistent with the equation of energy are not passed through in the actual path of the system from the phase A, then the value of C_1 may be different for such phases, the only necessary condition of steady motion being that C_1 must be the same for all phases belonging to the same path.

In what follows we shall assume that C_1 is constant for all phases consistent with the equation of energy and zero for all phases which that equation shows to be impossible and we shall deduce the number of systems which at a given instant are in a state less completely defined as for instance when the limits of $b_1 \ldots b_n \quad a_1 \ldots a_r$ are given but when the values of the remaining momenta may be any whatever consistent with the equation of energy.

If we write

$$\left.\begin{aligned} A_1^2 &= 2m_1(E - V) \\ A_2^2 &= \frac{m_2}{m_1}(A_1^2 - a_1^2) \\ A_3^2 &= \frac{m_3}{m_2}(A_2^2 - a_2^2) \\ A_{n-1}^2 &= \frac{m_{n-1}}{m_{n-2}}(A_{n-2}^2 - a_{n-2}^2) \end{aligned}\right\} \tag{25}$$

then the equation of energy (24) may be written

$$a_n^2 = \frac{m_n}{m_{n-1}} (A_{n-1}^2 - a_{n-1}^2) \tag{26}$$

We have next to integrate < $NC_1 a_n^{-1}$ with respect>

$$NC_1 m_n a_n^{-1} \, db_1 \dots db_n \, da_1 \dots da_{n-1}$$

the limits of the successive integrations being $\pm A_{n-1}$, $\pm A_{n-2}, \dots \pm A_1$. The general formula for integrations of this kind is

$$\int_{-A}^{+A} (A^2 - a^2)^2 da = \frac{\Gamma(1/2)\,\Gamma(q+1)}{\Gamma(q+3/2)} A^{2q+1} \tag{27}$$

Hence after r integrations, r being any number less than n, the result is

$$\begin{aligned} NC_1 \frac{(\Gamma(1/2))^{r+1}}{\Gamma(\frac{r+1}{2})} (m_n \, m_{n-1} \dots m_{n-r})^{1/2} \, m_{n-r}^{-(r-1)/2} \\ \times A_{n-r}^{r-1} \, db_1 \dots db_n \, da_1 \dots da_{n-r-1} \end{aligned} \tag{28}$$

Putting $r = n - 1$, this becomes

$$N_B = NC_1 \frac{(\Gamma(1/2))^n}{\Gamma(\frac{n}{2})} (m_n \dots m_1)^{1/2} \, [2(E-V)]^{\frac{n-2}{2}} \, db_1 \dots db_n \tag{29}$$

which is the number of systems in which the configuration is specified by the variables $b_1 \dots b_n$ while the momenta have any consistent with the equation of energy (24).

<Since C_1 is arbitrary, and since it was selected as a constant for an expression having special reference to a_n this expression, as it stands, is not symmetrical as to with respect to all the particles. We therefore>
If we write

$$C = C_1 \frac{\Gamma(1/2)^n}{\Gamma(\frac{n}{2})} (m_n \dots m_1)^{1/2} < \frac{1}{m_n} > 2^{\frac{n-2}{2}} \tag{30}$$

and the expression for the number of systems in the configuration B becomes

$$N_B = N\,C(E-V)^{\frac{n-2}{2}} \, db_1 \dots db_n \tag{31}$$

This equation is true only for configurations in which $E - V$ is positive, for since the kinetic energy is necessarily positive, the <while energy must be> potential energy must be less than the whole energy. For configurations specified in such a way that if they existed V would be greater than E, the value of N_B is zero.

N_B is also zero for configurations which though they make V less than E, cannot be reached by a continuous path from the original configuration without passing through configurations which make V greater than E.

To find the number of systems <in> which are in the configuration B and in which p_n is between a_1 and $a_1 + da_1$, we must make $r = n - 2$ in equation (28). The number is

$$N\,C_1\,\frac{\left(\Gamma(1/2)\right)^{n-1}}{\Gamma\left(\frac{n-1}{2}\right)}\,(m_n \ldots m_2)^{1/2}\,m_n^{-1}\,m_2^{-\frac{(n-3)}{2}}\;A_2^{n-3}\,db_1 \ldots db_n\,da_1 \qquad (32)$$

The whole number of systems in configuration B is given by (29). Hence the proportion of these systems in which p_1 is between a_1 and $a_1 + da_1$ is

$$\frac{\Gamma\left(\frac{n}{2}\right)\left(E - V - \frac{1}{2}m_1^{-1}a_1^2\right)^{\frac{n-3}{2}}}{\Gamma(1/2)\,\Gamma\left(\frac{n-1}{2}\right)(E-V)^{\frac{n-2}{2}}}\;2^{-1/2}\,m^{-1/2}\,da_1 \qquad (33)$$

If we write $k_1^2 = \frac{1}{2}\,m_1^{-1}a_1^2$, this becomes

$$\frac{\Gamma\left(\frac{n}{2}\right)\left(E - V - k_1^2\right)^{\frac{n-3}{2}}}{\Gamma(1/2)\,\Gamma\left(\frac{n-1}{2}\right)(E-V)^{\frac{n-2}{2}}}\,dk_1 \qquad (34)$$

If we write for any other particles $\qquad k_s^2 = \dfrac{1}{2}\,m_s^{-1/2}a_s^2 \qquad (35)$

the proportion of the system in configuration B in which k_s lies between k_s and $k_s + dk_s$ is

$$\frac{\Gamma\left(\frac{n}{2}\right)\left(E - V - k_s^2\right)^{\frac{n-3}{2}}}{\Gamma(1/2)\,\Gamma\left(\frac{n-1}{2}\right)(E-V)^{\frac{n-2}{2}}}\,dk_s \qquad (36)$$

Hence the law of distribution is the same for all the k's. Now k_s^2 is the kinetic energy of the particle m_s. Hence the law of distribution of kinetic energy is the same for all the particles. The maximum value of the kinetic energy of a particle is evidently $E - V$. The mean kinetic energy is found by multiplying (36) by k_1^2 <and> integrating with respect to k and dividing the result by the integral of (36) the limits in both integration being $\pm(E-V)^{1/2}$.

$$\int_{-(E-V)^{1/2}}^{(E-V)^{1/2}}(E - V - k^2)^{\frac{n-3}{2}}k^2\,dk = \frac{\Gamma\left(\frac{3}{2}\right)\Gamma\left(\frac{n-1}{2}\right)}{\Gamma\left(\frac{n+3}{2}\right)}\,(E-V)^{n/2} \qquad (37)$$

and

$$\int_{-(E-V)^{1/2}}^{(E-V)^{1/2}} (E - V - k^2)^{\frac{n-3}{2}}\, dk = \frac{\Gamma(\frac{1}{2})\,\Gamma(\frac{n-1}{2})}{\Gamma(\frac{n}{2})}\,(E-V)^{\frac{n-3}{2}} \tag{38}$$

Hence the mean value of k^2 is

$$\overline{k^2} = \frac{\Gamma(\frac{3}{2})\,\Gamma(\frac{n-1}{2})\,\Gamma(\frac{n}{2})}{\Gamma(\frac{n+2}{2})\,\Gamma(\frac{n-1}{2})\,\Gamma(1/2)}\,(E-V) = \frac{1}{n}\,(E-V) = \frac{1}{n}\,T \tag{39}$$

or the mean kinetic energy of every particle is the same, and is equal to the whole kinetic energy of the configuration divided by the number of particles.

Similarly we find that the mean value of k^4 is $\frac{1\cdot 3}{n\cdot(n+2)}\,T^2 = \frac{3n}{n+2}\,(\overline{k^2})^2$ and the mean value of k^{2r} is

$$\frac{1\cdot 3 \ldots 2r-1}{n\cdot n+2\ldots n+2r-2}\,T^{2r} = \frac{3\ldots 2r-1}{(1+\frac{2}{n})(1+\frac{4}{n})\ldots 1+\frac{2n-2}{n}}\,(\overline{k^2})^r$$

When n is very large, equation (36) approximates to

$$\frac{1}{\sqrt{2\pi}}\,\frac{1}{\sqrt{\overline{k^2}}}\,e^{-k^2/2\overline{k^2}}dk$$

which is the law of distribution when the number of particles is infinite.

[A page of notes may be missing here]

But[c] when the number of variables is very great, and when the potential energy of the specified configuration is very small compared with the total energy of the system, we may proceed as follows:–

Let the system consist of n' material particles, let V_{n-1} denote the potential energy of the first $n-1$ particles in their actual positions, the nth particle being removed to an infinite distance, and let W_n denote the work which must be done against the forces of the system in order to bring the nth particle from an infinite distance into its actual position, then the potential energy of the whole system is

$$V_n' = V_{n'-1} + W_{n'}$$

Writing, as in equation (52) K for the mean value of the kinetic energy of the particle we have

$$E - V_{n'} = n'K$$

$$< = E - V >$$

and

$$\left(E - V_{n'}\right)^{\frac{3n'-2}{2}} = \left(E - V_{n'-1}\right)^{\frac{3n'-2}{2}} \left(1 - \frac{W_{n'}}{E - V_{n'-1}}\right)^{3n'-2}$$
$$= \left(E - V_{n'-1}\right)^{\frac{3n'-2}{2}} e^{\frac{3n'-2}{2}} \log\left(1 - \frac{W_{n'}}{E - V_{n'-1}}\right)$$

and this, when n is very great and when $W_{n'}$ is small compared with $E - V_{n'}$ is approximately

$$\left(E - V_{n'-1}\right)^{\frac{3n'-2}{2}} e^{-\frac{3}{2}\frac{W_{n'}}{K}}$$

Hence the number of systems in which the last particle has a given <configuration> position, the configuration of the rest of the system being given, <def> is approximately proportional to exp. $-\frac{3}{2}W_{n'}/K$ where W_n is the potential energy of the particle with respect to the system in the given position.

If we suppose that there are a great many particles of the same kind, the average density of these particles <at any point> within any element $dx\, dy\, dz$ will be approximately proportional to

$$e^{-\frac{3}{2}\frac{W}{K}}$$

and the number of these whose kinetic energy is between k and $k + dk$ will, by equation 55, be approximately proportional to

$$\frac{1}{\sqrt{2\pi}} \frac{1}{K} e^{-\frac{3}{2}\frac{W+K}{K}} dk$$

We may express the result in words by saying that the average density, within a given element, of particles of a given kind having their kinetic energy between given limits is proportional to an exponential function, the index of which is three times the whole energy of the <given> particle in the given state divided by twice the average kinetic energy of a particle of the system.

This result agrees with that obtained by other methods in the case of an infinite number of particles. In the case of a finite system, however, it is <only> an approximation <and can be> to the true value only when the energy of a particle in the given state is very small compared with the whole <kinetic> energy of the system. <Indeed> If we suppose the energy of the particle equal to that of the whole system

a. This is part of a draft of Maxwell, "On Boltzmann's Theorem on the Average Distribution of Energy in a System of Material Points," *Trans. Cambridge Phil. Soc.* 12 (1879): 547–569, 550–558.

b. There is a break in the manuscript here when it is compared to the published paper.

c. The notes from here to the end are pps. 560-561 of the published paper.

3. [Energy of Internal Motion][a]

Cambridge University Library, Maxwell Manuscripts

To find the energy of internal motion of a system consisting of $p-1$ particles in terms of the seven constants of the system of p particles, together with the position and motion of the pth particle.

Since we may choose our axes of reference as we please we shall assume that the origin of coordinates is the centre of mass of the p particles, that the axis passes through the pth particle and that the axes of y and z are turned into such a position that

$$MN - AL = 0$$

We shall also for brevity write *in the present investigation*

$$u \text{ for } u - U \qquad v \text{ for } v - V \qquad \text{and } w \text{ for } w - W$$

that is to say we shall refer the velocity to the centre of mass of the p particles, and

$$\mu \text{ for } \frac{Mm}{M-m}$$

We shall distinguish quantities belonging to the system of $p-1$ particles <by an> accented letters.

We then have,

$$M' = M - m$$

$$X = Y = Z = U = V = W = 0$$

$$X' = -\frac{\mu x}{M} \qquad Y' = -\frac{\mu y}{M} \qquad Z' = -\frac{\mu z}{M}$$

$$U' = -\frac{\mu u}{M} \qquad V' = -\frac{\mu v}{m} \qquad W' = -\frac{\mu w}{M}$$

$$P' = P \qquad Q' = Q + \mu x w \qquad R' = R - \mu x v$$

$$A' = A \qquad b' = B - \mu x^2 \qquad C' = C - \mu \xi^2$$

$$L' = L \qquad M' = M \qquad N' = N$$

Similarly we may write,

$$M = \sum_1^n (m) \qquad M_p = \sum_1^p (m)$$

$$MX = \sum_1^n mx \qquad M_pX_p = \sum_1^p (mx)$$

$$MU = \sum_1^n mu \qquad M_pU_p = \sum_1^p (mu)$$

$$F = \sum_1^n (f) \qquad F_p = \sum_1^p (f)$$

for the mass, &c of the whole system and of the first p particles of the system. The coordinates of the centre of mass are $X\,Y\,Z$ and its velocity components are $U\,V\,W$. If we also write

$$P = F + M(ZV - YW)$$
$$Q = G + M(XW - ZU)$$
$$R = H + M(YU - XV)$$

P, Q, R are the components of angular momentum about the centre of mass,

$$\xi = x - X$$
$$\eta = y - Y$$
$$\zeta = z - Z$$

then for the rotation of a rigid system having the same configurations as the actual system and the same components of angular momentum we have the equations

$$P = Ap + Nq + Mr$$
$$Q = Np + Bq + Lr$$
$$R = Mp + Lq + Cr$$

where $p\,q\,r$ are the components of angular velocity and,

$$A = \sum m(\eta^2 + \zeta^2) \qquad B = \sum m(\zeta^2 + \xi^2) \qquad C = \sum m(\xi^2 + \eta^2)$$
$$L = -\sum m(\eta\zeta) \qquad M = -m(\zeta\xi) \qquad N = -m(\xi\eta)$$

are the moments of inertia and the products of inertia with their signs reversed.

Let $a\,b\,c\,\ell\,m\,n$ be such that,

$$p = aP + nQ + mR$$
$$q = nP + bQ + \ell R$$
$$r = mP + \ell Q + cR$$

The component in the direction of x of the velocity of m in the actual system relative to its velocity if the system were to become rigid is

$$u' = u - U - q\xi - r\eta$$

We are now able to express the three divisions of the kinetic energy and to prove that their sum is the whole kinetic energy.

Let K be the whole kinetic energy

$$K = \frac{1}{2}m(u^2 + v^2 + w^2)$$

Let K_T be the energy of translation

$$K_T = \frac{1}{2}M(U^2 + V^2 + W^2)$$

Let K_R be the energy of rotation

$$K_R = \frac{1}{2}(Pp + Qq + Rr)$$

K_R may be determined from the equation

$$\begin{vmatrix} A & N & M & P \\ N & B & L & Q \\ M & L & C & R \\ P & Q & R & 2K_R \end{vmatrix} = 0$$

Let K_I be the energy of internal motion

$$\begin{aligned}
K_I &= \sum \frac{1}{2}m(u'^2 + v'^2 + w'^2) \\
&= \frac{1}{2}\sum m(u - U - q\zeta + r\eta)^2 + \frac{1}{2}\sum m(v - V - r\xi + p\zeta)^2 \\
&\quad + \frac{1}{2}\sum m(w - W - p\eta + q\xi)^2 \\
&= \frac{1}{2}\sum m(u^2 + v^2 + w^2) + \frac{1}{2}M(U^2 + V^2 + W^2) \\
&\quad - \sum m(Uu + Vv + Ww) + p\sum m[\xi(v - V) - \eta(w - W)] \\
&\qquad + q\sum m[\xi(w - W) - \zeta(u - U)] + r\sum m[\eta(u - U) - \xi(v - V)] \\
&\qquad + \frac{1}{2}(p^2A + q^2B + r^2C) + qrL + rpM + pqN \\
&= K + K_T - 2K_T - 2K_R + K_R \\
K_I &= K - K_T - K_R
\end{aligned}$$

Since the energy of translation and the energy of rotation can be found in terms of the six constants when the configuratio [sic]
Since the energy of translation can be found from the components of the momentum and the energy of rotation can be found from <when> the components of angular momentum and the energy of rotation is given, the above equation enables us to express the energy of internal motion in terms of the seven constants $U\,V\,W\,F\,G\,H\,\&\,K$ when the configuration is given.

a. These are a series of notes for Maxwell, "On Boltzmann's Theorem on the average Distribution of Energy in a System of Material Points," *Trans. Cambridge Phil. Soc.* 12 (1879): 547–570, pps. 568–569, reprinted in *Scientific Papers,* vol. 2: 713–741.

4. [Internal Energy in a Free System]

Cambridge University Library, Maxwell Manuscripts

In a <conservative> system not acted on by external forces the sums of the momenta in the three coordinate directions and the sums of the angular momenta about the three coordinate axes remain constant. If the system is a conservative one, the total energy of the system also remains constant. We may therefore consider seven of the variables as functions of these seven constants and of the remaining variables.

We shall therefore eliminate from equation (8) the first seven components of momentum, namely m_1u_1, m_1v_1, m_1w_1 for the first particle, m_2u_2, m_2v_2, m_2w_2 for the second and m_3u_3 for the third, expressing these in terms of the remaining variables and the seven sums above mentioned. If we write

$$f = m(yw - zv)$$
$$g = m(zu - xw)$$
$$h = m(xv - yu)$$
$$k = 1/2m(u^2 + v^2 + w^2)$$

then

$$m_1u_1 + m_2u_2 + m_3u_3 = U_3$$
$$m_1v_1 + m_2v_2 + m_3v_3 = V_3$$
$$m_1w_1 + m_2w_2 + m_3w_3 = W_3$$

where $U_3\,V_3\,W_3$ are functions of the remaining momenta. Similarly,

$$f_1 + f_2 + f_3 = F_3$$
$$g_1 + g_2 + g_3 = G_3$$
$$h_1 + h_2 + h_3 = H_3$$
$$k_1 + k_2 + k_3 = K_3$$

where $U_3\, V_3\, W_3\, F_3\, G_3\, K_3$ are functions of the remaining momenta. $U_n \ldots H_n$ are constant and $K_n = E - V_n$.

We now have to express $du_1\, dv_1\, dw_1\, du_2\, dv_2\, dw_2\, du_3$ in terms of $dU\, dV\, dW\, dF\, dG\, dH\, dK$.

Differentiating U, V, W, F, G, H and K with respect to the components of momentum m_1u_1, m_1v_1, m_1w_1 m_2u_2, m_2v_2, m_2w_2 and m_3u_3 we obtain the functional determinant

$$\begin{vmatrix} 1, & 0, & 0, & 0, & z_1, & -y_1, & u_1 \\ 0, & 1, & 0, & -z_1, & 0, & x_1, & v_1 \\ 0, & 0, & 1, & y_1, & -x_1, & 0, & w_1 \\ 1, & 0, & 0, & 0, & z_2, & -y_2, & u_2 \\ 0, & 1, & 0, & -z_2, & 0, & x_2, & v_2 \\ 0, & 0, & 1, & y_2, & -x_2, & 0, & w_2 \\ 1, & 0, & 0, & 0, & z_3, & -y_3, & u_3 \end{vmatrix}$$

which we may write as $\alpha\Delta$ where

$$\alpha = (y_1 - y_2)(z_2 - z_3) - (y_2 - y_3)(z_1 - z_2)$$

and

$$\Delta = (u_1 - u_2)(x_1 - x_2) + (v_1 - v_2)(y_1 - y_2) + (w_1 - w_2)(z_1 - z_2)$$

We have therefore the following equation for the product of seven differentials

$$m_1^3\, m_2^3\, m_3\; du_1 dv_1 dw_1\; du_2 dv_2 dw_2\; du_3 = \frac{dU\, dV\, dW\, dF\, dG\, dH\, dE}{\alpha\Delta}$$

We may obtain a similar expression for the product of the seven corresponding differentials at the other extremity of the motion by accenting the variable symbols. The seven quantities $U \ldots E$ are the same at both extremities of the motion. Dividing out their differentials, we obtain for a system of n particles

$$dv_3 dw_3\, du_4 \ldots dw_n\, dx_1 \ldots dz_n \alpha\Delta = dv_3' \ldots dw_n'\, dx_1' \ldots dz_n' \alpha' \Delta'$$

[The rest of this page is blank. The following material is part of the same calculation but some of the pages seem to be missing.]

$$\begin{aligned}(m_1 + m_2)(f_1 + f_2) - (m_1y_1 + m_2y_2)(m_1w_1 + m_2w_2) \\ - (m_1z_1 + m_2z_2)(m_1v_1 + m_2v_2) \\ = m_1m_2\{(y_1 - y_2)(w_1 - w_2) - (z_1 - z_2)(v_1 - v_2)\}\end{aligned}$$

If we write,

$$(m_1 + m_2)(f_1 + f_2 + f_3) - (m_1 y_1 + m_2 y_2)\, m_1 w_1$$

M_2 for $m_1 + m_2$ $\quad$ X_2 for $m_1 x_1 + m_2 x_2$ $\quad$ and so on

then we may write the first side of the above equation

$$M_2 F_2 - Y_2 W_2 + Z_2 V_2$$

and if also we write $P = M_2 F_3 + Y_2 W_3 + Z_2 V_3$

$$M_2 F_2 - Y_2 W_2 + Z_2 V_2 = P + m_3\{v_3(M_3 z_3 - Z_3) - w_3(M_3 y_3 - Y_3)\}$$

Squaring the three terms of similar form and adding

$$\begin{aligned}
&m_1^2 m_2^2\{((x_1 - x_2)^2 + (y_1 - y_2)^2 + (z_1 - z_2)^2) \\
&\quad \times ((u_1^2 - u_2^2)^2 + (v_1^2 - v_2^2)^2 + (w_1^2 - w_2^2)^2) - \Delta^2\} \\
&\quad = P^2 + Q^2 + R^2 + 2m_3 u_3 \left[R(M_3 y_3 - Y_3) - Q(M_3 z_3 - Z_3)\right] \\
&\qquad + 2m_3 v_3 \left[P(M_3 z_3 - Z_3) - R(M_3 x_3 - X_3)\right] \\
&\qquad + 2m_3 w_3 \left[Q(M_3 x_3 - X_3) - P(M_3 y_3 - Y_3)\right] \\
&\qquad + m_3^2(u_3^2 + v_3^2 + w_3^2) \\
&\qquad \times \left[(M_2 x_3 - X_2)^2 + (M_2 y_3 - Y_2)^2 + (M_2 z_3 - Z_3)\right] \\
&\qquad - m_3^2\left[u_3(M_2 x_3 - X_2) + v_3(M_2 y_3 - Y_3) + w_3(M_3 z_2 - Z_2)\right]^2 \\
&\quad = \Big[(x_1 - x_2)^2 + (y_1 - y_2)^2 + (z_1 - z_2)^2\Big] \\
&\qquad \times m_1 m_2 \left[2K_3 M_3 - (U_3^2 + V_3^2 + W_3^2)\right. \\
&\qquad \left. - m_3\{M_2(u_3^2 + v_3^2 + w_3^2) - 2(U_3 u_3 + V_3 v_3 + W_3 w_3)\}\right] \\
&\qquad - m_1^2 m_2^2 \Delta^2
\end{aligned}$$

[The rest of the page is blank]

If we write

$$\begin{aligned}
m_1 + m_2 \ldots + m_r &= M_r \\
m_1 x_1 \ldots m_r x_r &= X_r
\end{aligned}$$

and so on, and

$$M_r F_r - Y_r W_r - Z_r V_r = P_r$$

then

$$\begin{aligned}
P_1 &= 0 \\
P_2 &= m_1 m_2 (y_1 - y_2)(w_1 - w_2) - (z_1 - z_2)(v_1 - v_2) \\
&= M_2 F_3 - Y_2 W_3 + Z_2 V_3 + m_3 v_3(M_3 z_3 - Z_3) - w_3(M_3 y_3 - Y_3)
\end{aligned}$$

[The rest of the page is blank]

To express Δ in terms of the selected variables

$$\begin{aligned}(m_1+m_2)(F-f_3)&-(m_1y_1+m_2y_2)(W-m_3w_3)\\&+(m_1z_1+m_2z_2)(V-m_3v_3)\\&=m_1m_2\{(y_1-y_2)(w_1-w_2)-(z_1-z_2)(v_1-v_2)\}\\&=M_2F_2-Y_2W_2+Z_2V_2\end{aligned}$$

Forming the other two similar expressions squaring them and adding, we find,

$$\begin{aligned}&m_1^2m_2^2\Big\{(x_1-x_2)^2+(y_1-y_2)^2+(z_1-z_2)^2\Big\}\\&\quad\times\Big\{(u_1-u_2)^2+(v_1-v_2)^2+(w_1-w_2)^2\Big\}-\Delta^2\\&\quad=M_2^2\Big\{(F_2^2+G_2^2+H_2^2)+2M_2(F_2V_2Z_2\\&\qquad+G_2W_2X_2+H_2U_2Y_2-F_2W_2Y_2-G_2U_2Z_2-H_2V_2X_2\Big\}\\&\qquad+(X_2^2+Y_2^2+Z_2^2)(U_2^2+V_2^2+W_2^2)-(U_2X_2+V_2Y_2+W_2Z_2)^2\end{aligned}$$

But

$$\begin{aligned}2K_2(m-1+m_2)&-(U_2^2+V_2^2+W_2^2)\\&=m_1m_2[(u_1-u_2)^2+(v_1-v_2)^2(w_1-w_2)^2]\end{aligned}$$

Substituting this in ( )

$$\begin{aligned}m_1^2m_2^2=\{(x_1-x_2)^2&+(y_1-y_2)^2+(z_1-z_2)^2\}\\&\times m_1m_2\big[2(m_1+m_2)K_2-(U_2^2+V_2^2+W_2^2)\big]\\&+(X_2^2+Y_2^2+Z_2^2)(U_2^2+V_2^2+W_2^2)\\&-(U_2X_2+V_2Y_2+W_2Z_2)^2\\&+2M_2\big[F_2V_2Z_2+G_2W_2X_2+H_2U_2Y_2\\&-F_2W_2Y_2-G_2U_2Z_2-H_2V_2X_2\big]\\&+M_2^2(F_2^2+G_2^2+H_2^2)\end{aligned}$$

We have now to express these quantities in terms of v_3, w_3 and the remaining variables. The only other variable explicitly involved is u_3 which is involved in $U_2\,G_2\,H_2$ and K_2.

If we multiply the three equations of the form ( ) by x_1-x_2, y_1-y_2 and z_1-z_2 respectively and add we find,

$$\begin{aligned}(M_2F_2-Y_2W_2+Z_2V_2)(x_1-x_2)&+(M_2G_2-Z_2U_2+X_2W_2)(y_1-y_2)\\&+(M_2H_2-X_2V_2+Y_2U_2)(z_1-z_2)=0\end{aligned}$$

a. These are a series of notes for Maxwell, "On Boltzmann's Theorem on the average Distribution of Energy in a System of Material Points," *Trans. Cambridge Phil. Soc.* 12 (1879): 547–570, 561–568, reprinted in *Scientific Papers,* vol. 2: 713–741.

5. [Evaluation of an Integral][a]

Cambridge University Library, Maxwell Manuscripts

To integrate $\iiint u^{q/2} dx\, dy\, dz$ where,

$$u = d - a^2x^2 - b^2y^2 - c^2z^2 - 2a'yz - 2b'zx - 2c'zy - 2a''x - 2b''y - 2c''z$$

and the integration is extended to all values of $x\, y$ and z which make u positive.

We may reduce each integration to the definite integral,

$$\int (P - p\xi)^{q/2}\, d\xi = [?] \frac{\Gamma(1/2)\,\Gamma\left(\frac{q+2}{2}\right)}{\Gamma\frac{q+3}{2}}\, p^{-1/2}\, P^{\frac{q+1}{2}}$$

the limits being $\pm P^{1/2}\, p^{-1/2}$.

We begin by writing u in the form

$$\begin{aligned} u &= d + (c'y + b'z + a'')^2 a^{-1} - by^2 - 2a'yz \\ &= 2b''y - cz^2 - 2cz'' - a(x + (c'y + b'z + a'')a^{-1})^2 \end{aligned}$$

whence

$$\int u^{q/2} dx = \frac{\Gamma(1/2)\,\Gamma\left(\frac{q+2}{2}\right)}{\Gamma\left(\frac{q+3}{2}\right)}\, a^{-1/2}\, P^{\frac{q+1}{2}}$$

where P

We begin by writing,

$$u = P - p\xi^2$$

where

$$P = d - by^2 - cz^2 - 2a'yz - 2b''yz - 2c''z + \frac{(c'y + b'z + a'')^2}{a}$$

$$p = a$$

$$\xi = x + \frac{c'y + b'z + a''}{a}$$

whence

$$\int u^{q/2}\, dx = \frac{\Gamma\frac{1}{2}\,\Gamma\left(\frac{q+2}{2}\right)}{\Gamma\frac{q+3}{2}}\, a^{-1/2}\, P^{\frac{q+1}{2}}$$

We next write

$$P = \langle p \rangle Q - \langle p \rangle q\eta^2$$

where

$$\langle p \rangle Q = d - cz^2 - 2c''z + \frac{(b'z + a'')^2}{a} + \frac{(aa' - b'c')z + ab'' - c'a''}{a(ab - c)}$$

$$\langle p \rangle q = \frac{ab - c^2}{a}$$

$$\eta = y + \frac{(aa' - b'c')z + ab'' - c'a''}{ab - c^2}$$

whence

$$\int P^{\frac{q+1}{2}}\, dy = \frac{\Gamma 1/2\, \Gamma(\frac{q+3}{2})}{\Gamma \frac{q+4}{2}} \left(\frac{ab - c}{a}\right)^{-1/2} Q^{\frac{q+2}{2}}$$

Finally we write

$$Q = R - r\zeta^2$$

where

$$R = d - \frac{a''^2}{a} - \frac{(ab'' - c'a'')^2}{a(ab - c'^2)}$$
$$- \frac{(ac'' - b'a'')(ab - c'^2) - (aa' - b'c')(ab'' - c'a'')}{a(ab - c'^2)\left[(ac - b'^2)(ab - c'^2) - (aa' - b'c'^2)\right]}$$

$$r = \frac{(ac - b'^2)(ab - c'^2) - (aa' - b'c')^2}{a(ab - c'^2)}$$

$$\xi = z + \frac{(ac'' - b'a'')(ab - c'^2) - (aa' - b'c')(ab'' - c'a'')}{(ac - b'^2)(ab - c'^2) - (aa' - b'c')^2}$$

whence

$$\int Q^{\frac{q+2}{2}}\, dz = \frac{\Gamma(1/2)\, \Gamma \frac{q+4}{2}}{\Gamma \frac{q+5}{2}}\, r^{-1/2}\, R^{\frac{q+3}{2}}$$

We thus find

$$\iiint u^{q/2}\, dx\, dy\, dz = \frac{(\Gamma(1/2)^3\, \Gamma(\frac{q+3}{2})}{\Gamma(\frac{q+5}{2})} \begin{vmatrix} a & -c' & b' \\ c' & b & a' \\ b' & a' & c \end{vmatrix}^{-1/2} R^{\frac{q+3}{2}}$$

where R is the maximum value of u, and is given by the equation,

$$\begin{matrix} a & c' & b' & a'' \\ c' & b & a' & b' \\ b' & a' & c & c'' \\ a'' & b'' & c'' & d \end{matrix} = \begin{vmatrix} a & c' & b' \\ c' & b & a' \\ b' & a' & c \end{vmatrix} R$$

In the same way we may <show that> find the double integral

$$\iint u^{q/2}\, dy\, dz$$

where u is the same function of $x\, y\, z$ as before with the condition

$$\alpha x + \beta y + \gamma z + \delta = 0$$

for all values of the variables which satisfy this condition and make u positive. Putting

$$x = -\frac{\beta y + \gamma z + \delta}{\alpha}$$

we find

$$\iint u^{q/2}\, dy\, dz = \frac{(\Gamma 1/2)^2\, \Gamma(\frac{q+2}{2})}{\Gamma(\frac{q+4}{2})} \left\{ -\frac{1}{\alpha^2} \begin{vmatrix} a & c' & b' & \alpha \\ c' & b & a' & \beta \\ b' & a' & c & \gamma \\ \alpha & \beta & \gamma & 0 \end{vmatrix} \right\}^{-1/2} Q^{\frac{q+2}{2}}$$

where Q is the maximum value of u under the given condition and is given by the equation,

$$\begin{vmatrix} a & c' & b' & a'' & \alpha \\ c' & b & a' & b'' & \beta \\ b' & a' & c & c'' & \gamma \\ a'' & b'' & c'' & d & \delta \\ \alpha & \beta & \gamma & \delta & 0 \end{vmatrix} = \begin{vmatrix} a & c' & b' & \alpha \\ c' & b & a' & \beta \\ b' & a' & c & \gamma \\ \alpha & \beta & \gamma & \end{vmatrix} Q\alpha^2$$

a. This integral is similar to one Maxwell evaluates in Maxwell, "On Boltzmann's Theorem on the Average Distribution of Energy in a System of Material Points," *Trans. Cambridge Phil. Soc.* 12 (1879): 547–570, 565, reprinted in *Scientific Papers,* vol. 2: 713–741. This again demonstrates that while Maxwell, in his published papers appears cavalier with respect to the evaluation of the mathematical details, in his notes they are worked out in detail.

6. "On the Available <Ener> Kinetic Energy of a Material System"

Cambridge University Library, Maxwell Manuscripts

The kinetic energy of a material system may be divided into three parts

The first part may be called the energy of Translation. It is that of a mass equal to the whole system moving with the velocity of the centre of mass of the system.

The second part may be called the energy of Rotation. It is equal to that of a rigid system having the same configuration as the real system and rotating about its centre of mass so that its angular momenta are equal to those of the real system.

The third part may be called the energy of internal motion and is equal to that which the system would have if the velocity of every particle were <the difference> its velocity relative to what it would be if at that instant the system were to become rigid.

Of these three parts of the kinetic energy, the first or the energy of translation is not affected by any mutual action between the parts of the system. It can be changed only by the action of external force.

If we define energy as available or unavailable accordingly as it can or cannot be converted into work under given circumstances, we may call the energy of translation unavailable as regards the material system considered. It can be rendered available only by the introduction of some other material system.

The second part of the energy, that of rotation, is not unavailable in the same sense, for by causing the system to expand so as to increase its moments of inertia, we may diminish the energy of rotation and covert any part of it less than the whole into work or potential energy.

If, however, we do not admit finite changes in the distances between the parts of the system the energy of rotation is unalterable and therefore unavailable as a source of work.

The third part of the energy, however, that of internal motion is available as a source of work even though we do not admit any sensible change in the configuration. For though work cannot be done without some change of configuration the amount of this change may be made as small as we please <by increasing> provided the forces be sufficiently intense.

If therefore we suppose the internal forces called into play by changes of configuration to be so great that the system is reduced from its actual motion to that of a rigid system in so short a time that the configuration is not sensibly altered, then the work done against these forces will be equal to the energy of internal motion. In this sense, therefore we may call the energy of internal motion the available kinetic energy, because it can be converted into work by means of internal forces without any sensible change of configuration.

To show that the kinetic energy is the sum of these three parts we shall calculate the value of these parts. Let $m_1, m_2, \ldots m_n$ be the masses of the particles composing the system.

Let the coordinates of the particle m_p be $x_p\, y_p\, z_p$

Let its velocity components be $u_p\, v_p\, w_p$

Let its angular momenta be

$$f_p = m_p(y_p w_p - z - p v_p)$$
$$g_p = m_p(z_p u_p - x_p w_p)$$
$$h_p = m_p(x_p v_p - y_p u_p)$$

Let the kinetic energy be $k_p = \frac{1}{2} m_p(u_p^2 + v_p^2 + w_p^2)$

Let the whole kinetic energy of the system be $E = \sum_1^n k_p$

When we have occasion to refer to the kinetic energy of the first p particles we shall denote it by

$$K_p = \sum_1^p k_p$$

K_P is thus explicitly a function of $k_1 \ldots k_p$ being the sum of these quantities. We may however consider it a function of K and of the motions of the particles whose indices are higher than p—thus

$$K_p = K - \sum_{p+1}^n k$$

For the whole kinetic energy we have

$$K' = K - \frac{1}{2} m(u^2 + v^2 + w^2)$$

The <kinetic> energy of translation of the system of p particles is zero. That of the $p - 1$ particles is,

$$K'_T = \frac{1}{2}\frac{m^2}{M'}(u^2 + v^2 + w^2)$$

Hence

$$K' - K_T = K'_R + K'_I = K_R + K_I - \frac{1}{2}\mu(u^2 + v^2 + w^2)$$

We have next to find K'_R from the equation[a]

$$2K'_R = a'P'^2 + b'Q'^2 + c'R'^2 + 2\ell' QR + 2m' RP + 2n' PQ$$
$$< = a\frac{D}{D'} >$$

$$
\begin{aligned}
2K'_R \cdot D' = {} & \left[aD - \mu x^2 \left(B + C - \mu x^2\right)\right] P^2 \\
& + \left[bD - \mu x^2 A\right] \left[Q^2 + 2xQw + \mu^2 x^2 w^2\right] \\
& + \left[cD - \mu x^2 A\right] \left[R^2 - 2\mu x R v + \mu^2 x^2 v^2\right] \\
& + 2\left[mD + \mu x^2 M\right] \left[PR - \mu x P v\right] \\
& + 2\left[nD + \mu x^2 N\right] \left[PQ + \mu x P w\right] \\
= {} & 2K_R D + 2\mu x w \left[Qb' + Pn'\right] D' + 2\mu x v \left[Rc' + Pm'\right] D' \\
& + \mu^2 x^2 w^2 b' D' + \mu^2 x^2 v^2 c' D' \\
& - \mu x^2 \left[(D + C - \mu x^2) P^2 + A(Q^2 + R^2) - 2PRM - 2PQM\right]
\end{aligned}
$$

Hence

$$
\begin{aligned}
K'_I &= K' - K'_T - K'_R \\
\Big\langle &= K_I - \frac{1}{2}\mu\left(1 + \mu x^2 c'\right)\left(v - x\frac{Rc' + Pm'}{1 + \mu x^2 c'}\right)^2 \\
&\quad - \frac{1}{2}\mu\left(1 + \mu x^2 b'\right)\left(w + x\frac{Qb' + Pn'}{1 + \mu x^2 b'}\right)^2 \Big\rangle \\
&= K_I - \frac{1}{2}\mu u^2 - \frac{1}{2}\mu\left(1 + \mu x^2 c'\right)\left(v - x\frac{r'}{1 + \mu x^2 c'}\right)^2 \\
&\quad - \frac{1}{2}\mu\left(1 + \mu x^2 b'\right)\left(w + x\frac{n'}{1 + \mu x^2 b'}\right)^2
\end{aligned}
$$

a. The square brackets in the remainder of this document are Maxwell's.

7. "On Boltzmann's Theorem on the Average Distribution of Energy in a System of material Points"

Trans. Cambridge Phil. Soc. 12 (1879): 547–570. Reprinted in *Scientific Papers*, vol. 2: 713–741.

[From the *Cambridge Philosophical Society's Transactions*, Vol. XII.]

XCIV. *On Boltzmann's Theorem on the average distribution of energy in a system of material points.*

DR LUDWIG BOLTZMANN, in his "Studien über das Gleichgewicht der lebendigen Kraft zwischen bewegten materiellen Punkten" [*Sitzb. d. k. Akad. Wien*, Bd. LVIII., 8 Oct. 1868], has devoted his third section to the general solution of the problem of the equilibrium of kinetic energy among a finite number of material points. His method of treatment is ingenious, and, as far as I can see, satisfactory, but I think that a problem of such primary importance in molecular science ought to be scrutinized and examined on every side, so that as many persons as possible may be enabled to follow the demonstration, and to know on what assumptions it rests. This is more especially necessary when the assumptions relate to the degree of irregularity to be expected in the motion of a system whose motion is not completely known.

Mr H. W. Watson, in his Treatise on the Kinetic Theory of Gases*, has developed with great clearness the steps of the investigation of the distribution of energy among a set of particles which are supposed to act on each other only at very small distances. The particles may be acted on by external forces such as gravity, but it is expressly stipulated that the time during which a particle is encountering other particles is very small compared with the time during which there is no sensible action between it and other particles; and also that the time during which a particle is simultaneously within the distance of molecular action of more than one other particle may be neglected.

Now this method of treating the question, however necessary it may be in the subsequent investigation of the processes of diffusion, &c. in gases, is inapplicable to the theory of the equilibrium of temperature in liquids and

* *Clarendon Press Series*, 1876.

solids, for in these bodies the particles are never free from the action of neighbouring particles. It is true that in following the steps of the investigation, as given either by Boltzmann or by Watson, it is difficult, if not impossible, to see where the stipulation about the shortness and the isolation of the encounters is made use of. We may almost say that it is introduced rather for the sake of enabling the reader to form a more definite mental image of the material system than as a condition of the demonstration. Be this as it may, the presence of such a stipulation in the enunciation of the problem cannot fail to leave in the mind of the reader the impression of a corresponding limitation in the generality of the solution.

In the theorem of Boltzmann which we have now to consider there is no such limitation. The material points may act on each other at all distances, and according to any law which is consistent with the conservation of energy, and they may also be acted on by any forces external to the system provided these also are consistent with that law.

The only assumption which is necessary for the direct proof is that the system, if left to itself in its actual state of motion, will, sooner or later, pass through every phase which is consistent with the equation of energy.

Now it is manifest that there are cases in which this does not take place. The motion of a system not acted on by external forces satisfies six equations besides the equation of energy, so that the system cannot pass through those phases, which, though they satisfy the equation of energy, do not also satisfy these six equations.

Again, there may be particular laws of force, as for instance that according to which the stress between two particles is proportional to the distance between them, for which the whole motion repeats itself after a finite time. In such cases a particular value of one variable corresponds to a particular value of each of the other variables, so that phases formed by sets of values of the variables which do not correspond cannot occur, though they may satisfy the seven general equations.

But if we suppose that the material particles, or some of them, occasionally encounter a fixed obstacle such as the sides of a vessel containing the particles, then, except for special forms of the surface of this obstacle, each encounter will introduce a disturbance into the motion of the system, so that it will pass from one undisturbed path into another. The two paths must both satisfy the equation of energy, and they must intersect each other in the phase for which the conditions of encounter with the fixed obstacle are satisfied, but they are not subject to the equations of momentum. It is difficult in a case of such extreme complexity to arrive at a thoroughly satisfactory conclusion, but we may with considerable confidence assert that except for particular forms of the surface of the fixed obstacle, the system will sooner or later, after a sufficient number of encounters, pass through every phase consistent with the equation of energy.

I shall begin with the case in which the system is supposed to be contained within a fixed vessel, and shall afterwards consider the case of a free system, or of a system contained in a vessel rotating uniformly about an axis which itself moves uniformly in a straight line.

I have found it convenient, instead of considering one system of material particles, to consider a large number of systems similar to each other in all respects except in the initial circumstances of the motion, which are supposed to vary from system to system, the total energy being the same in all. In the statistical investigation of the motion, we confine our attention to the *number* of these systems which at a given time are in a phase such that the variables which define it lie within given limits.

If the number of systems which are in a given phase (defined with respect to configuration and velocity) does not vary with the time, the distribution of the systems is said to be *steady*.

It is shewn that if the distribution is steady, a certain function of the variables must be constant for all phases belonging to the same path. If the path passes through all phases consistent with the equation of energy, this function must be constant for all such phases. If however there are phases consistent with the equation of energy, but which do not belong to the same path, the value of the function may be different for such phases.

But whether we are able or not to prove that the constancy of this function is a necessary condition of a steady distribution, it is manifest that if the function is initially constant for all phases consistent with the equation of energy, it will remain so during the motion. This therefore is one solution, if not the only solution, of the problem of a steady distribution.

Now we know from the empirical laws of the diffusion of heat that the problem of the equilibrium of temperature in an isolated material system has one and only one solution. But we have found one solution of the problem of equilibrium of energy in a system of material points in motion. If, therefore, the real material system in which the equilibrium of temperature takes place is capable of being accurately represented by a system of material points (as defined in pure dynamics) acting on each other according to determinate, though unknown, laws, then the mathematical condition of the equilibrium of energy must be the dynamical representative of the physical condition of the equality of temperature.

It appears from the theorem that in the ultimate state of the system the average kinetic energy of two given portions of the system must be in the ratio of the number of degrees of freedom of those portions. This, therefore, must be the condition of the equality of temperature of the two portions of the system.

Hence at a given temperature the total kinetic energy of a material system must be the product of the number of degrees of freedom of that system into a constant which is the same for all substances at that temperature, being in fact the temperature on the thermodynamic scale multiplied by an absolute constant.

If the temperature, therefore, is raised by unity, the kinetic energy is increased by the product of the number of degrees of freedom into the absolute constant.

The observed specific heat of the body, expressed in dynamical measure, is the increment of the *total* energy when the temperature is increased by unity The observed specific heat cannot therefore be less than the product of the number of degrees of freedom into the absolute constant, unless the potential energy diminishes as the temperature rises.

Dynamical Specification of the motion.

We shall begin by supposing the material system to be of the most general type, having its configuration determined by the n variables $q_1, q_2 \ldots q_n$, and its motion determined by the corresponding momenta $p_1, p_2 \ldots p_n$. The state of the system at any instant is completely defined if we know the values of these $2n$ variables for that instant.

We shall suppose the forces acting between the parts of the system to be of the most general kind consistent with the conservation of energy. This may be expressed by defining V, the potential energy of the system, as a function of $q_1 \dots q_n$, the variables which define the configuration.

The kinetic energy of the system is denoted by T. We shall suppose it to be expressed in terms of the q's and p's as in Hamilton's method. The total energy is denoted by

$$E = V + T \quad \text{.................................(1)},$$

and is a constant during the motion of the system.

Hamilton's equations of motion for this system are

$$\frac{\partial q_r}{\partial t} = \frac{dE}{dp_r} \quad \text{.................................(2)},$$

$$\frac{\partial p_r}{\partial t} = -\frac{dE}{dq_r} \quad \text{.................................(3)},$$

where q_r and p_r are the co-ordinate and the momentum corresponding to each other.

Let us now consider a finite motion of the system. Let the initial co-ordinates and momenta be distinguished by accented letters, and the final co-ordinates and momenta by the same letters unaccented.

To define completely such a motion requires $2n+1$ variables to be given. These may be the n initial co-ordinates, the n initial momenta, and the time occupied by the motion.

There is another method however in which the $2n+1$ variables are the n initial co-ordinates, the n final co-ordinates, and the total energy. When these quantities are given there are in general only a finite number of possible motions.

Definition of the "Action" of the system during the motion.

Twice the time integral of the kinetic energy, taken from the beginning to the end of the motion, and expressed in terms of the initial and final co-ordinates and of the total energy, is called the "Action" of the system during the motion. If we denote it by A,

$$A = \int 2T dt \quad \text{.................................(4)},$$

and is expressed as a function of $q_1' \dots q_n'$, $q_1 \dots q_n$, and E.

It is shewn in treatises on dynamics* that

$$\frac{dA}{dq_r'} = -p_r' \quad \text{.........(5)},$$

and

$$\frac{dA}{dq_r} = p_r \quad \text{.........(6)}.$$

Hence

$$\frac{dp_r'}{dq_s} = -\frac{d^2A}{dq_r'dq_s} = -\frac{dp_s}{dq_r'} \quad \text{.........(7)}.$$

The indices r and s in this equation may be the same or different.
Also if t' and t are the values of the time at the beginning and at the end of the motion,

$$\frac{dA}{dE} = t - t' \quad \text{.........(8)}.$$

Hence

$$\frac{dp_r}{dE} = -\frac{dt'}{dq_r} \quad (9) \quad \text{and} \quad \frac{dp_s'}{dE} = -\frac{dt}{dq_s'} \quad \text{.........(10)}.$$

In the course of our investigation we shall have to compare the product of the differentials of the co-ordinates and momenta at the beginning of the motion with the corresponding product at the end of the motion. We shall write for brevity $ds = dq_1 \ldots dq_n$ for the product of the differentials of the co-ordinates, and $d\sigma = dp_1 \ldots dp_n$ for the product of the differentials of the momenta, and we shall use the product $ds'\,ds\,dE$ as a middle term in comparing $ds'\,d\sigma'\,dt'$ with $ds\,d\sigma\,dt$.

Now

$$ds'\,d\sigma'\,dt' = ds'\,ds\,dE\,\Sigma \pm \left(\frac{dp_1'}{dq_1} \cdots\cdots \frac{dp_n'}{dq_n}\,\frac{dt'}{dE}\right) \quad \text{.........(11)},$$

where

$$\Sigma \pm \left(\frac{dp_1'}{dq_1} \cdots\cdots \frac{dp_n'}{dq_n}\,\frac{dt'}{dE}\right)$$

denotes the functional determinant

$$\begin{vmatrix} \frac{dp_1'}{dq_1}, & \cdots\cdots \frac{dp_1'}{dq_n}, & \frac{dp_1'}{dE} \\ \cdots & \cdots & \cdots \\ \frac{dp_n'}{dq_1}, & \cdots\cdots \frac{dp_n'}{dq_n}, & \frac{dp_n'}{dE} \\ \frac{dt'}{dq_1}, & \cdots\cdots \frac{dt'}{dq_n}, & \frac{dt'}{dE} \end{vmatrix} \quad \text{.........(12)}.$$

* Thomson and Tait's *Natural Philosophy*, § 330.

Substituting for the elements of this determinant their values as given by equations (7), (9), and (10) it becomes

$$\begin{vmatrix} -\frac{dp_1}{dq_1'}, & \ldots\ldots & -\frac{dp_n}{dq_1'}, & -\frac{dt}{dq_1'} \\ \ldots & \ldots & \ldots & \ldots \\ -\frac{dp_1}{dq_n'}, & \ldots\ldots & -\frac{dp_n}{dq_n'}, & -\frac{dt}{dq_n'} \\ -\frac{dp_1}{dE}, & \ldots\ldots & -\frac{dp_n}{dE}, & -\frac{dt}{dE} \end{vmatrix} \quad \ldots\ldots\ldots\ldots\ldots\ldots(13).$$

Now the rows in this determinant are the same as the columns in the former one; the accented and unaccented letters being exchanged and the signs of all the elements changed. We may therefore express the relation between the two determinants in the abbreviated form

$$\Sigma \pm \left(\frac{dp_1'}{dq_1} \ldots\ldots \frac{dp_n'}{dq_n}\frac{dt'}{dE}\right) = (-)^{n+1}\, \Sigma \pm \left(\frac{dp_1}{dq_1'} \ldots\ldots \frac{dp_n}{dq_n'}\frac{dt}{dE}\right) \ldots\ldots(14).$$

Hence

$$ds'\, d\sigma'\, dt' = ds'\, ds\, dE\Sigma \pm \left(\frac{dp_1'}{dq_1} \ldots\ldots \frac{dp_n'}{dq_n}\frac{dt'}{dE}\right)$$

$$= (-)^{n+1} ds'\, ds\, dE\Sigma \pm \left(\frac{dp_1}{dq_1'} \ldots\ldots \frac{dp_n}{dq_n'}\frac{dt}{dE}\right)$$

$$= (-)^{n+1} d\sigma\, ds\, dt$$

$$= ds\, d\sigma\, dt \quad \ldots\ldots\ldots\ldots\ldots\ldots\ldots\ldots(15).$$

If we suppose the time, $t-t'$, to be given, $dt=dt'$ and

$$ds'\, d\sigma' = ds\, d\sigma \ldots\ldots\ldots\ldots\ldots\ldots\ldots(16),$$

or

$$dq_1' \ldots\ldots dq_n'\, dp_1' \ldots\ldots dp_n' = dq_1 \ldots\ldots dq_n\, dp_1 \ldots\ldots dp_n \quad \ldots\ldots(17).$$

The initial state of the system is a function of $2n$ variables. We have hitherto supposed these to be the n co-ordinates and the n momenta, but since the total energy E is a function of these variables we may substitute for one of the momenta, say p_1', its value in terms of the n co-ordinates, the $n-1$ remaining momenta, and E, and thus express every quantity we have to deal with in terms of the latter set of variables. Then since by equation (2)

$$\frac{dE}{dp_1'} = \frac{\partial q_1'}{\partial t} = \dot{q}_1' \quad \ldots\ldots\ldots\ldots\ldots\ldots\ldots\ldots\ldots\ldots\ldots(18),$$

$$dq_1'\ldots\ldots dq_n'\, dp_1'\ldots\ldots dp_n' = dq_1'\ldots\ldots dq_n'\, dp_2'\ldots\ldots dp_n'\, dE\, \frac{1}{\dot{q}_1'} \quad \ldots.(19).$$

Similarly we find for the final state of the system

$$dq_1\ldots\ldots dq_n\, dp_1\ldots\ldots dp_n = dq_1\ldots\ldots dq_n\, dp_2\ldots\ldots dp_n\, dE\, \frac{1}{\dot{q}_1} \quad \ldots\ldots\ldots(20).$$

The left-hand members of these equations have been proved equal, and in the right-hand members dE is the same at the beginning and end of the motion. Dividing out dE we find

$$dq_1'\ldots\ldots dq_n' dp_2'\ldots\ldots dp_n' \frac{1}{\dot{q}_1'} = dq_1\ldots\ldots dq_n dp_2\ldots\ldots dp_n \frac{1}{\dot{q}_1} \quad \ldots\ldots.(21).$$

This equation is applicable to the case in which the total energy is supposed not to vary from one particular instant of the motion to another, and in which, therefore, the $2n$ variables are no longer independent, but, being subject to the equation of energy, are reduced to $2n-1$.

Statistical Specification.

We have hitherto, in speaking of a phase of the motion of the system, supposed it to be defined by the values of the n co-ordinates and the n momenta. We shall call the phase so defined the phase (pq). We shall now adopt a wider definition by saying that the system is in the phase (a,b) whenever the values of the co-ordinates are such that q_1 is between b_1 and b_1+db_1, q_2 between b_2 and b_2+db_2, and so on; also p_2 between a_2 and a_2+da_2, and so on. The limits of the first component of momentum, p_1, are not specified, because the value of p_1 is not independent of the other variables, being given in terms of E and the other $2n-1$ variables in virtue of the equation of energy.

The quantities a, b are of the same kind as p and q respectively, only they are not supposed to vary on account of the motion of the system. In the statistical method of investigation, we do not follow the system during its motion, but we fix our attention on a particular phase, and ascertain whether the system is in that phase or not, and also when it enters the phase and when it leaves it.

Boltzmann defines the probability of the system being in the phase $(a_1 b)$ as the ratio of the aggregate time during which it is in that phase to the whole time of the motion, the whole time being supposed to be very great. I prefer to suppose that there are a great many systems the properties of which are the same, and that each of these is set in motion with a different set of values for the n co-ordinates and the $n-1$ momenta, the value of the total energy E being the same in all, and to consider the number of these systems which, at a given instant, are in the phase $(a_1 b)$. The motion of each system is of course independent of the other systems.

Let N be the whole number of systems, and let the number of these which, at the time t, are in the phase $(a_1 b)$ be denoted by $N\ (a_1,\ b,\ t)$. The aim of the statistical method is to express $N\ (a_1,\ b,\ t)$ as a function of N, of the co-ordinates and momenta with their limits, and of t. It is manifest that N can only enter the function as a factor, for the different systems do not act on each other. Also any differential as da or db can only enter as a factor, for the number of systems within any phase must vary in the ratio of the interval between the limits of that phase. We may therefore write

$$N(a_1 bt) = Nf(a_2, \ldots\ldots a_n,\ b_1, \ldots\ldots b_n,\ t)\, da_2 \ldots\ldots da_n db_1 \ldots\ldots db_n \ldots\ldots (22),$$

where we have to determine the form of the function f.

We shall now follow the motion of these systems from the time t', when we begin to watch the motion, to the time t when we cease to watch it.

Since the systems which at the time t form the group $N(a_1,\ b,\ t)$ are individually the same systems which at the time t' formed the group $N\,(a_1',\ b',\ t')$ we have

$$N\,(a_1,\ b,\ t) = N\,(a_1',\ b',\ t) \ldots\ldots\ldots\ldots\ldots\ldots\ldots\ldots (23),$$

or $$Nf\,(a_2 \ldots\ldots t)\, da_2 \ldots\ldots db_n = Nf\,(a_2' \ldots\ldots t')\, da_2' \ldots\ldots db_n' \ldots\ldots\ldots (24).$$

But by equation (21)

$$da_2 \ldots\ldots db_n\, (\dot{b}_1)^{-1} = da_2' \ldots\ldots db_n'\, (\dot{b}_1')^{-1} \ldots\ldots\ldots\ldots\ldots (25).$$

Hence $$f\,(a_2 \ldots\ldots t)\, \dot{b}_1 = f\,(a_2' \ldots\ldots t')\, \dot{b}_1' = C \ldots\ldots\ldots\ldots\ldots\ldots\ldots (26),$$

where C is a constant for all phases of the same motion, and we may write

$$f\,(a_2 \ldots\ldots t) = C\,(\dot{b}_1)^{-1} \ldots\ldots\ldots\ldots\ldots\ldots\ldots\ldots\ldots\ldots\ldots (27),$$

and $$N\,(a_1,\ b,\ t) = NC\,(\dot{b}_1)^{-1}\, da_2 \ldots\ldots db_n \ldots\ldots\ldots\ldots\ldots\ldots\ldots (28).$$

If the distribution of the N systems in the different phases is such that the number in a given phase does not vary with the time, the distribution is said to be steady. The condition of this is that C must be constant for all phases belonging to the same path. It will require further investigation to determine whether or not this path necessarily includes all phases consistent with the equation of energy.

If, however, we assume that the original distribution of the systems according to the different phases is such that C is constant for all phases consistent with the equation of energy, and zero for all phases which that equation shows to be impossible, then the law of distribution will not change with the time, and C will be an absolute constant.

We have therefore found one solution of the problem of finding a steady distribution. Whether there may be other solutions remains to be investigated.

Let $N(b)$ denote the number of systems in which q_1 is between b_1 and b_1+db_1, q_2 between b_2 and b_2+db_2, and so on, and q_n between b_n and b_n+db_n, the momenta not being specified otherwise than by their being consistent with the equation of energy, then

$$N(b)=\int\ldots\int N(a_1,\ b)\,da_2\ldots da_n \quad\text{.....................(29)},$$

the integration being extended to all values of the momenta consistent with the equation of energy.

To simplify the integration let us suppose the variables transformed so that the kinetic energy is expressed in terms of the squares of the component momenta,

$$T=\tfrac{1}{2}(\mu_1 a_1^2+\mu_2 a_2^2+\ldots+\mu_n a_n^2) \quad\text{.....................(30)},$$

where $a_1\ldots a_n$ are the transformed momenta, and $\mu_1\ldots\mu_n$ are functions of the co-ordinates, which we may call moments of mobility, and which, in the case of material points, are the reciprocals of the masses.

Now let us assume

$$\tfrac{1}{2}\mu_n A_n^2=T=E-V \quad\text{..............................(31)},$$

$$\mu_{n-1}A_{n-1}^2=\mu_n(A_n^2-a_n^2) \quad\text{..........................(32)},$$

$$\mu_{n-2}A_{n-2}^2=\mu_{n-1}(A_{n-1}^2-a_{n-1}^2) \quad\text{.....................(33)},$$

....

$$\mu_2 A_2^2=\mu_3(A_3^2-a_3^2) \quad\text{..............................(34)}.$$

Then by the equation of energy

$$\mu_1 a_1^2 = \mu_2 (A_2^2 - a_2^2) \quad\text{..............................(35).}$$

Of these quantities, A_n is a function of the co-ordinates only, because E is given and V is a function of the co-ordinates, A_{n-1} is a function of the co-ordinates and a_n, A_{n-2} of the co-ordinates and of a_n and a_{n-1}, and so on.

Also by equation (2)

$$\dot{b}_1 = \frac{dT}{da_1} = \mu_1 a_1$$

$$= (\mu_1 \mu_2)^{\frac{1}{2}} (A_2^2 - a_2^2)^{\frac{1}{2}} \quad\text{..............................(36).}$$

To integrate the expression

$$\iiint \cdots \int C (\dot{b}_1)^{-1} da_2 \ldots da_n,$$

we begin by integrating with respect to a_2, thus

$$\int C (\dot{b}_1)^{-1} da_2 = \int C (\mu_1 \mu_2)^{-\frac{1}{2}} (A_2^2 - a_2^2)^{-\frac{1}{2}} da_2 \quad\text{..................(37),}$$

the limits of integration being $\pm A_2$. The result is

$$\frac{\Gamma(\frac{1}{2})\Gamma(\frac{1}{2})}{\Gamma(\frac{2}{2})} C (\mu_1 \mu_2)^{-\frac{1}{2}} A_2^0 \quad\text{...........................(38).}$$

For the next integration we have

$$\int (\mu_3 A_3^2)^0 da_3 = \int_{-A_3}^{A_3} \{\mu_3 (A_3^2 - a_3^2)\}^0 da_3 = \frac{\Gamma\frac{1}{2}\Gamma\frac{2}{2}}{\Gamma\frac{3}{2}} A_3^{\frac{1}{2}} \quad\text{................(39).}$$

Hence after r integrations, r being any number less than n, the result is

$$NC \frac{(\Gamma(\frac{1}{2}))^{r+1}}{\Gamma\frac{r+1}{2}} (\mu_1 \mu_2 \ldots \mu_{r+1})^{-\frac{1}{2}} [\mu_{r+1} A_{r+1}^2]^{\frac{r-1}{2}} da_{r+2} \ldots da_r db_1 \ldots db_n \quad\text{.......(40).}$$

Putting $r = n-1$ and remembering that $\mu_n A_n^2 = 2E - 2V$, we find

$$N(b) = NC \frac{[\Gamma(\frac{1}{2})]^n}{\Gamma\left(\frac{n}{2}\right)} (\mu_1 \mu_2 \ldots \mu_n)^{-\frac{1}{2}} [2E - 2V]^{\frac{n-2}{2}} \quad\text{....................(41).}$$

This is the number of systems whose configuration is specified by the variables $b_1 \ldots b_n$, while the momenta may have any values consistent with the equation of energy.

The quantity $E-V$, which occurs in this equation, is, by equation (1), equal in magnitude to T, the kinetic energy of the system. The quantity T, however, is defined explicitly in terms of the velocities or the momenta of the system, whereas $E-V$ does not involve these quantities explicitly, but is expressed as a function of the configuration.

We shall find it convenient, however, especially in the study of more complicated problems, to remember that the number of systems in a given configuration is a function of the kinetic energy corresponding to that configuration.

If the kinetic energy is not expressed as a sum of squares, but in the more general form,

$$T=\tfrac{1}{2}[11]\,a_1^2+[12]\,a_1a_2+\&c.$$
$$+\tfrac{1}{2}[22]\,a_2^2+[23]\,a_2a_3+\&c. \quad\ldots\ldots(42),$$

where the quantities denoted by [11] &c. are functions of the co-ordinates, which we may call the moments and products of mobility of the system; then since the discriminant

$$\Delta=\begin{vmatrix}[11], & [12], \ldots\ldots & [1n]\\ [21], & [22], \ldots\ldots & [2n]\\ \ldots & \ldots & \ldots\\ [n1], & [n2], \ldots\ldots & [nn]\end{vmatrix} \quad\ldots\ldots(43)$$

is an invariant, its value is the same when T is reduced to a sum of squares, in which case all the elements except those in the principal diagonal of the determinant vanish, and we have

$$\Delta=\mu_1\mu_2\ldots\mu_n \quad\ldots\ldots(44),$$

and we may write the value of $N(b)$,

$$N(b)=NC\frac{(\Gamma(\frac{1}{2}))^n}{\Gamma\left(\frac{n}{2}\right)}\Delta^{-\frac{1}{2}}(2E-2V)^{\frac{n-2}{2}}\,db_1\ldots db_n \quad\ldots\ldots(45).$$

If the system consists of n' material particles, whose masses are $m_1\ldots m_{n'}$, then the number of degrees of freedom is $n=3n'$ and

$$\mu_1=\mu_2=\mu_3=m_1^{-1},\quad \mu_4=\mu_5=\mu_6=m_2^{-1} \text{ and so on} \quad\ldots\ldots(46).$$

Hence in this case we may write

$$N(b)=NC'\frac{(\Gamma(\frac{1}{2}))^{3n'}}{\Gamma\left(\frac{3n'}{2}\right)}(m_1\ldots m_{n'})^{\frac{3}{2}}\,[2E-2V]^{\frac{3n'-2}{2}}\,db_1\ldots db_n \quad\ldots\ldots(47).$$

These expressions give the number of systems in a given configuration only when $E-V$ is positive for that configuration, for since the kinetic energy is necessarily positive, the potential energy cannot exceed the total energy. For configurations specified in such a way that if they existed V would be greater than E, the value of $N(b)$ is zero.

The value of $N(b)$ is also zero for configurations which, though they make V less than E, cannot be reached by a continuous path from the original configuration without passing through configurations which make V greater than E.

We shall return to this expression for the number of systems in a completely specified configuration, but in the mean time it will be useful to consider how many of these systems have one of their momenta, p_n, between given limits. In this way we shall be able to determine completely the average distribution of momentum among the variables without making any assumptions about the nature of the system which might limit the generality of our results.

In order to find the number of systems in the configuration (b) for which one of the momenta, say p_n, lies between a_n and a_n+da_n, we must stop before the last integration. Putting $r=n-2$ in equation (40)

$$N(b_1a_n)=NC\frac{\Gamma(\frac{1}{2})^{n-1}}{\Gamma\left(\frac{n-1}{2}\right)}(\mu_1\ldots\mu_{n-1})^{-\frac{1}{2}}(\mu_{n-1}A_{n-1}^2)^{\frac{n-3}{2}}da_n db_1\ldots db_n \ldots\ldots(48).$$

The whole number of systems in configuration (b) is given by (45). Hence the proportion of these systems for which a_n lies between a_n and a_n+da_n is

$$\frac{2^{-\frac{1}{2}}\Gamma\left(\frac{n}{2}\right)}{\Gamma(\frac{1}{2})\Gamma\left(\frac{n-1}{2}\right)}\frac{\left[E-V-\frac{1}{2}\mu_n a_n^2\right]^{\frac{n-3}{2}}}{[E-V]^{\frac{n-2}{2}}}\mu_n^{\frac{1}{2}}da_n \ldots\ldots\ldots\ldots\ldots\ldots(49).$$

If we write

$$\tfrac{1}{2}\mu_n a_n^2=k_n \ldots\ldots\ldots\ldots\ldots\ldots\ldots\ldots\ldots\ldots(50),$$

then k_n denotes the part of the kinetic energy arising from the momentum a_n. The proportion of the systems in configuration (b) for which k_n is between k_n and k_n+dk_n is

$$\frac{\Gamma\left(\frac{n}{2}\right)}{\Gamma(\frac{1}{2})\Gamma\left(\frac{n-1}{2}\right)}\frac{[E-V-k_n]^{\frac{n-3}{2}}}{[E-V]^{\frac{n-2}{2}}}k_n^{-\frac{1}{2}}dk_n \ldots\ldots\ldots\ldots\ldots\ldots(51).$$

Since any one of the variables may be taken for q_n, the law of distribution of values of the kinetic energy is the same for all the variables. The mean value of the kinetic energy corresponding to any variable is

$$K = \frac{1}{n}(E - V) = \frac{1}{n} T \quad \text{.........................(52).}$$

The maximum value is $$T = nK \quad \text{..................................(53).}$$

The mean value of k^r is

$$\frac{1 . 3 \ldots 2r-1}{n . n+2 \ldots n+2r-2} n^r K^r \text{..........................(54).}$$

When n is very large, the expression (51) approximates to

$$\frac{1}{\sqrt{2\pi}} \frac{1}{K} e^{-\frac{k}{2K}} dk \quad \text{...............................(55).}$$

Recapitulation.

The result of our investigation may therefore be stated as follows:

(α) We begin by considering a set of material systems which satisfy the general equations of dynamics (2) and (3), and the equation of energy (1). If in these systems the distribution of configurations satisfies equation (45), and the distribution of motion satisfies equation (51), these equations will continue to be satisfied during the subsequent motion of the system. One result of equation (51), to which we shall have to refer, is that the average kinetic energy corresponding to any one of the variables is the same for every one of the variables of the system.

(β) We now turn our attention to a system of real bodies enclosed in a rigid vessel impervious to matter and to heat. We know by experiment that in such a system the temperature cannot remain steady in every part unless the temperature of every part of the system is the same, and that this condition is necessary in whatever manner the configuration of the system may be varied by altering the position and mean density of the portions of sensible size into which we are able to divide it.

Now if the system of real bodies is a material system which satisfies the equations of dynamics, and if equations (45) and (51) are also satisfied, the condition of the system will, as we have shewn, (α), be steady in every respect, and therefore in respect of temperature. Hence by (β) the temperature of every part of the system must be the same.

Therefore if equations (45) and (51) are satisfied, the condition of equality of temperature is also satisfied.

But the condition of equality of temperature does not depend on the configuration of the system, for though we can alter the configuration by external constraint we cannot prevent the temperature from becoming equalized. It does not depend, therefore, on equation (45). We must therefore conclude, that if equation (51) is satisfied, the condition of equality of temperature is also satisfied, or, in other words, that equation (51) is the condition of equality of temperature.

Hence when two parts of a system have the same temperature, the average kinetic energy corresponding to any one of the variables belonging to these parts must be the same.

If the system is a gas or a mixture of gases not acted on by external forces, the theorem that the average kinetic energy of a single molecule is the same for molecules of different gases is not sufficient to establish the condition of equilibrium of temperature between gases of different kinds such as oxygen and nitrogen, because when the gases are mixed we have no means of ascertaining the temperature of the oxygen or of the nitrogen separately. We can only ascertain the temperature of the mixture by putting a thermometer into it.

We cannot legitimately assert that the temperatures of the oxygen and of the nitrogen must be equal because they are in contact with each other, for the only way in which we can conceive the oxygen or the nitrogen as existing in the mixture is by picturing the medium as a system of molecules, and as soon as we begin to see the molecules distinctly, heat becomes resolved into motion.

But since our investigation is equally applicable to a system of any kind, provided only it satisfies the equations of dynamics, we may suppose it to consist of pure oxygen and pure nitrogen separated by a solid diaphragm, the solid diaphragm consisting of molecules capable of motion, but acting on each other with forces which are sufficient to prevent any molecule from getting far apart from its neighbours except under the action of disturbing forces greater than any which would occur in a system at the given temperature. In this system, though the oxygen and the nitrogen cannot mix, each can make an exchange of molecular energy with the surface molecules of the diaphragm, and exchanges of energy can go on within the solid diaphragm itself without any exchange of molecules between distant parts of the diaphragm.

Hence, in this system, the average kinetic energy of a molecule of oxygen will become equal to that of a molecule of nitrogen in the final state of the system, that is to say, when the temperatures of all parts of the system have become equal, and since in that final state we have pure oxygen on one side and pure nitrogen on the other, we can verify the equality of temperature by means of a thermometer, and we can now assert that the temperatures, not only of oxygen and nitrogen, but of all bodies, are equal when the average kinetic energy of a single molecule of each of these substances is the same.

Approximate value of the probability when V is small compared with E.

To find the number of systems the configuration of which is specified as regards the limits of certain of the variables while the other variables are left undetermined, we should have to integrate the expressions in equations (41), (45), or (47) with respect to each of the undetermined variables in succession, the integrations being extended to all values of these variables which are consistent with the equation of energy.

These integrations cannot be performed unless the potential energy of the system is a known function of the variables which determine its configuration. We cannot therefore in general continue the integration so as to determine the number of systems in which the limits are specified for some, but not all, of the variables.

But when the number of variables is very great, and when the potential energy of the specified configuration is very small compared with the total energy of the system, we may obtain a useful approximation to the value of $[E-V]^{\frac{n-2}{2}}$ in an exponential form, for if we write, as in equation (53), $E=nK$,

$$[E-V]^{\frac{n-2}{2}}=E^{\frac{n-2}{2}}e^{\frac{n-2}{2}\log\left(1-\frac{V}{nK}\right)}$$

$$=E^{\frac{n-2}{2}}e^{-\frac{V}{2K}}\qquad\text{(56)},$$

nearly, provided n is very great and V is small compared with E. The expression is no longer approximate when V is nearly as great as E, and it does not vanish, as it ought to do, when $V=E$.

Hence when the potential energy of the system in the given configuration is very small compared with its kinetic energy, we may use the approximately correct statement, that the number of systems in a given configuration is inversely proportional to the exponential function, the index of which is half the potential energy of the system in the given configuration divided by the average kinetic energy corresponding to each variable of the system.

If we divide the system into any two parts, A and B, we may consider V, the potential energy of the whole system, as made up of three parts, V_A and V_B, the potential energy of A and B, each on itself, and W, that of B with respect to A.

When, as in the case of a gas, the parts of a system are in a great degree independent of each other, the average values of V_A and V_B may be treated as constants, and the variations of V will be the same as those of W, so that the variable part of the exponential function will be reduced to

$$e^{-\frac{W}{2K}} \quad\text{.......................................(57).}$$

If we suppose that A denotes a single molecule of a particular kind of gas, and that B denotes all the other molecules, of whatever kind, in the system, then, since there are many molecules similar to A, we may pass, from the number of systems in which A is within a given element of volume, to the average number of molecules similar to A which are within that element, or, in other words, the average density of the gas A within that element.

We may therefore interpret the expression (57) as asserting that the density of a particular kind of gas at a given point is inversely proportional to the exponential function whose index is half the potential energy of a single molecule of the gas at that point, divided by the average kinetic energy corresponding to a variable of the system.

We must remember that since the centre of mass of a molecule is determined by *three* variables, the mean kinetic energy of agitation of the centre of mass of a molecule is *three* times the quantity K which denotes the mean kinetic energy of a single variable.

PART II. *A Free system.*

In a material system not acted on by external forces the motion satisfies six equations besides the equation of energy, so that we must not include in our integration all the phases which satisfy the equation of energy, but only those of them which also satisfy these six equations.

In what follows, we shall suppose the system to consist of n particles, whose masses are $m_1 \dots m_n$, and whose co-ordinates x, y, z, and velocity-components u, v, w, are distinguished by the same suffix as the particle to which they belong.

Let us now consider a system consisting of s of these particles, and write

$$m_1 + m_2 + \&\text{c.} + m_s = M_s \quad \text{.........................(58)},$$

$$\left.\begin{aligned} m_1x_1 + m_2x_2 + \&\text{c.} + m_sx_s &= M_sX_s, \\ m_1y_1 + m_2y_2 + \&\text{c.} + m_sy_s &= M_sY_s, \\ m_1z_1 + m_2z_2 + \&\text{c.} + m_sz_s &= M_sZ_s, \end{aligned}\right\} \quad \text{....................(59)},$$

then M_s will be the mass of the minor system and X_s, Y_s, Z_s the co-ordinates of its centre of mass. If we also write

$$\left.\begin{aligned} m_1u_1 + \&\text{c.} + m_su_s &= M_sU_s, \\ m_1v_1 + \&\text{c.} + m_sv_s &= M_sV_s, \\ m_1w_1 + \&\text{c.} + m_sw_s &= M_sW_s, \end{aligned}\right\} \quad \text{.........................(60)},$$

$$\left.\begin{aligned} m_1(y_1w_1 - z_1v_1) + \&\text{c.} + m_s(y_sw_s - z_sv_s) &= F_s + M_s(Y_sW_s - Z_sV_s), \\ m_1(z_1u_1 - x_1w_1) + \&\text{c.} + m_s(z_su_s - x_sw_s) &= G_s + M_s(Z_sU_s - X_sW_s), \\ m_1(x_1v_1 - y_1u_1) + \&\text{c.} + m_s(x_sv_s - y_su_s) &= H_s + M_s(X_sV_s - Y_sU_s), \end{aligned}\right\} \text{...(61)},$$

then U_s, V_s, W_s will be the velocity-components of the centre of mass, and F_s, G_s, H_s the components of angular momentum round this point.

We shall also write

$$\tfrac{1}{2}m_1(u_1^2 + v_1^2 + w_1^2) + \&\text{c.} + \tfrac{1}{2}m_s(u_s^2 + v_s^2 + w_s^2) = T_s \text{............(62)}.$$

The seven conditions satisfied by the whole system are that the seven quantities U_n, V_n, W_n, F_n, G_n, H_n and E are constant during the motion.

Under these conditions the $3n$ momentum-components are not independent. We shall therefore transform equation (17) into one in which the differentials of the first seven velocity-components are replaced by the differentials of the seven constants.

The functional determinant is found by differentiating the seven quantities U_n, V_n, W_n, F_n, G_n, H_n and E with respect to the momenta m_1u_1; m_1v_1, m_1w_1; m_2u_2, m_2v_2, m_2w_2; and m_3u_3. We thus obtain

$$\begin{vmatrix} 1, & 0, & 0, & 0, & z_1, & -y_1, & u_1 \\ 0, & 1, & 0, & -z_1, & 0, & x_1, & v_1 \\ 0, & 0, & 1, & y_1, & -x_1, & 0, & w_1 \\ 1, & 0, & 0, & 0, & z_2, & -y_2, & u_2 \\ 0, & 1, & 0. & -z_2, & 0, & x_2, & v_2 \\ 0, & 0, & 1, & y_2, & -x_2, & 0, & w_2 \\ 1, & 0, & 0, & 0, & z_3, & -y_3, & u_3 \end{vmatrix} = \Delta \ldots\ldots\ldots\ldots(63),$$

which we may write $$\Delta = a\, r_{12}\, \dot{r}_{12} \ldots\ldots\ldots\ldots(64),$$

where $$a = (y_1 - y_2)(z_2 - z_3) - (y_2 - y_3)(z_1 - z_2) \ldots\ldots\ldots\ldots(65),$$

or twice the projection on the plane of yz of the triangle whose vertices are m_1, m_2, and m_3, and

$$r_{12}\dot{r}_{12} = (u_1 - u_2)(x_1 - x_2) + (v_1 - v_2)(y_1 - y_2) + (w_1 - w_2)(z_1 - z_2) \ldots\ldots(66),$$

or the rate of increase of the distance between m_1 and m_2 multiplied into that distance.

In a system composed of material particles, each component of momentum is equal to the corresponding velocity-component multiplied into the mass of the particle. We may therefore write $p_1 = m_1u_1$ and so on, and since the masses are invariable we may omit them from both members of equation (17), and write it

$$dx_1' \ldots dz_n' \, du_1' \ldots dw_n' = dx_1 \ldots dz_n du_1 \ldots dw_n \ldots\ldots\ldots\ldots(67).$$

But $$dUdVdWdFdGdHdE = m_1^3 m_2^3 m_3 a' r'_{12} \dot{r}'_{12} du_1' dv_1' dw_1' du_2' dv_2' dw_2' du_3'$$
$$= m_1^3 m_2^3 m_3 a r_{12} \dot{r}_{12} du_1 \ldots\ldots du_3 \ldots\ldots\ldots\ldots(68).$$

Hence $$\frac{dx_1' \ldots dz_n' dv_3' \ldots dw_n'}{m_1^3 m_2^3 m_3 a' r'_{12} \dot{r}'_{12}} = \frac{dx_1 \ldots dz_n dv_3 \ldots dw_n}{m_1^3 m_2^3 m_3 a r_{12} \dot{r}_{12}} = C \ldots\ldots\ldots\ldots(69),$$

and equation (29) becomes

$$N(b) = \int^{8n-7} C(m_1^3 m_2^3 m_3 a r_{12} \dot{r}_{12})^{-1} dv_3 \ldots dw_n \quad \ldots\ldots\ldots\ldots(70).$$

We shall find it useful in what follows to define the energy of internal motion as the excess of the whole kinetic energy of the system over that which it would have if it were moving like a rigid body with the same configuration, and the same components of momentum and of angular momentum.

If we suppose the internal motion of the system to be destroyed in a very short time by internal forces, so that the configuration is not sensibly altered during the process, then the work done by the system against these forces is the measure of the energy of internal motion.

Writing T for the kinetic energy referred to the origin, K for that of the mass moving with the velocity of the centre of mass, J for the kinetic energy due to the rotation of the system as a rigid body, and I for the energy of internal motion, we have

$$I=T-K-J \quad\text{.......(71)},$$

where

$$T=\Sigma\left[\tfrac{1}{2}m\,(u^2+v^2+w^2)\right] \quad\text{.......(72)},$$

$$K=\tfrac{1}{2}M(U^2+V^2+W^2) \quad\text{.......(73)},$$

$$J=\tfrac{1}{2}(Fp+Gq+Hr) \quad\text{.......(74)},$$

where p, q, r are the components of angular velocity with respect to the axes of x, y, z and are related to F, G, H by the equations

$$\left.\begin{array}{ll} Ap-Nq-Mr=F, & aF-nG-mH=p, \\ -Np+Bq-Lr=G, & -nF+bG-lH=q, \\ -Mp-Lq+Cr=H, & -mF-lG+cH=r, \end{array}\right\} \quad\text{.......(75)}.$$

where

$$\left.\begin{array}{ll} A=\Sigma m\left[(y-Y)^2+(z-Z)^2\right] & L=\Sigma m\,(y-Y)(z-Z) \\ B=\Sigma m\left[(z-Z)^2+(x-X)^2\right] & M=\Sigma m\,(z-Z)(x-X) \\ C=\Sigma m\left[(x-X)^2+(y-Y)^2\right] & N=\Sigma m\,(x-X)(y-Y) \end{array}\right\} \text{...(76)}.$$

Writing for the sake of brevity

$$D=\begin{vmatrix} A, & -N, & -M \\ -N, & B, & -L \\ -M, & -L, & C \end{vmatrix}, \qquad d=\begin{vmatrix} a, & -n, & -m \\ -n, & b, & -l \\ -m, & -l, & c \end{vmatrix} \text{.......(77)},$$

the relations between the moments and products of mobility and those of inertia will be given by equations of the forms

$$\left.\begin{array}{ll} aD=BC-L^2 & Ad=bc-l^2 \\ lD=-MN-AL, & Ld=-mn-al, \\ \multicolumn{2}{c}{Dd=1.} \end{array}\right\} \text{.........(78)}.$$

If we write

$$\left.\begin{aligned}\xi &= u - U + qz - ry\\ \eta &= v - V + rx - pz\\ \zeta &= w - W + py - qx\end{aligned}\right\} \quad \ldots\ldots\ldots\ldots\ldots\ldots\ldots\ldots(79),$$

then ξ, η, ζ will be the velocity-components of a particle with respect to axes passing through the centre of mass of the system and rotating with the angular velocity whose components are p, q, r. We may therefore call ξ, η, ζ the velocity-components of the internal motion. If the system were to become rigid, the internal motion would become zero. The energy of internal motion may be expressed in terms of ξ, η, ζ, thus:—

$$I = \Sigma \tfrac{1}{2} m (\xi^2 + \eta^2 + \zeta^2) \ldots\ldots\ldots\ldots\ldots\ldots\ldots\ldots(80).$$

We have now to express the energy of internal motion of a system of $s-1$ particles in terms of the quantities U, V, W, F, G, H and T belonging to the system of s particles, together with the position and velocity of the sth particle.

To avoid the repetition of suffixes we shall distinguish quantities belonging to the minor system of $s-1$ particles by accented letters, and quantities belonging to the complete system of s particles and the particle m, by unaccented letters. We shall also write

$$\left.\begin{aligned}\mu &= \frac{Mm}{M'}\,.\\ \text{We thus find}\qquad M' &= M - m\\ M'X' &= MX - mx\\ M'U' &= MU - mu,\\ F' &= F - \mu(y-Y)(w-W) + \mu(z-Z)(v-V)\\ A' &= A - \mu(y-Y)^2 - \mu(z-Z)^2\\ L' &= L - \mu(y-Y)(z-Z)\\ T' &= T - \tfrac{1}{2}m(u^2+v^2+w^2)\\ K' &= K + \tfrac{1}{2}\mu\left[(U-u)^2 + (V-v)^2 + (W-w)^2\right] - \tfrac{1}{2}m(u^2+v^2+w^2).\end{aligned}\right\} \ldots(81).$$

Since the choice of the axes of reference is arbitrary, we may simplify the expressions by taking for origin the centre of mass of the system M, and for the axis of z the line passing through the particle m. We may also turn the axes of x and y about that of z till A becomes a maximum, the condition of which is

$$LM + CN = 0.$$

We shall also reckon velocities with reference to the centre of mass of the system M.

With these simplifications we find

$$\left.\begin{array}{lll} F' = F + \mu vz & G' = G - \mu uz & H' = H \\ A' = A - \mu z^2 & B' = B - \mu z^2 & C' = C \\ L' = L & M' = M & N' = N \\ a' = \dfrac{a}{1 - a\mu z^2}, & & l' = \dfrac{l'}{1 - b\mu z^2}, \\ b' = \dfrac{b}{1 - b\mu z^2}, & & m' = \dfrac{m}{1 - a\mu z^2}, \\ c' = c + \mu z^2\left(\dfrac{l^2}{1 - b\mu z^2} + \dfrac{m^2}{1 - a\mu z^2}\right), & & n' = n = 0, \\ & D' = D(1 - a\mu z^2)(1 - b\mu z^2). & \end{array}\right\} \ldots\ldots(82).$$

We are now able to calculate the energy of rotation, J', of the minor system:

$$2J' = a'F'^2 + b'G'^2 + c'H'^2 - 2l'G'H' - 2m'H'F' - 2n'F'G' \ldots\ldots\ldots(83),$$

$$\left.\begin{array}{l} = 2J + \dfrac{1}{1 - a\mu z^2}[v^2 a\mu z^2 - 2v\mu z(Fa - Hm) + \mu z^2(Fa - Hm)^2] \\ + \dfrac{1}{1 - b\mu z^2}[u^2 b\mu z^2 + 2u\mu z(Gb - Hl) + \mu z^2(Gb - Hl)^2] \end{array}\right\} (84).$$

Combining these results and reducing we find for the energy of internal motion of the system M'

$$I' = I - \tfrac{1}{2}\mu(1 - b\mu z^2)^{-1}(u - Gb + Hl)^2 - \tfrac{1}{2}\mu(1 - a\mu z^2)^{-1}(v - Fa + Hl)^2 - \tfrac{1}{2}\mu w^2 \ldots(85).$$

$$\text{Hence} \iiint I'^{\frac{q}{2}}\, du\, dv\, dw = \frac{(\Gamma(\frac{1}{2}))^3\, \Gamma\left(\frac{q+2}{2}\right)}{\Gamma\left(\frac{q+5}{2}\right)} \left(\frac{2}{\mu}\right)^{\frac{3}{2}} (1 - a\mu z^2)^{\frac{1}{2}}(1 - b\mu z^2)^{\frac{1}{2}} I^{\frac{q+3}{2}} \ldots(86),$$

the integration being extended to all values of u, v, and w which make I' positive.

Now $(1 - a\mu z^2)(1 - b\mu z^2) = \dfrac{D'}{D}$, and this is an invariant.

Hence in general, whatever axes we choose,

$$\iiint [M_{s-1}^3 D_{s-1}]^{-\frac{1}{2}} I_{s-1}^{\frac{q}{2}} du_s dv_s dw_s = \frac{(\Gamma(\frac{1}{2}))^3 \Gamma\left(\frac{q+2}{2}\right)}{\Gamma\left(\frac{q+5}{2}\right)} [\tfrac{1}{2} m_s]^{-\frac{3}{2}} [M_s^3 D_s]^{-\frac{1}{2}} I_s^{\frac{q+3}{2}} \ldots\ldots(87).$$

For the system consisting of the two particles m_1 and m_2 the energy of rotation is

$$J_2 = \tfrac{1}{2} \frac{M_2}{m_1 m_2 r_{12}^2} (F_2^2 + G_2^2 + H_2^2) \ldots\ldots(88),$$

and the energy of internal motion is

$$I_2 = \tfrac{1}{2} \frac{m_1 m_2}{M_2} \dot{r}_{12}^2 \ldots\ldots(89).$$

Hence we may write equation (70)

$$N(b) = \int^{3n-7} C(m_1^3 m_2^3 m_3 a_1 r_2)^{-1} \left(2 \frac{M_2}{m_1 m_2}\right)^{-\frac{1}{2}} I_2^{-\frac{1}{2}} dv_3 \ldots dw_n \ldots\ldots(90).$$

We have first to express I_2 in terms of quantities having the suffix $_3$.

If we make the plane of yz pass through the three particles m_1, m_2, m_3, so that the origin coincides with their centre of mass and has the same velocity, and the axis of z passes through m_3, then a is twice the area of the triangle whose vertices are m_1, m_2 and m_3,

$$F_2 = F_3 + \frac{M_3 m_3}{M_2} z_3 v_3, \qquad G_2 = G_3 - \frac{M_3 m_3}{M_2} z_3 u_3, \qquad H_2 = H_3 \ldots\ldots(91),$$

$$a m_3 u_3 = G_3 (y_1 - y_2) + H_3 (z_1 - z_2) \ldots\ldots(92),$$

$$I_2 = I_3 - \tfrac{1}{2} \frac{M_3 m_3}{M_2} \left(1 + \frac{M_3 m_3}{m_1 m_2} \frac{z_3}{r_{12}^3}\right)$$

$$\left[\left(u - \frac{M_2 G_3 z_3}{m_1 m_2 r_{12}^2 + M_3 m_3 z_3^2}\right)^2 + \left(v + \frac{M_2 F_3 z_3}{m_1 m_2 r_{12}^2 + M_3 m_3 z_3}\right)^2\right] - \tfrac{1}{2} \frac{M_3 m_3}{M_2} w^2 \ldots(93).$$

We have now to integrate

$$\iint I^{-\frac{1}{2}} dv_3 dw_3,$$

extending the integration to all values of v_3 and w_3 which make I_2 positive, and remembering that equation (92) shews that u_3 is independent of v_3 and w_3. The result is

$$\iint I_3^{-\frac{1}{2}}\, dv_3 dw_3 = \frac{(\Gamma(\frac{1}{2}))^3}{\Gamma(\frac{3}{2})}\left[\frac{1}{4}\frac{M_3^2 m_3^2}{M_2^2}\cdot\frac{m_1 m_2 r_{12}^2 + M_3 m_3 z_3^2}{m_1 m_2 r_{12}^2}\right]^{-\frac{1}{2}} I_3^{\frac{1}{2}} \text{ ....(94).}$$

Now for the three particles m_1, m_2, m_3,

$$D_3 = \frac{m_1 m_2 m_3}{M_3^2}\left[r_{23}^2\, m_2 m_3 + r_{31}^2\, m_3 m_1 + r_{12}^2\, m_1 m_2\right] a^2 \text{..............(95),}$$

where r_{23}, r_{31} and r_{12} are the distances between the particles, and a is the area of the triangle $m_1 m_2 m_3$.

Also $$r_{23}^2 m_2 m_3 + r_{31}^2 m_3 m_1 + r_{12}^2 m_1 m_2 = \frac{M_3}{M_2}(m_1 m_2 r_{12}^2 + M_3 m_3 z_3^2) \text{.........(96).}$$

We may now write equation (90) in the form

$$N(b) = \int^{3n-9} C \frac{(\Gamma\frac{1}{2})^3}{\Gamma(\frac{3}{2})}\left[\tfrac{1}{2} m_1^3 m_2^3 m_3^3 M_3^3 D_3\right]^{-\frac{1}{2}} I_3^{\frac{1}{2}}\, du_4 \ldots dw_n \text{...........(97).}$$

Continuing the integration by equation (87) we find

$$N(b) = 2^{\frac{3n-8}{2}} C \frac{(\Gamma\frac{1}{2})^{3n-6}}{\Gamma\left(\frac{3n-6}{2}\right)} (m_1 \ldots m_n)^{-\frac{3}{2}} M_n^{-\frac{3}{2}} D_n^{-\frac{1}{2}} I_n^{\frac{3n-8}{2}} \text{..........(98),}$$

where I_n is what we have defined as the energy of internal motion of the system, or the work which the system would do, in virtue of its motion, against the system of internal forces which would be called into play if the distances between the parts of the material system were in an insensibly small time to become invariable.

In order to determine the number of systems in a given configuration for which the velocity-components of the particle m_n lie between the limits $u \pm \frac{1}{2}du$, $v \pm \frac{1}{2}dv$, $w \pm \frac{1}{2}dw$, we must form the expression for $N(b, u_n, v_n, w_n)$ by stopping short before the last triple integration.

We thus find $N(b, u_n, v_n, w_n)$

$$= 2^{\frac{3n-11}{2}} C \frac{\{\Gamma(\frac{1}{2})\}^{3n-9}}{\Gamma\left(\frac{3n-9}{2}\right)} (m_1 \ldots m_{n-1})^{-\frac{3}{2}} M_{n-1}^{-\frac{3}{2}} D_{n-1}^{-\frac{1}{2}} I_{n-1}^{\frac{3n-11}{2}}\, du_n dv_n dw_n \text{...(99).}$$

If, as in equations (82) to (86), we suppose the origin of co-ordinates to be the centre of mass of the whole system, the axis of z to pass through the particle m_n, and the axes of x and y to be in the directions of the principal axes of the section of the momental ellipsoid normal to z, then writing

$$\xi = u - qz, \quad \eta = v + pz, \quad \zeta = w \text{..........................(100)},$$

so that ξ, η, ζ are the velocity-components of m_n relative to axes moving as the system would do if it were then to become rigid, with the angular velocity whose components are p, q, r, we may write

$$I_{n-1} = I_n - \tfrac{1}{2}\mu\,(1 - b\mu z^2)^{-1}\xi^2 - \tfrac{1}{2}\mu\,(1 - a\mu z^2)^{-1}\eta^2 - \tfrac{1}{2}\mu z^2 \text{.........(101)}.$$

The sum of the last three terms of this expression, with its sign taken positive, represents the part of the internal motion of the system which is due to the fact that the particle m_n is moving with the relative velocity whose components are ξ, η, ζ.

We may also define it as the work which would be done by the particle m_n against the internal forces of the system, if these forces were suddenly to become such as to render the whole system rigid in an infinitely short time.

Comparing this result with that obtained in equation (48), we see that the law of distribution of the velocities of the particle m_n is the same as what it would be in a fixed vessel containing $n - 2$ particles, provided that we substitute for u^2, v^2, w^2 the quantities $(1 - b\mu z^2)^{-1}\xi^2$, $(1 - a\mu z^2)^{-1}\eta^2$, ζ^2 respectively.

Hence the mean square of the velocity in the direction of the line joining the particle with the centre of mass is the same at all points of the system, but the mean square of the velocity in other directions is less than this in the ratio of $1 - a\mu z^2$ to 1, where z is the perpendicular from the centre of mass on the line of relative motion of the particle, and a is the moment of mobility of the system about an axis through the centre of mass and normal to the plane through that centre and the line of motion.

When the product of the mass of the particle into the square of its distance from the centre is so small that it may be neglected in comparison with the moments of inertia of the system, then quantities like $a\mu z^2$ and $b\mu z^2$ may be neglected in respect of unity, and we may assert that the mean square of the relative velocity, for a particle of given mass, is the same in all directions and at all points of the system; but that for different particles it varies inversely as their masses; so that the average energy of motion relative to the moving axes is the same for particles of all kinds throughout the system.

We have already learned from equation (98) that in a free system of n particles the number of cases in which the system is in a given configuration, or, in other words, the probability of that configuration, is proportional to the $\frac{3n-8}{2}$ power of the energy of internal motion corresponding to that configuration.

We have next to consider the manner in which this probability depends on the position of a particular particle, say of the last particle, m_n.

Let $I_n^{(0)}$ denote the energy of internal motion of the complete system when m_n is at the centre of mass of the system and is without any velocity relative to that centre. It is manifest that in this case m_n contributes nothing towards the energy of internal motion.

Now let m_n be carried from the centre of mass to the point $(0, 0, z)$ and left there without any velocity (that is, let $u=v=w=0$).

Let W be the work which must be done against the forces of the system to effect this transference, then since the total energy of the system and the three angular momenta must be maintained constant, we shall have after this displacement, for the energy of internal motion of the remaining $n-1$ particles,

$$I_{n-1}=I_n^{(0)}-W \quad\text{..............................(102).}$$

But by equation (85)

$$I_n=I_{n-1}+\tfrac{1}{2}\mu\,(1-b\mu z^2)^{-1}(u-qz)^2+\tfrac{1}{2}\mu\,(1-a\mu z^2)^{-1}(v+pz)^2+\tfrac{1}{2}\mu w^2\text{...(103).}$$

Substituting the value of I_{n-1} from equation (102), and remembering that $u=v=w=0$, we find for the energy of internal motion in the new configuration

$$I_n=I_n^{(0)}-W+\tfrac{1}{2}\mu\,(1-b\mu z^2)^{-1}q^2z^2+\tfrac{1}{2}\mu\,(1-a\mu z^2)\,p^2z^2\text{...........(104).}$$

The probability, therefore, of a configuration in which, the positions of all the other particles being given, that of m_n is varied, is proportional to $I_n^{\frac{3n-8}{2}}$, I_n being given by equation (104).

When, as in the case of a gas, there are a great many particles similar to m_n, we may speak of the density of the medium consisting of such particles in the element $dxdydz$. In this case, however, for reasons already given, neglect the quantities $a\mu z^2$ and $b\mu \dot{z}^2$, and we may write m for μ. We may also choose our axes in the manner which is most convenient. We shall therefore make the axis of z that round which the system, if it were rendered rigid, would rotate with velocity ω, and we shall suppose this axis to be vertical, as otherwise a steady motion under the action of gravity could not exist, and we shall denote the horizontal distance from this axis by r.

We may now write for the density of the gas at the point (z, r)

$$\rho = \rho_0 \left[1 + (2I_n^{(0)})^{-1}(m\omega^2 r^2 - 2mgz)\right]^{\frac{3n-8}{2}} \text{ ....(105)},$$

where ρ_0 is the density at the origin.

When n is a very large number and when the second term of the binomial is very small compared with unity, we may write for this the exponential expression

$$\rho = \rho_0 e^{\frac{3}{4}\frac{mn}{T}(\omega^2 r^2 - 2gz)} \text{ ....(106)}.$$

If m_0 is the mass of a molecule of hydrogen, μm_0 will be the mass of a molecule of the kind of gas considered, where μ is the chemical equivalent of the gas.

Also if T is the temperature on the centigrade scale, and α the coefficient of dilatation of a perfect gas, then since the "velocity of mean square" of agitation of the molecules of hydrogen at 0°C. is $1{\cdot}844 \times 10^5$ centimetres per second, the kinetic energy of agitation of a system containing n molecules of any kind will be

$$\tfrac{3}{2} m_0 n (1{\cdot}844)^2 10^{10} (1 + \alpha T),$$

and the difference between this and the energy of internal motion may be neglected.

We thus find for the density at any point

$$\rho = \rho_0 e^{\frac{\mu}{2}\cdot\frac{\omega^2 r^2 - 2gz}{(1{\cdot}844)^2 10^{10}(1+\alpha T)}} \text{ ....(107)}.$$

Let us now consider a tube of uniform section placed on a whirling table so that one end, A, of the tube coincides with the axis while the other end, B, revolves about the axis with the angular velocity ω. The linear velocity of B is ωr, and we shall suppose, for the sake of easy calculation, that this velocity is one-tenth of the velocity of agitation of the molecules of hydrogen.

The velocity of the end B would be 184·4 metres per second. If the tube contains hydrogen at 0°C., the ratio of the density of the gas at B to the density at A will be $e^{\frac{1}{200}}$, or approximately $1+\frac{1}{200}$.

If it contains a gas whose chemical equivalent is μ, the ratio will be

$$1+\frac{\mu}{200}.$$

If the tube contains hydrogen and carbonic acid, and if a certain volume of the tube at A contains 200 parts of hydrogen and 200 of carbonic acid, then an equal volume of the tube at B will contain 201 parts of hydrogen and 222 parts of carbonic acid.

The time during which the experiment would require to be continued in order to obtain a given degree of approximation to the ultimate distribution of the mixed gases varies as the square of the length of the tube.

Thus in Loschmidt's experiments on the diffusion of gases he used a tube about a metre long, and continued his experiments from half an hour to an hour in order to obtain the results from which he could best deduce the coefficient of diffusion.

In these experiments the inequalities of distribution of hydrogen and carbonic acid were reduced to less than a third part of their original value in half an hour, and if the experiment had gone on for two hours the differences from the ultimate distribution would have been reduced to a hundredth part of their original value.

We may therefore consider two hours as ample time for an experiment on the ultimate distribution of these two gases in a tube one metre in length.

But if we make the whirling tube 20 centimetres long, the differences of distribution from the ultimate distribution would be reduced to a hundredth part of their original value in a twenty-fifth part of the time, that is to say in 4 minutes 48 seconds.

If it were found more convenient to have bulbs on the ends of the tubes, so as to be able to secure the gas at each end before it got mixed up by the violent commotion arising from the stopping of the whirling tube, we should have to allow a longer time for the whirling.

In order to obtain a similar distribution of the two gases in a vertical tube by the action of gravity the tube would require to be 1720 metres high, and in order to obtain the same degree of approximation to the ultimate distribution we should have to let the experiment go on for 675 years, carefully preserving the tube during that time from all inequalities of temperature, which, by causing convection-currents, would continually mix up the gases and prevent their partial separation.

VI Documents: Radiometer & Rarified Gas Dynamics

1. Letter from Maxwell to William Huggins, October 13, 1868[a]

Lewis Campbell and William Garnett, *The Life of James Clerk Maxwell* (London, 1884): 260–261.[b]

Ardhallow, Dunoon, Oct. 13/68.

MY DEAR SIR—I sympathise with you in your great sorrow. Though my own mother was only eight years with me, and my father became my companion in all things, I felt her loss for many years, and can in some degree appreciate your happiness in having so long and so complete fellowship with your mother. I have little fear, however, that the nearness to the other world which you must feel will in any way unfit you for the work on which you have been engaged, for the higher powers of the intellect are strengthened by the exercise of the nobler emotions.

. . . . . .

Your identification of the spectrum of comet II with that of carbon is very wonderful.[c] The dynamical state of comets' tails is most perplexing, but the chemistry and activity of their heads leads to new questions. With respect to the transparency of a heavenly body, I think it indicates scattered condition rather than gasity. A cloud of large blocks of stone is much more transparent than air of the same average density. Such blocks in a nebula would never be themselves seen, but perhaps if they were often to encounter each other, the results of the collision would be incandescent gases, and might be the only visible part of the nebula.

... Any opinion as to the form in which the energy of gravitation exists in space is of great importance, and whoever can make his opinion probable will have made an enormous stride in physical speculation. The apparent universality of gravitation, and the equality of its effects on matter of all kinds are most remarkable facts, hitherto without exception; but they are purely experimental facts, liable to be corrected by a single observed exception. We cannot conceive of matter with negative inertia or mass; but we see no way of *accounting* for the proportionality of gravitation to mass by any legitimate method of demonstration. If we can see the tails of comets fly off in the direction opposed to the sun with an accelerated velocity, and if we believe these tails to be matter and not optical illusions or mere tracks of vibrating disturbance, then we must admit a force in that direction, and we may establish that it is caused by the sun if it always depends upon his position and distance. I therefore admit that the proposition that the sun repels comets' tails is capable of proof; but whether he does so by his ordinary attractive power being changed into repulsion by a change of state of the matter of the tail is another question. Now, it seems ascertained by simple observations with telescopes that the coma is formed by successive explosions out of the nucleus, mostly on the side of the sun, and that the formation of the tail depends on the coma, though the substance is invisible in the state of passing from the coma to the tail. Then, by your observations, the nucleus and coma have light of their own, probably due to carbon in some gaseous form; but the tail's light being polarised in the plane of the sun is due to him. Hence the head is fire and the tail smoke. The head obeys gravitation, which is exerted on it with precisely the same intensity as on all other known matter, solid or gaseous. The tail appears to be acted on in a contrary way. If the comet consisted of a mixture of gravitating and levitating matter, and is analysed by the sun, then before the emission of the tail the acceleration due to gravitation should be less than on a planet at the same distances; the more complete the discharge of tail the greater the intensity of gravitation on the remaining head.

N.B.—To understand the dynamics of the tail, the motion in space of particular portions of it must be studied.

a. This letter is also reprinted in Campbell and Garnett, *Life*, with new preface and appendix and letters from the 1884 edition by Robert Kargon (New York: Johnson Reprint, 1969) xxiii–xxiv.

b. William Huggins (1824–1910) was an early astrophysicist whose observations on the spectra of stars, nebulae and comets were published from the 1860s until his death. According to H. Dingle, "William Huggins," *Dictionary of Scientific Biography*, vol. 6, 542, Huggins, after having "consulted with Clerk Maxwell on the theory of matter" made the first estimates of the radial velocities of stars from the Doppler shifts of their spectral lines.

c. His observation on the comet mentioned were published as, Huggins,"On the Spectrum of Comet II," *Proc. R. Soc. London* 16 (1868): 393–394, that followed a short note, "Note on the Spectrum of Comet II," *Monthly Not. Astron. Soc.* 27 (1867): 288.

2. Letter from Maxwell to Peter Guthrie Tait, 1873[a]

Cambridge University, Tait Letters.

O T[b]

I have received your acceptance of Addle[c] and I and others are delighted. The Board meets on Friday, not exactly to nominate you, but to announce our nomination to the V. C.[d]

Election by Senate on May 7.

To read $\theta\Delta$cs at one gulp is impossible. I am all my time either at the Lab or designing plans therefor.[e]

Experimental Class on Magnetic measurements began on Wednesday. Elementary practical class on geometrical optics begins on Monday.

Measurements of electric capacity to follow.

$$\frac{dp}{dt}$$

I saw Crookes experiments at the R. S. Soi ... Ree. They whip spirits all to pieces.[f] A candle at 3 inches acts on a pith disk as promptly as a magnet does on a compass needle. No time for air currents and the force far greater than the weight of all the air left in the vessel. Attraction by a bit of ice very lively. All this at the best attainable vacuum. At 12 mm pressure—no effects, at 760 mm pressure reverse effects but not nearly as prompt or even so strong as in a vacuum.

On Thursday I saw conductivity of Selenium as affected by light.[g] It is most sudden. Effect of a copper heater insensible. That of the sun great.

a. This letter is dated by its reference to William Crookes' experiments with the radiometer, first exhibited at the Royal Society in 1873.

b. O T was Maxwell's usual salutation to William Thomson, or, Peter Guthrie Tait. We take this as addressed to Tait from the contents of the letter, the reference to $\theta\Delta$cs [thermodynamics] and the details of his teaching. He shared these interests with Tait elsewhere, but not with Thomson.

c. Additional Examiner at Cambridge University, sometimes written as Addles-eggs-aminer.

d. Vice Chancellor.

e. Maxwell was planning and supervising the construction of the new Cavendish Laboratory at Cambridge during 1872 and 1873.

f. An allusion to Crookes' interest in mediums and psychic phenomena in the years just before he took up the radiometer. See W. H. Brock, "William Crookes," *Dictionary of Scientific Biography,* ed. Charles C. Gillispie (New York: Scribners, 1971) vol. 3, 474–482.

g. Photoconductivity was discovered in 1873 by Willoughby Smith through experiments on selenium. See Smith, "Effect of Light on Selenium during the passage of an electric Current," *Am. J. Sci.* 5 (1873); *Nature* 7 (1873): 303, 361. Maxwell is referring to experiments done at the same time by Lieutenant Sale, also demonstrated before the Royal Society. See Sale, "The Action of Light on the Electrical Resistance of Selenium," *Proc. R. Soc. London* 21 (1873): 283–285. There appears to have been some dispute over the priority of this discovery.

3. "Report on Mr. Crookes' Paper on the Action of Heat on Gravitating Masses," 24 February, 1874[a]

Royal Society of London Archives, no. 295.

In this paper the author describes a series of researches which seem to have conducted him to a remarkable discovery. This discovery, if it really <represents> indicates a repulsion due to the radiation of heat, is of such transcending scientific value, that the record of the steps by which the discoverer was led to it becomes worthy of a permanent place in the history of science. If, on the other hand, the phenomena observed are due, not to the cause assigned by the author, but to some peculiarity of his instrumental method, the nature of which he has not discovered, then the paper must take its place beside the many other laborious and delicate investigations which have not resulted in any gain to science. My own impression from reading the paper, is that the author has made a great discovery with respect to the mode in which a body placed in air of different densities is acted on when bodies of different temperature are placed near it.

In a space exhausted as much as is possible the body is repelled from bodies warmer than the average temperature, and attracted towards bodies colder than the average temperature of surrounding bodies.

In air <highly rarified> of greater density, but still highly rarified, the reverse is the case; hot bodies attract and cold bodies repel the suspended body. This action, however, does not seem to be so prompt and decided as the opposite action in a more perfect vacuum.

Most of the recorded experiments were made with a balance arrangement moving in a vertical plane. In such an arrangement the expansion or distortion of the beam by heat might introduce an unknown force arising from the displacement of the centre of gravity. It is therefore satisfactory to learn from Art. 69 that the author has examined cases in which the motion is in a horizontal plane, and has made use of the torsion-balance suspension, and obtained results of the same kind as with the balance. I think the result of an experiment with this torsion balance ought to be explicitly stated in the paper.

I am of [the] opinion, therefore, that if any doubt has hitherto been thrown, by those who have witnessed the experiments, upon the constancy and regularity of the results, the paper which describes the discovery is an important scientific document, and ought to be printed in the Philosophical Transactions.

The Title of the paper appears to indicate rather what the author intended at first to investigate than what he actually discovered. He seems at first to have examined the question whether the gravitation of bodies depends in any way on their temperature. (John Herapath, in his Mathematical Physics, Vol. II, supposes that the attraction of the gravitation increases as the temperature of either of the bodies increases.)[b] But as the investigation goes on the results obtained are seen to depend neither on the gravitating mass of the movable body, nor on its "heat" or temperature, but on the sectional area which it exposes to the radiation of neighbouring hot bodies. The author tells us that it is a surface action and does not depend on the interior mass, and also that it is an instantaneous action, which does not wait till the body has changed its temperature. Above all, it is a *directed* action, being a repulsion from a heat-emitting body.

If, as I suppose, this repulsion is due to radiation, it ought to depend, not on the excess of temperature of the repelling body over that of the moveable body, but on the excess of the sum of the radiations exchanged between the hot and the moveable body over the sum of the radiations between the moveable body and that which lies on the opposite side of it.

In my treatise on Electricity, Arts. 792 & 793,[c] I have pointed out a probable repulsive action of radiation. But the effects observed by Mr Crookes seem to indicate forces of much larger value. Mr Crookes seems to have taken the proper precautions against electric action, but a coil or net of wire *inside* the exhausted tube would be a perfect safeguard against electrification, without impeding vision or radiation. Finally, the opposite effects observed in denser air appear to me to be, if they are regular, the most inexplicable of all the phenomena.

J. Clerk Maxwell
24 Feb. 1874

a. William Crookes (1832–1919) had been a student of A. W. Hoffmann at the Royal College of Chemistry in London (later evolving into Imperial College). He discovered Thallium. His work on the radiometer effect followed from experiments in which he tried to improves the accuracy of the chemical balance by placing it in an evacuated chamber. His later physical researches included some notable experiments on electrical discharges in gases and radioactivity. The paper Maxwell is here refereeing was read before the Royal Society on December 11, 1873. An abstract appeared in *Proc. R. Soc. London* 22 (1873):37–41 and published in full as William Crookes, "On Attraction and Repulsion resulting from Radiation,"

Phil. Trans. R. Soc. London 164 (1874): 501–527.

b. John Herapath, *Mathematical Physics* (London: Whittaker and Co., 1847), vol. 2; reprinted in *Mathematical Physics and Selected Papers by John Herapath*, ed. Stephen G. Brush (New York: Johnson Reprint, 1972).

c. Maxwell,*A Treatise on Electricity and Magnetism* (New York: Dover reprint of 3rd edition 1954) vol. 2, 440–441; the section is headed "Energy & States of Radiation."

4. "Report on Mr. Crookes paper On the Attraction and Repulsion resulting from Radiation," [1875][a]

Royal Society of London Archives, no. 370.

This is a continuation of Mr Crookes' former paper on the same subject and he promises additional matter when he had made more progress in his researches.

Any doubts which might remain in the mind after reading the former paper as to the reality of the results in a vacuum may be considered as completely dissipated by the experiments described in this paper. The apparatus now described is better designed than that in the former paper, and indeed is quite capable of being made to furnish numerical results in absolute measure, such as <indeed> Mr Crookes himself seems to contemplate giving in a future paper.

The paper therefore is an important one in the present state of science, and deserves immediate publication. The question of its permanent value as a contribution to science depends a good deal on what is to follow. The history of the original discovery is valuable, however imperfect the methods used in the final experiments may have been. <The details of the method by which> If the original discoverer, or any one else, completes the work by making numerical determinations, the details of the method by which the best results were obtained is also of great value. The intermediate steps, in which various forms of apparatus are devised, has less scientific value in the future than it does in the present.

Hence there are reasons for publishing a paper like the one before us in the Proceedings, where it would appear earlier than in the Transactions. At the same time I regard the whole research as so important, and the probability of the author obtaining still more interesting results as so great, that if the author prefers to have it in the Transactions as a continuation of his former paper, I should think it a very suitable contribution to the Transactions.

It is perhaps premature to say anything about the evidence furnished in this paper as to the cause of the phenomena in a vacuum. As for those in air and gas of various densities, which are of a far more complicated kind, I shall not attempt to speak of them.

But one or two points may be noted

At p. 29 it appears that the effect of a uniform radiation on a surface of pith is to produce an effect which increases continuously and does not reach its maximum for more than 150 seconds. This seems to point to the rise of temperature of pith as an important element in the experiment, as Mr Crookes points out at p. 27.

Another observation of Mr Crookes is contained in the supplement to the abstract of this paper, namely, that a black surface is more energetically repelled than a white one. Now if the cause of the repulsion were radiation alone, there would be greater repulsion of the white surface where there is a reflected radiation than at the black where there is none. But if the result depends on the heating of the surface, we know that the black surface is heated by luminous rays more than the white one.

If, according to Prof. Reynold's theory, the force is due to evaporation, it is plain that the greatest force which the vapour can exert on a surface is the pressure corresponding to the maximum density at the temperature of that surface. This pressure will not attain its maximum till there is complete thermal and mechanical equilibrium. The recent experiments of Kundt and Warburg show that even in a good vacuum the viscosity of the remaining medium is nearly the same as at atmospheric pressure, the difference mainly arising from the different quality of the vapours which form the residium.[b] Hence any currents formed in a rare gas will subside very rapidly, and any phenomenon which would be disturbed by such currents will take place more regularly.

Mr Crookes at p. 33 states his belief that the repulsion due to radiation always exists, and is only masked by the phenomena observed in denser air. But the numbers which he gives at pp. 16, 17 shew that the repulsion in a good vacuum is far more powerful than any other forces observed, and when the admission of a little air has once destroyed this effect the variations of the phenomena when more air is admitted are very trifling. It seems therefore that the admission of a little air *destroys* the vacuum effect. The experiments of Kundt and Warburg (Berlin Sitzungsberichte 1875) show that the properties of the most perfect vacuum are very different from those of a good Sprengel vacuum.[c]

One test is that the rate of cooling of a thermometer bulb surrounded by a thin shell of vacuum is at least twice as slow for the best vacuum as for an ordinary one. This test, combined with the non-discharge of the electrical spark and Mr Crookes own results show that the research into the properties of a real vacuum is likely to yield valuable results. Finally I would hope that Mr Crookes, who determines with such care the period of vibration of his torsion beams, would also determine the moment of inertia of the beam, or what would do as well, give us the *weight* of this rod of glass, two inches long, which he uses to test his glass fibres (p. 23). The forces of repulsion would then be expressed in absolute measure.

a. The paper Maxwell is here refereeing was read before the Royal Society on April 22, 1875, (received March 20th, supplement added April 20th 1875). An abstract was published in *Proc. R. Soc. London* 23 (1875): 373–378 and published in full as William Crookes, "On the Repulsion resulting from Radiation Part II," *Phil. Trans. R. Soc. London* 165 (1875): 510–547.

b. August Kundt and Emil Warburg,"Ueber Reibung und Wärmeleitung verdünnter Gase," *Monats. Preuss. Akad. Berlin* (1875): 160–173; *Phil. Mag.* 50 (1875): 53–62; *Ann. Phy.* 155 (1875): 337-366, 525–555; 156 (1875): 177–211.

c. This is the same article as footnote b.

5. [Report on Crookes paper on Repulsion resulting from Radiation, 1876][a]

Memoir and Scientific Correspondence of the late George Gabriel Stokes, ed. Joseph Larmor, (Cambridge: At the University Press, 1909) vol. 2, 433-436.

The experiments recorded in these papers relate to bodies placed in a vessel as completely exhausted as possible. These phenomena are more regular than those observed by Mr Crookes in less highly rarified air, and the forces are more powerful.

Mr Crookes has now succeeded in making the phenomena so regular that they can be subjected to the test of measurement and comparison. He has therefore made several different kinds of instruments for the measurement of repulsion force.

Sect. (128). A differential radiometer, consisting of two pith disks, one white and one black, at the ends of a horizontal rod suspended by a silk fibre.

(135) The same with a bar of pith, one end white and one end black, with magnet and mirror, suspended by a silk fibre and controlled by a magnet.

In both these instruments light is incident on both surfaces, but the black surface is repelled more than the white.

In the first instrument there is equilibrium when the stronger specific action of the black surface is counteracted by its greater distance from the source of radiation and the greater obliquity of the rays. This can take place only when the source is very near the instrument.

In the second, the directive force of the magnet tends to bring the instrument back to zero, and enables it to be used when the source is at a considerable distance, so that the rays may be considered as parallel.

(143) The rotating radiometer, or "Light-Mill." This is also a differential instrument, the sides of the vanes being alternately black and white, so that the effect of radiation is to drive the mill round. The final velocity of rotation depends on the friction of the pivot, and whatever other resistances may act on the exhausted chamber.

It appears from the experiments that the velocity is inversely as the square of the distance of the source; and as it has been already shown that this is the law of the pressure on a surface at rest, it follows either that the resistance to the motion of the "light-mill" is proportional to its velocity, or that the pressure resulting from radiation is different for surfaces at rest and in motion.

The most important experiments with the light-mill are those which show the effect of the change of temperature of the instrument.

When the instrument is cold, and luminous radiation falls on it, each vane goes with its white surface foremost. This Mr Crookes calls the normal or positive direction.

Under other circumstances the black surface goes first. This may be called the negative rotation.

(166) An aluminium light-mill, while being heated, went	+
(167) Afterwards, while cooling, it went	-
When the bulb was cooled by evaporation of ether	+
When the bulb was surrounded by a heated glass shade	-
When the bulb was surrounded by hot air from Bunsen's burner	-
When the bulb was heated by experimenter's body	-

(At (165) the effect of the hot glass shade or the brass light-mill seemed reversed, but the statement is not quite clear.)

These results lead me to think, what indeed Mr Crookes at Section (195) seems to hint at, namely, that the effect depends on the difference of temperature between the white and the black surface.

If we admitted the corpuscular theory of light, we might suppose that the force was due to the impact of the corpuscles. This force would be greater at the white surface, from which the corpuscles recoil, than at the black surface, where they are merely stopped. According to the electromagnetic theory of light there is a pressure in the line along which light is propagated. This would be greater if (as near a reflecting surface) light is propagated both ways, than near a black surface, where there is no reflected light. Hence neither of these hypotheses explains the phenomenon of the black surface being most repelled.

But if we suppose that the force depends on the temperature of the surface (that is, not on radiation, but on "thermometric heat"), I think we can explain most of the facts.

For a black surface is known to absorb and to radiate more than a white one, even when the radiation is not luminous.

Hence when the radiometer is exposed to radiation from sources of high temperature, the black surfaces absorb more, become warmer, and are "repelled."

When the radiometer is hot the black surfaces radiate most, and become cooler than the white ones, and the motion is negative. When a hot glass shade is put over the instrument, or when heated air rises round it, the heating effect acts on both white and black surfaces, but probably most on black; but the radiation of the black surfaces through the transparent hot body cools them more than the white ones, and the motion is negative.

Cooling the instrument by ether renders it less able to radiate, and the black surfaces are now most warmed by radiation from without, so that the rotation is positive.*[b]

Mr Crookes's view, as expressed in (195), agrees with this up to a certain point. He admits that in normal rotation the black surface is hotter than the white, but he attributes the motion to the *radiation from* the black surface, owing to its higher temperature. Now if the same radiation falls on a white and a black surface, the total radiation from the white surface will be greater than that from the black, because some of the radiation is reflected directly by the white surface, whereas what falls on the black is absorbed, and the radiation of the black is only that due to its own temperature.

It is not, therefore, because the black surface radiates more, but because it is hotter, that it is more repelled.

With respect to the cause of this repulsion Mr Crookes (130) has shown clearly that it is not evaporation of water, as in Prof. Reynolds's *illustrative* experiment (Proc. R. S.). Indeed Prof. Reynolds himself shows that at the low pressure obtained by the mercury pump, and at ordinary temperatures, liquid water cannot permanently exist.[c]

But the experiments of Kundt and Warburg (Pogg., 1874–5) show that in the best exhausted vessel there is still enough matter to offer considerable resistance to motion and to conduct heat.[d] (They seem, however, to have succeeded in exhausting a vessel of small capacity till the passage of heat was not affected by the kind of gas with which it was originally filled.)

Hence any inequality in the pressure, arising from evaporation of any other cause, may continue to exist in an exhausted vessel as long as it would in a less rarified gas.

But whatever may be the cause of the phenomenon, the determination of its magnitude, as compared with the radiation which occasions it, is the greatest value as a means of testing the various hypotheses. Mr Crookes has made great progress in the construction of measuring apparatus. He has employed three methods–the "Torsion Rod" of Michell and Coulomb,[e] the Horizontal Pendulum of Zollner,*[f] and the Torsion Balance of Ritchie.[g]

Of these the best adapted for quantitative measurement is the Torsion Rod proper, in which the force to be measure is balanced against the torsional elasticity of the suspension fibre. The instrument described at the beginning of Part IV is of this kind.

[Some calculation follows, showing that the data of this kind in the paper do not suffice to deduce the pressure on the disks.][h]

The horizontal pendulum, in which gravity is the principal force, is liable to great alterations of zero if the instrument is not rigidly fixed. Hence Mr Crookes abandoned it.

The torsion balance described at the end of Part IV is a common gravitation balance, in which the final adjustment is made by the torsion of a fibre instead of by a rider. The consistency of Mr Crookes's results is a proof of his wonderful skill in the construction of such apparatus.

He has ascertained that the pressure due to weak sunlight, on a certain day, was 0.009 grain per square inch. This is $\frac{1}{381000}$ of an inch of mercury, or 0.089 dynes per square centimetre.

This is an important measurement, but what would give it more value would be a simultaneous measurement of the heating power of the sun's light by an actinometer, the light being subjected to the same absorbing media in its path to the actinometer as in its path *to the pith.*

The experiments in which the instrument itself was brought to various temperatures are important, and should be extended.

But if the disks themselves could be constructed of two substances separated by a small interval, the temperatures of the two surfaces might be separately altered.

Thus if a pith disk with a hole in it had a copper disk at its back, the back of which is blackened,—then if the copper disk were heated by means of a burning glass, the rays passing through the hole in the pith, the copper would be hotter than the pith, and the whole might move towards the light; and other experiments might be devised with compound disks to determine whether temperature or radiation is the cause of repulsion.

These remarks are not intended to depreciate the value of Mr Crookes's papers, which I consider worthy to be printed in the *Transactions,* and likely to become important documents for the history of Science.

No explanation is given in the paper of the mode in which the Morse instrument was made to register the revolutions of the radiometer as shown in the drawings.

a. The paper Maxwell is here refereeing was published as William Crookes, "On Repulsion resulting from Radiation. Parts III and IV," both read before the Royal Society February 10, 1876. Abstracts were published in *Proc. R. Soc. London* 24 (1876): 276–279, 279–283 and the paper published in full in *Phil. Trans. R. Soc. London* 166 (1876): 325–376.

b. This refers to a footnote of Larmor's at the bottom of the page [*On these points, cf. Prof. Stokes' remarks, supra, pp. 377, 386 seq., and his paper reprinted in *Math. and Phys. Papers,* vol. v, pp. 24–35.]

c. Osborne Reynolds, "On the Force caused by the Communication of Heat between a Surface and a Gas and a New Photometer," *Proc. Roy. Soc. London* 24 (1876): 388–391, and *Phil. Mag.* 2 (1876): 231–233. The paper was published in full in *Phil. Trans. R. Soc. London* 166 (1877): 725–735. Maxwell, himself refereed this paper also, see, document VI-6.

d. Kundt and Warburg, "Ueber Reibung und Wärmeleitung verdünnter Gase," *Monats. Preuss. Akad. Berlin* (1875): 160–173x; *Phil. Mag.* 50 (1875): 53–62; *Ann. Phys.* 155 (1875): 337–366, 525–555, 156 (1875): 177–211.

e. There were many claimants to the "invention" of the torsion balance in the seventeenth and the eighteenth centuries, the last two being John Michell and Charles Coulomb. Maxwell will not decide the issue here. It is discussed in C. Stewart Gilmour, *Coulomb and the Evolution of Physics and Engineering in eighteenth century France* (Princeton: Princeton University Press, 1971).

f. The * refers to Maxwell's footnote at the bottom of the page "I think it was first used in America." Johann Karl Friedrich Zollner describes his horizontal pendulum in, Zollner, "Untersuchung über die Bewegungen Strahlender und bestrahlter Körper," *Ann. Phys.* 160 (1877): 154–169, 296–317, 459–466.

g. William Ritchie (died 1837) wrote a series of papers on various instruments. Those on the torsion balance include, Ritchie, "Description and Application of a Torsion Galvanometer," *J. R. Inst.* 1 (1821): 29–38; "Elastic Threads of Glass with Some of the Most Useful Applications of the Properties to Torsion Balances," *Phil. Trans. R. Soc. London* 120 (1830): 215–222.

h. This remark is Larmor's. We have been unable to recover the original of this document.

6. Report on Prof. Reynolds' paper "On the forces caused by the communication of heat between a surface & a gas; and on a new Photometer," 7 April, 1876[a]

Royal Society of London, Archives.

The experiments in this paper were made with Mr Crookes' Radiometer or Light Mill as constructed by Dr Geissler of Bonn.[b] The first argument as

to the nature of the forces concerned is drawn from the relation between the velocity of the mill and the amount of radiation which falls on it.
Since the velocity is nearly proportional to the number of candles, and since the force is also proportional to the number of candles, the resistance to the motion must be nearly as the velocity. Now the resistance arising from the friction of the pivot is nearly proportional of the velocity.
That arising from the viscosity of the medium is proportional to the velocity and independent of the density. That arising from the communication of motion to the medium is proportional to the square of the velocity and to the density. In a very rare medium, therefore, the resistance will increase at first slower than the velocity, but for very high velocities the term involving the square of the velocity will become sensible, and will make the resistance increase faster than the velocity.
But in the ordinary experiments with the radiometer the main part of the resistance is as the velocity, and this shows that it is mainly due to the viscosity of the medium.

Hence what remains in the vessel is capable of exerting a force of sensible magnitude.

The argument about the inequality of the action and reaction of a hot and a cold body is founded on statements in Mr Crookes first memoir which are not very clear. All the experiments show that a body moves in obedience to the radiations which fall on it from *all sides* and that the attraction towards a cold body arises from the radiation of the warmer bodies on the other side. The interpretation of Dr. Schuster's experiment is a more satisfactory argument.[c]
It is shown that the <force> stress which drives the light mill is a stress *internal* to the instrument, that it is inserted between the vanes of the mill and the glass vessel in which it is contained, and not between the vanes and the source of radiation. The actual difference of pressure on the black and white surfaces was calculated from the recoil deflexion to be .00000042 *lb* or 0.00294 grains weight per square inch.
This is less than Mr Crookes' value of 0.009 grains weight per square inch for the total pressure due to weak sunlight and this is to be expected as it is only the difference between black and white.
Prof. Reynolds next assumes that the pressure of the medium is equal to that of the vapour of mercury at $60° F$, and finds that the pressure due to the differential action of the black and white surfaces is only one thousandth of the whole pressure of the medium.
Can this be explained by the difference of temperature of the black and

white surfaces?

If the volume of a portion of gas is constant and if p be the pressure and θ the absolute < *value* > temperature

$$\frac{dp}{p} = \frac{d\theta}{\theta}$$

(Prof. Reynolds by writing 1/2 for 2, at p. 15 makes $\frac{dp}{p} = \frac{1}{4}\frac{d\theta}{\theta}$ at p. 15. This only makes it harder to explain the phenomenon.) If therefore the temperature of a film of air close to a surface is *suddenly* increased from θ to $\theta + d\theta$, the pressure will be *at first* $= p + \frac{p}{\theta} d\theta$.

But this excess of pressure will soon cause the air to move away from the surface and become rarified, and so reduced in pressure. But the force required to keep up a motion of given velocity in the air is for rare air, independent of its density. Whereas the rate at which a given supply of heat can increase the volume of the gas varies inversely as the density. Hence the rarer the gas the less will the pressure on the hot surface be relieved by the gas moving round to the cold surface. If there were no yielding of the gas the increment of pressure required to produce the observed force would be according to Prof. Reynolds 1.6640 P or, since he has made a numerical mistake, it would be 0.416 P.

To test whether the radiation of the actual source of light would produce this difference of temperature between a white and a black surface, Prof. Reynolds has constructed a differential air thermometer with one hemisphere of each bulb coloured, one white, the other black on the inside*,[d] so that the source of light may shine through the clear part of each bulb and strike the black and white surfaces <where they> in immediate contact with the air in the bulb.

In this way he estimates that the constant radiation of the source of light employed by Dr Schuster would cause a difference of $24°F$ in the temperatures of the black and the white bulbs. Of course, in the light mill there is a rapid equalization of temperature between the two faces of the vanes by conduction, owing to the thinness of the films of which they are made; but in spite of this, it is probable that the difference of temperature is sufficient to account for the difference of pressure. Prof. Reynolds suggests a light-mill with its vanes set like those of a windmill, and acted on by light incident parallel to the axis of radiation. The going of such a mill would not, I think, settle the question as against the direct effect of radiation, because the <reflected> radiation reflected from the vanes would (on the radiation theory) make the mill to go.

But the following experiment would I think be decisive against radiation. Let one of Mr Crookes' "torsion rods" carry at one end a film of mica or a very thin slice of rock salt, and let a black pith surface be *fixed* in the vessel behind it, inside the vacuum chamber. Then let a beam of light be passed through the moveable transparent disks, so as to fall on the black surface behind it. According to the radiation theory there is more radiation in front than behind the disk, so that it ought to be repelled by the incident light.

According to Reynolds' theory, there is more generation of heat behind it than in front, so that the disk ought to advance against the incident ray.

I think the latter result would follow.

To conclude, I consider that Prof. Reynolds' paper is well designed and clearly expressed, that his results are well established as far as they go, and that the paper deserves a place in the Transactions.

J. Clerk Maxwell

7th April 1876

a. The paper Maxwell is refereeing is, Osborne Reynolds, "On the Forces caused by the Communication of Heat between a Surface and a Gas; and on a New Photometer," read before the Royal Society March 24, 1876, abstracted in *Proc. R. Soc. London* 24 (1876): 388–391 and published in full in *Phil. Trans. R. Soc. London* 166 (1877): 725–735. Osborne Reynolds (1842–1912) was apprenticed to a mechanical engineer for three years before entering Cambridge where he graduated 7th wrangler in 1867, a year in which Maxwell was the examiner. His work on the radiometer was his first really important research. It contributed to his interest in dimensional methods which led, afterwards, to the "Reynolds number" in the study of turbulence.

b. Johann Heinrich Wilhelm Geissler (1815–1879) was an expert glass blower and at the University of Bonn he made standard thermometers of very high precision. He invented a high vacuum pump and the "Geissler tube" for electrical discharge experiments. His "light mill" was a precursor of Crookes's radiometer.

c. Arthur Schuster, "On the Nature of the Force producing the Motions of a Body Exposed to Rays of Heat and Light," *Proc. R. Soc. London* 24 (1876): 391–392; *Phil. Mag.* 2 (1876): 313–314, published in full in *Phil. Trans. R. Soc. London* 166 (1877): 715–724. Maxwell also refereed this paper, see document VI-7.

d. The * refers to Maxwell's footnote at the bottom of the page "I cannot remember whether any of Leslie's photometers were coloured on the *inside*. They were on the *outside*."

7. Report on Dr. Schuster's Paper "On the Nature of the force producing the motion of a body exposed to rays of heat and light," 1876[a]

Royal Society of London, Archives, no. 473

The experiments described in this paper are those referred to in Prof. Reynolds' paper on the same subject.

A "Light mill," in its exhausted glass chamber, was suspended by two fibres so as to be capable of rotatory vibrations about a vertical axis. The lower part of the vessel was dipped in oil, in order to damp the vibrations. The apparatus being at rest and in equilibrium, at zero, a beam of light was thrown on it, and the vessel was observed to be deflected in the opposite direction <from> to the motion of the light-mill. After several oscillations, when the motion of the mill was steady, <and> the vessel was again found to be in equilibrium and at rest at zero, showing that no external force was acting on the suspended system consisting of the mill and its envelope.

The actual pressure on the vanes of the mill was then deduced from the values of the first few elongations. This method is inferior to in accuracy to that which might be applied to Mr Crookes' observations, if he had given *all* the data required, but it was the only method available with the instrument as <it existed> possessed by Prof. Reynolds. As it might be worth while for the author to deduce the force from the equations of motion of the system, I will give them here—

Let I and i be the moments of inertia of the glass vessel and the light-mill.

x and y their angular displacements,

k the coefficient of damping between mill and glass,

K the coefficient of damping between glass and oil-vessel,

L moment of force due to light between mill and vessel

HY moment of force of restitution due to bifilar suspension

Equations of motion

$$i\ddot{x} + k(\dot{y} - \dot{x}) = L \tag{1}$$

$$I\ddot{y} + k(\dot{y} - \dot{x}) + K\dot{y} + H\dot{y} = -L \tag{2}$$

Since i is insensible compared to I, $\dot{y}$ will be insensible compared to $\dot{x}$, and we may solve the equations separately, thus

$$\dot{x} = \frac{L}{k}(1 - e^{-\frac{k}{i}t}) \tag{3}$$

which gives the velocity of the mill beginning at zero and tending towards $\frac{i}{k}$.

$$I\ddot{y} + (K + k)\dot{y} + Hy = -L\,e^{-\frac{k}{i}t} \tag{4}$$

The solution of this equation of motion of the vessel is

$$y = A\,e^{-\lambda t}\sin(nt + \alpha) + B\,e^{-\frac{k}{i}t} \tag{5}$$

where α and n can be deduced from direct observation of the oscillations not influenced by light, and the other quantities may be found by substitution to fulfill the following equations

$$I(\lambda^2 - n^2) - (K + k)\lambda + H = 0 \tag{6}$$

$$2I\lambda n - (K + k)n = 0 \tag{7}$$

$$B(I\frac{k^2}{i^2} - (K + k)\frac{h}{i} + M = L \tag{8}$$

Since when $t = 0$, $\dot{y} = 0$ and $y = 0$,

$$A \sin\alpha + B = 0 \tag{9}$$

$$A(\lambda \sin\alpha - n\cos\alpha) + B\frac{k}{i} = 0 \tag{10}$$

Of these quantities,
n and $\frac{k}{i}$ may be determined by special experiment.
H is known from the data of the bifilar suspension and the weight of the instrument.
I is determined from this and from n.
From this and equation 7 we get $K + k$.
A and B are found from 9 & 10 combined with the first and second elongations.
Finally substituting in 8 we get L, the quantity sought.
The forces observed in these experiments are far greater than the pressure which according to the electromagnetic (and also the corpuscular) theory of light exists in the direction of radiation. Besides, the true pressure of light would be greater on a reflecting or a white surface than an absorbing or a black surface, and this is the opposite of what is observed.
I think Dr Schuster's experiments so important that his paper should be printed in the Transactions. I think, however, that the discussion on p. 6 might be made clearer by avoiding the symbolical notation and speaking of the driving stress, external and internal, and the resisting stress, external and internal.

J. Clerk Maxwell [b]

a. The paper Maxwell is refereeing here is, Arthur Schuster, "On the Nature of the Force Producing the Motion of a Body Exposed to Rays of Heat and Light," which was read before the Royal Society March 23rd 1876, abstracted in *Proc. R. Soc. London* 24 (1876): 391–392, and published in full in *Phil. Trans. R. Soc. London* 166 (1877):715–724. Arthur Schuster (1851–1934) was of German-Jewish extraction and came to Manchester in 1870. He worked with Balfour Stewart and Osborne Reynolds, then joined Maxwell at the Cavendish Laboratory in 1877 where he worked on electrical standards. In 1881 he returned to Manchester as professor of Applied Mathematics, later professor of Physics. His subsequent researches included work on terrestrial magnetism, spectroscopy, ionized gases, radiation transfer, and a near miss at the discovery of the electron. Schuster, *Progress of Physics During Thirty-Three Years* (Cambridge: Cambridge University Press, 1911) and "The Clerk-Maxwell Period," in *A History of the Cavendish Laboratory, 1870–1910* (London: Longmans, 1910), 14–39 are of more than ordinary interest. In 1907 Schuster turned over his chair at Manchester to Ernest Rutherford.

b. In a second, very short report to Stokes on the revised version of this paper Maxwell wrote,

> Dear Stokes,
>
> I consider the part about stresses in the new edition p. 4 not so good as the original mode of statement. After p. 8 the new solution becomes the best again. There are also some emendations here and there on the new edition some of which are manifestly improvements and others are probably so being considered by the author.
>
> I use the word stress as short for action—and reaction. Schuster seems to use it as a new-fashioned word for force.
>
> Yours truly
> J. Clerk Maxwell

8. Letter from Maxwell to Robert Cay, 15 May 1876[a]

Peterhouse Library, Cambridge.

Dear Uncle Robert, I enclose receipt signed. I forgot to dispatch it as I was sent for to London, to be ready to explain to the Queen why Otto von Guericke devoted himself to the discovery of nothing, and to show her the two hemispheres in which he kept it, and the picture of the 16 horses who could not separate the hemispheres, and how after 200 years W. Crookes has come much nearer to nothing and has sealed it up in a glass globe for public inspection.[b] Her Majesty however let us off very easily and did not make much ado about nothing, as she had much heavy work cut out for her all the rest of the day ...

a. Robert Dundas Cay was Maxwell's paternal uncle. He was a Writer to the Signet (a member of the Scottish Bar). From 1844 until 1853 he held an appointment in Hong Kong before returning to Edinburgh after the death of his wife. He was the father of Charles Hope Cay, see Garber, Brush, and Everitt *Maxwell on Molecules and Gasses*, 350, n. a. For William Dyce Cay, see Brush, Everitt, and Garber *Maxwell on Saturn's Rings*, 39, 40, 58, a well-known civil engineer, and Elizabeth Dunn, see C. W. F. Everitt *James Clerk Maxwell* (New York: Scribners, 1975), 55. About twenty letters from Maxwell to R. D. Cay, mostly on business matters, are preserved at Peterhouse, Cambridge along with the photographs of Maxwell collected by Elizabeth Dunn.

b. This is a reference to the Loan Exhibition of Scientific Apparatus, held in London in 1876. Queen Victoria visited the exhibition where Maxwell represented "Molecular Physics" and explained the operation of various pieces of apparatus.

9. Report on Part V "Repulsion resulting from Radiation" by Mr Crookes, 23 January 1878[a]

Royal Society of London, Archives, no. 87.

In this long paper Mr Crookes describes several most extensive series of experiments. The labour, skill and ingenuity displayed in these experiments are marvellous and many of the individual results are valuable, but I think the improvement of the methods of making, measuring, and preserving a vacuum is likely to be still more valuable to science than any of the actual experiments. The title of the paper is continued from Mr Crookes' former researches, though he has long given up any theory of the direct action of radiation in moving his apparatus. The only danger of retaining the title is that it may tend to suggest enquiries likely to be fruitless.

To avoid this it is desirable to distinguish between the different kinds of action in the experiments.

It is convenient to use the word radiation to denote energy propagated after the manner of light, whether the wave-length is such as to excite the sensation of light or not.

It is better not to call radiation by the name of heat, though heat-rays may be spoken of as a particular kind of rays, the principal effect of which is to generate heat.

Radiations are of different kinds, and these kinds may be sifted from each other by glass, water &c screens.

When radiation is absorbed by a body heat is generated at the place where the absorption takes place, and the temperature of the body is raised. Heat is of one kind only, and temperature has only one meaning.

At the surface of a hot body heat is communicated to the gas in contact with it at a rate depending on superficial conductivity. The body also loses heat by radiation. Both these effects cool the body but the first only is the cause of the motion in Mr Crookes experiments.

Hence in the experiments with good conducting metal vanes p. 29 one side of which is covered with lamp black, the effect of the lamp black is to make the black surface hotter when the radiation falls on it, but when the radiation falls on the other side the black surface is cooler than if it had not been blackened, 1st because it radiates faster, 2nd because it does not conduct as well as metal.

In the last experiment the negative rotation of the vanes and of the disk at 1 mm pressure is very curious. If the principal action was that between the vanes and the disk, negative rotation would take place but it is difficult to see how this action could be greater than that between the vanes and the lower mica plate which is heated by the hot wire. On the whole I think the paper contains a valuable series of experiments and deserves a place in the Philosophical Transactions.

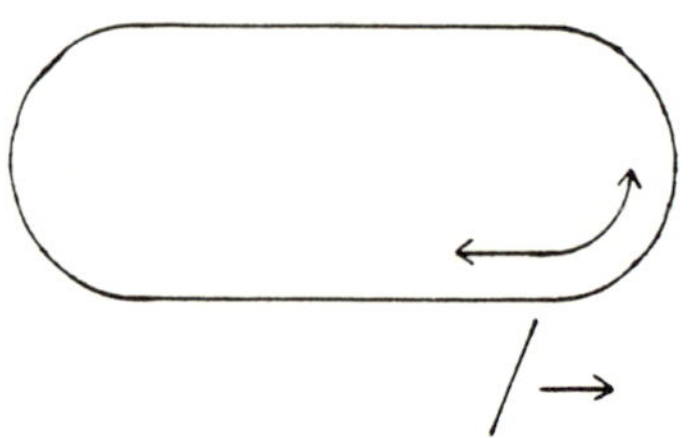

J. Clerk Maxwell

Cambridge, 23 Jan. 1878

a. The paper Maxwell is refereeing is William Crookes, "Repulsion Resulting from Radiation Part V," which was the Bakerian Lecture, read December 13, 1878 (it was received December 3rd 1877), abstracted in *Proc. R. Soc. London* 27 (1878): 29–33 and published in full in *Phil. Trans. R. Soc. London* 169 (1878): 243–313.

10. Letter from Maxwell to William Thomson, March 7, 1878

"Origins of Clerk Maxwell's Electrical Ideas, as Described in Familiar Letters to William Thomson," *Proc. Cambridge Phil. Soc.* 32 (1936): 695–750, 743–744.

Cavendish Laboratory
7 March 1878

Dear Thomson

Stress in gas of non-uniform temperature [in *Phil. Trans.* Memoir of same year 1878][a]

$$P = p - \frac{2}{3}\frac{R\theta}{C}\left(\frac{du}{dx} - \frac{1}{3}\left(\frac{du}{dx} + \frac{dv}{dy} + \frac{dw}{dz}\right)\right) + \frac{R^2\theta}{3C^2\rho}\left(\frac{d^2\theta}{dx^2} - \frac{1}{3}\left(\frac{d^2\theta}{dx^2} + \frac{d^2\theta}{dy^2} + \frac{d^2\theta}{dz^2}\right)\right)$$

Q & R similar

$$S = -\frac{R\theta}{3C}\left(\frac{dw}{dy} + \frac{dv}{dz}\right) + \frac{R^2\theta}{3c^2\rho}\frac{d^2\theta}{dz\,dy}$$

$p =$ hydrostatic pressure, $\frac{R\theta}{3C} = \mu =$ coefft of viscosity.

Hence $\frac{R^2\theta}{3C^2\rho} = 3\frac{\mu^2}{\rho\theta}$ (N.B. The approximations assume that p is large compared with the terms in the last set. When ρ is zero they do not really become ∞. At a millionth of an atmosphere $\frac{\mu^2}{\rho} = 25.9$ for air in CGS measure, so it is not small. Hence if there are two hot bodies in a cold vessel there will be a pressure between them because the $\frac{d^2\theta}{dx^2}$ between them is +. If they are colder than the vessel vice versa. Two infinite flat parallel plates would *not* repel each other whether hot, cold; or one hot & one cold. There would only be an effect near the edges [which has recently been formulated into a theory].

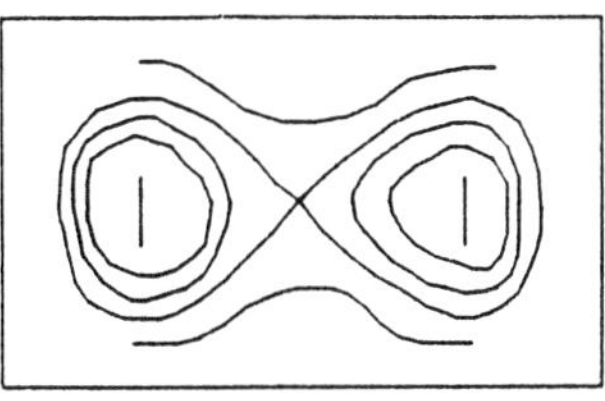

But the most remarkable thing is the equation of motion

$$\rho\frac{\partial u}{\partial t} + \frac{dp}{dx} - \mu\left(\frac{d^2u}{dx^2} + \frac{d^u}{dy^2} + \frac{d^2u}{dz^2}\right)$$
$$-\frac{1}{3}\mu\frac{d}{dx}\left(\frac{du}{dx} + \frac{dv}{dy} + \frac{dw}{dz}\right)$$
$$+2\frac{\mu^2}{\rho\theta}\frac{d}{dx}\left(\frac{d^2\theta}{dx^2} + \frac{d^2\theta}{dy^2} + \frac{d^2\theta}{dz^2}\right) = \rho X.$$

From the form of this equation it appears that the inequality of temperature will not produce any current (as indeed might be predicted if gravity were absent) but there will be a state of steady flow of heat without current, hydrostatic pressure uniform but yet stresses different in different directions.

Here then we have a kind of stress increasing as the density decreases (but not ∞ at zero) which does not produce currents but may act on solid bodies, and which is nonexistent when the temperature is a linear function of x, y, z. For similar systems it is inversely as the square of the dimensions. Hence small radiometers answer best. The isothermal surfaces for a hot cup are like the electric equipotentials of the same. Hence $\frac{d^2\theta}{dx^2}$ is greatest on the convex side and the cup is driven mouth foremost. [But see *Phil. Trans.]*

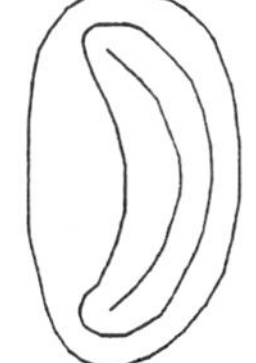

Yours very truly,
J. Clerk Maxwell

a. The notes in square brackets are Joseph Larmor's. In the first he refers to Maxwell, "Stresses in Rarified Gases Arising from Inequalities of Temperature," *Phil. Trans. R. Soc. London* 170 (1880): 231–256, document VI-31. While the paper did not appear until 1880 an abstract was published in 1878, document VI-11.

11. Maxwell, "On Stresses in Rarified Gases arising from Inequalities of Temperature"

Proc. R. Soc. London 27 (April 11, 1878): 304-308, received, 19 March, 1878

(Abstract.)

1. In this paper I have followed the method given in my paper "On the Dynamical Theory of Gases" (Phil. Trans., 1867, p. 49).[a] I have shown that when inequalities of temperature exist in a gas, the pressure at a given point is not the same in all directions, and that the difference between the maximum and the minimum pressure at a point may be of considerable magnitude when the density of the gas is small enough, and when the inequalities of temperature are produced by small solid bodies at a higher or lower temperature than the vessel containing the gas.

2. The nature of this stress may be thus defined: let the distance from the given point, measured in a given direction, be denoted by h, and the absolute temperature by θ; then the space-variation of the temperature for a point moving along this line will be denoted by $\frac{d\theta}{dh}$, and the space-variation of this quantity along the same line by $\frac{d^2\theta}{dh^2}$. There is in general a particular direction of the line h, for which $\frac{d^2\theta}{dh^2}$ is a maximum, another for which it is a minimum, and a third for which it is a maximum-minimum. These three directions are at right angles to each other and are the axes of principal stress at the given point; and the part of the stress arising from inequalities of temperature is in each of these principal axes a pressure equal to—

$$3\frac{\mu^2}{\rho\theta}\frac{d^2\theta}{dh^2}$$

where μ is the coefficient of viscosity, ρ the density, and θ the absolute temperature.

3. Now, for dry air at $15°C.$, $\mu = 1.9X10^{-4}$ in centimetre-gramme-second measure, and $\frac{3\mu^2}{\rho\theta} = \frac{1}{p}0.315$, where p is the pressure, the unit of pressure being one dyne per square centimetre, or nearly one-millionth part of an atmosphere.

If a sphere of one centimetre in diameter is T degrees centigrade hotter than the air at a distance from it, then, when the flow of heat has become steady, the temperature at a distance of r centimetres will be,

$$\theta = T_0 + \frac{T}{2r}, \qquad \text{and} \qquad \frac{d^2\theta}{dr^2} = \frac{T}{r^3}$$

Hence, at a distance of one centimetre from the centre of the sphere, the pressure in the direction of the radius arising from inequality of temperature will be

$$\frac{T}{p} = 0.315 \text{ dynes per square centimetre.}$$

4. In Mr. Crookes' experiments the pressure, p, was often so small that this stress would be capable, if it existed alone, of producing rapid motion in small masses.

Indeed, if we were to consider only the normal part of the stress exerted on solid bodies immersed in the gas, most of the phenomena observed by Mr. Crookes could be readily explained.

5. Let us take the case of two small bodies symmetrical with respect to the axis joining their centres of figure. If both bodies are warmer than the air at a distance from them, then in any section perpendicular to the axis joining their centres, the point where it cuts this line will have the highest temperature, and there will be a flow of heat outwards from this axis in all directions.

Hence $\frac{d^2\theta}{dh^2}$ will be positive for the axis, and it will be a line of maximum pressure, so that the bodies will repel each other.

If both bodies are colder than the air at a distance, everything will be reversed; the axis will be a line of minimum pressure, and the bodies will attract each other.

If one body is hotter, and the other colder, than the air at a distance, the effect will be smaller; and it will depend on the relative sizes of the bodies, and on their exact temperatures, whether the action is attractive or repulsive.

6. If the bodies are two parallel disks, very near to each other, the central parts will produce very little effect, because between the disks the temperature varies uniformly and $\frac{d^2\theta}{dh^2} = 0$. Only near the edges will there be any stress arising from inequality of temperature in the gas.

7. If the bodies are encircled by a ring having its axis in the line joining the bodies, then the repulsion between the two bodies, when they are warmer than the air in general, may be converted into attraction by heating the ring, so as to produce a flow of heat inwards towards the axis.

8. If a body in the form of a cup or bowl is warmer than the air, the distribution of temperature in the surrounding gas is similar to the distribution of electric potential near a body of the same form, which has been investigated by Sir W. Thomson.*[b] Near the convex surface the value of $\frac{d^2\theta}{dh^2}$ is nearly the same as if the body had been a complete sphere, namely $2T\frac{1}{\alpha^2}$, where T is the excess of temperature, and a is the radius of the sphere. Near the concave surface the variation of temperature is exceedingly small. Hence the normal pressure on the convex surface will be greater than on the concave surface, as Mr. Crookes has shown by the motion of his radiometers.

Since the expressions for the stress are linear as regards the temperature, everything will be reversed when the cup is colder than the surrounding air.

9. In a spherical vessel, if the two polar regions are made hotter than the equatorial zone, the pressure in the direction of the axis will be greater than that parallel to the equatorial plane, and the reverse will be the case if the polar regions are made colder than the equatorial zone.

10. All such explanations of the observed phenomena must be subjected to careful criticism. They have been obtained by considering the normal stresses alone, to the exclusion of the tangential stresses; and it is much easier to give an elementary exposition of the former than of the latter.

If, however, we go on to calculate the forces acting on any portion of the gas in virtue of the stresses on its surface, we find that when the flow of heat is steady, these forces are in equilibrium. Mr. Crookes tells us that there is no molar current, or wind, in his radiometer vessels. It may not be easy to prove this by experiment, but it is satisfactory to find that the system of stresses here described as arising from inequalities of temperature will not, when the flow of heat is steady, generate currents.

11. Consider, then, the case in which there are no currents of gas, but a steady flow of heat, the condition of which is

$$\frac{d^2\theta}{dx^2} + \frac{d^2\theta}{dy^2} + \frac{d^2\theta}{dz^2} (= -\nabla^2\theta) = 0.$$

(In the absence of external forces, such as gravity, and if the gas in contact with solid bodies does not slide over them, this is always a solution of the equations, and it is the only permanent solution.) In this case the equations of motion show that every particle of the gas is in equilibrium under the stresses acting on it.

Hence any finite portion of the gas is also in equilibrium; also, since the stresses are linear functions of the temperature, if we superpose one system of temperatures on another, we also superpose the corresponding systems of forces. Now the system of temperatures due to a solid sphere of uniform temperature immersed in the gas, cannot of itself give rise to any force tending to move the sphere in one direction rather than in another. Let the sphere be placed within the finite portion of gas which, as we have said, is already in equilibrium. The equilibrium will not be disturbed. We may introduce any number of spheres at different temperatures into the portion of gas, and when the flow of heat has become steady the whole system will be in equilibrium.

12. How, then, are we to account for the observed fact that forces act between solid bodies immersed in rarified gases, and this, apparently, as long as inequalities of temperature are maintained?

I think we must look for an explanation in the fact discovered in the case of liquids by Helmholtz and Piotrowski,*[c] and for gases by Kundt and Warburg,† [d] that the fluid in contact with the surface of a solid must slide over it with a finite velocity in order to produce a finite tangential stress.

The theoretical treatment of the boundary conditions between a gas and a solid is difficult, and it becomes more difficult if we consider that the gas close to the surface is probably in an unknown state of condensation. We shall, therefore, accept the results obtained by Kundt and Warburg on their experimental evidence.

They have found that the velocity of sliding of the gas over the surface due to a given tangential stress varies inversely as the pressure.

The coefficient of sliding for air on glass was found to be $\lambda = \frac{10}{p}$ centimetres, where p is the pressure in millionths of an atmosphere. Hence at ordinary pressures λ is insensible, but in the vessels exhausted by Mr. Crookes it may be considerable.

Hence if close to the surface of a solid there is a tangential stress, S, acting on a surface parallel to that of the body, in a direction, h, parallel to that surface, there will also be a sliding of the gas in contact with the solid over its surface in the direction h, with a finite velocity $= S\frac{\lambda}{\mu.}$

13. I have not attempted to enter on the calculation of the effect of this sliding motion, but it is easy to see that if we begin with the case in which there is no sliding, the effect of permission being given to the gas to slide must be in the first place to diminish the action of all tangential stresses on the surface without affecting the normal stresses; and, in the second place to set up currents sweeping over the surfaces of solid bodies, thus completely destroying the simplicity of our first solution of the problem.

14. When external forces, such as gravity, act on the gas, and when the thermal phenomena produce differences of density in different parts of the vessel, then the well-known convection currents are set up. These also interfere with the simplicity of the problem and introduce very complicated effects. All that we know is that the rarer the gas and the smaller the vessel, the less is the velocity of the convection currents; so that in Mr. Crookes' experiments they play a very small part.

a. Maxwell, "On the Dynamical Theory of Gases," *Phil. Trans. R. Soc. London* 157 (1867): 49–88; *Phi. Mag.* 35 (1868): 129–145, 185–217, reprinted in *Scientific Papers*, vol. 2, 26–78, and Garber, Brush, and Everitt, *Maxwell on Molecules and Gases*, 419–472.

b. The * is to a footnote at the bottom of the page "Reprint of Papers on Electrostatics, p. 178." This is William Thomson, *Reprint of Papers on Electrostatics and Magnetism* (London: Macmillan, 1873). The paper referred to is Thomson, "Determination of the Distribution of Electricity on a Circular Segment of Plane or Spherical Conducting Surface, under any given Influence," in

Papers on Electrostatics and Magnetism, 178–191.

c. The * refers to a footnote at the bottom of the page, 'Wiener Sitzb., xl (1860), p. 607." The paper referred to is H. von Helmholtz and G. von Piotrowski, "Ueber Reibung tropfbauer Flüssigkeiten," *Sitz. Math.-Naturwiss. Cl. Akad. Wiss. Wien* 40 (1860): 607–658.

d. † refers to the footnote "Pogg. Ann., clv (1875), p. 337." The paper referred to is August Kundt and Emil Warburg, " Ueber Reibung und Wärmeleitung verdünnte Gase," *Ann. Phy.* 155 (1875): 337–366, 525–550; 156 (1876): 177–211.

12. [Report of William Thomson on Maxwell, "Stresses in Rarified Gases arising from Inequalities in Temperature,"] June 1878[a]

Royal Society of London, Archives, no. 123.

June 15, 1878

Professor Clerk Maxwell's paper "On Stress in Rare[?i]fied Gases arising from Inequalities of Temperature" is an exceedingly valuable contribution to the kinetic theory of gases and is in my opinion suitable for publication in the Transactions of the Royal Society.

I <have> made several rough notes in pencil on the paper on first reading of it, and since that time have had an opportunity of talking over the subject with the author and calling his attention to them. So I need not (as I at first intended) enter on them at length in this report but merely say that[b]

(1) It seems desirable that a linear dimension definable from the gas itself <at> when of the particular <condition as to> density considered should be pointed out, as a standard with reference to which the solid bodies referred to in p. 1 shall be called "small" or not "small."

(2) The only lineal dimension definable from the gas itself at any density <included in> for which the investigation is applicable is the average <length> distance from one atom to <its> the one which is nearest <neighbour at any> to it at any instant; or something similarly dependent on the distances between <the> atoms, as for example the average length of path throughout which the radius of curvature of the path exceeds a stated large multiple of this length, or, more simple, the average length of path travelled by an atom before experiencing more than a stated small amount of change of direction. For the investigation seems not to take into account any properties dependent on the ratio of the sum of the volumes[c] of the atoms (or of their spheres of sensible influence) being other than infinitely small in proportion to the whole volume of the space in which they move; that is to say it supposes the average length of path within which there is sensible change of direction to be infinitely small in comparison with the whole path. Hence (as far as I can see) there is *no* lineal dimension involved in the investigation, than the average distance from a particle to its next

neighbour for the instant. But I do not quite see how *this* is involved in the formulas though no doubt if I had been able to *learn* them thoroughly from the paper I should have seen it. The author might be able to help the reader in this respect by some slight additional explanation.

(3) It is desirable that the author should somewhere give a Table of "dimensions" of the various numerical <values> quantities representing viscosity, thermal conductivity, diffusivity, and *slipperiness* (or "coefficient" (!) of sliding) which appear in the theory, as the author has done with excellent effect in many other cases (following the excellent example of Fourier, who, after Newton, seems to have been the first scientific writer who cared for such things).

(4) Referring to pencil note on p. 2, how will the law of proportion in similar systems stand when the "slipperiness" of the surfaces [of] solids is taken into account?

(5) It is desirable that in some way the fact that the radiometer action depends (as *first* I believe pointed out by Tait and Dewar) on the lineal dimensions of the moveable vanes being not infinitely great in comparison with the distance from atom to nearest neighbour, should be made apparent in the present investigation.

(5) Referring to pencil notes on p. 5 or on § 11 of printed abstract—the mention of "sphere" tends to put the reader on a wrong track. What follows (which is of enormous importance) will be more readily understood if the remark here is made for a solid of any shape instead of for the particular case of a sphere.

(6) § § 11, 12, 13 are of exceedingly great importance and interest. It is probable that Prof. Stokes may have known the principle before, but infinitely improbable that any one else did.[d] It is desirable that some slight additional information should be given to help the reader.

< *I* suppose> If the surface were like the sketch so that all collisions would be against the projecting gills and only an infinitely small proportion of them against the bottoms of the hollows (β in the sketch) <it> would it act as would the same space occupied by the gas? Suppose the surface of the solid to be covered with square pyramids or with simple harmonic hills ($z = h \sin mx \sin ny$) as in the second sketch how would it be if the vertical angle (θ) of each pyramid is infinitely small? (The same answer here of course as to the preceding question.) If $\theta = 180°$, and the surface of the solid is supposed to be frictionless (that is to say if we have simply a smooth frictionless solid) how will it be in this case? (I suppose in this, as in the other extreme, the resultant action on the solid will be equilibrant.) Will a very fine powdery deposit over the whole surface or over part of the surface, of a solid concerned in the radiometer influence its action on

account of this property of slipperiness?

In conclusion I can only say that I think the author should be invited to add anything he can or will to the paper, without delaying its publication, and to make further investigations in continuation.

William Thomson

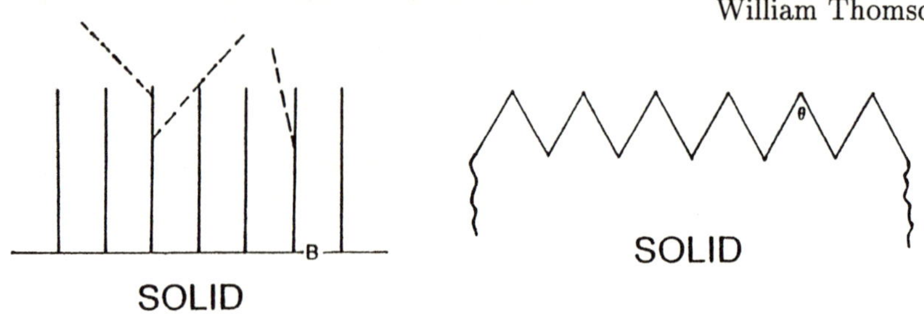

a. The paper under discussion is Maxwell, "On Stresses in Rarified gases arising from Inequalities of Temperature," *Phil. Trans. R. Soc. London* 170 (1880): 231–256. Abstract published in *Proc. R. Soc. London* 27 (1878): 304–308, and *Nature* 18 (1878): 54–55.

b. On 18 June 1878 Stokes wrote to Maxwell a letter which begins: "I enclose a long extract from the report of the first referee of your paper. I thought I might presume on his leave. I did not think it needful to wait till your paper had been brought before the committee of papers in order to have authority for the communication." The extract begins with the following paragraphs numbered 1, 2, etc. in the above document, and is identical with our transcription except for the omission of deleted words and some minor variations indicated below.

c. Stokes's extract originally had this word "values" but was corrected later; in the same sentence and elsewhere in his extract Stokes substituted dashes for parentheses. (The latter were apparently not available on his typewriter.)

d. The principle referred to is presumably that "the system of temperatures due to a solid sphere of uniform temperature immersed in gas, cannot of itself give rise to any force tending to move the sphere in one direction rather than another" unless one introduces the assumption that the gas may slide along the surface.

13. Letter from Peter Guthrie Tait to Maxwell, June 26, 1878

Cambridge University, Tait Letters.

38 George Square
Edinburgh
26/6/78

O $\frac{dp}{dt}$,

I got a *proof* of Macfarland[a] & sent it you this afternoon—but I was too busy to write.

1) As to Dewar[b] & Tait. Their joint paper, which I still believe to be not only the *first correct,* but the only *correct,* explanation of Crookes &c. was prevented from appearing before the public in proper guise by the efforts of those foes to knowledge the Secretaries R. S., to wit Stokes & Huxley! True, I assure thee it was as thus—I was busy with an eloge of $\sum\tau\omega\xi$ for Nature,[c] so Dewar undertook to give the paper, w[h] he did uncommonly well. Scott Lang took notes and made an article for "Nature," which appeared.[d] Dewar had *one* evening only to correct the press of this, & at 4 P. M. he got a note from Huxley (then *Survival's* locum tenens) saying that *that* evening would suit him to come & see Dewar & McKendrick on Electric Currents by light falling on frog's eyes.[e] So Lang's *undigested* notes went out to the world. Dewar was engaged drawing out his lecture (and mine) for Trans. R. S. E.[f] when $\sum\tau\omega\xi$ took it into his head to start him for the Jacksonian, and kept us all alive with telegrams. So no more has been done. Still, unsophisticated as is Scott Lang's version, and full of absolute absurdities,*it shows that the right thing was said.* Dixi.

As to Tennent all I know is that at the Council meeting a letter was read from him saying that his paper was in refutation of a remark in a letter from *you,* to *himself.* Whereupon someone asked, Does Maxwell allow the objection? Balfour[g] then said "Oh, I'll ask *him* T.[t] that." I suppose he has done so, and that this has brought Tennent on you. But *nothing,* not coming from an official of the Society, has an official authority. Dixi_2

Co is not brittle, and gives strong currents (getting *wildly* stronger by rise of temperature) with Cu. A supplementary ΘH diagram is one of the plates I hope to send you on Friday.[h]

R. U. Disposed to look at prooves? I like those pencillings of yours so much that I cannot have too many. But answer *straight;* I don't want to be indiscreet.

What is the intensity of plane sound waves coming through a hole in a wall—in terms of aperture?

Yrs.
G.[i]

a. Alexander Macfarlane, see document III-43, n. e.

b. James Dewar (1842–1923), inventor of the dewar vessel for research in low temperature physics. He never bothered to patent his invention and so made no money on it when it was marketed for holding coffee of tea. Dewar was assistant in chemistry at Edinburgh to Crum Brown while working with Tait. He was elected Jacksonian professor of natural philosophy at Cambridge in 1875; hence the reference to "start for the Jacksonian." Most of his subsequent research was done in the Royal Institution, London, where he held a chair from 1877.

c. P. G. Tait, "Scientific Worthies, V.–George Gabriel Stokes," *Nature* 12 (July 15, 1875): 201–203.

d. J. C. Scott Lang was Tait's assistant at Edinburgh and subsequently a professor at St. Andrews University.

e. Thomas Henry Huxley (1825-1895) was at this time a Secretary of the Royal Society. "*Survival's* locum tenens" refers to his role as public defender of Darwin's theory of evolution. His famous debate with Bishop Wilberforce took place at the 1860 Oxford meeting of the British Association, one of the meetings at which Maxwell presented some of his early results on kinetic theory. See Garber, Brush, and Everitt, *Maxwell on Molecules and Gases*, Document II-8. Despite their theological disagreements Tait and Huxley were personal friends, making common cause on the golf-course. For the experiment on frogs's eyes, see John Gray McKendrick and James Dewar, "On the Physiological Action of Light," *Proc. R. Inst.* 7 (1875): 360–367, 8 (1879): 137–149; *Nature* 15 (1877): 433–435, 453–454. McKendrick was later professor of physiology at Edinburgh.

f. Transactions of the Royal Society of Edinburgh.

g. Probably J. H. Balfour secretary of the Royal Society of Edinburgh.

h. This refers to Tait's work on thermoelectricity (ΘH) in particular on cobalt (Co) and copper (Cu).Θ denoting thermo-, and H capital eta for electricity.

i. The signature, an elaborate G, stands for Tait's pseudonym "Guthrie Headstone," where Head = Tête and Stone = Peter. Tait was Guthrie to his intimates, he strongly disliked Peter. A facsimile is reproduced in Knott, *Life of Tait,* facing p. 89.

14. "Report on Mr. W. Crookes paper 'On Repulsion resulting from Radiation, Part VI'," 23 October 1878[a]

Royal Society of London, Archives, no. 38.

In this paper, as in former papers with the same title, the first thing that we have to notice is the inexhaustible patience and skill with which such multitudes of delicate instruments have been prepared, every result having been verified and tested by a whole series of new forms of radiometer specially designed and constructed for the occasion.

Indeed this profusion in construction of apparatus tends in some degree to mask and obscure the actual path of research pursued by the author, in spite of the general lucidity with which every step of that path is described. The most important observations in the present paper are thus made by the help of a small fly with clear mica vanes which can be placed in different parts of the radiometer bulb to detect differences of pressure on its vanes. This is a most valuable instrument of research, <only> but it appears from this paper [?] that Mr Crookes has succeeded in constructing a fly like a common windmill with screw propeller turning on a horizontal axis. Such a fly would be a still better explorer of the field, because it indicates the pressure in the direction of its axis, whereas the fly used only measures differences of pressure at its right & left sides.

Nevertheless the exploration of the field near a hemicylinder of metal shows clearly the state of things near the edge of the metal, which may be compared to a wind sweeping along the surface *from* the edge like the water near the cutwater of a ship, and the opposite way from the motion of the electrical wind near the edge or point of an electrified body.

Now suppose the temperature of the hemicylinder to be uniform, and the air to be kept at rest, the isothermal surface would be something like the curves in the figure—nearly circular on the converse side, like those for a complete cylinder, <but> except near the edge where the nearer to the metal.

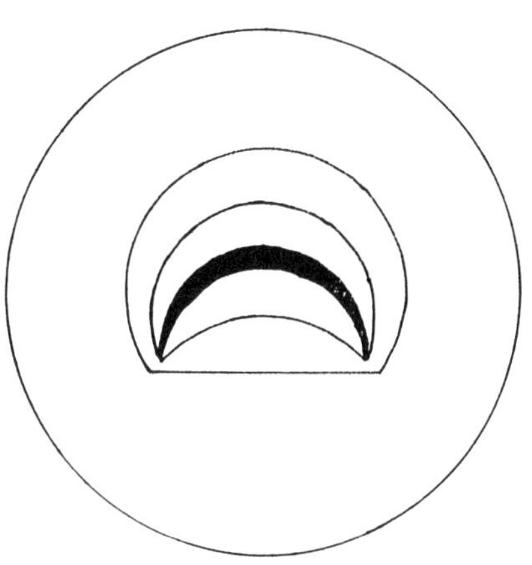

On the concave side the isothermals are very much spread out, showing a great uniformity of temperature except near the edge, where the variation is exceedingly rapid.

Now according to my theory of stress in a rarified gas the tangential force at a surface in the direction of s, a line drawn on the surface is $C\,\frac{d^2\theta}{ds\,dv}$ where θ is the temperature and dv is an element of the normal. This shows that if the <consective> isothermal surface consecutive to the surface of the solid is parallel to the surface of the solid, there is no tangential stress, but if not there is a tangential stress of the gas on the solid *from* places where the consecutive surfaces are close together *to* places where they are further apart.

At the surface of the solid the gas cannot support this stress, but must yield to it. This causes a current sweeping over the surface <just as if a superficial tangential stress stress equal to the differences between the stresses due to variation of temperature> The motion of the gas will extend itself to a finite distance from the surface, but still it will be a wind arising from a tangential surface action.

This wind will disturb the distribution of temperature, so that it would be difficult except by actual exploration to ascertain the final result.

It is satisfactory, however, to learn from Mr Crookes observations that the wind is distributed in a manner not very different from that arising <from the> at the instant when sliding is supposed to begin.

The wind blows along the surface from the edge on the curved side as well as the concave side, but it is much stronger near the edge on the concave side than on the converse side. Now the value of $\frac{d^2\theta}{ds\,dv}$ is greatest on the concave side of the edge and therefore the tangential force is greatest there. If we were to assume as an alternative hypothesis that the normal force is the primary force, and that this produces a wind which is fed by an indraught at the side, this side wind would be stronger on the convex side of the edge because the normal force is great<est>er on that side.

Hence Mr Crookes' result is in favour of the slipping of the gas over the surface near an edge being a primary part of the phenomenon. The experiments with moveable mica screens fastened to the fly show that the stress is between two surfaces, and that if both these surfaces are rigidly connected, the stress produces no tendency of the whole fly to move.

On the whole I consider that this paper should be printed in the Transactions. One reason is that it is stated to be the last of this series. If another paper describing numerous variations of form in radiometers were presented, I should not recommend it to be placed in the Transactions unless it contained some new principle not already established in the former papers.

At art. 435 there is a comparison of a certain radiometer to Carnot's reversible heat engine which I think might be left out with advantage to the paper.

In the radiometer, when the direction of the flow of heat through the bulb is reversed, the fly goes in the reverse direction. Still, however, it is the flow of heat which drives the fly, just as if the wind suddenly blows from the opposite point, a fixed windmill would turn in the opposite direction.

In Carnot's engine the engine is worked backwards by an external agent and this causes <the heat to pass from> heat to be abstracted from a cold body and a larger quantity of heat to be communicated to a hot body. Nothing of this kind has as yet been done with a radiometer.

J. Clerk Maxwell

Cambridge 23 Oct. 1878

a. The paper under discussion is, William Crookes, "On Repulsion Resulting from Radiation. Part VI," read before the Royal Society November 21, 1878, abstracted in *Proc. R. Soc. London* 28 (1878): 35, published in full in *Phil. Trans. R. Soc. London* 170 (1878): 87–134.

15. "Report on a paper by Prof. Osborne Reynolds 'On Certain Dimensional Properties of Matter in the Gaseous State'," 28 March, 1879[a]

Royal Society of London, Archives, no. 188.

---

(1) Report on the Experimental Part

This paper consists of two parts. In the first the author describes how he was led by certain theoretical considerations to devise and perform experiments of three distinct kinds, thereby obtaining results of great scientific value and throwing some light on the constitution of gases. In the second part the kinetic theory of gases is applied to the phenomena observed in the first.

As the difficulty of reporting on the second part is much greater than in the case of the first, I shall report on the first part separately.

In the first place the author has established the fact of what he calls Thermal Transportation, namely that when two portions of the same gas are separated by a porous plug, the two surfaces of which are at different temperatures, the condition of equilibrium is no longer that the pressures of the gas on the two sides of the plug should be equal, but that the pressure on the hotter side should exceed that on the colder side by a certain quantity, and that if this is not the case the gas will transpire through the plug till this condition is satisfied. This phenomenon, so far as I am aware, has neither been observed nor suspected by any one, and by itself it would be a discovery the account of which ought to be printed in the Transactions. But this discovery was suggested by a train of thought which connects it with Mr Crookes' experiments, and it was followed up by an extensive series of measurements which throw an entirely new light on some of the experiments of Graham which have hitherto seemed to yield no exact results to physical theory.[b]

The author has also devised an experiment of the kind discovered by Mr Crookes in which a silk fibre was acted on in a gas so dense that Mr. Crookes' force would have been insensible on a vane of larger dimensions.

All three results and reasonings have this idea in common—that the dynamical similarity of two gaseous systems, bounded by solid surfaces geometrically similar, depends on the ratio of the <linear dimensions> homologous lines of the system to a certain quantity, having the dimensions of a line, and which varies inversely as the density of the gas.

Hence if a certain phenomenon occurs when a gas passes through a plug the pores of which are of a certain size, a similar phenomenon will occur in a plug with larger pores, if the density of the gas be diminished in a certain ratio.

It is the application of this idea to a well devised series of experiments on plates of stucco and meerschaum that has enabled Prof. Reynolds to make the first great step in reducing to order the very perplexing results of Graham's experiments on the passage of gases through porous septa.

The method adopted for comparing the series of observations, by plotting curves whose coordinates are proportional to the logarithms of the observed times, and then ascertaining the amount of displacement required to make the curve <of one series> corresponding to one porous plate coincide with that corresponding to another plate, is admirably fitted to detect the conditions of dynamical similarity in plates whose pores are of different size. The result shows clearly that two portions of the same gas bounded by solid surfaces which are geometrically similar are not dynamically similar unless the densities are inversely as the homologous lines in the two surfaces.

Professor Reynolds thinks that this is a proof that a gas is not a continuous and homogeneous substance. I think that it is an argument tending in the direction of such a proof, but that what is established by the experiments is neither more nor less than this—It being granted that the total energy of a material system is capable of being expressed as a function of all the variables which determine the configuration and velocities of the system and that the behaviour of the system under the action of any external forces may be determined in terms of the derivatives of this function with respect to the variables—the experiments show that this function for air and for hydrogen, is a function of at least one linear magnitude, besides those which determine the form of the containing vessel.

Thus if a gas conformed strictly to Boyle's law, it might not be easy to prove by experiments on its density, pressure, and temperature that its properties are related to an absolute linear magnitude, but if it departs from Boyle's law and becomes liquefied, the circumstances under which this occurs furnish an additional equation by means of which a linear magnitude can be defined absolutely.

The <properties of> coefficients of diffusion, viscosity and conduction of heat are of the dimensions $\left[\frac{L^2}{T}\right]$ and therefore are capable, along with the equation, $p = R\rho\theta$ of furnishing the definition of an absolute length.

The capillary constants of liquids furnish the same kind of evidence. I think therefore that the argument from Prof. Reynolds' results is one of many in favour of a certain heterogeneity of structure in a gas, and that it belongs to a particular class of these arguments, namely that derived from the phenomena of diffusion, conduction, of heat, and viscosity.

The only suggestion I have to make as to the experiments on thermal transpiration is, that if these experiments are repeated it would be desirable to test the temperature of the surfaces of the porous plate by putting one junction of a thermoelectric current in contact with the surface, the other junction being in a vessel of water of variable temperature. One junction of iron and german silver wire would be sufficiently sensitive. There should be two circuits, one for each side of the plate.

(2) Report on the Mathematical Part

I find it necessary to consider the mathematical part of the paper separately. The author tells us that it was not begun till late last summer, and the M.S̃. extends to 93 pages, and consists for the most part of investigations, the foundations of which have to be laid with extensive care. I am sure that the experience he has gained in the construction of his theory would enable him to make very considerable improvements in his foundations and definitions if he were now to revise them, and I would suggest that this course should be pointed out to him with respect to the mathematical part of his paper. I do *not* think that it would be desirable to have a new edition of the experimental part, revised by the light of his mature knowledge, because I think that such a faithful record of the process of sound scientific thought is more valuable than a textbook, however well digested.

There is one serious error in definition which appears in equation (1) and is repeated in equation (64), p. 100.
All previous writers have defined the mean velocity of the gas as the velocity of the centre of mass of all the molecules which at a given instant are within the element of volume, or what comes to the same thing, they define its components in terms of the quantity of gas carried through three fixed planes. Professor Reynolds, on the other hand, writing, as I do,

$$u = U + \xi, \qquad v = V + \eta, \qquad w = W + \zeta,$$

where u, v, w	are the velocity components of a molecule
$U\ V\ W$	are the velocity components of the gas
ξ, η, ζ	are the velocity components of the agitation of a molecule
$u - U$, $v - V$, $w - W$	the velocity components relatively to the mean velocity

defines the *three* quantities U, V, W by means of the *two* equations

$$\overline{\xi^2} = \overline{\eta^2} = \overline{\zeta^2}$$

or $$U^2 + 2Uu + u^2 = V^2 - 2Vv + v^2 = W^2 - 2Ww + w^2$$

The correction of this blunder would do no harm to any other part of the work.

I think however that more care ought to be taken in the enunciation of the Theorem I.

The difficulty, which Prof. Reynolds has attacked with great courage and with some success, is the treatment of the theory of a gas near a rigid surface.

At an element of the gas which is not near the boundary of the medium, we take advantage of the fact that the molecules which at any instant are within the element have come from places in the neighbourhood in which the properties of the gas vary in a continuous manner from place to place. We may therefore assume that the distribution of the molecules in the element according to their velocities is expressed by a continuous function of their velocities, and the actual form of the functions in certain cases can be ascertained.

But supposing the properties of the gas to vary in a continuous manner within a region bounded by a solid surface, the distribution of velocities in an element near the surface must not be the same as if the surface had been removed, and the properties of the gas had varied according to the same continuous law in the region beyond. For the velocities of those molecules which rebound from the surface are quite different from those of the hypothetical molecules which would have come from the gas supposed to exist in the region beyond.

Hence the function which expresses the distribution of velocities in such an element will not be a continuous one, for the law of the velocities of the molecules which have rebounded from the surface is quite different from that of the molecules which come from neighbouring regions in the gas.

The method which I have used in my papers presented to the Royal Society in 1866 and 1878 depends entirely on the assumption that the distribution of velocities is continuous.[c] It deals with the mean values of quantities of the forms ξ, $\xi^2\,\xi^3$, ξ^4, &c. and in this way contrives to avoid the consideration of <molecules> the place from which the molecules come.

In fact I suppose that everything depends on these mean values of powers and products of ξ, η, ζ and that these mean values vary in a continuous manner, and I proceed to consider the variations of these quantities within the element, arising partly from the mutual actions of the molecules within the element, and partly from the exchange of molecules which is going on through the surface of the element. But I carefully abstain from asking the molecules which enter where they last started from. I only count them and register their mean velocities, avoiding all personal enquiries which would only get me into trouble.

Hence this method is totally inapplicable to the treatment of the gas near a solid surface, and in my paper of 1878 I have kept clear of the question. It has been treated, both theoretically and experimentally, by M. M. Kundt and Warburg [Pogg. Ann. CLV (1875) p. 337].[d] Prof. Reynolds does not refer to this investigation but it would be interesting to compare them. Kundt & Warburg deal only with friction and not with thermal transpiration. They test the results of the hypothesis that the velocity of the rebounding molecules resolved parallel to the surface is, on the average, zero. This lead to the conclusion that there is a certain finite amount of sliding between the surface and the gas in contact with it, but their experiments show that the amount of sliding is much greater than the calculated value, so that it is probable that the average velocity of the rebounding molecules, resolved parallel to the surface, is in the same direction as that of the impinging molecules. Prof. Reynolds has apparently independently arrived at this result that there must be slipping at a solid surface. See Art. 98, p. 107.

Professor Reynolds' theory of what happens at the surface is not very definitely stated.

To arrive at a set of conditions having a certain degree of probability we might frame the hypothesis that the surface has on it small prominences of various shapes from which the molecules rebound, with a velocity which is greater or less than that of the impinging molecules, accordingly as the temperature of the solid is greater or less than that of the gas.

Thus if ℓ, m, n are the direction cosines of the normal to the element dS of the surface, which we may suppose sensibly at rest (though vibrating on account of its heat) then the velocity before impact resolved along this normal is

$$p = \ell u + mv + nw$$

Now let the impact change p into $-p(1 + k\frac{\delta\theta}{\theta})$

where θ is the temperature of the gas and $\theta \pm d\theta$ that of the solid then, after impact,

$$u' = u - p\ell(l + k\frac{\delta\theta}{\theta})$$

with two similar equations.

The number of molecules which in unit of time impinge on dS is $Np\,dS$, where N is the number in unit of volume.

Now let the surface nearly coincide with the plane of $x\,y$, and let the slopes of the prominences be turned indifferently in all directions, then if we write

< If we put > $$\iint (l - u^2)\,dxdy = q\,x\,y$$

and $$\iint (l - u^2)^2\,dxdy = r\,x\,y$$

we find for the mean value of u, v, w, u^2, &c.

$$u' = u\left(1-\left(2+k\frac{\delta\theta}{\theta}\right)q\right)$$

$$v' = v\left(1-\left(2+k\frac{\delta\theta}{\theta}\right)q\right)$$

$$w' = w\left(1-\left(2+k\frac{\delta\theta}{\theta}\right)\right)\left[1-q+\frac{1}{2}q\frac{u^2+v^2}{w^2}\right]$$

$$u'^2 = u^2 - 2\left(2+k\frac{\delta\theta}{\theta}\right)q\,u^2 + \left(2+k\frac{\delta\theta}{\theta}\right)^2\left[\frac{3}{8}r\,(3u^2+v^2)+\frac{1}{2}(q-r)\,w^2\right]$$

$$u'v' = uv\left[1-2\left(2+k\frac{\delta\theta}{\theta}\right)q + \frac{3}{4}\left(2+k\frac{\delta\theta}{\theta}\right)^2 r\right]$$

See below

[The two equations following are inserted at the bottom of the page:]

$$u'w' = uw\left[1-\left(2+k\frac{\delta\theta}{\theta}\right)\left(1+\frac{1}{2}\frac{u^2+v^2}{w^2}\right) + \left(2+k\frac{\delta\theta}{\theta}\right)^2\left(\frac{3}{2}(q-r)+\frac{3}{2}\frac{u^2+v^2}{w^2}\right)\right]$$

$$w'^2 = w^2\left[1-\left(2+k\frac{\delta\theta}{\theta}\right)(1-q)+\left(2+k\frac{\delta\theta}{\theta}\right)^2(1-2q+r)\right] - (u^2+v^2)\left(2+k\frac{\delta\theta}{\theta}\right)\left[q-\frac{3}{4}(q-r)\left(2+k\frac{\delta\theta}{\theta}\right)\right]$$

Conditions of this kind might perhaps assist Prof. Reynolds in forming a theory about the rebounding molecules.

There is another point in which I think Prof. Reynolds has been misled by my authority, namely in asserting that the distribution of velocities is that given by the expression

$$e^{-\left(\frac{\xi^2+\eta^2+\zeta^2}{\alpha^2}\right)}d\xi\,d\eta\,d\zeta$$

This can be true only in a gas at rest and in thermal equilibrium. If the pressures are different in different directions the distribution of molecules must be different from this. Another law of distribution has been investigated by Boltzmann and adopted by me in my paper of 1878.

I do not think that anything in Prof. Reynolds' paper depends on the exactness of this distribution, but it should be recognised that the law given is not exact for the cases he considered.

I think that Theorem I and all that depends on it would be improved by drawing the distinction more clearly between the mean velocity of the molecules which, at a given instant, are within a given element, and the mean velocity of the molecules which, during unit time, are projected from the element after encounters with other molecules. Let us confine our attention to molecules whose velocity makes an angle less than a given small angle with the line whose direction-cosines are ℓ, m, n, and of these let us consider only those the direction of whose motion is positive.

Of these we have to consider first those which in unit of time pass through unit of area of the plane normal to ℓmn. These have all come from the neighbourhood of the plane and have been projected from places of a greater or less distance on the negative side of the plane. It is therefore a legitimate assumption that their mean velocity is equal to the mean velocity of molecules projected from an element of a certain distance (called the mean range) on the negative side of the plane. By making this distinction between the molecules projected from an element in a given time, and the molecules which pass through a plane in a given time, we express what Prof. Reynolds means in Theorem I, without introducing the phrase "if the condition of the gas were uniform" which is not very precise, and, I rather think, is not true in the case of a gas with a uniform shearing velocity. ($U = bz$) We may use this method directly to explain the fact of thermal transpiration & c.

Let P be an element of the surface of a solid and let us suppose that the temperature is the same for gas and solid along any vertical line but that it increases from left to right.

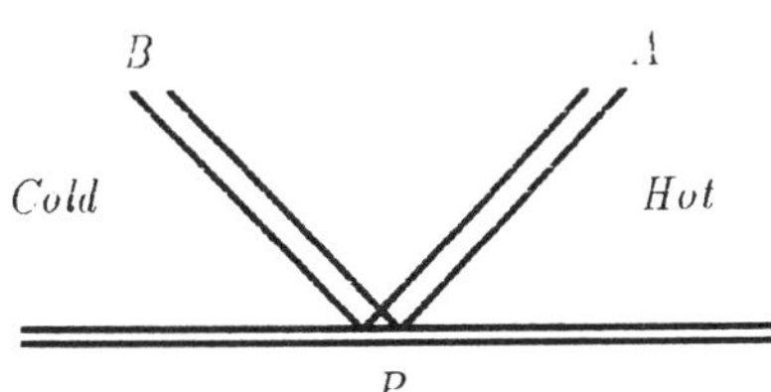

Let us also suppose that the pressure is at first the same everywhere. Consider the molecules moving in the direction AP. They come from regions hotter than P and therefore have a greater mean velocity than the particles near P.

Similarly the particles coming in the direction BP will have a smaller mean velocity than the molecules near P.

If the molecules approaching along AP returned along PB and vice versa, then the total mean velocity of molecules near P going and coming along

AP will be the same as along BP. This would be the case of perfect smoothness of the plane surface, a case which no one ever supposed possible.

In every other case, the difference in the mean velocity of the rebounding molecules in the two directions fails to make up for the difference in the mean velocities of the approaching molecules, and the pressure on P is greater in the direction AP than in the direction BP, that is to say, there is a tangential stress, urging the gas from cold to hot, and the surface from hot to cold.

This, I think, is satisfactory as an explanation to the nonmathematical mind, and I do not think there is any error in it, though it is very far from being a satisfactory introduction to a numerical calculation of the resulting force.

I have found it necessary to spend a good deal of time in examining Prof. Reynolds' paper. Perhaps if I had spent more time I should have been able to make a shorter report, but my principal anxiety has been to satisfy myself of the validity of a method which depends on so many assumptions and which grapples with difficulties which I have made it a point of strategy to avoid. I now think that certain imperfections which I have mentioned do not affect the truth of the conclusions, and that the author might easily put them right. I think however that pains would be well bestowed in explaining the steps of the mathematical calculations in the same lucid style as the descriptions of the general reasonings and of the general principles of the mathematics.

I therefore recommend that the second part of the paper by printed in the Transactions after the author has had the opportunity to make certain improvements in it.

James Clerk Maxwell

11 Scroope Terrace,
Cambridge
28th March 1879

a. The paper under discussion is, Osborne Reynolds, "On Certain Dimensional Properties of Matter in the Gaseous State, Part I and Part II," read February 5, 1879 before the Royal Society and abstracted in *Proc. R. Soc. London* 28 (1879): 304–321 and *Nature* 19 (1879): 435–437, printed in full in *Phil. Trans. R. Soc. London* 170 (1880): 727–845, and reprinted in Reynolds, *Papers on Mechanical and Physical Subjects*, (Cambridge: At the University Press, 1900–1903), vol. 1, 257.

b. Thomas Graham's experiments on gases had puzzled Maxwell before as he could not explain them with his theory of transport phenomena, see Maxwell, "Illustrations of the Dynamical theory of Gases," *Phil. Mag.* 19 (1860): 19–32, 20 (1860): 21–37, reprinted in *Scientific Papers*, vol. 1, 377–409 and Garber, Brush, and Everitt, *Maxwell on Molecules and Gases*, 285-318, 309.

c. The paper referred to is Maxwell, "On the Dynamical Theory of Gases," read to the Royal Society 31 May, 1866, abstracted in *Proc. R. Soc. London* 15 (1867): 167–171 and *Phil. Mag.* 32 (1866): 390-393, published in full in *Phil. Trans. R. Soc. London* 157 (1867): 49–88, reprinted in *Scientific Papers* vol. 2, 26–78, and Garber, Brush, and Everitt, *Maxwell on Molecules and Gases*, 419–471.

d. The paper referred to is Maxwell, "On Stresses in Rarified Gases Arising from Inequalities in Temperature," read before the Royal Society 11 April, 1878, reproduced here as document VI-31.

e. August Kundt and Emil Warburg, "Ueber Reibung und Wärmeleitung verdünnte Gase," *Ann. Phys.* 155 (1875): 337–366, 525–550; 156 (1876): 177–211.

16. Letter from Maxwell to George Gabriel Stokes, 1879[a]

Royal Society Archive, no. 188, filed with Maxwell's report, document VI-15

Reynolds is numerically right according to his own dynamical theory. What I said was independent of any dynamical theory of molecules and rested on ordinary gaseous laws.

Reynolds' increase of temperature is that of the recoiling molecules only. But it appears impossible to measure this temperature by means of any kind of thermometer, for the thermometer will be acted on by the molecules which have not yet impinged as well as those which have already impinged, so that it is difficult to form a practical notion of Reynolds' rise of temperature except as a mathematical deduction from the increase of velocity of a certain set of the molecules, namely those which have just impinged.

I have been examining Lorentz (of Leyden) theory of Reflection & Refraction <on> by the Electromagnetic hypothesis. He gives outlines of Fresnel, Cauchy, &c.[b] Is there any thing since Cauchy & Green on this subject? I have Cornu[c] on Crystalline Reflexion but I do not known whether Jamin or Quincke have themselves attacked the theory.[d]

Lorentz makes the electric displacement perp.[endicular] to plane of polarization and gets on very well with Fresnel's formula and total reflexion but I have to take more pains to understand his metallic reflexion where metallic conduction comes in to cause rapid absorption.[e]

Has any thing been done about reflexion from a stratum in which the properties of the medium change continuously from one state to another within a fraction of a wave-length? The advantage of the electromagnetic theory is that there are only 3 coefficients for a crystal instead of 6 as in the elastic solid theory and that no forced ? relations are required to "save appearances."[f] Also the same hypothesis about the direction of electric vibrations being perp.[endicular] to plane of polarization serves for reflexion & for double refraction.

Yours truly,
J. Clerk Maxwell

a. This letter is dated by its contents and clearly relates to the points Maxwell raised in his referee report, document VI-15. In addition both the report and letter are filed together in the Royal Society archives.

b. Henrik Antoon Lorentz (1853–1928) developed an electromagnetic theory of reflection and refraction in his doctoral dissertation, see Lorentz, "Sur la théorie de la réflexion et de la réfraction de la lumière," Leiden, 1875. This french translation and the dutch version were published in Lorentz, *Collected Papers* (The Hague: M. Nijhoff, 1934–39), vol. 1, 193–382, and 1–192, respectively.

c. Marie Alfred Cornu (1843–1902) entered the École Polytechnique in 1860 and wrote his thesis on crystalline reflection in 1867. It was published in several parts, Cornu, " Théorème sur la réflexion crystalline," *Comptes Rendus* 60 (1863): 47–49; *Ann. Phy.* 126 (1863): 466–469; "Théorème géométrique relatifs à la réflexion crystalline, " *Comptes Rendue* 62 (1866): 1327–1329; "Théorie nouvelle de la réflexion crystalline d'après les idées de Fresnel," *Comptes Rendus* 63 (1866): 1058–1061; "Recherche sur la réflexion crystalline," *Ann. Chim.* 11 (1867): 283–389.

d. Jules Celestin Jamin (1818–1886) worked on double refraction and total reflection in transparent substances and liquids but seems to have left the problem of the reflection of light by crystals to his student Cornu. Georg Hermann Quincke (1834–1934) wrote a series of papers on light during the 1870s. What probably alerted Maxwell to this work was Quincke, "Extracts from Two Letters on Refractive Index of Flass and Quartz as Tested by Refraction," *Proc. R. Soc. London* 9 (1878): 567–569.

e. Earlier in his career Maxwell had purposely avoided any attempt at deriving Fresnel's laws for the reflection and refraction of polarized light from electromagnetic theory as he was uncertain about the boundary conditions to use for high frequencies. See his letter to Stokes, October 15, 1864,

> I am trying to understand the conditions at a surface for reflexion and refraction but they may not be the same for the period of vibration of light and for experiments at leisure.

Memoir and Scientific Correspondence of the Late Sir George Gabriel Stokes ed. J. Larmor (Cambridge: The University Press, 1909), vol. 2, 26. The letter explains some of Maxwell's difficulties. However, Lorentz showed that the problem was easier than Maxwell supposed and the static boundary condition used by Maxwell in Maxwell, "On Faraday's Lines of Force," read in December 1855 and February 1856, published in *Trans. Cambridge Phil. Soc.* 10 (1864): 27–65, reprinted in *Scientific Papers*, vol. 1, 155–229 applied. It is likely that Maxwell was led to Lorentz's work when he reviewed G. F. FitzGerald, "On the Electromagnetic Theory of Reflexion and Refraction," for the Royal Society. His review, dated February 6, 1879, is reprinted in Larmor, *Memoir and Correspondence of Stokes*, 40–43. At Cambridge University Library there is also an undated fragment on the same problem, reprinted in Ian Hopley, " Clerk Maxwell's Contribution to Physics," unpublished PhD. dissertation, London University, 1965.

Maxwell's caution about "conditions at boundary surfaces" in any physical situation is reflected in his report on Reynolds's paper, document VI-15. Only in the last revisions to Maxwell, "On Stresses in Rarified Gases Arising from Inequalities in Temperature," did he finally throw caution to the wind and attack a problem of this kind.

f. For the innumerable "forced relations" in elastic solid theories of light, see Stokes, "Report on Double Refraction," *Rep. 32nd Meeting BAAS* (1862): 253–282, reprinted in Stokes, *Mathematical and Physical Papers* (Cambridge: The University Press, 1880–1905) vol. 4, 157–202, and the papers by Rayleigh, "On Double Refraction," *Phil. Mag.* 41 (1871): 519–528, and, "Notes, Chiefly Historical on some Fundamental Propositions in Optics," *Phil. Mag.* 21 (1886): 466-476, reprinted in Rayleigh, *Scientific Papers* (Cambridge: Cambridge University Press, 1899–1920) vol. 1, 111–134, and 518–536, respectively, and J. W. Gibbs, "A Comparison of the Elastic and Electrical Theories of Light with Respect to the Law of Double Refraction and the Dispersion of Colors," *Am. J. Sci.* 35 (1888): 467–475, reprinted in Gibbs, *Scientific Papers*, 223–246. These papers are discussed in Everitt, *Maxwell*, 144 and n. 44.

17. Letter from Osborne Reynolds to Maxwell, 7 July, 1879

Cambridge University Library, Maxwell Collection

66 Marina
St. Leonards
7 July 1879

My dear Sir[a]

Thank you for the paper and for the additions on page 45 and 81. I have been trying the reduction necessary to obtain the extensions to equations (126) and (127) but I have so far failed and I therefore venture to ask if you will kindly put me in the way.

It seems to me that equation (122) is obtained by putting in equation 45—

$$\frac{Q}{\mu} = \xi^2 + 2u\xi - \frac{\xi^2 + \eta^2 + \zeta^2 + 2u\xi + 2v\eta + 2w\zeta}{3}$$

and that for the extension u, v, w may be supposed zero.
Whence, I get

$$\frac{1}{3}\Big[\frac{d}{du}\rho\left(\xi^3 + \xi\eta^2 + \xi\zeta^2\right) + \frac{d}{du}\rho\left(\eta\xi^2 + \eta^3 + \eta\zeta^2\right) + \frac{d}{du}\rho\left(\zeta\xi^2 + \zeta\eta^2 + \zeta^3\right)\Big] - \frac{d}{dx}\rho\,\xi^3 - \frac{d}{dy}\rho\,\eta\xi^2 - \frac{d}{dz}\rho\,\zeta\xi^2 = 3k\,A_1\,\rho\,q$$

Then substituting from (147) and neglecting terms involving $\dfrac{dp^2}{\rho^2\theta}\dfrac{d\theta}{du}$

$$\frac{5}{3}\frac{\mu^2}{\rho\theta}\left\{\frac{d^2\theta}{dx^2}+\frac{d^2\theta}{dy^2}+\frac{d^2\theta}{dz^2}\right\}$$
$$+\frac{1}{3kA_2}\left\{\frac{d}{du}\rho\xi^3+\frac{d}{dy}\rho\eta\xi^2+\frac{d}{dz}\rho\zeta\xi^2\right\}=-q$$

In order to reduce this further, if I am right so far, it appears necessary to obtain the relative values of ξ^3, $\xi\eta^2$, $\xi\zeta^2$ as they occur in equation (147). It is here that I am stuck and I find the same difficulty with the equations (127).

Also, you have added the term $\frac{3\mu^2}{\rho\theta}\frac{d^2\theta}{dx\,dy}$ to the first of the equations (127) the left hand side of which is $\rho\eta\xi$ and there appears to me to be a want of symmetry about his which suggests that there should be another term

$$\frac{3\mu^2}{\rho\theta}\frac{d^2\theta}{dz\,dx}$$

or that the one term should be

$$\frac{3\mu^2}{\rho\theta}\frac{d^2\theta}{dy\,dz}$$

I hope that you will forgive me for thus troubling you,

Yours sincerely
Osborne Reynolds

Prof. Maxwell F. R. S.

a. The equations to which Reynolds refers are in a draft of Maxwell, "On Stresses in Rarified Gases Arising from Inequalities in Temperature," *Phil. Trans. R. Soc. London* 170 (1880): 231–256, abstract in *Proc. R. Soc. London* 27 (1878): 304–308, reprinted in *Scientific Papers*, vol. 2, 682–707, and document VI-31. Maxwell obviously sent Reynolds a copy and asked for comments.

18. Letter from George Gabriel Stokes to Maxwell, August 18, 1879[a]

Cambridge University Library

LENSFIELD COTTAGE, CAMBRIDGE, 18 AUGUST 1879.

MY DEAR MAXWELL,

I HAVE HAD MY HANDS VERY FULL WITH OTHER MATTERS, AND I FEAR I HAVE UNREASONABLY PUT OFF LOOKING OVER YOUR PAPER.

I DON'T MUCH LIKE THE LOOK OF THE SEPARATE LINE SUCH AS "NOTE ADDED MAY, 1879" ON P. 6. IT SEEMS TO ME TO BREAK UP THE SUBJECT MATTER IN AN INCONVENIENT WAY. I THINK IT WILL BE MUCH BETTER TO PUT <TO PUT> THE THING IN ITALICS AT THE END OF THE SQUARE BRACKET. THIS I THINK IS WHAT IS GENERALLY DONE IN SIMILAR CASES. WHERE WE HAVE A SUBSTANTIAL ADDITION OF THE NATURE OF AN APPENDIX, AS AT P. 10, THE DATE OF ADDITION COMES VERY WELL AT THE HEAD, AND THEN WE DON'T WANT THE SQUARE BRACKET. I MAY NOTICE THAT A FINAL SQUARE BRACKET APPEARS AT THE END OF THE PAPER, THOUGH THERE IS NO INITIAL BRACKET TO READ TO IT. I WILL STRIKE IT OUT. THE NOTE ON P. 18 MIGHT STAND AS IT IS, AS IT OCCURS AT THE END OF THE PAPER, BUT I THINK IT WOULD BE BETTER TO MAKE IT MATCH THE OTHERS, AND I WILL DO SO IF YOU DON'T OBJECT.

I SEE THERE IS A SMALL PINCH OF THE SOLIDUS NOTATION IN P.18. SO FAR AS I KNOW IT HAS ONLY JUST COME TO THE BIRTH, AND I AM AFRAID THAT MANY A READER MIGHT NOT KNOW WHAT IT MEANT. IT IS ALL VERY WELL IN A WORK, WHERE IT IS EXPLAINED ONCE FOR ALL IN THE PREFACE, AND USED REPEATEDLY. BUT IT HARDLY DOES I THINK FOR IT TO DROP IN CASUALLY AS IT WERE IN A PAPER IN THE PHILOSOPHICAL TRANSACTIONS. IF YOU HAD EVEN USED IT REGULARLY WHEN FRACTIONS OCCURRED IN THE TEXT IT WOULD HAVE BEEN DIFFERENT. BUT YOU GENERALLY USE ONE OF THE OLD NOTATIONS. HENCE THOUGH I LIKE THE SOLIDUS AND AM RESPONSIBLE FOR THE STARTING OF THE THING, I THINK THAT IN THE PRESENT CASE WE HAD BETTER STICK TO ONE OF THE OLD NOTATIONS. I SHOULD PROPOSE THE OLD ÷ IF YOU DON'T OBJECT TO ITS SOMEWHAT CLUMSY LOOK.

I FEEL A GOOD DEAL OF DOUBTS ABOUT ART. 8, P. 3. YOU SEEM TO IMPLY THAT A CUP INITIALLY AT REST IN AN EXPANSE OF RARIFIED GAS, WITHDRAWN WE WILL SUPPOSE FROM THE INFLUENCE OF GRAVITY , WOULD BEGIN TO MOVE AND GO ON MOVING CONCAVITY FOREMOST, THE CUP BEING SUPPOSED HOTTER THAN THE GAS.

NOW I THINK ALL THE EXPERIMENTS THAT HAVE BEEN TRIED INDICATE THAT THE MOTION ESSENTIALLY DEPENDS UPON THE PRESENCE OF ANOTHER BODY AT A REASONABLE DISTANCE. I AM NOT SURE WHETHER YOU MEAN TO GO FURTHER THAN TO SAY THAT WHAT IS STATED IN 8 IS MERELY WHAT WOULD BE SUPPOSED AT FIRST SIGHT, AND MEAN SECTION 10 TO CORRECT SECTION 8. IF SO, IT IS NOT VERY CLEAR. FOR YOU APPEAR TO OFFER 8 AS THE EXPLANATION OF WHAT CROOKES OBTAINED. NOW I THINK THAT THE EXPERIMENTS SHOW VERY CLEARLY THAT THE REACTION DEPENDS ON WHAT TAKES PLACE BETWEEN THE FLY AND THE WALL OF THE CASE. IF A GIVEN FLY WERE PUT INSIDE A SUFFICIENTLY LARGE EXHAUSTED VESSEL, THERE WOULD APPARENTLY BE NO MOTION.

MY WIFE, ARTHUR AND ISABELLA STARTED FOR IRELAND THE LAST DAY OF JULY.[b] I INTEND TO START SOME DAY THIS WEEK. YOU HAD BEST ADDRESS HERE, AND IF I SHOULD HAVE GONE THE LET-

TER WILL BE FORWARDED.

PRAY GIVE MY KINDEST REMEMBRANCES TO MRS. MAXWELL, AND BELIEVE ME

YOURS SINCERELY,
G. G. Stokes

THE RAIN IT RAINETH EVERY DAY

a. This letter is typed in block capitals. All comments in this paper refer to a draft of Maxwell's, "On Stresses in Rarified Gases Arising from Inequalities in Temperature," *Phil. Trans. R. Soc. London* 170 (1880): 231–256, abstract in *Proc. R. Soc. London* 27 (1878): 304–308, reprinted in *Scientific Papers,* vol. 2, 681–707, and document VI-31.

b. "Arthur and Isabella" refers to Stokes's daughter and son-in-law.

19. Letter from Maxwell to George Gabriel Stokes, 25 August, 1879[a]

Memoir and Scientific Correspondence of the Late George Gabriel Stokes, ed. Joseph Larmor, (Cambridge: The University Press, 1909) vol. 2, 43–44.

Glenlair, Dalbeattie.
21 August 1879.

MY DEAR STOKES,

I have not been able for work of any kind for some time, so that it is with difficulty that I can answer your letter, though it is about my own paper, and you have been taking so much trouble to make it somewhat more like a paper.

At p. 3 I would read—

Hence the normal pressure will be greater on the convex surface than on the concave surface, and if we were to neglect the tangential pressures we might think this an explanation of the motion of Mr Crookes' cups.

At p. 6. Note added May 1879] should be put at p. 7 in italics before the square bracket].

At p. 12 transfer Added May 1879] to the end of line 12 of p. 13 *in italics* with square bracket thus

$$\frac{2}{3}\frac{p}{mu}. \qquad \textit{Added May 1879}]$$

At p. 18 transfer Note added June 1879 to foot of page and put it in italics.

At foot of p. 26 delete]

in p. 18, line	6	from bottom	*for*	$\mu^4/\rho^2 p\theta$	*read*	$\mu^4 \div \rho^2 p\theta$
"	7	"	"	$\mu^2/\rho\theta$	"	$\mu^2 \div \rho\theta$

"	9	"	"	$\mu^2/\rho p$	"	$\mu^2 \div \rho p$
"	13	"	"	μ/p	"	$\mu \div p$.

I hope the R. S. will not send me any papers to report on for I could not do it.

My wife got caught in the rain on the 13th, and on Monday I was afraid of bronchitis, but the doctor thinks that it has taken a better turn now, but she is very much distressed with neuralgia in the face and with toothache. She was much pleased by getting a letter from Mrs Stokes this morning.

Among the subscribers to Dr Smith's Optics, 1738, appears the name of "Mr Gabriel Stocks." I dare say however your optical studies were already somewhat advanced (from a heredity point of view) in 1738.

Yours very truly,
J. CLERK MAXWELL

a. These corrections are for Maxwell, "On Stresses in Rarified Gases arising from Inequalities in Temperature," *Phil. Trans. R. Soc. London* 170 (1880): 231–256, abstract in *Proc. R. Soc. London* 27 (1878): 304–308, reprinted in *Scientific Papers,* vol. 2, 681–707, and document VI-31. The specific points addressed are those raised by Stokes in document VI-18.

20. Letter from Maxwell to William Thomson, 25 August, 1879[a]

Joseph Larmor, "Origins of Clerk Maxwell's Electrical Ideas, as Described in Familiar Letters to William Thomson," *Proc. Cambridge Phil. Soc.* 32 (1936): 744–745

Glenlair, Dalbeattie.
1879 August 25

DEAR THOMSON

Professor Stokes has sent me again our reports on Osborne Reynolds, together with two long letters from O. R. in answer to the advice culled by Stokes from said reports and sent as suggestions to O. R.

The labour O. R. has had to go through in reperusing his own MS. seems to have impressed him with new views as to the trouble likely to be experienced by a Referee, and also by the Sec. R. S., who has to harmonize two referees and one author to the satisfaction of a committee of papers.

The author has also given up his method of determining the 3 components U, V, W, of mean velocity from the two quadric equations

$$\sum(\xi - U)^2 = \sum(\eta - V)^2 = \sum(\zeta - W)^2$$

This was decidedly the most striking mathematical conception in the paper, so that when it is removed we shall be less likely to come into collision with him.

His reason for not quoting Stoney, viz that his dispute with [G. J.] Stoney is in the *Phil. Mag.*, can be settled by a foot note giving the reference. I certainly do not believe in the Stoney Stratum though the Crookes Layer is a thing capable of being explored.[b]

If a hot plate discharged all molecules out of a broadside of guns normal to its surface it might produce a Stoney Stratum, but then it would require an army of demons to glean the molecules as they recoiled and aim them normal to the plate again. [Cf. modern emission of electrons.][c] This is the only way to maintain a uniform aelotropic stress throughout a finite region of gas.

Otherwise the stress is uniform in all directions with a correction of the form

$$\frac{d^2\theta}{dh^2}\left(\text{not } \frac{d\theta}{dh}\right)$$

I never sailed through a plate of meerschaum on a molecule, so that I do not known whether the passage is like the Forth & Clyde canal or like the Caledonian canal but I should not expect to have many long reaches, but rather plenty of locks and sluices so that it would not be a very like a tube but more like a cullender. Nevertheless I do not see how O. R. could detect all this between the lines of my paper of 1867.

My principal question is, What are your coordinates? and shall I send Stokes' bundle in that direction with any chance of its being reflected back with suitable reflexions to the Observatory, Armagh within a reasonable time?

As for me I was for 6 weeks or so in a state of Thermal Transpiration in Feb. & March, and if any more papers are sent me I shall be an exhausted receiver, for I am hardly able to correct proofs of Electricity & Magnetism [2 ed].[d]

When I saw you in Cambridge I did not know that W. Garnett, my demonstrator was going in for Edinburgh. He only made up his mind to do so at the last moment, just before the election. If he should go in for Glasgow I may say that his strength is as the strength of 10 because his heart is pure. Whatever he undertakes he does it with his might. He has out of his own energies developed a course of Nat. Phil for elementary men, in which the experiments are not only exhibited but are reasoned from, and the intellectual status of those who attended them was raised pretty considerably, so that the results contrasted favourably with those obtained by mathematical coaching.[e]

I do not think he has done much in the Theory of Numbers and that sort of mathematics but he is up in all the practical part, and knows how to use it himself and how to inculcate it into others. He has also great power of grinding up a new subject for himself and boiling it down for his pupils. He quite sees the difficulties of things, but if they have to be done he always does them, and that within the specified time. In short he is one of the few men I know who has a strong will and knows how to govern it. He has just married and Mrs Maxwell's opinion of the lady is very favourable.

a. These first paragraphs are Maxwell's further comments on Reynolds' response to his own and Thomson's referee reports on what was published as, Reynolds, "On Certain Dimensional Properties of Matter in the Gaseous States. Part I and Part II," read February 5, 1879; abstracted in *Proc. R. Soc. London* 28 (1879): 304–321, and *Nature* 19 (1879): 435, printed in full in *Phil. Trans. R. Soc. London* 170 (1880): 727–845, reprinted in Reynolds, *Papers* vol. 1, 257. The revised version of the paper was received at the Royal Society, October 25th 1879.

b. The controversy between Osborne Reynolds and George Johnstone Stoney (1826–1911) appears in Stoney, "The Radiometer and Its Lessons," *Nature* 17 (1878): 181; Reynolds's reply is in Reynolds, "The Radiometer and its Lessons," *Nature* 17 (1878): 220, and Stoney's rejoinder follows on p. 301. Maxwell is mistaken in the journal in which the exchanged appeared.

c. The remark in [ ]'s is from Joseph Larmor.

d. The second edition of Maxwell, *Treatise on Electricity and Magnetism* was not published until 1881. Maxwell revised the first eight chapters extensively but the rest was really a reprint of the first edition. The proofs were corrected by W. D. Niven, the editor of Maxwell's collected papers.

e. William Garnett (850–1932) was Demonstrator at the Cavendish Laboratory from 1874 to 1880. For more details see Garber, Brush, and Everitt, *Maxwell on Molecules and Gases*, 515–516. Garnett's "going in for Edinburgh" means applying for the chair of mathematics at Edinburgh University. Maxwell had previously written a strong letter in support of George Chrystal. According to one of Tait's later, unpublished letters, Maxwell threw the electors into confusion by sending along an even stronger letter in support of Garnett. This uncharacteristically erratic action was one of the first things that made Tait realize that Maxwell's illness was serious. Chrystal got the job.

21. Letter from Maxwell to William Thomson, 1 September, 1879

Joseph Larmor, "Origins of Clerk Maxwell's Electrical Ideas, as Described in Familiar letters to William Thomson," *Proc. Cambridge Phil. Soc.* 32 (1936): 745–748

Glenlair, Dalbeattie.
1879 Sept. 1

DEAR THOMSON!

Peacocks as Gardeners. We got our original stock from Mrs McCunn, Ardhallow.[a] At that time (1860) the garden there was the finest on the coast and the peacocks sat on the parapets & banks near the house. After those days there came a gardener who found that the peacocks destroyed the garden, so the peacocks were destroyed, and the garden ceased to produce the commonest vegetables, and became quite a desert. M^r McCunn was very fond of his garden and very particular about it, but he also cared for his peacocks and whenever he went out he had bits of bread &c for them, by which he generated a centre of attraction near the house and the peacocks were the great beauty of the place. When he died, of course M^{rs} McCunn let the gardener be master, so the peacocks died and the garden is nothing worth to see.

Mrs Maxwell says that if you were constantly at home and combined the feeding of peacocks with your morning smoke and meditation, the birds would soon be concentrated near the centre of attraction thus provided and would not trouble the garden. Mrs Maxwell always gets the peacocks to choose the gardener and they have chosen one who has now been seven years with us.

At seed time they are confined in an outhouse where they have some Indian corn and a vessel of water. When the hen is sitting she is not shut up, for she keeps to her nest and nobody is supposed to know where that is, but she comes once a day to the house, and calls for her dinner, and eats it, and goes back to her nest at once.

The peacocks will eat the young cabbages but the gardener tells them to go out again and they find it pleasanter to be about the house and to sit on either side of the front door.

Mrs Maxwell will not send them unless on consideration they would be acceptable.

I send O. R.'s reports and his letters. "Thermal Transpiration" would be called by Graham "Thermal Diffusion" or *even* "Effusion." I note some references to properties of bodies in recent publications.

Specific heat of gases. Eilhardt Wiedemann *Phil. Mag.* 1876.[b]

Calorimetric data. A. Schulier un[d] V. Wartha (Budapest).[c]

Unit of heat= $\frac{1}{100}$ (gramme H_2O at $100°C - -d^o$ at $0°C$).

Latent heat of water 79.86. Regnault 78.39.

Heat of combination of H_2 and 0 at $0°C$.

Andrews	1845	33613[d]	Wiedemann's Annalen
Than	1877	33867[e]	B^d II s 357.

Thomsen	1872	33971[f]
Favre & Silb.	1845	34065[g]
Schuller & W.	1877	34126[h]

Diffusion calculated from Graham's experiments *Phil. Trans.* 1861 by Stefan,[i] *Wiener Sitz.* 23 Jan 1879 $\frac{(\text{centimeters})^2}{\text{Day}}$

Caramel	$10°C$	k= 0.047
Albumin	$13°C$	0.063
Cane sugar	9°	0.312
NaCl	5°	0.765
	9°	0.910
HCl	5°	1.742

Zinc Sulphate in water by H. F. Weber, *Züricher naturforschenden Ges.* 25 Nov 1878[j]

Percentage	Temp	k	$(\text{cm})^2$
0.3120	1°.20	0.1252	
—	18°.55	0.2421	
—	44°.70	0.4146	

$$\{k = 0.1187\,[1 + 0.0557\,t]\,\}$$

Stefan *Wiener Sitz,* March 1878[k]

CO_2 in water at 16°.5 $\quad k = 1.41\frac{cm^2}{\text{day}}$

In alcohol at 16°.6

	CO_2	O		N
absorption	3.122	0.284		0.121
diffusion	2.772	1.243	x	2.66

J. Schuhmeister *Wiener Sitz.* 3 April 1879.[ℓ]

Diffusion of salts increases with temperature also with concentration.

10 per cent solutions at $10°C\,\frac{\text{cm}^2}{\text{day}}$

KCL	1.10	NHaCl	1.04		
KBr	1.13	NaBr	0.86	LiBr	0.80
KI	1.12	NaI	0.80	LiI	0.80
KNO_3	0.80	NaCl	0.84	LiCl	0.70
K_2CO_3	0.60	NaNO	0.60	CaCl	0.68
K_2SO_4	0.75	NaCO	0.40	$CuCl_2$	0.43
		Na_2SO_4	0.66	$CuSO_4$	0.21
				$ZnSO_4$	0.20
				$MgSO_4$	0.28

Baumgarten, *Wiener Sitz.* 8 Feb 1877.[m]

diffusion of vapours in gases.

	H	Coal gas	Air	CO_2
temperatures Schefelalhers	(16.8) 0.244	19.9 0.165	16.1 0.110	18.3 0.089
Schwefelkohlenstoff	17.3 0.261	19.9 0.165	17.9 0.078	17.3 0.063
Chloroform	18.1 0.270	17.4 0.125	18.4 0.074	17.9 0.047
Alkohol	17.9 0.344	16.5 0.173	17.1 0.096	

Your objection to the phrase "centre of gravity" of bodies in general is thus expressed by Lord Mahon (afterwards Stanhope of the Printing Press, a pupil of Le Sage)[n] Principles of Electricity 1779 "So that mathematically & correctly speaking the *common Definition of the Centre of Gravity* of a Body (viz that Point, upon which, if the Body be suspended, in *any* position whatever it will always remain in perfect *aequilibrio*) is a compleat absurdity, and no less than a contradiction in terms."

$$\frac{1}{\frac{dp}{dt}}$$

a. Mrs McCunn was Mrs Maxwell's sister.

b. Ernest Eilhard Gustav Wiedemann, "Ueber die specifische Wärme der Gase," *Ann. Phy.* 157 (1876): 1–42; *Phil. Mag.* 2 (1876): 81–108.

c. Alois Schuller and Vincze Wartha, "Calorimetrische Untersuchungen," *Ann. Phy.* 2 (1877): 559–383; *Ber. Chem. Gesell. Berlin* 16 (1877): 129–135.

d. Thomas Andrews divided his time between teaching, administration and research. His work on gases was first published in W. A. Miller, *Elements of Chemistry* (London: J. W. Parker and Sons, 1863). The remark may refer to the paper by Schuller and Wartha n. b, and their use of Andrews's value in their own paper.

e. Carl von Than, "Die Verbrennungswärme der Krallgases in geschlossen gefas sen," *Ber. Chem. Gesell. Berlin* 10 (1877): 947–952, 2141.

f. Julius Thomsen (1826–1909) was a prolific Danish chemist. His papers on the heat generated from chemical reactions stretched over many years and he published several every year. Of the several published in 1872 the reference is probably to Thomsen, "Ueber die Bildungswärme der Sauren des Stickstoffs," *Ber. dtsch. Chem. Gesell.* 5 (1872): 508–510.

g. Pierre Antoine Favre and J. T. Silberman, "Sur les chaleurs dégagées pendant la combinasion chimiques," *Comptes Rendus* 20 (1845): 156–159; 21 (1845); 944–950.

h. This reference is that in n. b.

i. Josef Stefan, "Über Diffusion der Flussigkeiten," *Sitz. Math.-Naturwiss. Cl. Akad. Wiss. Wien* 78 (1879): 957–979; 79 (1879): 161–214.

j. H. F. Weber, "Untersuchung über der Elementargesetz der Hydrodiffusion," *Vierteljahrsschriften, Zurich* 23 (1878): 325–366; *Ann. Phy.* 7 (1879): 469–487, 536-552; *Phil. Mag.* 8 (1879): 487–500, 523–536.

k. Josef Stefan, "Über die Diffusion der Kohlensaure durch Wasser und Alkohol," *Sitz. Math.-Naturwiss. Cl. Akad. Wiss. Wien* 77 (1878): 371–409.

ℓ. J. Schuhmeister, "Untersuchung über die Diffusion des Salzlösungen," *Sitz. Math.-Naturwiss. Cl. Akad. Wiss. Wien* 75 (1877): 603–626.

m. Georg Baumgartner, "Ueber den Einfluss der Temperatur auf die Verdampfungsgeschwindkeit von Flussigkeiten," *Sitz. Math.-Naturwiss. Cl. Akad. Wiss. Wien* 75 (1877): 679–688.

n. "Lord Mahon, afterwards, Stanhope of the Printing Press," refers to Charles Stanhope (1753–1816) third Earl of Stanhope who invented machines for printing and typefounding that revolutionized the industry in Britain. George-Louis LeSage (1724–1803) published his *Essai de chimie mécanique* (Rouen, 1758). According to J. B. Gough, "George-Louis LeSage," in *Dictionary of Scientific Biography* vol. 8, 260, although his work was not popularized until the publication of Pierre Prevost's treatises in 1805 and 1818, LeSage's wide correspondence with "the great natural philosophers and mathematicians of his age," presumably extended to Lord Stanhope. For Maxwell on LeSage see Garber, Brush, and Everitt, *Maxwell on Molecules and Gases,* 45, 105, 128, 170–173, 204–208, 422.

22. Maxwell to George Gabriel Stokes, 2 September, 1879[a]

Royal Society of London, Archives

Glenlair, Dalbeattie.
Sept. 2, 1879

Dear Stokes

Thomson's moorings till October are at Netherhall, Largs, Ayrshire. I have sent him the reports and O. R. letters and he will send them to you at Armagh.

O. R.[b] has abandoned his equation 1 whereby he deduced 3 single valued quantities from 2 quadratic equations. He seems also, on reading over his paper, to have appreciated some of the difficulties which would occur to a stranger in reading it. Of course I cannot profess to follow with minute attention the course of an acrobat who drives 24 in hand, but as on more than one occasion he throws up the reins and starts a new team, it is probable that the results will be sufficiently flexible to adapt themselves to the facts, whatever the facts may be.[c] But O. R. says he has improved all this, and I hope he has. Thomson thinks Stoney should be acknowledged. Of course in the historical part it is an inexcusable omission to represent O. R.'s paper as the first attempt to assign dimensions to the structure of gases, for Loschmidt, Stoney, and Thomson had all assigned numerical values to the diameter of a molecule, a very definite thing, and Clausius & Maxwell had still earlier measured a mean path, something not quite as definite, whereas all that O. R. has done is to give a hardly numerical estimate of a still less definite quantity call the "mean range."

But <in order> to criticise Stoney when he comes to the Radiometer and the Stoney Stratum which he calls the Crookes' Layer requires hermeneutic powers of an order considerably above $\frac{dp}{dt}$[d] or even O. R. and this fact is a powerful temptation to the feeble minded to pass the whole thing over in silence, though to the eagle eye of Thomson a happy <thought> expression, even though in the midst of erroneous words, may illuminate the whole conglomeration of blunders with a meaning which the author himself could never be made to recognise.

With respect to Graham's experiments, O. R. is right and Thomson wrong, for there is considerable difficulty, perceived by Graham himself, in getting any simple law out of them. The result was that Graham was driven to the excessively fine pores of graphite plate, whereby he approximated to the law of effusion and got rid of transpiration almost entirely. On the other hand O. R. is impervious to Thomson's lucid statement of the difference between effusion and transpiration, namely that what restrains the flow of gas <in effusion> is collisions but that in effusion the collisions between two molecules of gas are very few compared with those between a molecule and a solid surface whereas in transpiration the collisions between two molecules mostly preponderate, so that in effusion the velocity of the molecules governs the velocity of effusion, whereas in transpiration viscosity is the ruling consideration.

I am afraid I have not answered your letter at all, except about O. R. being the discoverer of dimensional properties in gases. I have always felt inclined to give him leave to practise at his "mean range" till he has qualified himself to go in among the all comers for the R. S. meetings. ...

a. These are a continuation of Maxwell's comments of Osborne Reynolds paper, Reynolds, "On Certain Dimensional Properties of Matter in the Gaseous State. Part I and Part II," read February 5, 1879, abstract in *Proc. R. Soc. London* 28 (1879): 304–321 and *Nature* 19 (1879): 435, printed in full in *Phil. Trans. R. Soc. London* 170 (1880): 727–845, reprinted in Reynolds, *Papers* vol. 1, 257. It is clear that Thomson had returned his comments to Maxwell, and Maxwell is commenting upon both Reynolds and Thomson here.

b. O. R. is Osborne Reynolds

c. "24 in had," refers to the 24 terms in Reynolds's stress formulae. See also the letter from Maxwell's father to Maxwell, 3 November 1851, Campbell, and Garnett, *Life*, 160.

d. Maxwell

23. Letter from William Thomson to Maxwell, 7 September, 1879[a]

Cambridge University Library, Tait Letters

Sep. 7/79
Netherhall.
Largs.
Ayrshire

Dear $\frac{1}{\frac{dt}{dt}}$

Yes please send the peacocks. I think we will like them very much. Their coming to be fed and becoming tame (? attached–? affactionate?) is a real merit. Say when and how they are to come.

I have returned the OR[b] papers to Stokes with a few remarks. Transpiration, effusion, and diffusion leave still a want of a word for diffusion when the resistance to explosion into a vacuum is that due to collisions on solid and not terribly to mutual collisions betw.[een] molecules of gas. I don't think O. R. quite understands this: and with respect to the *"dimensional"* he does not take up the idea that the essence of the true theory of the radiometer is that the dimensions of the movable body must not be infinitely great in comparison with the length of the free path and that the *discoverer* of the theory of the radiometer was discovering that. As to thermal effusion I take the solid as if absolutely impermeable to heat and simply reckon the velocities at <the two> orifices on one side and at orifices on the other. Then it follows with general obviousness and particularly from your effusion formula in your 1866 paper. I don't approve of your investigation of it in your report on O. R.

Wild haste not to be late for church. Post this on the way.

W. T.

a. These are William Thomson's further comments on Osborne Reynolds, "On Certain Dimensional Properties of Matter in the Gaseous State. Part I and Part II," read February 5, 1879, abstract in *Proc. R. Soc. London* 28 (1879): 304–321; *Nature* 19 (1879): 435, printed in full in *Phil. Trans. R. Soc. London* 170 (1880): 727–845, reprinted in Reynolds, *Papers,* vol. 1, 257.

b. OR is Osborne Reynolds.

24. [Draft of Section 1 of "Stresses in Rarified Gases"] [a]

Cambridge University Library, Maxwell Manuscripts.

Let the masses of the molecules be M_1 and M_2 and their velocity-components ξ_1, η_1, ζ_1 and ξ_2, η_2, ζ_2 respectively. Let the velocity-compo nents of the centre of mass of the two molecules be a, b, c and <let those> let the velocity of M_1 relative to M_2 be V < $\alpha\,\beta\,\gamma$ then> and its direction cosines < $\alpha\,\beta\,\gamma$ > $\ell\,m\,n$

$$\xi_1 = a + \frac{M_2}{M_1 + M_2}\,\ell\,V \qquad \xi_2 = a - \frac{M_1}{M_1 + M_2}\,\ell\,V$$

<Let $\alpha^2 + \beta^2 + \gamma^2 = V$ >
then V is the velocity of M_1 relative to M_2.

The values of $a\,b\,c$ will <not be changed> be constant during the encounter. Before the encounter let a straight line be drawn through M_1 parallel to V and let a perpendicular b be drawn from M_2 to this line. The magnitude and direction of b and V will be constant as long as no force acts on the molecules. During the encounter the two molecules act on each other. If the force acts in the line joining their centres of mass, the product bV will remain constant, and the sum of the internal energies of the molecules is not changed by the encounter, V, and therefore b will be of the same magnitude after the encounter as before it, but their directions will be turned in the plane of V and b through an angle 2θ, this angle being a function of b and V which vanishes for values of b greater than the limit of molecular action. The plane of this angle must pass through V, but all positions of the plane <under> satisfying this condition are equally probable.

If we distinguish by an accent the values of the velocity components after the encounter, we may express the result by saying that a, b, c are unchanged, but that ℓ, m, n become ℓ', m', n' where ℓ', m', n' are the <components> direction cosines of a vector V' of the same magnitude as V but turned through an angle 2θ from the <posit> direction of V.

Let us next consider a function of the direction cosines $\ell\,m\,n$ and let us determine its value after an encounter. Considering $\ell\,m\,n$ as the coordinates of a point on a sphere, any function of these coordinates can be expressed in the form of a series of spherical harmonics.

Let $Y^{(n)}$ be the harmonic of order n belonging to this series. We may consider it as a function of $\ell\,m\,n$ or of the corresponding point say P on the sphere before the encounter. The effect of the encounter is to change ℓ, m, n into ℓ', m', n', the coordinates of another point P' such that $PP' = 2\theta$.

After the encounter, therefore <we have to substitute> instead of the value of

a. Draft of Maxwell, "Stresses in Rarified Gases Arising from Inequalities in Temperature," *Phil. Trans. R. Soc. London* 170 (1880): 231–256, abstract in *Proc. R. Soc. London* 27 (1878): 304–308, reprinted in Maxwell, *Scientific Papers,* vol. 2, 681–707 and as document VI-31.

25. [Expanded version of "Application of Spherical Harmonics to the Theory of Gases"][a]

Cambridge University Library, Maxwell Manuscripts.

Let the masses of the two molecules be M_1 and M_2 and their velocity-components $\xi_1\,\eta_1\,\zeta_1$ and $\xi_2\,\eta_2\,\zeta_2$ respectively.

Let the velocity components of the centre of mass of the two molecules be $a\,b\,c$ and let the velocity of M_1 relative to M_2 be V and let the direction cosines of V be $\ell\,m\,n$.

Let us also consider the direction of V as determined by the position of a point P on a sphere, so that the radius through P is parallel to V. We shall call this sphere the sphere of reference. It is, in fact, a diagram of velocity.

Before the encounter let a straight line be drawn through M_1 parallel to V and let a perpendicular b be drawn from M_2 to this line. The magnitude and direction of b and V will be constant as long as no force acts on the molecules. If, during the encounter, the force between the molecules acts in the line joining their centres of mass, the product bV will remain constant, and if the sum of the internal energies of the two molecules is not changed by the encounter, V, and therefore b will be of the same magnitude after the encounter, but their directions will be turned in the plane of V and b through an angle 2θ, this angle being a function of b and V which vanishes for values of b greater than the limiting distance of molecular action.

The effect of the encounter is therefore to change the point P on the sphere of reference into P', where $PP' =< 2\theta >$ is drawn in the plane of V and b and is equal to 2θ.

The values of a, b, c and V are not changed by the encounter.

Let us next consider a function, F, of ℓ, m, n, or what is the same thing, of the position of P, the point of reference, and let us determine what it becomes when $\ell\,m\,n$ become $\ell'\,m'\,n'$ or in other words, when P becomes P'. Let us call the new value F'.

Also, since every direction in which PP' can be drawn is equally probable, the mean value of F' is the arithmetical mean of all the values of F' for the circle of which the centre is P and the radius 2θ.

Let us begin by supposing that F is a spherical harmonic of order n which we may write $A_n Y^{(n)}$, then F' is the mean value of $Y^{(n)}$ for all points of the circle described about P with a radius 2θ. Let us write $\mu = \cos 2\theta$.

< F' is therefore a function of μ and since>

Since F' is found by a taking the mean of a system of spherical harmonics of the order n, it is itself an harmonic of order n and since it is a function of μ only, it must be a zonal harmonic. Also when $\mu = 1$ P' coincides with

P and $F' = F$. Hence if $F = AY^{(n)}$, $F' = AY_n^{(n)}\, P^{(n)}(\mu)$, where $P^{(n)}$ is the zonal harmonic, or Legendre Function, of order n.

[The following fragment is on a separate loose page but seems to be related to the preceding material]

If we distinguish by an accent the values of the velocity components after the encounter, we may express the result by saying that a, b, c and V remain the same but that

[rest of the page is blank]

Now any function F of $\ell\, m\, n$ may be expressed in the form of a series of spherical harmonics

$$F = A_0\, Y^{(0)} + \&c + A_n\, Y^{(n)}$$

The mean value of F after an encounter in which the cosine of the angle of deviation is μ will be

$$F' = A_0\, Y^0\, P^{(0)}(\mu) + \&c + A_n\, Y^{(n)}\, P^{(n)}(\mu)$$

in which each term is multiplied by the zonal harmonic of corresponding order whose argument is μ.

To find the effect of an encounter on the value of any function of ξ_1, η_1, ζ_1, ξ_2, η_2, ζ_2 we begin by substituting for these quantities

$$a + \tfrac{M_2}{M_1+M_2}\ell\, V, \qquad b + \tfrac{M_2}{M_1+M_2} m\, V, \qquad c + \tfrac{M_2}{M_1+M_2} n\, V$$

$$a - \tfrac{M_1}{M_1+M_2}\ell\, V, \qquad b - \tfrac{M_1}{M_1+M_2} m\, V, \qquad c - \tfrac{M_1}{M_1+M_2} n\, V$$

respectively. We thus obtain a function of a, b, c, V, ℓ, m, n and <arranging> if we express it as a series of spherical harmonics in ℓ, m, n each multiplied by a function of a, b, c and V the mean value after the encounter may be found by multiplying each <term> of the spherical harmonics by the zonal harmonic of the same order whose argument is μ.

We may then express a, b, c, ℓV, mV, nV in terms of ξ_1, η_1, ζ_1, ξ_2, η_2, ζ_2, and thus obtain the function as modified by the encounter in terms of the quantities.

$$M_1\xi_1 + M_2\xi_2, \qquad M_1\eta_1 + M_2\eta_2, \qquad M_1\zeta_1 + M_2\zeta_2$$

$$\xi_1 - \xi_2, \qquad \eta_1 - \eta_2, \qquad \zeta_1 - \zeta_2,$$

When as in the present investigation we are dealing with a single gas, and we are considering the value of a function of ξ, η, ζ, the <component> velocity-components of any one of the molecules, we shall have to take into account the effect of the encounter on both molecules, so that the function of which we have to find the variation is $f(\xi_1, \eta_1, \zeta_1) + f(\xi_2, \eta_2, \zeta_2)$.

If we write

$$\xi_1 = a + x \qquad \eta_1 = b + y \qquad \zeta_1 = c + z$$

$$\xi_2 = a - x \qquad \eta_2 = b - y \qquad \zeta_2 = c - z$$

<then $\xi_1 + \xi_2 = 2a$ >
<and since this is not altered by the encounter>

Let us being with $\sum(M\xi)$, or the sum of all the momenta of the molecules resolved in the direction of x. The part of this quantity which is <affected by the> we have to consider is that which belongs to the two molecules between which the encounter takes place

$$\text{or} \qquad M_1\xi_1 + M_2\xi_2$$

and since the value of this quantity is not altered by the encounter we find for the variation of $\sum(M\xi)$ due to the encounter

$$\delta \sum(M\xi) = 0$$

Let us next take a solid harmonic function of ξ, η, ζ of the second order, such as $\sum(M\xi\eta)$

$$M_1\xi_1\eta_1 + M_2\xi_2\eta_2 = (M_1 + M_2)\,ab + \frac{M_1 + M_2}{M_1 + M_2} V^2 \ell m$$

After the encounter, ℓm becomes $\ell m P^{(2)}(\mu)$, but ab does not vary. Substituting for $V^2\ell m$ its value $(\xi_1 - \xi_2)(\eta_1 - \eta_2)$ or

$$\xi_1\eta_1 + \xi_2\eta_2 - \xi_1\eta_2 - \xi_2\eta_1$$

we observe that <if as the product of> a function of the motion of one molecule is independent of the motion of another, so that the mean value of $\xi_1\,\eta_2$ is $\overline{\xi_1}\,\overline{\eta_2}$

We thus find

$$\delta \sum(M\xi\eta) = \frac{M_1 M_2}{M_1 + M_2}\left[\xi_1\eta_1 + +\xi_2\eta_2 - \overline{\xi_1\eta_2} - \overline{\xi_2}\,\overline{\eta_1}\right]\left(P^{(2)}(\mu) - 1\right)$$

When, as in the present investigation there is only one kind of gas, there is no distinction between the molecules M_1 and M_2 <and when we define the component velocities with reference> and when we define $\xi\,\eta\,\zeta$ as the velocity-components relative to the centre of mass of the system of molecules the mean values of these quantities are zero, so that the equation may be written

$$\delta \sum(M\xi\eta) = M\xi\eta\left(P^{(2)}(\mu) - 1\right)$$

Since, as we have shown, the

We might have taken any solid harmonic in ξ, η, ζ of the second degree, say $H^{(2)}$ and found

$$\delta \sum(MH^{(2)}) = MH^{(2)}\left(P^{(2)}(\mu) - 1\right)$$

Let us now consider the function $\sum M(\xi^2+\eta^2+\zeta^2)$

$$M_1(\xi_1^2+\eta_1^2+\zeta_1^2)+M_2(\xi_2^2+\eta_2^2+\zeta_2^2)=(M_1+M_2)(a^2+b^2+c^2)+\frac{M_1M_2}{M_1+M_2}V^2$$

Since none of the quantities a, b, c, V is altered by the encounter we find that

$$\delta\big(M(\xi^2+\eta^2+\zeta^2)\big)=0$$

<Since all fun>

We are now able to express the effect of the encounter on any function of the second degree. We resolve the function into a solid harmonic in ξ, η, ζ, and a multiple of $\xi^2\,\eta^2\,\zeta^2$. The variation of the function is that due to the solid harmonic only, for the other part does not vary.

We have next to consider a function of three dimensions in ξ, η, ζ. We begin by resolving it into a solid harmonic of the third order and a solid harmonic of the first order multiplied by $(\xi^2+\eta^2+\zeta^2)$

$$F(M\xi\eta\zeta)=MH^{(3)}+MH^{(1)}(\xi^2+\eta^2+\zeta^2)$$

Let us begin with the solid harmonic of the third order.

$$\begin{aligned}H^{(3)}(\xi\,\eta\,\zeta)=H^{(3)}&(a,\,b,\,c)\\&+\frac{M_2}{M_1+M_2}V\left[\ell\frac{dH}{da}+m\frac{dH}{db}+n\frac{dH}{dc}\right]^b\\&+\frac{M_2^2}{(M_1+M_2)^2}V^2\left[a\frac{dH}{d\ell}+b\frac{dH}{dm}+c\frac{dH}{dn}\right]\\&+\frac{M_2^3}{(M_1+M_2)^3}V^3\,H(\ell m,n)\end{aligned}$$

The first term of the second member of this equation is not altered by the encounter. The second term is cancelled by the corresponding term for the second molecule. If $M_1=M_2$ the fourth term also vanishes so that the variation depends on the third term, in which the first derivatives of H are solid harmonics of the second order. Remembering that the mean values of ξ, η, ζ are zero the third term reduces to $\frac{3}{2}\,MH^{(2)}(\xi\,\eta\,\zeta)$ and we find

$$\delta\sum MH^{(3)}(\xi,\,\eta,\,\zeta)=\frac{3}{2}MH^{(2)}(\xi,\,\eta,\,\zeta)\,\left(P^{(2)}(\mu)-1\right)$$

If the function is of the form $M\xi(\xi^2+\eta^2+\zeta^2)$ we <find> put the sum of the <two> values for the two molecules in the form

$$2Ma(a^2+b^2+c^2)+\frac{1}{2}MaV^2+MV^2(a\ell^2+b\ell m+c\ell n)$$

The first and second terms are not altered by the encounter and we may write the third term in the form

$$\left\langle \frac{1}{3}MV^2 a + \frac{2}{3}MV^2\, a\left(\frac{3}{2}\ell^2 - \frac{1}{2}\right)\right\rangle$$

$$\frac{1}{3}MV^2 a + \frac{2}{3}MV^2\left[a\left(\ell^2 - \frac{1}{2}m^2 - \frac{1}{2}n^2\right) + \frac{3}{2}b\,\ell\, m + \frac{3}{2}c\,\ell\, n\right]$$

in which the first term does not vary and the second term is a harmonic of the second order, which must be multiplied by $P^{(2)}(\mu) - 1$.

<Substituting the> Expressing the last term as a function of ξ, η, ζ we find

$$\delta \sum M\xi(\xi^2 + \eta^2 + \zeta^2) = \frac{2}{3}M\xi(\xi^2 + \eta^2 + \zeta^2)$$

[the rest of the page is blank]

<We are now able to es>

Now let h_n be a function of x, y, z which is a solid harmonic. After an encounter in which $cos 2\theta = \mu$ it becomes

$$h_n P(n)(\mu)$$

If there are N molecules and if of these δN have encounters of this kind in the time δt then

$$N\frac{\delta}{\delta t}n_n =$$

$$\frac{\delta}{\delta t}\sum h_n = \delta N h_n\left[P^{(2)}(\mu) - 1\right]$$

[the rest of the page is blank]

Let us suppose that at a given instant there are in unit of volume dN_1 molecules of the first kind whose velocity components are between ξ_1 and $\xi_1 + d\xi_1$, η_1 and $\eta_1 + d\eta_1$ ζ_1 and $\zeta_1 + d\zeta_1$ and dN_2 of the second kind whose velocity components have their limits distinguished by the suffix $_{(2)}$.

Let us first suppose that they do not interfere with each others motion and let us determine the number of the molecules of the first kind which would in the time δt pass a given molecule of the second kind at a distance between b and $b + db$.

Each of these molecule must pass through an annular area $2\pi\, bdb$ and the number in the time δt would be $dN_1 r\delta t\, 2\pi\, bdb$.

The whole number of pairs which pass one another within the given limits is therefore

$$dN_1\, dN_2\, 2\pi\, bdb\, r\delta t$$

This would be the number of encounters which would take place in the time δt if each encounter did not take place till the first molecule had reached the plane through the second molecule normal to r

But the encounter begins before <this> the second molecule reaches this plane, so that the actual number of encounter in <unit of time> the time δt is somewhat greater than this when the gas is dense. In rarefied gases however we may consider this as a first approximation to the number of encounters of a given kind.

Now let μ be the cosine of the angle through which r is turned by an encounter of the given kind, μ is a function of b and r only, and let h_n be a solid harmonic function of $x\, y\, z$ of the order n multiplied by a constant or by a power of r^2.

If H_n is any homogeneous function of x, y, and z of positive integral degree n then the homogeneous function of the same degree

$$h_n = H_n + \frac{n \cdot n-1}{2(n+3)} \nabla^2 H_n\, r^2 + \&c$$
$$+ \frac{\nabla^{2p}\, H_n\, r^{2p}}{2 \cdot 4 \cdot 2p \cdot (n+3)(n+5) \cdot (n+2p+1)} + \&c.$$

is a solid harmonic of order n.

For if we perform on h_n the operation ∇^2, it will vanish.

We may therefore express H_n in the form

$$H_n = h_n + H_{n-2}\, r^2$$

where < H_{n-2} > h_n is a solid harmonic of order n, and H_{n-2} is a homogeneous function of n-2 degrees.

Proceeding in the same way we may express H_{n-2} in the form

$$h_{n-2} + H_{n-4}\, r$$

where h_{n-2} is a solid harmonic of the order $n-2$.

We may therefore express any homogeneous function H_n in the form <of a series of terms each of which is a solid harmonic>

$$H_n = h_n + h_{n-2}\, r^2 + \&c$$

where each term is a solid harmonic multiplied by a power of r.

a. See Maxwell, "Stresses in Rarified Gases Arising from Inequalities in Temperature. Note added to Section 1, May 1879," *Phil. Trans. R. Soc. London* 170 (1880): 231–256.

b. The square brackets are Maxwell's in this and following equations.

26. [Draft of Section 2 of "Stresses in Rarified Gases"][a]

Cambridge University Library, Maxwell Manuscripts.

In the gas

The number of molecules which at a given instant are within the element of volume $dx\ dy\ dz$, and whose velocity-components lie between the limits $\xi \pm 1/2 d\xi$, $\eta \pm 1/2 d\eta$, $\zeta \pm 1/2 d\zeta$ being denoted by

$$dN + f(\xi\,\eta,\ \zeta,\ x,\ y,\ z,\ t)\, d\xi d\eta,\, d\zeta\, dx\, dy\, dz$$

Boltzmann[b] has shown that the function f must satisfy the equation

$$\frac{df}{dt} + \xi\frac{df}{dx} + \eta\frac{df}{dy} + \zeta\frac{df}{dz} + X\frac{df}{d\xi} + Y\frac{df}{d\eta} + Z\frac{df}{d\zeta} + \iiint d\xi_1 d\eta_1 d\zeta_1 \int bdb \int \rho d\rho\, V(f\, f_1 - f' f_1') = 0$$

<where f_1 denotes what f becomes when> where f_1 f' and f_1' denote what f becomes when in place of the velocity components $\xi\,\eta\,\zeta$ we write, respectively, ξ_1, η_1, ζ_1 those of another molecule, < $\xi'\,\eta'\,\zeta'$ these> and $\xi'\,\eta'\,\zeta'$ and $\xi_1'\,\eta_1'\,\zeta_1'$ the <values> velocity component of the two molecules after the encounter.

If the medium is enclosed in a vessel with perfectly elastic sides, the ultimate value of f is

$$f = A\, e^{-h[\chi + 1/2\,(\xi^2 + \eta^2 + \zeta^2)]}$$

where χ is the potential of the external force whose components are X, Y, Z and h is a constant depending on the ultimate temperature of the medium.

This is the only case in which the problem can be solved without assuming <the particular form> an expression for the force between two molecules in terms of their distance.

<I have shown that by the assumption that the <force> repulsion between two molecules is inversely as the fifth power of the distance, the integrations in the last term of equation ( ) can be easily performed.>

If we suppose the action between two molecules to be like that of two "rigid elastic" sphere, the <nature> angle of deflexion depends only on b, the <distance> shortest distance between the molecules on their undisturbed paths. Hence the number of encounters of a given kind is proportional to the relative velocity of the encountering molecules, and the quantity to be integrated involves this relative velocity V, <which> so that the integration cannot be performed unless we <know> assume a particular form for the function.

If, however, we suppose that the force between the molecules is <exerted at various distances> a continuous function of the distance between them, the angle of deflexion will depend on the relative velocity and will be less when the relative velocity is greater. In particular, when the force is inversely as the fifth power of the distance <between the> the angle of deflexion will be the same when V is increased and be diminished in the proportion, so that the quantity $Vb^2\,db$ is in this case independent of V.

a. Maxwell, "Stresses in Rarified Gases Arising from Inequalities in Temperature," *Phil. Trans. R. Soc. London* 170 (1880): 231–256.

b. Ludwig Boltzmann, "Weitere Studien über das Wärmegleichgewicht unter Gasmolekülen," *Sitz. Math.-Naturwiss. Cl. Akad. Wiss. Wien* 66 (1872): 275–370. English translation in Brush, *Kinetic Theory,* vol. 2.

27. [Draft of parts of Sections 7–11 of "Stresses in Rarified Gases"][a]

Cambridge University Library, Maxwell Manuscripts.

We shall employ the method given in my paper on the Dynamical Theory of Gases Phil. Trans. 1867, but <as we shall> for the sake of simplicity we shall assume that all the molecules are of the same kind.

Let M be the mass of a single molecule and let its velocity components be $u+\xi$, $v+\eta$, $w+\zeta$, u, v, w, being those of the centre of mass of all the molecules within a certain small distance.

Let N be the number of molecules in unit of volume and ρ their density then $\rho = NM$.

Let the symbol $\sum$ prefixed to any expression denote the sum of all the values of this expression within unit of volume.

Thus $\qquad \sum M = \rho \qquad \sum M = 0 \qquad \sum M\xi^2 = p \qquad$ &c

Let Q be any property of a molecule, we have to consider the variation of $\sum Q$ in an element of volume arising from various causes.

In the first place molecules may enter the element or pass out of it and so alter the value of $\sum Q$.

In the second place the value of Q for molecules within the element may be changed, and this either by encounters with other molecules or by the action of external forces.

We shall use the symbol $\frac{\partial}{\partial t}$ to denote the actual time-variation in an element moving with velocity-components U, V, W, and $\frac{\delta}{\delta t}$ to denote that which arises from the <action of forces > encounters of the molecules and $\frac{\partial'}{\partial t}$ to denote the effect of external forces.

Let the element under consideration travel with the component velocities U, V, W, then if Q be an property of a molecule

$$\frac{\partial \sum Q}{\partial t} + \frac{d}{dx}\sum(u+\xi-U)\,Q + \frac{d}{dy}\sum(v+\eta-V)\,Q + \frac{d}{dz}\sum(w+\zeta-W)\,Q = \frac{\delta \sum Q}{\delta t} \tag{1}$$

<But $Q = M$, Since>

If after performing the differentiations we make $U = u$, $V = v$, $W = w$ the equation becomes

$$\frac{\partial \sum(Q)}{\partial t} + (\frac{du}{dx} + \frac{dv}{dy} + \frac{dw}{dz})\sum Q + \frac{d}{dx}\sum Q\xi + \frac{d}{dy}\sum Q\eta + \frac{d}{dz}\sum Q\zeta = \frac{\delta \sum Q}{\delta t} \tag{2}$$

Let us begin by putting $Q = M$. Since M is not altered by the action of forces, $\frac{\delta M}{\delta t} = 0$ and we find

$$\frac{\partial \rho}{\partial t} + \rho(\frac{du}{dx} + \frac{dv}{dy} + \frac{dw}{dz}) = 0 \tag{3}$$

which is the ordinary "equation of continuity"

Let us next put $Q = M(u+\xi)$ the momentum of a molecule parallel to x.

$$\frac{\partial(\rho u)}{\partial t} + (\frac{du}{dx} + \frac{dv}{dy} + \frac{dw}{dz})\rho u + \frac{d}{dx}\sum M\xi^2 + \frac{d}{dy}\sum M\eta\zeta + \frac{d}{dz}\sum M\xi\zeta = \frac{\delta \sum Q(u+\xi)}{\delta t} \tag{4}$$

Multiplying (3) by $-u$ and adding we obtain the ordinary equation of motion

$$\rho\frac{\partial u}{\partial t} + \frac{d}{dx}\sum M\xi^2 + \frac{d}{dy}\sum M\xi\eta + \frac{d}{dz}\sum M\xi\zeta = \rho X \tag{5}$$

For since the encounters of the molecules cannot change the sum of their momenta parallel to x, the only cause of variation is the external force X. There are two other equations of motion of a similar form.

Eliminating[c] by means of this equation the second term of (2) we obtain the more convenient form

$$\left\langle \frac{\partial \sum Q}{\partial t} - \frac{\sum Q}{\rho}\frac{\partial \rho}{\partial t} + \frac{d}{dx}\sum(Q\xi) + \frac{d}{dy}\sum(Q\eta) + \frac{d}{dz}\sum(Q\zeta) = \frac{\delta \sum Q}{\delta t} \quad (4) \right\rangle$$

$$\rho\frac{\partial}{\partial t}\left(\frac{\sum Q}{\rho}\right)+\frac{d}{dx}\sum(Q\xi)+\frac{d}{dy}\sum Q\eta+\frac{d}{dz}\sum Q\zeta=\frac{\delta\sum Q}{\delta t} \tag{4}$$

Let us next put $Q = M(u+\xi)$, then $\sum Q = \rho u$ and since the mutual encounters of the molecules do not alter the sum of their momenta parallel to x the only cause of variation of this quantity is the external force X, hence

$$\frac{\partial\sum M(u+\xi)}{\delta t}=\rho X$$

also $\dfrac{\partial(\rho u)}{\partial t}-\dfrac{\rho u}{\rho}\dfrac{\partial\rho}{\partial t}=\rho\dfrac{\partial u}{\partial t}$ and the equation becomes

$$\rho\frac{\partial u}{\partial t}+\frac{d}{dx}\sum(M\xi^2)+\frac{d}{dy}\sum(M\xi\eta)+\frac{d}{dz}\sum(M\xi\zeta)=\rho X \tag{5}$$

This is one of the three ordinary equations of motion.

Let us next put $Q = M(u+\xi)^2$ then $\sum Q = \rho u^2 + \sum M\xi^2$

$$\begin{aligned}\rho\frac{\partial}{\partial t}(u^2+\frac{\sum(M\xi^2)}{\rho}&+\frac{d}{dx}(2u\sum(M\xi^2)+\sum M\xi^3)\\&+\frac{d}{dy}(2u\sum M\xi\eta+\sum M\xi^3\eta)+\frac{d}{dz}(2u\sum M\xi\zeta+\sum M\xi^2\zeta)\\&=\frac{\delta}{\delta t}(\rho u^2+\sum M\xi^2)\end{aligned}$$

Multiplying 5 by $2u$ and subtracting

$$\begin{aligned}\left\langle\rho\frac{\partial}{\partial t}\left(\frac{\sum M\xi^2}{\rho}\right)+\frac{d}{dx}\sum(M\xi^2)+\frac{d}{dy}\sum(M\xi^2\eta)+\frac{d}{dz}\sum(\xi^2\zeta)\right.\\\left.=\frac{\delta}{\delta t}\sum M\xi^2\right\rangle\end{aligned}$$

$$\begin{aligned}\rho\frac{\partial\xi^2}{\partial t}+2\rho\,\xi^2\frac{du}{dx}+2\rho\,\xi\eta\frac{du}{dy}+2\rho\,\xi\zeta\frac{du}{dz}\\+\frac{d}{dx}\rho\,\xi^2+\frac{d}{dy}\rho\,\xi\eta+\frac{d}{dz}\rho\,\xi\zeta=\frac{\delta}{\delta t}\sum M\xi^2\end{aligned}$$

$$\begin{aligned}\rho\frac{\partial}{\partial t}\xi\eta+\rho\,\xi\eta\frac{du}{dx}+\rho\,\eta^2\frac{du}{dy}+\rho\,\eta\zeta\frac{du}{dz}+\rho\,\xi^2\frac{dv}{dx}+\rho\,\xi\eta\frac{dv}{dy}+\rho\,\xi\zeta\frac{dv}{dz}\\+\frac{d}{dx}\rho\,\xi^2\eta+\frac{d}{dy}\rho\,\xi\eta^2+\frac{d}{dz}\rho\,\xi\eta\zeta=\frac{\delta}{\delta t}\rho\,\xi\eta\end{aligned}$$

Lastly let us put $Q = M(u+\xi)^3$

$$Q=\rho u^3+3u\sum(M\xi^2)+\sum(M\xi^3)$$

$$\rho\frac{\partial}{\partial t}\Big(u^3+3\frac{u}{\rho}\sum(M\xi^2)+\frac{1}{\rho}\sum(M\xi^3)+\frac{d}{dx}\sum M(3u^2\xi^2+3u\xi^3+\xi^4)$$

$$+\frac{d}{dy}\sum M(3u^2\xi\eta+3u\xi^2\eta+\xi^3\eta)+\frac{d}{dz}\sum M(3u^2\xi\zeta+3u\xi^2\zeta+\xi^3\zeta)\Big)$$

$$=\frac{\delta}{\delta t}(\rho u^3+3u\sum(M\xi^2)+\sum M\xi^3)$$

Eliminating by means of the former equations

$$\Big\langle\rho\frac{\partial}{\partial t}\sum(\frac{M\xi^3}{\rho})+\frac{d}{dx}\sum(M\xi^4)+\frac{d}{dy}\sum(M\xi^3\eta)+\frac{d}{dz}\sum M\xi^3\zeta$$

$$=\frac{\delta}{\delta t}\sum(M\xi^3)\Big\rangle$$

$$\rho\frac{\partial}{\partial t}\xi^3+3\rho\,\xi^3\frac{du}{dx}+3\rho\,\xi^2\eta\frac{du}{dy}+3\rho\,\xi^2\zeta\frac{du}{dz}-3\xi^2\Big(\frac{d}{dx}\rho\,\xi^2+\frac{d}{dy}\rho\,\xi\eta+\frac{d}{dz}\rho\,\xi\zeta\Big)$$

$$+\frac{d}{dx}\rho\xi^4+\frac{d}{dy}\rho\xi^3\eta+\frac{d}{dz}\rho\xi^3\zeta=\frac{\delta}{\delta t}\rho\xi^3$$

a. Maxwell, "Stresses in Rarified Gases Arising from Inequalities in Temperature." *Phil. Trans. R. Soc. London* 170 (1880): 231–256.

b. Maxwell, "On the Dynamical Theory of Gases," *Phil. Trans. R. Soc. London* 157 (1867): 49–88, reprinted in *Scientific Papers,* vol. 2, 26–78, and Garber, Brush, and Everitt, *Maxwell on Molecules and Gases,* 419–472.

c. This paragraph, which begins a new page in Maxwell's notes, seems to be a revision of part of the preceding material.

28. [Draft of "Note added June, 1879" to Section 15 of "Stresses in Rarified Gases"][a]

Cambridge University Library, Maxwell Manuscripts.

I have recently examined the functions of the <fourth fifth and> four five and six dimensions, in order to find <whether they give rise> the nature of the terms arising from them in the expressions <we have already> for the conduction of heat, the stresses, and the equations of motion.

If we neglect those terms of six dimensions which would be zero in a gas in equilibrium, and suppose that the other terms have the same values as in a gas in equilibrium and if we use these values to determine the rate of decay of functions of five dimensions, we find that if the function of five dimensions is either a solid harmonic of <five dimensions> the fifth order or a solid harmonic of the third order multiplied by the square of the velocity, the rate of decay is zero. But we know that the rate of decay of such functions is <in a defin> equal to the function itself divided by a definite time-modulus. Hence as there is no decay these functions themselves must be zero.

If, however we consider an harmonic of the first order multiplied by the fourth power of the velocity we find that the rate of decay depends on the very same integral as that which determines the rate of decay of harmonics of the second order so that we may express it in terms of the viscosity. We thus obtain for the approximate mean values of terms of five dimensions

$$\overline{\xi^5} = -30\frac{\mu}{\rho}R^3\theta^2\frac{d\theta}{dx}$$

$$\overline{\xi^3\eta^2} = \overline{\xi^3\zeta^2} = \overline{\xi\eta^4} = \overline{\xi\zeta^4} = -6\frac{\mu}{\rho}R^3\theta^2\frac{d\theta}{dx}$$

$$\overline{\xi\eta^2\zeta^2} = -2\frac{\mu}{\rho}R^3\theta^2\frac{d\theta}{dx}$$

Terms in which η or ζ only is of odd dimensions may be similarly expressed in terms of the derivatives of θ with respect to y and z respectively.

The values of terms in which ξ, η and ζ are all of odd dimensions may be neglected in comparison with these. If we trace the effect of these terms on terms of lower dimensions, we find that they introduce, into the terms of four dimensions, corrections depending on second derivatives of the temperature multiplied into $\mu^2/\rho\,\theta$.

<Into the terms of the three dimensions which already contain>

<We thus find that wherever in the expressions of the present paper we have terms of the form $(\frac{\mu}{\rho\,\theta}\frac{d}{dx})^2\theta$, the <addi> correction which must be added will consist of terms of the form $(\frac{\mu}{\rho\,\theta}\frac{d}{dx})^{n+2}\theta$ >

We thus find that wherever in the present paper we have terms depending on θ arising from the principal terms of the fourth order, we shall have as the next approximation terms depending in the same way upon the second derivatives of θ multiplied by λ^2 or μ^2/ρ

Hence in the equations of motion, in which we have here the <second der> third derivatives of θ multiplied by $\mu^2/\rho\theta$, we should have, as additional terms, the fifth derivatives of θ multiplied by $\mu^4/\rho^3\theta$.

It is only in very highly rarified gases <that these terms> and only when the <changes of> temperature varies considerably within distances comparable to λ or μ/ρ, that these additional terms becomes of sensible value.

In precisely the same way I have found that the terms of <the fourth> four dimensions contain terms depending on the motion of the gas, so that, if we consider these terms only, we have

$$\xi^4 = 3R^2\theta^2 + \frac{48}{7}\frac{\mu}{\rho} R^2\,\theta^2\,(2\frac{du}{dx} - \frac{dv}{dy} - \frac{dw}{dz})$$
$$\eta^2\zeta^2 = R^2\,\theta^2 + \frac{8}{7}\frac{\mu}{\rho} R^2\,\theta^2\,(2\frac{du}{dx} - \frac{dv}{dy} - \frac{dw}{dz})$$

$$\xi^3\eta = \zeta\eta^3 = \frac{36}{7}\frac{\mu}{\rho} R^2\,\theta^2\left(2\frac{du}{dy} + \frac{dv}{dx}\right) = 3\xi\eta\zeta^2$$

In the equations of motion, in addition to the terms involving the second derivatives of u, v, w multiplied by μ, <these terms> there will <introduce> be additional terms involving the fourth derivatives of u, v, w multiplied by μ^3/ρ^2.

Hence the corrections which must be applied to the formulae of this paper are zero <unless> whenever the velocities can be expressed as functions of three dimensions and the temperature as a function of four dimensions in x, y, z, and in all cases they are very small provided the <rate of> variation of the velocity and temperature within distances comparable to λ is small.

a. Maxwell, "Stresses in Rarified Gases Arising from Inequalities in Temperature." *Phil. Trans. R. Soc. London* 170 (1880): 231–256.

29. [Expanded version of document 28]

Cambridge University Library, Maxwell Manuscripts.

We shall not consider functions of more than six dimensions in $\xi\,\eta\,\zeta$, and of these only the principal terms which are, to the degree of approximation we adopt,

$$\xi^6 = 15\,R^3\,\theta^3$$
$$\xi^4\,\eta^2 = 3\,R^3\,\theta^3$$
$$\xi^2\,\eta^2\,\zeta^2 = R^3\,\theta^3$$

In the equations for terms of five dimensions we shall consider that part only which involves the principal terms of six dimensions neglecting the others, as they involve the small factor μ/ρ.

We thus find

$$\rho\,\frac{\delta}{\delta t}\xi^5 = 5\xi^4\rho\,\frac{\partial u}{\partial t} + \frac{d}{dx}\rho\,\xi^6 = 15\rho\,R^3\,\theta\frac{d}{dx}\theta^2$$

or

$$\frac{\delta}{\delta t}\xi^5 = 30\,R^3\,\theta^2\,\frac{d\theta}{dx}$$

Similarly we obtain

$$\frac{\delta}{\delta t}\xi^3\eta^2 = \frac{\delta}{\delta t}\xi^3\zeta^2 = \frac{\delta}{\delta t}\xi\eta^4 = \frac{\delta}{\delta t}\xi\zeta^4 = 6\,R^3\,\theta^2\frac{d\theta}{dx}$$
$$\frac{\delta}{\delta t}\xi\eta^2\zeta^2 = 2\,R^3\,\theta^2\frac{d\theta}{dx}$$

To determine the values of the terms of five dimensions we observe that the solid harmonics of the fifth order in which the dimensions of ξ are odd and those of η and ζ are even are

$$\xi^5 - 10\xi^3\eta^2 + 5\xi\eta^4 \qquad (\;= A)$$
$$\xi^5 - 10\xi^3\eta^2 + 5\xi\zeta^4 \qquad (\;= B)$$
$$\xi\eta^4 - 6\xi\eta^2\zeta^2 + \xi\zeta^4 \qquad (\;= C)$$

Now $\frac{\delta}{\delta t}H_5 =$ [space] H
differentiating each of these expressions with respect to t and substituting the values in [space] we find that all three disappear, and since the time-variation of any solid harmonic is proportional to the solid harmonic itself, it follows that each of the solid harmonics is zero.

We have next to consider terms of the form of solid harmonics of the third order multiplied by $(\xi^2 + \eta^2 + \zeta^2)$. These are

$$(\xi^3 - 3\xi\eta^2)\,(\xi^2 + \eta^2 + \zeta^2)$$

and

$$(\xi^3 - 3\xi\zeta^2)\,(\xi^2 + \eta^2 + \zeta^2)$$

Multiplying these out, differentiating with respect to t and substituting the values in [space] we find that these terms also disappear, and we conclude as before that each of the products of solid harmonics of the third order into $(\xi^2 + \eta^2 + \zeta^2)$ is zero.

We have therefore only to consider the term

$$\xi(\xi^2 + \eta^2 + \zeta^2)^2 \qquad (= F)$$

We find by differentiation and substitution as before, that

$$\frac{\delta F}{\delta t} = 70\,R^3\,\theta^2\frac{d\theta}{dx}$$

but

$$\frac{\delta F}{\delta T} = -\frac{p}{\mu}F$$

We therefore have the following equations to determine the functions of <the> five dimensions

$$F = -70\,\frac{\mu}{p}\,R^3\,\theta^2\,\frac{d\theta}{dx}$$
$$A = B = C = D = E = 0$$

whence we find

$$\xi^5 = -30\,\frac{\mu}{p}\,R^3\,\theta^2\,\frac{d\theta}{dx}$$
$$\xi^3\eta^2 = \xi^3\zeta^2 = \xi\eta^4 = \xi\zeta^4 = -6\,\frac{\mu}{p}\,R^3\theta^2\,\frac{d\theta}{dx}$$
$$\xi\eta^2\zeta^2 = -2\,\frac{\mu}{p}\,R^3\,\theta^2\,\frac{d\theta}{dx}$$

The functions of five dimensions in which η or ζ only is odd are of the same form, but depend on $\frac{d\theta}{dy}$ and $\frac{d\theta}{dz}$ respectively. Those in which the dimensions of all the components are odd are of a higher order of small quantities and are to be neglected.

We come next to the equations for the variation of terms of four dimensions. We shall arrange these in two classes, according to whether all the dimensions are even, or two of them are odd. In the terms in $\frac{\partial}{\partial t}$ and in the terms involving the differential coefficients of u, v, w, we retain only the principal terms of four dimensions.

$$\rho\,\frac{\delta}{\delta t}\xi^4 = \rho\,\frac{\partial}{\partial t}\xi^4 + 4\xi^3\rho\,\frac{\partial u}{\partial t} + \frac{d}{dx}\rho\,\xi^5 + \frac{d}{dy}\rho\,\xi^4\eta + \frac{d}{dz}\rho\,\xi^4\zeta + 4\xi^4\rho\,\frac{du}{dx}$$

But

$$4\xi^3\rho\,\frac{\partial u}{\partial t} = -4\xi^3\,\frac{d}{dx}\rho\,\xi^2 - 4\xi^3\,\frac{d}{dy}\rho\,\xi\eta - 4\xi^3\,\frac{d}{dz}\rho\,\xi\zeta$$

Hence, substituting the <principal> first approximate values as follows

$$\langle\xi^4 = 3R^3\,\theta^2\rangle$$
$$\zeta^2 = R\theta(1+\alpha^2) \qquad \xi\eta = R\,\theta\,\alpha\beta$$
$$\xi^3 = (R\theta)^{3/2}\alpha^3 \qquad \xi\eta^2 = (R\,\theta)^{3/2}\alpha\beta^2$$
$$\xi^4 = 3R^2\theta^2(1+2\alpha^2) \qquad \xi^2\eta^2 = R^2\theta^2(1+\alpha^2+\beta^2)$$
$$\xi^3\eta = 3R^2\theta^2\,\alpha\beta \qquad \xi\eta\zeta^2 = R^2\theta^2\,\alpha\beta$$
$$\xi^5 = 10(R\,\theta)^{5/2}\alpha^3 \qquad \xi^3\eta^2 = (R\,\theta)^{5/2}(\alpha^3+3\alpha\beta^2),$$
$$\xi\eta^2\zeta^2 = (R\,\theta)^{5/2}(\alpha\beta^2+\alpha\gamma^2)$$

30. [Notes for Appendix to "Stresses in Rarified Gases"] [a]

Cambridge University Library, Maxwell Manuscripts.

We shall suppose that part of the fixed surface is such that in whatever manner the molecules strike it, they are afterwards projected from the surface with their velocities distributed as they would be in a gas in complete equilibrium with the surface.

We may suppose that this is effected by the gas being condensed on these parts of the surface and then, <evapor> after having lost all traces of its original distribution of velocities, evaporating quietly from the surface.

On this hypothesis, the molecules close to the surface which at any instant are moving away from it have their velocities distributed in the same way as if they had come from a mass of gas on the other side of the surface free from currents and having the temperature of the <surface> solid, the only other condition being that the <pressure> density of the imaginary gas must be such that as many molecules pass through the surface in the one direction as in the other.

The hypothesis of MM Kundt & Warburg[b] was <something> of this kind, but they found by experiment that though it accounted for the phenomenon of a finite slipping between the gas and the surface the observed slipping was greater than that calculated.

The easiest way to alter the hypothesis to suit the case is to suppose that certain parts of the surface act like a perfectly reflecting smooth plane, that is to say, they reverse the normal component of these molecules which fall on it and leave the components parallel to the surface unaltered.

We shall suppose that in every unit of surface, an area f is such as to turn the medium into ordinary gas, and an area $1-f$ is perfectly reflecting.

Let us suppose the surface to be a plane surface normal to the axis of x, and that the medium is on the negative side of the surface. Let the suffix (1) be applied to all molecules moving in the positive direction and the suffix (2) to all those moving in the negative direction, and let those which are projected from the surface as ordinary gas be marked with an accent.

The mass of all the molecules which in unit of time strike unit of area of the surface is $\rho_1\xi_1$. The portion of these which falls on the perfectly reflecting part of the surface is $(1-f)\rho_1\xi_1$. These are reflected with the sign of ξ reversed thereby contributing a portion equal to $(f-1)\rho_1\xi_1$ to the value of $\rho_2\xi_2$.

The portion which falls on the absorbing part of the surface is $f\rho_1\xi_1$, which <becomes must> and that which evaporates from the same surface is $f\rho_2'\xi_2'$ and there is no loss of molecules, so that

$$f\rho_1\xi_1 + f\rho_2'\xi_2' = 0$$

Now the total mass which leaves the surface is $\rho_2\xi_2$ so that

$$\rho_2\xi_2 = (f-1)\rho_1\xi_1 + f\,\rho_2'\xi_2'$$

Similarly if we consider the momentum in the direction of y the incident momentum is $\rho_1\xi_1\eta'$ <the reflected part> $1-f$ of this is reflected and becomes $(f-1)\rho_1\xi\eta$ and f of it is absorbed and is evaporated with η changed into $-v$ so that it becomes $-f\,\rho_2'\xi_2'v$ We thus find

$$\rho_2\xi_2\eta_2 = (f-1)\rho_1\xi_1\eta_1 - f\rho_2'\xi_2'v$$

Eliminating $f\rho_2'\xi_2'$ from these equations we find

$$(1-f)\rho_1\xi_1\eta_1 + \rho_2\xi_2\eta_2 = v[(1-f)\rho_1\xi_1 + \rho_2\xi_2]$$

If we now assume that the motion of the medium as a whole is expressed by the equation

$$\begin{aligned}
\rho_1\xi_1 + \rho_2\xi_2 &= 0\\
\rho_1\xi_1 - \rho_2\xi_2 &= 2\rho\,(2\pi)^{-1/2}\,(R\theta)^{1/2}\left(1+\frac{1}{2}\alpha^2\right)\\
\rho_1\xi_1\eta_1 + \rho_2\xi_2\eta_2 &= \rho\,R\theta\,\alpha\beta\\
\rho_1\xi_1\eta_1 - \rho_2\xi_2\eta_2 &= \rho(2\pi)^{-1/2}R\theta\,\alpha^2\beta
\end{aligned}$$

Substituting these values in

$$(2-f)\rho\,R\theta\,\alpha\beta - f(2\pi)^{-1/2}\rho\,R\theta\,\alpha^2\beta - 2f(2\pi)^{-1/2}\left(1+\frac{1}{2}\alpha^2\right)(R\theta)^{1/2}\,v = 0$$

In the first term of this equation the quantity $\rho R\theta\alpha\beta$ represents the stress in the direction of y on unit of area of the surface, or as we may write it p_{xy}.

In the second term $\rho\,R\theta\,\alpha^2\beta = -\frac{3}{2}\mu\left(\frac{R}{\theta}\right)^{1/2}\frac{d\theta}{dy}$

[the rest of the page is blank]

In this equation the surface is assumed to be normal to the axis of x and the motion to be in the direction of y but it is easy to write down the general equations for a point on the surface at which the normal drawn into the gas has the direction-cosines $\ell\,m\,n$. Writing

$$\frac{d}{d\nu} = \ell\frac{d}{dx} + m\frac{d}{dy} + n\frac{d}{dz} \tag{70}$$

for the operation of differentiating with respect to the normal ν we find as the conditions close to the surface of the solid

$$\left.\begin{aligned}
&2(p\rho)^{1/2}u - g\mu\frac{d}{d\nu}\left[(1-\ell^2)u - \ell m v - \ell n w\right] \\
&\qquad +\frac{3}{2}\mu\,(p/\rho)^{1/2}\frac{1}{\theta}\left(\frac{d}{dx} - \ell\frac{d}{d\nu}\right)\left(\theta + 2g\mu(\rho p)^{-1/2}\frac{d\theta}{d\nu}\right) = 0 \\
&2(p\rho)^{1/2}v - g\mu\frac{d}{d\nu}\left[(1-m^2)\,v - mnw - m\ell\,u\right] \\
&\qquad +\frac{3}{2}\mu\,(p/\rho)^{1/2}\frac{1}{\theta}\left(\frac{d}{dy} - m\frac{d}{d\nu}\right)\left(\theta + 2g\mu\,(\rho p)^{-1/2}\frac{d\theta}{d\nu}\right) = 0 \\
&2(p\rho)^{1/2}w - g\mu\frac{d}{d\nu}\left[(1-n^2)\,w - n\,\ell\,u - nmv\right] \\
&\qquad +\frac{3}{2}\mu\,(p/\rho)^{1/2}\frac{1}{\theta}\left(\frac{d}{dz} - n\frac{d}{d\nu}\right)\left(\theta + 2g\mu\,(\rho p)^{-1/2}\frac{d\theta}{d\nu}\right) = 0
\end{aligned}\right\} \tag{71}$$

These conditions satisfy of themselves the condition

$$\ell\,u + mv + nw = 0 \tag{72}$$

When there is no inequality of temperature we see by equation (69) that <the>

$$v = \frac{1}{2}g\mu(p\rho)^{-1/2}\frac{dv}{dx} \tag{73}$$

If we put

$$G = \frac{1}{2}g\mu(p\rho)^{-1/2} \tag{74}$$

we may call G the coefficient of slipping (Gleitungscoefficient, Helmholtz & Piotrowski, Kundt & Warburg).[c] G is a line which is related to λ and ℓ by the equations

$$G = -\frac{1}{4}(\pi/2)^{1/2}g\lambda = -\frac{1}{3}(2/\pi)^{1/2}g\ell \tag{75}$$

The motion of the gas takes place in very nearly the same manner as if a stratum of depth G had been removed from the solid and replaced by gas <and that the > there being now no slipping between the new surface and the gas in contact with it.

Kundt and Warburg found that the coefficient of slipping for air on a surface of glass was nearly equal to twice the mean free path of a molecule. It follows from this that $f = 1/2$ or the surface behaves as if half of its area were a perfect reflector and the other half absorbed the gas and then allowed it to evaporate. If we suppose the surface acts entirely by way of absorption and evaporation, $f = 1$ and $G = \frac{2}{3}\ell$.

The last term of equation (67) shows that if the temperature of the gas close to the surface increases in passing along the surface in the direction of y there will be a tendency of the gas to flow along the surface in the direction of y, that is to say, from colder to hotter parts of the surface. The ultimate result of this tendency will depend on whether the gas is perfectly free to move or is constrained by neighbouring surfaces.

Let us therefore <take> suppose that the gas is contained in a cylindrical tube of circular section and internal radius equal to a. Let us also suppose that the temperature is uniform in any one section of the tube, but varies from section to section.

Let us take the axis of the tube for that of z then the equation of steady motion in the interior of the gas is

$$\frac{dp}{dz} - \mu\left(\frac{d^2w}{dx^2} + \frac{d^w}{dy^2}\right)$$

Since everything is symmetrical about the axis the solution of this equation is

$$w = A + \frac{1}{4\mu}\frac{dp}{dz}r^2$$

where

$$r^2 = x^2 + y^2$$

If Q denotes the quantity of gas which passes through the section of the tube in unit of time

$$Q = 2\pi\int \rho w\, r\, dr$$
$$= \pi\rho\, a^2\left(A + \frac{1}{8\mu}\frac{dp}{dz}a^2\right)$$

At the inner surface of the tube the velocity of the gas is

$$w = A + \frac{1}{4\mu}\frac{dp}{dz}a^2 = \frac{Q}{\pi\rho\, a^2} + \frac{1}{8\mu}\frac{dp}{dz}a^2$$

<The ratio of> and

$$\frac{dw}{dr} = \frac{1}{2\mu}\frac{dp}{dz}u$$

Equation (67) may therefore be written

$$\frac{Q}{\rho\pi a^2} + \frac{1}{8\mu}\frac{dp}{dz}a^2 + G\frac{1}{2\mu}\frac{dp}{dz}a - \frac{3}{4}\frac{\mu}{\rho\theta}\frac{d\theta}{dz}$$

which <is the equation> gives the relation between the quantity of gas which passes through the tube, the rate of variation of pressure, and the rate of variation of temperature.

When there is no flow of gas on the whole, $Q = 0$ and

$$\frac{dp}{dz}(a^2 + 4Ga) = 6\frac{\mu^2}{\rho\theta}\frac{d\theta}{dz}$$

The pressure therefore increases from the colder to the hotter parts of the tube, and if we neglect the variation of μ^2/θ the difference of the pressures at the two ends of the tube will be proportional to the difference of the temperatures. Let us suppose p to be measured in millionths of an atmosphere, then for dry air at $15°C$

$$\frac{dp}{d\theta} = \frac{0.63}{p(a^2 + 4Ga)}$$

a. Maxwell, "Stresses in Rarified Gases Arising from Inequalities in Temperature." *Phil. Trans. R. Soc. London* 170 (1880): 231–256, reprinted in *Scientific Papers*, vol. 2, 682–707.

b. August Kundt and Emil Warburg, "Ueber Reibung und Wärmeleitung verdünnte Gase," *Ann. Phys.* 155 (1875): 337–366, 525–555; 156 (1875): 177–211, *Phil. Mag.* 50 (1875): 53–62. H./ von Helmholtz and G. von Piotrowski, "Ueber Reibung tropfbarer Flussigkeiten," *Sitz. Math.- Naturwiss. Cl. Akad. Wiss. Wien* 40 (1860): 607–658,

31. "On Stresses in Rarefied Gases Arising from Inequalities in Temperature"

Phil. Trans. R. Soc. London 170 (1880): 231–256. Reprinted in *Scientific Papers,* vol. 2, 681–712.

[From the *Philosophical Transactions of the Royal Society*, Part I. 1879.]

XCIII. *On Stresses in Rarified Gases arising from Inequalities of Temperature.*

1. IN this paper I have followed the method given in my paper "On the Dynamical Theory of Gases" (*Phil. Trans.*, 1867, p. 49). I have shewn that when inequalities of temperature exist in a gas, the pressure at a given point is not the same in all directions, and that the difference between the maximum and the minimum pressure at a point may be of considerable magnitude when the density of the gas is small enough, and when the inequalities of temperature are produced by small * solid bodies at a higher or lower temperature than the vessel containing the gas.

2. The nature of this stress may be thus defined:—Let the distance from a given point, measured in a given direction, be denoted by h; then the space-

* The dimensions of the bodies must be of the same order of magnitude as a certain length λ, which may be defined as the distance travelled by a molecule with its mean velocity during the time of relaxation of the medium.

The time of relaxation is the time in which inequalities of stress would disappear if the rate at which they diminish were to continue constant. Hence

$$\lambda = 2\left(\frac{2p}{\pi\rho}\right)^{\frac{1}{2}} . \frac{\mu}{p} = 2\mu\left(\frac{2}{\pi p\rho}\right)^{\frac{1}{2}} .$$

On the hypothesis that the encounters between the molecules resemble those between "rigid elastic" spheres, the free path of a molecule between two successive encounters has a definite meaning, and if l is its mean value,

$$l = \frac{3}{2}\mu\left(\frac{\pi}{2p\rho}\right)^{\frac{1}{2}} = \frac{3\pi}{8}\lambda = 1{\cdot}178\lambda.$$

So that the mean path of a molecule may be taken as representing what we mean by "small".

If the force between the molecules is supposed to be a continuous function of the distance, the free path of a molecule has no longer a definite meaning, and we must fall back on the quantity λ, as defined above.

variation of the temperature for a point moving along this line will be denoted by $\frac{d\theta}{dh}$, and the space-variation of this quantity along the same line by $\frac{d^2\theta}{dh^2}$.

There will, in general, be a particular direction of the line h for which $\frac{d^2\theta}{dh^2}$ is a maximum, another for which it is a minimum, and a third for which it is a maximum-minimum. These three directions are at right angles to each other, and are the axes of principal stress at the given point; and the part of the stress arising from inequalities of temperature is, in each of these principal axes,

$$3\frac{\mu^2}{\rho\theta}\frac{d^2\theta}{dh^2},$$

where μ is the coefficient of viscosity, ρ the density, and θ the absolute temperature.

3. Now for dry air at 15° C., $\mu = 1{\cdot}9 \times 10^{-4}$ in centimetre-gramme-second measure, and $\frac{3\mu^2}{\rho\theta} = \frac{1}{p}\,0{\cdot}315$, where p is the pressure, the unit of pressure being one dyne per square centimetre, or nearly one-millionth part of an atmosphere.

If a sphere of $2a$ centimetres in diameter is T degrees centigrade hotter than the air at large distances from it, then, when there is a steady flow of heat, the temperature at a distance of r centimetres from the centre will be

$$\theta = \theta_0 + \frac{Ta}{r}, \quad \text{and} \quad \frac{d^2\theta}{dr^2} = \frac{2Ta}{r^3}.$$

Hence, at a distance of r centimetres from the centre of the sphere, the pressure in the direction of the radius arising from inequality of temperature will be

$$\frac{Ta}{pr^3}\,0{\cdot}63 \text{ dynes per square centimetre.}$$

4. In Mr Crookes' experiments the pressure, p, was often so small that this stress would be capable, if it existed alone, of producing rapid motion in a radiometer.

Indeed, if we were to consider only the normal part of the stress exerted on solid bodies immersed in the gas, most of the phenomena observed by Mr Crookes could be readily explained.

5. Let us take the case of two small bodies symmetrical with respect to the axis joining their centres of figure. If both bodies are warmer than

the air at a distance from them, then, in any section perpendicular to the axis joining their centres, the point where it cuts this line will have the highest temperature, and there will be a flow of heat outwards from this axis in all directions.

Hence $\frac{d^2\theta}{dh^2}$ will be positive for the axis, and it will be a line of maximum pressure, so that the bodies will repel each other.

If both bodies are colder than the air at a distance, everything will be reversed; the axis will be a line of minimum pressure, and the bodies will attract each other.

If one body is hotter and the other colder than the air at a distance, the effect will be smaller, and it will depend on the relative sizes of the bodies, and on their exact temperatures, whether the action is attractive or repulsive.

6. If the bodies are two parallel disks very near to each other, the central parts will produce very little effect, because between the disks the temperature varies uniformly, and $\frac{d^2\theta}{dh^2}=0$. Only near the edges will there be any stress arising from inequality of temperature in the gas.

7. If the bodies are encircled by a ring having its axis in the line joining the bodies, then the repulsion between the two bodies, when they are warmer than the air in general, may be converted into attraction by heating the ring so as to produce a flow of heat inwards towards the axis.

8. If a body in the form of a cup or bowl is warmer than the air, the distribution of temperature in the surrounding gas is similar to the distribution of electric potential near a body of the same form, which has been investigated by Sir W. Thomson. Near the convex surface the value of $\frac{d^2\theta}{dh^2}$ is nearly the same as if the body had been a complete sphere, namely $2T\frac{1}{a^2}$, where T is the excess of temperature, and a is the radius of the sphere. Near the concave surface the variation of temperature is exceedingly small.

Hence the normal pressure will be greater on the convex surface than on the concave surface, and if we were to neglect the tangential pressures we might think this an explanation of the motion of Mr Crookes' cups.

Since the expressions for the stress are linear as regards the temperature, everything will be reversed when the cup is colder than the surrounding air.

9. In a spherical vessel, if the two polar regions are made hotter than the equatorial zone, the pressure in the direction of the axis will be greater than that parallel to the equatorial plane, and the reverse will be the case if the polar regions are made colder than the equatorial zone.

10. All such explanations of the observed phenomena must be subjected to careful criticism. They have been obtained by considering the normal stresses alone, to the exclusion of the tangential stresses, and it is much easier to give an elementary exposition of the former than of the latter. If, however, we go on to calculate the forces acting on any portion of the gas in virtue of the stresses on its surface, we find that when the flow of heat is steady, these forces are in equilibrium. Mr Crookes tells us that there is no molar current or wind in his radiometer vessels. It is not easy to prove this by experiment, but it is satisfactory to find that the system of stresses here described as arising from inequalities of temperature will not, when the flow of heat is steady, generate currents.

11. Consider, then, the case in which there are no currents of gas but a steady flow of heat, the condition of which is

$$\frac{d^2\theta}{dx^2}+\frac{d^2\theta}{dy^2}+\frac{d^2\theta}{dz^2}=0.$$

(In the absence of external forces such as gravity, and if the gas in contact with solid bodies does not slide over them, this is always a solution of the equations, and it is the only permanent solution.) In this case the equations of motion shew that every particle of the gas is in equilibrium under the stresses acting on it. Hence, any finite portion of the gas is also in equilibrium; also, since the stresses are linear functions of the temperature, if we superpose one system of temperatures on another, we also superpose the corresponding systems of forces.

Now the system of temperatures due to a solid sphere of uniform temperature immersed in the gas, cannot of itself give rise to any force tending to move the sphere in one direction rather than in another. Let the sphere be placed within the finite portion of gas which, as we have said, is already in equilibrium. The equilibrium will not be disturbed. We may introduce any number of spheres at different temperatures into the portion of gas, so as to form a body of any shape, heated in any manner, and when the flow of heat has become steady the whole system will be in equilibrium.

12. How, then, are we to account for the observed fact that forces act between solid bodies immersed in rarified gases, and this, apparently, as long as inequalities of temperature are maintained?

I think we must look for an explanation in the phenomenon discovered in the case of liquids by Helmholtz and Piotrowski*, and for gases by Kundt and Warburg†, that the fluid in contact with the surface of a solid must slide over it with a finite velocity in order to produce a finite tangential stress.

The theoretical treatment of the boundary conditions between a gas and a solid is difficult, and it becomes more difficult if we consider that the gas close to the surface is probably in an unknown state of condensation. We shall therefore accept the results obtained by Kundt and Warburg on their experimental evidence.

They have found that the velocity of sliding of the gas over the surface due to a given tangential stress varies inversely as the pressure.

The coefficient of sliding for air on glass was found to be $G = \frac{8}{p}$ centimetres, where p is the pressure in millionths of an atmosphere. Hence at ordinary pressures G is insensible, but in the vessels exhausted by Mr Crookes it may be considerable.

Hence, if close to the surface of a solid there is a tangential stress S, acting on a surface parallel to that of the body in a direction h parallel to that surface, there will also be a sliding of the gas in contact with the solid over its surface in the direction h with a finite velocity $= \frac{SG}{\mu}$.

13. I have not attempted to enter on the calculation of the effect of this sliding motion, but it is easy to see that if we begin with the case in which there is no sliding, the instantaneous effect of permission being given

* *Wiener Sitzb.*, xl. 1860, p. 607.
† Pogg. *Ann.*, clv. 1875, p. 337.

to the gas to slide must be to diminish the action of all tangential stresses on the surface, without affecting the normal stresses, and in course of time to set up currents sweeping over the surfaces of solid bodies, thus completely destroying the simplicity of our first solution of the problem.

14. When external forces, such as gravity, act on the gas, and when the thermal phenomena produce differences of density in different parts of the vessel, then the well-known convection currents are set up. These also interfere with the simplicity of the problem and introduce very complicated effects. All that we know is that the rarer the gas and the smaller the vessel the less is the effect of the convection currents, so that in Mr Crookes' experiments they play a very small part.

We now proceed to the calculations:—

(1) *Encounter between Two Molecules.*

The motion of the two molecules after an encounter depends on their motion before the encounter, and is capable of being determined by purely dynamical methods. If the encounter of the molecules does not cause rotation or vibration in the individual molecules, then the kinetic energy of the centres of mass of the two molecules must be the same after the encounter as it was before.

This will be true on the average, even if the molecules are complex systems capable of rotation and internal vibration, provided the temperature is constant. If, however, the temperature is rising, the internal energy of the molecules is, on the whole, increasing, and therefore the energy of translation of their centres of mass must be, on an average, diminishing at every encounter. The reverse will be the case if the temperature is falling.

But however important this consideration may be in the theory of specific heat and that of the conduction of heat, it has only a secondary bearing on the question of the stresses in the medium; and as it would introduce great complexity and much guesswork into our calculations, I shall suppose that the gas here considered is one the molecules of which do not take up any sensible amount of energy in the form of internal motion. Kundt and Warburg* have shewn that this is the case with mercury gas.

* Pogg. *Ann.*, clvii. 1876, p. 353.

Let the masses of the molecule be M_1 and M_2, and their velocity-components ξ_1, η_1, ζ_1, and ξ_2, η_2, ζ_2 respectively. Let V be the velocity of M_1 relative to M_2.

Before the encounter let a straight line be drawn through M_1 parallel to V, and let a perpendicular b be drawn from M_2 to this line. The magnitude and direction of b and V will be constant as long as the motion is undisturbed.

During the encounter the two molecules act on each other. If the force acts in the line joining their centres of mass, the product bV will remain constant, and if the force is a function of the distance, V and therefore b will be of the same magnitude after the encounter as before it, but their directions will be turned in the plane of V and b through an angle 2θ, this angle being a function of b and V, which vanishes for values of b greater than the limit of molecular action. Let the plane through V and b make an angle ϕ with the plane through V parallel to x, then all values of ϕ are equally probable.

If ξ_1' be the value of ξ_1 after the encounter,

$$\xi_1' = \xi_1 + \frac{M_2}{M_1+M_2}\left((\xi_2-\xi_1)\, 2\sin^2\theta + [(\eta_2-\eta_1)^2+(\zeta_2-\zeta_1)^2]^{\frac{1}{2}} \sin 2\theta \cos\phi\right)\ldots(1).$$

When the two molecules are of the same kind, $\frac{M_2}{M_1+M_2} = \frac{1}{2}$, and in the present investigation of a single gas we shall assume this to be the case.

If we use the symbol δ to indicate the increment of any quantity due to an encounter, and if we remember that all values of ϕ are equally probable, so that the average value of $\cos\phi$ and of $\cos^3\phi$ is zero, and that of $\cos^2\phi$ is $\frac{1}{2}$, we find

$$\delta(\xi_1+\xi_2) = 0 \ldots\ldots(2)$$

$$\delta(\xi_1^2+\xi_2^2) = -[3(\xi_2-\xi_1)^2 - V^2]\sin^2\theta\cos^2\theta \ldots\ldots(3)$$

$$\delta(\xi_1^3+\xi_2^3) = -\tfrac{3}{2}(\xi_1+\xi_2)[3(\xi_2-\xi_1)^2 - V^2]\sin^2\theta\cos^2\theta \ldots\ldots(4).$$

From these by transformation of coordinates we find

$$\delta(\xi_1\eta_1+\xi_2\eta_2) = -3(\xi_2-\xi_1)(\eta_2-\eta_1)\sin^2\theta\cos^2\theta \ldots\ldots(5)$$

$$\delta(\xi_1\eta_1^2+\xi_2\eta_2^2) = -\tfrac{1}{2}[9(\xi_1\eta_1^2+\xi_2\eta_2^2) - 3(\xi_1\eta_2^2+\xi_2\eta_1^2) - (\xi_1+\xi_2)(6\eta_1\eta_2+V^2)]\sin^2\theta\cos^2\theta \ldots\ldots(6)$$

$$\delta(\xi_1\eta_1\zeta_1+\xi_2\eta_2\zeta_2) = -\tfrac{1}{2}[9(\xi_1\eta_1\zeta_1+\xi_2\eta_2\zeta_2) - 3(\xi_1\eta_1\zeta_2+\xi_1\eta_2\zeta_1+\xi_1\eta_2\zeta_2+\xi_2\eta_1\zeta_1 + \xi_2\eta_1\zeta_2+\xi_2\eta_2\zeta_1)]\sin^2\theta\cos^2\theta \ldots\ldots(7).$$

[*Application of Spherical Harmonics to the Theory of Gases.*

If we suppose the direction of the velocity of M_1 relative to M_2 to be indicated by the position of a point P on a sphere, which we may call the sphere of reference, then the direction of the relative velocity after the encounter will be indicated by a point P', the angular distance PP' being 2θ, so that the point P' lies in a small circle, every position in which is equally probable.

We have to calculate the effect of an encounter upon certain functions of the six velocity-components of the two molecules. These six quantities may be expressed in terms of the three velocity-components of the centre of mass of the two molecules (say u, v, w), the relative velocity of M_1 with respect to M_2 which we call V, and the two angular coordinates which indicate the direction of V. During the encounter, the quantities u, v, w and V remain the same, but the angular coordinates are altered from those of P to those of P' on the sphere of reference.

Whatever be the form of the function of ξ_1, η_1, ζ_1, ξ_2, η_2, ζ_2, we may consider it expressed in the form of a series of spherical harmonics of the angular coordinates, their coefficients being functions of u, v, w, V, and we have only to determine the effect of the encounter upon the value of the spherical harmonics, for their coefficients are not changed.

Let $Y^{(n)}$ be the value at P of the surface harmonic of order n in the series considered.

After the encounter, the corresponding term becomes what $Y^{(n)}$ becomes at the point P', and since all positions of P' in a circle whose centre is P are equally probable, the mean value of the function after the encounter must depend on the mean value of the spherical harmonic in this circle.

Now the mean value of a spherical harmonic of order n in a circle, the cosine of whose radius is μ, is equal to the value of the harmonic at the pole of the circle multiplied by $P^{(n)}(\mu)$, the zonal harmonic of order n, and amplitude μ.

Hence, after the encounter, $Y^{(n}$ becomes $Y^{(n)}P^{(n)}(\mu)$, and if F_n is the corresponding part of the function to be considered, and δF_n the increment of F_n arising from the encounter, $\delta F_n = F_n(P^{(n)}(\mu) - 1)$.

This is the mean increment of F_n arising from an encounter in which $\cos 2\theta = \mu$. The rate of increment is to be found from this by multiplying it by the number of encounters of each molecule per second in which μ lies between μ and $\mu+d\mu$, and integrating for all values of μ from -1 to $+1$.

This operation requires, in general, a knowledge of the law of force between the molecules, and also a knowledge of the distribution of velocity among the molecules.

When, as in the present investigation, we suppose both the molecules to be of the same kind, and take both molecules into account in the final summation, the spherical harmonics of odd orders will disappear, so that if we restrict our calculations to functions of not more than three dimensions, the effect of the encounters will depend on harmonics of the second order only, in which case $P^{(2)}(\mu)-1=\frac{3}{2}(\mu^2-1)=-\frac{3}{2}\sin^2 2\theta$.—Note added May, 1879.]

(2) *Number of Encounters in Unit of Time.*

We now abandon the dynamical method and adopt the statistical method. Instead of tracing the path of a single molecule and determining the effects of each encounter on its velocity-components and their combinations, we fix our attention on a particular element of volume, and trace the changes in the average values of such combinations of components for all the molecules which at a given instant happen to be within it. The problem which now presents itself may be stated thus: to determine the distribution of velocities among the molecules of any element of the medium, the current-velocity and the temperature of the medium being given in terms of the coordinates and the time. The only case in which this problem has been actually solved is that in which the medium has attained to its ultimate state, in which the temperature is uniform and there are no currents.

Denoting by

$$dN=f_1(\xi,\eta,\zeta,x,y,z,t)\,d\xi d\eta d\zeta dx dy dz$$

the number of molecules of the kind M_1 which at a given instant are within the element of volume $dxdydz$, and whose velocity-components lie between the limits $\xi\pm\frac{1}{2}d\xi$, $\eta\pm\frac{1}{2}d\eta$, $\zeta\pm\frac{1}{2}d\zeta$, Boltzmann has shewn that the function f_1 must satisfy the equation

$$\frac{df_1}{dt}+\xi_1\frac{df_1}{dx}+\eta_1\frac{df_1}{dy}+\zeta_1\frac{df_1}{dz}+X\frac{df_1}{d\xi_1}+Y\frac{df_1}{d\eta_1}+Z\frac{df_1}{d\zeta_1}+$$

$$+\iiint d\xi_2 d\eta_2 d\zeta_2\int b\,db\int d\phi\, V(f_1f_2-f_1'f_2')=0 \;\ldots\ldots\ldots\ldots(8)$$

where f_2, f_1', f_2' denote what f_1 becomes when in place of the velocity-components of M_1 before the encounter we put those of M_2 before the encounter, and those of M_1 and M_2 after the encounter, respectively, and the integration is extended to all values of ϕ and b and of ξ_2, η_2, ζ_2, the velocity-components of the second molecule M_2.

It is impossible, in general, to perform this integration without a knowledge, not only of the law of force between the molecules, but of the form of the functions f_1, f_2, f_1', f_2', which have themselves to be found by means of the equation.

It is only for particular cases, therefore, that the equation has hitherto been solved.

If the medium is surrounded by a surface through which no communication of energy can take place, then one solution of the equation is given by the conditions

$$f_1 f_2 - f_1' f_2' = 0,$$

and

$$\xi \frac{df}{dx} + \eta \frac{df}{dy} + \zeta \frac{df}{dz} + X \frac{df}{d\xi} + Y \frac{df}{d\eta} + Z \frac{df}{d\zeta} = 0,$$

which give

$$f_1 = A_1 e^{-h(2\psi_1 + \xi_1^2 + \eta_1^2 + \zeta_1^2)} \quad \text{.........................(9)}$$

where ψ_1 is the potential of the force whose components are X_1, Y_1, Z_1, and A_1 is a constant which may be different for each kind of molecules in the medium, but h is the same for all kinds of molecules.

This is the complete solution of this problem, and is independent of any hypothesis as to the manner in which the molecules act on each other during an encounter. The quantity h which occurs in this expression may be determined by finding the mean value of ξ^2, which is $\frac{1}{2h}$. Now in the kinetic theory of gases,

$$\rho \overline{\xi^2} = p = R\rho\theta \quad \text{................................(10)}$$

where p is the pressure, ρ the density, θ the absolute temperature, and R a constant for a given gas. Hence

$$\frac{1}{2h} = R\theta \quad \text{..................................(11).}$$

We shall suppose, however, with Boltzmann, that in a medium in which there are inequalities of temperature and of velocity

$$dN = N\{1 + F(\xi, \eta, \zeta)\} f_0(\xi, \eta, \zeta)\, d\xi d\eta d\zeta \quad\text{.................(12)}$$

where F is a rational function of ξ, η, ζ, which we shall suppose not to contain terms of more than three dimensions, and f_0 is the same function as in equation (9).

Now consider two groups of molecules, each defined by the velocity-components, and let the two groups be distinguished by the suffixes ($_1$) and ($_2$). We have to estimate the number of encounters of a given kind between these two groups in a unit of volume in the time δt, those encounters only being considered for which the limits of b and ϕ are $b \pm \frac{1}{2}db$ and $\phi \pm \frac{1}{2}d\phi$.

Let us first suppose that both groups consist of mere geometrical points which do not interfere with each other's motion. The group dN_1 is moving through the group dN_2 with the relative velocity V, and we have to find how many molecules of the first group approach a molecule of the second group in a manner which would, if the molecules acted on each other, produce an encounter of the given kind. This will be the case for every molecule of the first group which passes through the area $bdbd\phi$ in the time δt. The number of such molecules is $dN_1 Vbdbd\phi\delta t$ for every molecule of the second group, so that the whole number of pairs which pass each other within the given limits is

$$Vbdbd\phi dN_1 dN_2 \delta t,$$

and if we take the time δt small enough, this will be the number of encounters of the real molecules in the time δt.

(3) *Effect of the Encounters.*

We have next to estimate the effect of these encounters on the average values of different functions of the velocity-components. The effect of an individual encounter on these functions for the pair of molecules concerned is given in equations (3), (4), (5), (6), (7), each of which is of the form

$$\delta P = Q \sin^2\theta \cos^2\theta \quad\text{..............................(13)}$$

where P and Q are functions of the velocity-components of the two molecules, and if we write $\bar{P}$ for the average value of P for the N molecules in unit of volume, then taking the sum of the effects of the encounters—

$$\Sigma\delta P = N\delta\bar{P} \quad\text{.................................(14).}$$

We thus find

$$\frac{\delta P}{\delta t} = N\iiiint\!\!\iiiint Q \sin^2\theta\cos^2\theta\, Vbdbd\phi f_1 f_2 d\xi_1 d\eta_1 d\zeta_1 d\xi_2 d\eta_2 d\zeta_2 \quad\text{...(15).}$$

Now, since θ is a function of b and V, the definite integral

$$V\int_0^{2\pi}\int_0^{\infty} b\sin^2\theta\cos^2\theta dbd\phi = B \quad\text{........................(16)}$$

will be a function of V only.

If the molecules are "rigid-elastic" spheres of diameter s,

$$B = \tfrac{1}{8}\pi s^2 V \quad\text{....................................(17).}$$

If they repel each other with a force inversely as the fifth power of the distance, so that at a distance r the force is κr^{-5}, then

$$B = \left(\frac{2\kappa}{M}\right)^{\frac{1}{2}} A_2 \quad\text{...................................(18)}$$

where A_2 is the numerical quantity 1·3682. In this case B is independent of V.

The experiments of O. E. Meyer*, Kundt and Warburg†, Puluj‡, Von Obermayer§, Eilhard Wiedemann‖, and Holman¶, shew that the viscosity of air varies according to a lower power of the absolute temperature than the first, probably the 0·77 power. If the viscosity had varied as the first power of the absolute temperature, B would have been independent of V. Though this is not the case, we shall assume, for the sake of being able to effect the integrations, that B is independent of V.

We shall find it convenient to write for B,

$$B = \frac{p}{3N\mu} \quad\text{....................................(19)}$$

where p is the hydrostatic pressure, N the number of molecules in unit of

* Pogg. *Ann.*, 1873, Bd. 148, p. 222.

† *Ibid.* 1876, Bd. 159, p. 403.

‡ *Wiener Sitz.*, 1874 and 1876.

§ *Ibid.* 1875.

‖ *Arch. des Sci. Phys. et Nat.*, 1876, t. 56, p. 273.

¶ *American Academy of Arts and Sciences*, June 14, 1876. *Phil. Mag.*, s. 5, vol. 3, No. 16, Feb., 1877.

volume, and μ a new coefficient which we shall afterwards find to be the coefficient of viscosity.

Equation (15) may now be written

$$\frac{\delta P}{\delta t} = \frac{p}{3\mu}\iiiint\!\!\iint Q f_1 f_2 \, d\xi_1 d\eta_1 d\zeta_1 d\xi_2 d\eta_2 d\zeta_2 \quad \text{..................(20)}$$

where the integrations are all between the limits $-\infty$ and $+\infty$, and f_1 and f_2 are of the form

$$f = \{1 + F(\xi, \eta, \zeta)\} h^{\frac{3}{2}} \pi^{-\frac{3}{2}} e^{-h(\xi^2+\eta^2+\zeta^2)} \quad \text{.....................(21)}$$

$F(\xi, \eta, \zeta)$ being small compared with unity.

We may write F in the form

$$\begin{aligned} F = (2h)^{\frac{1}{2}}(\alpha\xi + \beta\eta + \gamma\zeta) &+ 2h\left(\tfrac{1}{2}\alpha^2\xi^2 + \tfrac{1}{2}\beta^2\eta^2 + \tfrac{1}{2}\gamma^2\zeta^2 + \beta\gamma\eta\zeta + \gamma\alpha\zeta\xi + \alpha\beta\xi\eta\right) \\ &+ (2h)^{\frac{3}{2}}\left(\tfrac{1}{6}\alpha^3\xi^3 + \tfrac{1}{6}\beta^3\eta^3 + \tfrac{1}{6}\gamma^3\zeta^3 + \tfrac{1}{2}\alpha^2\beta\xi^2\eta + \tfrac{1}{2}\alpha^2\gamma\xi^2\zeta + \tfrac{1}{2}\beta^2\gamma\eta^2\zeta \right. \\ &\left. + \tfrac{1}{2}\beta^2\alpha\eta^2\zeta + \tfrac{1}{2}\gamma^2\alpha\zeta^2\xi + \tfrac{1}{2}\gamma^2\beta\zeta^2\eta + \alpha\beta\gamma\xi\eta\zeta\right) \quad \text{.................................(22)} \end{aligned}$$

where each combination of the symbols $\alpha\beta\gamma$ is to be taken as a single independent symbol, and not as a product of the component symbols.

(4) *Mean Values of Combinations of* ξ, η, ζ.

To find the mean value of any function of ξ, η, ζ for all the molecules in the element, we must multiply this function by f, and integrate with respect to ξ, η, and ζ.

If the non-exponential factor of any term contains an odd power of any of the variables, the corresponding part of the integral will vanish, but if it contains only even powers, each even power, such as $2n$, will introduce a factor

$$R^n\theta^n (2n-1)(2n-3)\ldots\ldots 3\cdot 1$$

into the corresponding part of the integral.

First, let the function be 1, then

$$1 = \iiint f \, d\xi d\eta d\zeta \quad \text{................................(23)}$$

or

$$1 = 1 + \tfrac{1}{2}(\alpha^2 + \beta^2 + \gamma^2) \quad \ldots\ldots(24)$$

which gives the condition

$$\alpha^2 + \beta^2 + \gamma^2 = 0 \quad \ldots\ldots(25).$$

Let us next find the mean value of ξ in the same way, denoting the result by the symbol $\bar{\xi}$,

$$\bar{\xi} = (R\theta)^{\frac{1}{2}}[\alpha + \tfrac{1}{2}(\alpha^3 + \alpha\beta^2 + \alpha\gamma^2)] \quad \ldots\ldots(26).$$

Since in what follows we shall denote the velocity-components of each molecule by $u+\xi$, $v+\eta$, $w+\zeta$, where u, v, w are the velocity-components of the centre of mass of all the molecules within the element, it follows that the mean values of ξ, η, ζ are each of them zero. We thus obtain the equations

$$\left.\begin{array}{l} \alpha + \tfrac{1}{2}(\alpha^3 + \alpha\beta^2 + \alpha\gamma^2) = 0 \\ \beta + \tfrac{1}{2}(\alpha^2\beta + \beta^3 + \beta\gamma) = 0 \\ \gamma + \tfrac{1}{2}(\alpha^2\gamma + \beta^2\gamma + \gamma^3) = 0 \end{array}\right\} \quad \ldots\ldots(27).$$

Remembering these conditions, we find that the mean values of combinations of two, three, and four dimensions are of the forms

$$\left.\begin{array}{ll} \overline{\xi^2} &= R\theta(1+\alpha^2) \\ \overline{\xi\eta} &= R\theta\alpha\beta \end{array}\right\} \quad \ldots\ldots(28)$$

$$\left.\begin{array}{ll} \overline{\xi^3} &= (R\theta)^{\frac{3}{2}}\alpha^3 \\ \overline{\xi\eta^2} &= (R\theta)^{\frac{3}{2}}\alpha\beta^2 \\ \overline{\xi\eta\zeta} &= (R\theta)^{\frac{3}{2}}\alpha\beta\gamma \end{array}\right\} \quad \ldots\ldots(29)$$

$$\left.\begin{array}{ll} \overline{\xi^4} &= 3R^2\theta^2(1+2\alpha^2) \\ \overline{\xi^3\eta} &= 3R^2\theta^2\alpha\beta \\ \overline{\xi^2\eta^2} &= R^2\theta^2(1+\alpha^2+\beta^2) \\ \overline{\xi^2\eta\zeta} &= R^2\theta^2\beta\gamma \end{array}\right\} \quad \ldots\ldots(30).$$

(5) *Rates of Decay of these Mean Values.*

If any term of Q in equation (20) contains symbols belonging to one group alone of the molecules, the corresponding term of the integral may be found from the above table, but if it contains symbols belonging to both groups we must consider the sextuple integral (20). But we shall not find it necessary to do this for terms of not more than three dimensions, for in these, if both groups of symbols occur, the index of one of them must be odd, and the integral vanishes.

We thus find from equations (3), (4), (5), (6), and (7)

$$\frac{\delta}{\delta t}\alpha^2 = -\frac{p}{\mu}\alpha^2 \quad\text{.............(31)}$$

$$\frac{\delta}{\delta t}\alpha\beta = -\frac{p}{\mu}\alpha\beta \quad\text{.............(32)}$$

$$\frac{\delta}{\delta t}\alpha^3 = \frac{1}{2}\frac{p}{\mu}(-2\alpha^3 + \alpha\beta^2 + \alpha\gamma^2) \quad\text{.............(33)}$$

$$\frac{\delta}{\delta t}\alpha\beta^2 = \frac{1}{6}\frac{p}{\mu}(\alpha^3 - 8\alpha\beta^2 + \alpha\gamma^2) \quad\text{.............(34)}$$

$$\frac{\delta}{\delta t}\alpha\beta\gamma = -\frac{3}{2}\frac{p}{\mu}\alpha\beta\gamma \quad\text{.............(35).}$$

[Any rational homogeneous function of $\xi\ \eta\ \zeta$ is either a solid harmonic, or a solid harmonic multiplied by a positive integral power of $(\xi^2 + \eta^2 + \zeta^2)$, or may be expressed as the sum of a number of terms of these forms.

If we express any one of these terms as a function of u, v, w, V and the angular coordinates of V, we can determine the rate of change of each of the spherical harmonics of the angular coordinates.

If we then transform the expression back to its original form as a function of ξ_1, η_1, ζ_1, ξ_2, η_2, ζ_2, and if we add the corresponding functions for both molecules, we shall obtain an expression for the rate of change of the original function.

Thus among the terms of two dimensions we have the five conjugate solid harmonics

$$\frac{1}{3}(2\xi^2 - \eta^2 - \zeta^2),$$

$$\xi\eta, \qquad \xi\zeta,$$

$$\eta^2 - \zeta^2, \quad \eta\zeta.$$

The rate of increase of each of these arising from the encounters of the molecules is found by multiplying it by $-\frac{p}{\mu}$. We may therefore call $\frac{p}{\mu}$ the "modulus of the time of relaxation" of this class of functions.

The function $\xi^2+\eta^2+\zeta^2$ is not changed by the encounters.

Homogeneous functions of three dimensions are either solid harmonics of the third order or solid harmonics of the first order multiplied by $\xi^2+\eta^2+\zeta^2$, or combinations of these.

The time modulus for solid harmonics of the third order is $\frac{3}{2}\frac{p}{\mu}$.—Note added May, 1879.]

That of ξ, η, or ζ, multiplied by $\xi^2+\eta^2+\zeta^2$ is $\frac{2}{3}\frac{p}{\mu}$.

(6) *Effect of External Forces.*

The only effect of external forces is expressed by equations of the form

$$\frac{\partial u}{\partial t}=X \qquad (36).$$

The average values of ξ, η, ζ and their combinations are not affected by external forces.

(7) *Variation of Mean Values within an Element of Volume.*

We have employed the symbol δ to denote the variation of any quantity within an element, arising either from encounters between molecules or from the action of external forces.

There is a third way, however, in which a variation may occur, namely, by molecules entering the element or leaving it, carrying their properties with them.

We shall use the symbol ∂ to denote the actual variation within a specified element.

If MQ is the average value of any quantity for each molecule within the element, then the quantity in unit of volume is ρQ. We have to trace the variation of ρQ.

We begin with an element of volume moving with the velocity-components U, V, W, then by the ordinary investigation of the "equation of continuity"

$$\frac{\partial}{\partial t}[Q\rho]+\frac{d}{dx}[Q(u+\xi-U)]+\frac{d}{dy}[Q(v+\eta-V)]+\frac{d}{dz}[Q(w+\zeta-W)]=\rho\frac{\delta}{\delta t}Q\ldots(37).$$

If after performing the differentiations we make $U=u$, $V=v$, $W=w$, the equation becomes for an element moving with the velocity (u, v, w)

$$\frac{\partial}{\partial t}(Q\rho)+\rho Q\left(\frac{du}{dx}+\frac{dv}{dy}+\frac{dw}{dz}\right)+\frac{d}{dx}(\rho Q\xi)+\frac{d}{dy}(\rho Q\eta)+\frac{d}{dz}(\rho Q\zeta)=\rho\frac{\delta}{\delta t}Q\ \ldots(38).$$

(8) *Equation of Density.*

Let us first make $Q=1$, then, since the mass of a molecule is invariable, the equation becomes

$$\frac{\partial\rho}{\partial t}+\rho\left(\frac{du}{dx}+\frac{dv}{dy}+\frac{dw}{dz}\right)=0\ \ldots\ldots\ldots\ldots\ldots\ldots\ldots\ldots(39),$$

which is the ordinary "equation of continuity."

Eliminating by means of this equation the second term of the general equation (38) we obtain the more convenient form—

$$\rho\frac{\partial Q}{\partial t}+\frac{d}{dx}(\rho Q\xi)+\frac{d}{dy}(\rho Q\eta)+\frac{d}{dz}(\rho Q\zeta)=\rho\frac{\delta Q}{\delta t}\ \ldots\ldots\ldots\ldots(40).$$

(9) *Equations of Motion.*

Putting $Q=u+\xi$, this equation becomes

$$\rho\frac{\partial u}{\partial t}+\frac{d}{dx}(\rho\xi^2)+\frac{d}{dy}(\rho\xi\eta)+\frac{d}{dz}(\rho\xi\zeta)=\rho X\ \ldots\ldots\ldots\ldots\ldots(41)$$

where any combination of the symbols ξ, η, ζ is to be taken as the average value of that combination.

Substituting their values as given in (28)

$$\rho\frac{\partial u}{\partial t}+R\frac{d}{dx}(\rho\theta)+R\left[\frac{d}{dx}(\rho\theta\alpha^2)+\frac{d}{dy}(\rho\theta\alpha\beta)+\frac{d}{dz}(\rho\theta\alpha\gamma)\right]=\rho X\ldots..(42),$$

which is one of the three ordinary equations of motion of a medium in which stresses exist.

(10) *Terms of Two Dimensions.*

Put $Q=(u+\xi)^2$. Since the resulting equation is true whatever be the values of u, v, w, we may, after differentiation, put each of these quantities equal to zero. We shall thus obtain the same result which we might have obtained by elimination between this and the former equations. We find

$$\rho\frac{\partial}{\partial t}\overline{\xi^2}+2\rho\overline{\xi^2}\frac{du}{dx}+2\rho\overline{\xi\eta}\frac{du}{dy}+2\rho\overline{\xi\zeta}\frac{du}{dz}+\frac{d}{dx}(\rho\overline{\xi^3})+\frac{d}{dy}(\rho\overline{\xi^2\eta})+\frac{d}{dz}(\rho\overline{\xi^2\zeta})=\rho\frac{\delta}{\delta t}\overline{\xi^2}\ldots(43),$$

or by substituting the mean values of these quantities from (29)

$$\rho\frac{\partial\theta}{\partial t}+\rho\frac{\partial}{\partial t}(\theta\alpha^2)+2\rho\theta\frac{du}{dx}+2\rho\theta\left(\alpha^2\frac{du}{dx}+\alpha\beta\frac{du}{dy}+\alpha\gamma\frac{du}{dz}\right)$$

$$+R^{\frac{1}{2}}\left[\frac{d}{dx}(\rho\theta^{\frac{3}{2}}\alpha^3)+\frac{d}{dy}(\rho\theta^{\frac{3}{2}}\alpha^2\beta)+\frac{d}{dz}(\rho\theta^{\frac{3}{2}}\alpha^2\gamma)\right]=-\frac{R\rho^2\theta^2}{\mu}\alpha^2\ldots(44)$$

with two other equations of similar form.

Similarly we obtain by putting $Q=(u+\xi)(v+\eta)$

$$\rho\frac{\partial}{\partial t}(\theta\alpha\beta)+\rho\theta\left(\frac{dv}{dx}+\frac{du}{dy}\right)$$

$$+\rho\theta\left(\alpha^2\frac{dv}{dx}+\alpha\beta\frac{dv}{dy}+\alpha\gamma\frac{dv}{dz}+\alpha\beta\frac{du}{dx}+\beta^2\frac{du}{dy}+\beta\gamma\frac{du}{dz}\right)$$

$$+R\left[\frac{d}{dx}(\rho\theta^{\frac{3}{2}}\alpha^2\beta)+\frac{d}{dy}(\rho\theta\,\alpha\beta^2)+\frac{d}{dz}(\rho\theta^{\frac{3}{2}}\alpha\beta\gamma)\right]=-\frac{R\rho^2\theta^2}{\mu}\alpha\beta\ \ldots(45)$$

with two other equations of like form for $\beta\gamma$ and $\gamma\alpha$.

(11) *Terms of Three Dimensions.*

Putting $Q=(u+\xi)^3$ and in the final equation making $u=v=w=0$ and eliminating $\frac{\partial u}{\partial t}$ by (41) we find

$$\rho\frac{\partial}{\partial t}\overline{\xi^3}+3\rho\overline{\xi^3}\frac{du}{dx}+3\rho\overline{\xi^2\eta}\frac{du}{dy}+3\rho\overline{\xi^2\zeta}\frac{du}{dz}$$

$$+\frac{d}{dx}(\rho\overline{\xi^4})+\frac{d}{dy}(\rho\overline{\xi^3\eta})+\frac{d}{dz}(\rho\overline{\xi^3\zeta})$$

$$-3\overline{\xi^2}\left[\frac{d}{dx}(\rho\overline{\xi^2})+\frac{d}{dy}(\rho\overline{\xi\eta})+\frac{d}{dz}(\rho\overline{\xi\zeta})\right]=\rho\frac{\delta}{\delta t}\overline{\xi^3}\ldots\ldots\ldots\ldots(46).$$

which gives

$$R^{\frac{3}{2}}\rho\frac{\partial}{\partial t}(\theta^{\frac{3}{2}}\alpha^2)+3\rho(R\theta)^{\frac{3}{2}}\left(\alpha^3\frac{du}{dx}+\alpha^2\beta\frac{du}{dy}+\alpha^2\gamma\frac{du}{dz}\right)$$

$$+3R^2\rho\theta\frac{d\theta}{dx}+3R^2\rho\theta\left(\alpha^2\frac{d\theta}{dx}+\alpha\beta\frac{d\theta}{dy}+\alpha\gamma\frac{d\theta}{dz}\right)+3R^2\rho\theta\frac{d}{dx}(\alpha^2\theta)$$

$$-3R^2\theta\alpha^2\left[\frac{d}{dx}(\rho\theta\alpha^2)+\frac{d}{dy}(\rho\theta\alpha\beta)+\frac{d}{dz}(\rho\theta\alpha\gamma)\right]=\rho(R\theta)^{\frac{3}{2}}\frac{1}{2}\frac{p}{\mu}(-2\alpha^3+\alpha\beta^2+\alpha\gamma^2)\ldots(47).$$

Since the combinations of $\alpha\beta\gamma$ represent small numerical quantities, we may at this stage of the calculation, when we are dealing with terms of the third order, neglect terms involving them, except when they are multiplied by the large coefficient p/μ. The equation may then be written approximately :—

$$3R^2\rho\theta\frac{d\theta}{dx}=\rho(R\theta)^{\frac{3}{2}}\frac{1}{2}\frac{p}{\mu}(-2\alpha^3+\alpha\beta^2+\alpha\gamma^2)\ldots\ldots\ldots\ldots\ldots\ldots(48).$$

Similarly, by putting $Q=(u+\xi)(v+\eta)^2$, we obtain the approximate equation

$$R^2\rho\theta\frac{d\theta}{dx}=\rho(R\theta)^{\frac{3}{2}}\frac{1}{6}\frac{p}{\mu}(\alpha^3-8\alpha\beta^2+\alpha\gamma^2)\ \ldots\ldots\ldots\ldots\ldots\ldots(49),$$

and in the same way we find

$$R^2\rho\theta\frac{d\theta}{dx}=\rho(R\theta)^{\frac{3}{2}}\frac{1}{6}\frac{p}{\mu}(\alpha^3+\alpha\beta^2-8\alpha\gamma^2)\ \ldots\ldots\ldots\ldots\ldots\ldots(50).$$

(12) *Approximate Values of Terms of Three Dimensions.*

From equations (48), (49), and (50), we find

$$\alpha^3=-\frac{9}{2}\frac{\mu}{p}\left(\frac{R}{\theta}\right)^{\frac{1}{2}}\frac{d\theta}{dx},\quad \alpha\beta^2=\alpha\gamma^2=-\frac{3}{2}\frac{\mu}{p}\left(\frac{R}{\theta}\right)^{\frac{1}{2}}\frac{d\theta}{dx}$$

From which by substitution we obtain

$$\beta^3=-\frac{9}{2}\frac{\mu}{p}\left(\frac{R}{\theta}\right)^{\frac{1}{2}}\frac{d\theta}{dy},\quad \alpha^2\beta=\beta\gamma^2=-\frac{3}{2}\frac{\mu}{p}\left(\frac{R}{\theta}\right)^{\frac{1}{2}}\frac{d\theta}{dy}$$

$$\gamma^3=-\frac{9}{2}\frac{\mu}{p}\left(\frac{R}{\theta}\right)^{\frac{1}{2}}\frac{d\theta}{dz},\quad \alpha^2\gamma=\beta^2\gamma=-\frac{3}{2}\frac{\mu}{p}\left(\frac{R}{\theta}\right)^{\frac{1}{2}}\frac{d\theta}{dz}\ \ldots\ldots\ldots\ldots\ldots(51).$$

The value of $\alpha\beta\gamma$ is of a smaller order of magnitude, and we do not require it in this investigation.

(13) *Equation of Temperature.*

Adding the three equations of the form (44), and omitting terms containing small quantities of two dimensions, and also products of differential coefficients such as $\frac{d\mu}{dx}\frac{d\theta}{dx}$, we find

$$\frac{\partial\theta}{\partial t}=\frac{5}{2}\frac{\mu}{\rho}\left(\frac{d^2\theta}{dx^2}+\frac{d^2\theta}{dy^2}+\frac{d^2\theta}{dz^2}\right)+\frac{2}{3}\frac{\theta}{\rho}\frac{\partial\rho}{\partial t}\quad\ldots\ldots\ldots(52).$$

The first term of the second member represents the rate of increase of temperature due to conduction of heat, as in Fourier's Theory, and the second term represents the increase of temperature due to increase of density. We must remember that the gas here considered is one for which the ratio of the specific heats is $1{\cdot}\dot{6}$.

(14) *Stresses in the Gas.*

Subtracting one-third of the sum of the three equations from (44), we obtain

$$p\alpha^2=-2\mu\frac{du}{dx}+\frac{2}{3}\mu\left(\frac{du}{dx}+\frac{dv}{dy}+\frac{dw}{dz}\right)+3\frac{\mu^2}{\rho\theta}\frac{d^2\theta}{dx^2}+\frac{3}{2}\frac{\mu^2}{\rho\theta}\left(\frac{d^2\theta}{dx^2}+\frac{d^2\theta}{dy^2}+\frac{d^2\theta}{dz^2}\right)\ldots\ldots(53).$$

This equation gives the excess of the normal pressure in x above the mean hydrostatic pressure p. The first two terms of the second member represent the effect of viscosity in a moving fluid, and are identical with those given by Professor Stokes (*Cambridge Transactions*, Vol. VIII., 1845, p. 297). The last two terms represent the part of the stress which arises from inequality of temperature, which is the special subject of this paper.

There are two other equations of similar form for the normal stresses in y and z.

The tangential stress in the plane xy is given by the equation

$$p\alpha\beta=-\mu\left(\frac{du}{dy}+\frac{dv}{dx}\right)+3\frac{\mu^2}{\rho\theta}\frac{d^2\theta}{dxdy}\quad\ldots\ldots\ldots(54).$$

There are two other equations of similar form for the tangential stresses in the planes of yz and zx.

(15) *Final Equations of Motion.*

We are now prepared to complete the equations of motion by inserting in (42) the values of the quantities α^2, $\alpha\beta$, $\alpha\gamma$, and we find for the equation in x

$$\rho\frac{\partial u}{\partial t}+\frac{dp}{dx}-\mu\left(\frac{d^2u}{dx^2}+\frac{d^2u}{dy^2}+\frac{d^2u}{dz^2}\right)+\frac{1}{3}\mu\frac{d}{dx}\left(\frac{du}{dx}+\frac{dv}{dy}+\frac{dw}{dz}\right)$$

$$+\frac{9}{2}\frac{\mu^2}{\rho\theta}\frac{d}{dx}\left(\frac{d^2\theta}{dx^2}+\frac{d^2\theta}{dy^2}+\frac{d^2\theta}{dz^2}\right)=\rho X \quad\text{.....................(55).}$$

If we write

$$p'=p+\frac{1}{3}\mu\left(\frac{du}{dx}+\frac{dv}{dy}+\frac{dw}{dz}\right)+\frac{9}{2}\frac{\mu^2}{\rho\theta}\left(\frac{d^2\theta}{dx^2}+\frac{d^2\theta}{dy^2}+\frac{d^2\theta}{dz^2}\right) \quad\text{..........(56)}$$

$$=p+\frac{9}{5}\frac{\mu}{\theta}\frac{\partial\theta}{\partial t}-\frac{23}{15}\frac{\mu}{\rho}\frac{\partial\rho}{\partial t}\quad\text{.............................(57)}$$

or, if the pressure p is constant, so that $\rho\partial\theta+\theta\partial\rho=0$

$$p'=p+\frac{10}{3}\frac{\mu}{\theta}\frac{\partial\theta}{\partial t}\quad\text{...................................(58)}$$

then the equation (55) may be written

$$\rho\frac{\partial u}{\partial t}+\frac{dp'}{dx}-\mu\left(\frac{d^2u}{dx^2}+\frac{d^2u}{dy^2}+\frac{d^2u}{dz^2}\right)=\rho X\quad\text{.................(59).}$$

If there are no external forces such as gravity, then one solution of the equations is

$$u=v=w=0,\ p'=\text{constant},$$

and if the boundary conditions are such that this solution is consistent with them, it will become the actual solution as soon as the initial motions, if any exist, have subsided. This will be the case if no slipping is possible between the gas and solid bodies in contact with it.

But if such slipping is possible, then wherever in the above solution there is a tangential stress in the gas at the surface of a solid or liquid, there cannot be equilibrium, but the gas will begin to slide over the surface till the velocity of sliding has produced a frictional resistance equal and opposite to the tangential stress. When this is the case the motion may become steady.

I have not, however, attempted to enter into the calculation of the state of steady motion.

[I have recently applied the method of spherical harmonics, as described in the notes to sections (1) and (5), to carrying the approximations two orders higher. I expected that this would have involved the calculation of two new quantities, namely, the rates of decay of spherical harmonics of the fourth and sixth orders, but I found that, to the order of approximation required, all harmonics of the fourth and sixth orders may be neglected, so that the rate of decay of harmonics of the second order, the time-modulus of which is $\mu \div p$, determines the rate of decay of all functions of less than 6 dimensions.

The equations of motion, as here given (equation 55) contain the second derivatives of u, v, w, with respect to the coordinates, with the coefficient μ. I find that in the more approximate expression there is a term containing the fourth derivatives of u, v, w, with the coefficient $\mu^3 \div \rho p$.

The equations of motion also contain the third derivatives of θ with the coefficient $\mu^2 \div \rho\theta$. Besides these terms, there is another set consisting of the fifth derivatives of θ with the coefficient $\mu^4 \div \rho^2 p\theta$.

It appears from the investigation that the condition of the successful use of this method of approximation is that $l\frac{d}{dh}$ should be small, where $\frac{d}{dh}$ denotes differentiation with respect to a line drawn in any direction. In other words, the properties of the medium must not be sensibly different at points within a distance of each other, comparable with the "mean free path" of a molecule. —Note added June, 1879.]

APPENDIX.

(Added May, 1879.)

In the paper as sent in to the Royal Society, I made no attempt to express the conditions which must be satisfied by a gas in contact with a solid body, for I thought it very unlikely that any equations I could write down would be a satisfactory representation of the actual conditions, especially as it is almost certain that the stratum of gas nearest to a solid body is in a very different condition from the rest of the gas.

One of the referees, however, pointed out that it was desirable to make the attempt, and indicated several hypothetical forms of surfaces which might be tried. I have therefore added the following calculations, which are carried to the same degree of approximation as those for the interior of the gas.

It will be seen that the equations I have arrived at express both the fact that the gas may slide over the surface with a finite velocity, the previous investigations of which have been already mentioned;* and the fact that this velocity and the corresponding tangential stress are affected by inequalities of temperature at the surface of the solid, which give rise to a force tending to make the gas slide along the surface from colder to hotter places.

This phenomenon, to which Professor Osborne Reynolds has given the name of Thermal Transpiration, was discovered entirely by him. He was the first to point out that a phenomenon of this kind was a necessary consequence of the Kinetic Theory of Gases, and he also subjected certain actual phenomena, of a somewhat different kind, indeed, to measurement, and reduced his measurements by a method admirably adapted to throw light on the relations between gases and solids.

It was not till after I had read Professor Reynolds' paper that I began to reconsider the surface conditions of a gas, so that what I have done is simply to extend to the surface phenomena the method which I think most suitable for treating the interior of the gas. I think that this method is, in

* Sect. 12 of introduction.

some respects, better than that adopted by Professor Reynolds, while I admit that his method is sufficient to establish the existence of the phenomena, though not to afford an estimate of their amount.

The method which I have adopted throughout is a purely statistical one. It considers the mean values of certain functions of the velocities within a given element of the medium, but it never attempts to trace the motion of a molecule, not even so far as to estimate the length of its mean path. Hence all the equations are expressed in the forms of the differential calculus, in which the phenomena at a given place are connected with the space variations of certain quantities at that place, but in which no quantity appears which explicitly involves the condition of things at a finite distance from that place.

The particular functions of the velocities which are here considered are those of one, two, and three dimensions. These are sufficient to determine approximately the principal phenomena in a gas which is not very highly rarified, and in which the space-variations within distances comparable to λ are not very great.

The same method, however, can be extended to functions of higher degrees, and by a sufficient number of such functions any distribution of velocities, however abnormal, may be expressed. The labour of such an approximation is considerably diminished by the use of the method of spherical harmonics as indicated in the note to Section I. of the paper.

On the Conditions to be Satisfied by a Gas at the Surface of a Solid Body.

As a first hypothesis, let us suppose the surface of the body to be a perfectly elastic smooth fixed surface, having the apparent shape of the solid, without any minute asperities.

In this case, every molecule which strikes the surface will have the normal component of its velocity reversed, while the other components will not be altered by impact.

The rebounding molecules will therefore move as if they had come from an imaginary portion of gas occupying the space really filled by the solid, and such that the motion of every molecule close to the surface is the optical reflection in that surface of the motion of a molecule of the real gas.

In this case we may speak of the rebounding molecules close to the surface as constituting the *reflected* gas. All directed properties of the incident gas are reflected, or, as Professor Listing might say, *perverted* in the reflected gas; that is to say, the properties of the incident and the reflected gas are symmetrical with respect to the tangent plane of the surface.

The incident and reflected gas together constitute the actual gas close to the surface. The actual gas, therefore, cannot exert any stress on the surface, except in the direction of the normal, for the oblique components of stress in the incident and reflected gas will destroy one another.

Since gases can actually exert oblique stress against real surfaces, such surfaces cannot be represented as perfectly reflecting surfaces.

If a molecule, whose velocity is given in direction and magnitude, but whose line of motion is not given in position, strikes a fixed elastic sphere, its velocity after rebound may with equal probability be in any direction.

Consider, therefore, a stratum in which fixed elastic spheres are placed so far apart from one another that any one sphere is not to any sensible extent protected by any other sphere from the impact of molecules, and let the stratum be so deep that no molecule can pass through it without striking one or more of the spheres, and let this stratum of fixed spheres be spread over the surface of the solid we have been considering, then every molecule which comes from the gas towards the surface must strike one or more of the spheres, after which all directions of its velocity become equally probable.

When, at last, it leaves the stratum of spheres and returns into the gas, its velocity must of course be *from* the surface, but the probability of any particular magnitude and direction of the velocity will be the same as in a gas at rest with respect to the surface.

The distribution of velocity among the molecules which are leaving the surface will therefore be the same as if, instead of the solid, there were a portion of gas at rest, having the temperature of the solid, and a density such that the number of molecules which pass from it through the surface in a given time is equal to the number of molecules of the real gas outside which strike the surface.

To distinguish the molecules, which, after being entangled in the stratum of spheres, afterwards return into the surrounding gas, we shall call them, collectively, the *absorbed and evaporated* gas.

If the spheres are so near together that a considerable part of the surface of each sphere of the outer layer is shielded from the direct impact of the incident molecules by the spheres which lie next to it, then if we call that point of each sphere which lies furthest from the solid the *pole* of the sphere, a greater proportion of molecules will strike any one of the outer layer of spheres near its pole than near its equator, and the greater the obliquity of incidence of the molecule, the greater will be the probability that it will strike a sphere near its pole.

The direction of the rebounding molecule will no longer be with equal probability in all directions, but there will be a greater probability of the tangential part of its velocity being in the direction of the motion before impact, and of its normal part being opposite to the normal part before impact.

The condition of the molecules which leave the surface will therefore be intermediate between that of evaporated gas and that of reflected gas, approaching most nearly to evaporated gas at normal incidence and most nearly to reflected gas at grazing incidence.

If the spheres, instead of being hard elastic bodies, are supposed to act on the molecules at finite, though small distances, and if they are so close together that their spheres of action intersect, then the gas which leaves the surface will be still more like reflected gas, and less like evaporated gas.

We might also consider a surface on which there are a great number of minute asperities of any given form, but since in this case there is considerable difficulty in calculating the effect when the direction of rebound from the first impact is such as to lead to a second or third impact, I have preferred to treat the surface as something intermediate between a perfectly reflecting and a perfectly absorbing surface, and, in particular, to suppose that of every unit of area a portion f absorbs all the incident molecules, and afterwards allows them to evaporate with velocities corresponding to those in still gas at the temperature of the solid, while a portion $1-f$ perfectly reflects all the molecules incident upon it.

We shall begin by supposing that the surface is the plane yz, and that the gas is on that side of it for which x is positive.

The incident molecules are those which, close to the surface, have their normal component of velocity negative. We shall distinguish these molecules by the suffix $({}_1)$. For these, and these only, ξ_1 is negative.

The rebounding molecules are those which have ξ positive. We shall distinguish them by the suffix $({}_2)$. Those which are evaporated will be further distinguished by an accent.

Symbols without any mark refer to the whole gas, incident, reflected, and evaporated, close to the. surface.

The quantity of gas which is incident on unit of surface in unit of time, is $-\rho_1\xi_1$.

Of this quantity the fraction $1-f$ is reflected, so that the sign of ξ is reversed, and the fraction f is evaporated, the mean value of ξ in evaporated gas being ξ', where the accent distinguishes symbols belonging to unpolarized gas at rest relative to the surface, and having the temperature, θ', of the solid.

Equating the quantity of gas which is incident on the absorbing part of the surface to that which is evaporated from it, we have

$$f\rho_1\xi_1+f\rho_2'\xi_2'=0 \qquad (60).$$

Equating the whole quantity of gas which leaves the surface to the reflected and evaporated portions

$$\rho_2\xi_2=(f-1)\,\rho_1\xi_1+f\rho_2'\xi_2' \qquad (61).$$

If we next consider the momentum of the molecules in the direction of y, that of the incident molecules is $\rho_1\xi_1\eta_1$. A fraction $(1-f)$ of this is reflected and becomes $(1-f)\,\rho_1\xi_1\eta_1$, and a fraction f of it is absorbed and then evaporated, the mean value of η being now $-v$, namely, the velocity of the surface relatively to the gas in contact with it.

The momentum of the evaporated portion in the direction of y is therefore $-f\rho_2'\xi_2'v$, and this, together with the reflected portion, makes up the whole momentum which is leaving the surface, or

$$\rho_2\xi_2\eta_2=(f-1)\,\rho_1\xi_1\eta_1-f\rho_2'\xi_2'v \qquad (62).$$

Eliminating $f\rho_2'\xi_2'$ between equations (61) and (62)

$$(1-f)\,\rho_1\xi_1\eta_1+\rho_2\xi_2\eta_2+v\left[(1-f)\,\rho_1\xi_1+\rho_2\xi_2\right]=0 \qquad (63).$$

The values of functions of ξ, η and ζ for the incident molecules are to be found by multiplying the expression in equation (22) by the given function, and integrating with respect to ξ between the limits $-\infty$ and 0, and with respect to η and ζ between the limits $\pm\infty$.

The values of the same functions for the molecules which are leaving the surface are to be found by integrating with respect to ξ from 0 to ∞.

We must remember, however, that since there is an essential discontinuity in the conditions of the gas at the surface, the expression in equation (22) is a much less accurate approximation to the actual distribution of velocities in the gas close to the surface than it is in the interior of the gas. We must, therefore, consider the surface conditions at which we arrive in this way as liable to important corrections when we shall have discovered more powerful methods of attacking the problem.

For the present, however, we consider only terms of three dimensions or less, and we find

$$\left.\begin{array}{l}\rho_1\xi_1 = -\rho\,(2\pi)^{-\frac{1}{2}}\,(R\theta)^{\frac{1}{2}}\,(1+\frac{1}{2}\alpha^2)\\ \rho_2\xi_2 = \rho\,(2\pi)^{-\frac{1}{2}}\,(R\theta)^{\frac{1}{2}}\,(1+\frac{1}{2}\alpha^2)\end{array}\right\}\quad\ldots\ldots\ldots\ldots(64)$$

$$\left.\begin{array}{l}\rho_1\xi_1\eta_1 = \frac{1}{2}\rho R\theta\alpha\beta - \frac{1}{2}\rho\,(2\pi)^{-\frac{1}{2}}\,R\theta\alpha^2\beta\\ \rho_2\xi_2\eta_2 = \frac{1}{2}\rho R\theta\alpha\beta + \frac{1}{2}\rho\,(2\pi)^{-\frac{1}{2}}\,R\theta\alpha^2\beta\end{array}\right\}\quad\ldots\ldots\ldots\ldots(65).$$

Substituting these expressions in equation (63), and neglecting α^2 in comparison with unity, we find

$$(2-f)\,\rho R\theta\alpha\beta + f\,(2\pi)^{-\frac{1}{2}}\,\rho R\theta\alpha^2\beta + 2f\,(2\pi)^{-\frac{1}{2}}\,(1+\tfrac{1}{2}\alpha^2)\,(R\theta)^{\frac{1}{2}}\,\rho v = 0 \quad\ldots(66).$$

If we write

$$G = \tfrac{1}{2}\mu\,(2\pi)^{\frac{1}{2}}\,(p\rho)^{-\frac{1}{2}}\left(\frac{2}{f}-1\right)\quad\ldots\ldots\ldots\ldots(67)$$

and substitute for $\alpha\beta$ and $\alpha^2\beta$ their values as given in equations (54) and (51), and divide by $2\,(p\rho)^{\frac{1}{2}}$, equation (66) becomes

$$v - G\left(\frac{dv}{dx} - \frac{3}{2}\frac{\mu}{\rho\theta}\frac{d^2\theta}{dx\,dy}\right) - \frac{3}{4}\frac{\mu}{\rho\theta}\frac{d\theta}{dy} = 0 \quad\ldots\ldots\ldots\ldots(68).$$

If there is no inequality of temperature, this equation is reduced to

$$v = G\frac{dv}{dx}\quad\ldots\ldots\ldots\ldots(69).$$

If, therefore, the gas at a finite distance from the surface is moving parallel to the surface, the gas in contact with the surface will be sliding over it with the finite velocity v, and the motion of the gas will be very nearly the same as if the stratum of depth G had been removed from the solid and filled with the gas, there being now no slipping between the new surface of the solid and the gas in contact with it.

The coefficient G was introduced by Helmholtz and Piotrowski under the name of *Gleitungs-coefficient*, or coefficient of slipping. The dimensions of G are those of a line, and its ratio to l, the mean free path of a molecule, is given by the equation

$$G=\frac{2}{3}\left(\frac{2}{f}-1\right)l \quad \text{.................................(70)}.$$

Kundt and Warburg found that for air in contact with glass, $G=2l$, whence we find $f=\frac{1}{2}$, or the surface acts as if it were half perfectly reflecting and half perfectly absorbent. If it were wholly absorbent, $G=\frac{2}{3}l$.

It is easy to write down the surface conditions for a surface of any form.

Let the direction-cosines of the normal ν be l, m, n, and let us write

$$\frac{d}{d\nu} \text{ for } l\frac{d}{dx}+m\frac{d}{dy}+n\frac{d}{dz}.$$

We then find as the surface conditions

$$\left.\begin{aligned} u-G\frac{d}{d\nu}[(1-l^2)u-lmv-lnw]+\frac{3}{4}\frac{\mu}{\rho\theta}\left(\frac{d}{dx}-l\frac{d}{d\nu}\right)\left(\theta+4G\frac{d\theta}{d\nu}\right)=0 \\ v-G\frac{d}{d\nu}[(1-m^2)v-mnw-mlu]+\frac{3}{4}\frac{\mu}{\rho\theta}\left(\frac{d}{dy}-m\frac{d}{d\nu}\right)\left(\theta+4G\frac{d\theta}{d\nu}\right)=0 \\ w-G\frac{d}{d\nu}[(1-n^2)w-nlu-nmv]+\frac{3}{4}\frac{\mu}{\rho\theta}\left(\frac{d}{dz}-n\frac{d}{d\nu}\right)\left(\theta+4G\frac{d\theta}{d\nu}\right)=0 \end{aligned}\right\} \text{...(71)}.$$

In each of these equations the first term is one of the velocity-components of the gas in contact with the surface, which is supposed fixed; the second term depends on the slipping of the gas over the surface, and the third term indicates the effect of inequalities of temperature of the gas close to the surface, and shows that in general there will be a force urging the gas from colder to hotter parts of the surface.

Let us take as an illustration the case of a capillary tube of circular section, and for the sake of easy calculation we shall suppose that the motion is so slow, and the temperature varies so gradually along the tube that we may suppose the temperature uniform throughout any one section of the tube.

Taking the axis of the tube for that of z, we have for the condition of steady motion parallel to the axis

$$\frac{dp}{dz}=\mu\left(\frac{d^2w}{dx^2}+\frac{d^2w}{dy^2}\right) \quad \text{..............................(72)}.$$

Since everything is symmetrical about the axis, if we write r^2 for x^2+y^2 we find as the solution of this equation

$$w = A + \frac{1}{4\mu}\frac{dp}{dz}r^2 \quad \text{.................................(73).}$$

If Q denotes the quantity, of gas which passes through a section of the tube in unit of time

$$Q = 2\pi \int \rho w r dr$$

$$= \pi\rho a^2 \left(A + \frac{1}{8\mu}\frac{dp}{dz}a^2\right) \quad \text{...........................(74).}$$

At the inner surface of the tube we have $r = a$, and

$$w = A + \frac{1}{4\mu}\frac{dp}{dz}a^2$$

$$= \frac{Q}{\pi\rho a^2} + \frac{1}{8\mu}\frac{dp}{dz}a^2 \quad \text{.............................(75)}$$

also

$$\frac{dw}{d\nu} = -\frac{1}{2\mu}\frac{dp}{dz}a \quad \text{.................................(76).}$$

The last of equations (71) may therefore be written

$$\frac{Q}{\pi\rho a^2} + \frac{1}{8\mu}(a^2 + 4Ga)\frac{dp}{dz} - \frac{3}{4}\frac{\mu}{\rho\theta}\frac{d\theta}{dz} = 0 \quad \text{......................(77).}$$

Equation (77) gives the relation between the quantity of gas which passes through any section of the tube, the rate of variation of pressure, and the rate of variation of temperature in passing along the axis of the tube.

If the pressure is uniform there will be a flow of gas from the colder to the hotter end of the tube, and if there is no flow of gas the pressure will increase from the colder to the hotter end of the tube.

These effects of the variation of temperature in a tube have been pointed out by Professor Osborne Reynolds as a result of the Kinetic Theory of Gases, and have received from him the name of Thermal Transpiration: a name in strict analogy with the use of the word Transpiration by Graham.

But the phenomenon actually observed by Professor Reynolds in his experiments was the passage of gas through a porous plate, not through a capillary tube; and the passage of gases through porous plates, as was shown by Graham, is of an entirely different kind from the passage of gases through capillary tubes, and is more nearly analogous to the flow of a gas through a small hole in a thin plate.

When the diameter of the hole and the thickness of the plate are both small compared with the length of the free path of a molecule, then, as Sir William Thomson has shown, any molecule which comes up to the hole on either side will be in very little danger of encountering another molecule before it has got fairly through to the other side.

Hence the flow of gas in either direction through the hole will take place very nearly in the same manner as if there had been a vacuum on the other side of the hole, and this whether the gas on the other side of the hole is of the same or of a different kind.

If the gas on the two sides of the plate is of the same kind but at different temperatures, a phenomenon will take place which we may call *thermal effusion.*

The velocity of the molecules is proportional to the square root of the absolute temperature, and the quantity which passes out through the hole is proportional to this velocity and to the density. Hence, on whichever side the product of the density into the square root of the temperature is greatest, more molecules will pass from that side than from the other through the hole, and this will go on till this product is equal on both sides of the hole. Hence the condition of equilibrium is that the density must be inversely as the square root of the temperature, and since the pressure is as the product of the density into the temperature, the pressure will be directly proportional to the square root of the absolute temperature.

The theory of thermal effusion through a small hole in a thin plate is therefore a very simple one. It does not involve the theory of viscosity at all.

The finer the pores of a porous plate, and the rarer the gas which effuses through it, the more nearly does the passage of gas through the plate correspond to what we have called effusion, and the less does it depend on the viscosity of the gas.

The coarser the pores of the plate and the denser the gas, the further does the phenomenon depart from simple effusion, and the more nearly does it approach to transpiration through a capillary tube, which depends altogether on viscosity.

To return to the case of transpiration through a capillary tube. When the temperature is uniform

$$Q = -\frac{\pi\rho a^4}{8\mu}\frac{dp}{dz}\left(1 + 4\frac{G}{a}\right) \text{ ....(78).}$$

By experiments on capillary tubes of glass, MM. Kundt and Warburg found* for the value of G for air at different pressures and at from 17° C. to 27° C.,

$$G = \frac{8}{p} \text{ centimetres..............................(79)}$$

where p is the pressure in dynes per square centimetre, which is nearly the same as in millionths of an atmosphere. For hydrogen on glass

$$G = \frac{15}{p} \text{ centimetres ....(80).}$$

When there is no flow of gas in a tube in which the temperature varies from end to end, the pressure is greater at the hot end than at the cold end. Putting $Q = 0$ we have

$$\frac{dp}{d\theta} = 6\frac{\mu^2}{\rho\theta}\frac{1}{a^2 + 4Ga} \text{ ....(81).}$$

The quantity $6\frac{\mu^2}{\rho\theta}$ is just double of that calculated in section (3) of the introduction, and is therefore in C.G.S. measure $0{\cdot}63 \times p$ for dry air at 15° C. Let us suppose $a = 0{\cdot}01$ centimetre, and the pressure 40 millimetres of mercury, then $G = {\cdot}00016$ centimetre.

If one end of the tube is kept at 0° C. and the other at 100° C., the pressure at the hot end will exceed that at the cold end by about 1·2 millionths of an atmosphere.

The difference of pressure might be increased by using a tube of smaller bore and air of smaller density, but the effect is so small that though the theoretical proof of its existence seems satisfactory, an experimental verification of it would be difficult.

* Pogg. *Ann.*, July, 1876.

A Maxwell Bibliography

Maxwell's Published Works

This bibliography includes Maxwell's published papers, reprinted in these three volumes, and his other publications on related topics. The entries are listed in the order, and with the paper number assigned to them in Maxwell, *Scientific Papers,* ed. W. D. Niven. If there is no number noted the paper was not included in Maxwell's collected works. The full citation for an entry cited here as Davies (1871) is in the bibliography of secondary sources. The following abbreviations are used for the three texts in this series.

MSR = *Maxwell on Saturn's Rings*

MMG = *Maxwell on Molecules and Gases*

MH = *Maxwell on Heat & Statistical Mechanics*

XVIII. "On Theories of the Constitution of Saturn's Rings," *Proc. R. Soc. Edinburgh* 4 (1862): 99–101.

XIX. *On the Stability of the Motion of Saturn's Rings. An Essay, which gained the Adams Prize for the Year 1856, in the University of Cambridge.* Comments by Cook and Franklin (1964), Davies (1871), Deslandres (1895), Faye (1872), Goldsbrough (1951), Hirn (1872), Keeler (1895), Pendes (1933), Proctor (1865, 1869), Taylor (1884), Tremaine (1985), Webb (1867). Reprinted in MSR, 68–158.

— "On the Dynamical Theory of Gases," *Rep. 29th Meeting BAAS* (1859): 9. Reprinted in MMG, document III-2, 281–282.

XX. "Illustrations of the Dynamical Theory of Gases," *Phil. Mag.* 19 (1860): 19–32; 20 (1860): 21–37. Excerpts in Brush (1965) and in Boorse and Motz (1966). Report by

Jochmann (1860). Comments by Clausius (1860), Jochmann (1860). MMG document III-6, 285–318.

— "On the Results of Bernoulli's Theory of Gases as Applied to Their Internal Friction, Their Diffusion, and Their Conductivity for Heat," *Rep. 30th Meeting BAAS* (1860): 15–16. MMG, document III-8, 320–321.

XXVII. "On the Viscosity or Internal Friction of Air and Other Gases," *Phil. Trans. R. Soc. London* 156 (1866): 249–268; abstract in *Proc. R. Soc. London* 15 (1867): 14–17. Report by Bertram (1866). Comments by Meyer (1871). MMG documents III-18, 359–388 and III-17, 354–358, respectively.

XXVIII. "On the Dynamical Theory of Gases,"*Phil. Trans. R. Soc. London* 157 (1867): 49–88; *Phil. Mag.* 32 (1866): 390–393; 35 (1868): 129–145, 185–217; abstract in *Proc. R. Soc. London* 15 (1867): 167–171. Report by Bertram (1866) and Grossmann (1866). Reprinted in Brush (1966). MMG documents III-30, 419–472 and III-29, 415–418.

— Discussion Remark on paper by Brodie. *Chemical News* 15 (1867): 303.

XLI. "Address to the Mathematical and Physical Sections of the British Association," *Rep. 40th Meeting BAAS* (1870): 1–9. MMG document III-9, 89–104.

XLIV. "Introductory Lecture on Experimental Physics," (Cambridge, 1871). MMG document II-11, 110–125.

— *Theory of Heat.* London: Longmans, Green & Co., 1871; 2d ed. 1872; 3d ed. 1872; 4th ed. 1875; 5th ed. 1877; 6th ed. 1880; 10th ed. with corrections and additions by Lord Rayleigh, 1891. 3rd ed. reprinted, Westport, Conn.: Greenwood Press, 1970. German translation *Theorie der Wärme*, translated from the 4th English ed. by F. Auerbach, Breslau, 1877. Second translation by F. Neeson, Braunschweig: Viewig, 1878. Russian translation *Teoriya Teploty v Elementarnoi Obrabotky* translated by A. L. Korol'kov, Kiev: Kumnerev, 1888. Reviews by Howard (1892), Stewart (1872), [Anonymous] (1872). Extracts in MH documents II-6, 119–120, III-22, III-23, III-32.

LII. "On the Proof of the Equations of Motion of a connected System," *Proc. Cambridge Phil. Soc.* 2 (1873): 292–294.

— "Clerk-Maxwell's Kinetic Theory of Gases," *Nature* 8 (1873): 84. MH document II-7, 121–123.

LIX. "On Loschmidt's Experiments on Diffusion in Relation to the Kinetic Theory of Gases," *Nature* 8 (1873): 298–300. MMG document III-42, 496–504.

— "On the Equilibrium of Temperature of a Gaseous Column Subject to Gravity," *Nature* 8 (1873): 527–528. MH document II-10, 125–126.

— "On the Final State of a System of Molecules in Motion Subject to Forces of any Kind," *Rep. 43rd Meeting BAAS* (1873): 29–32. MH document II-14, 138–143.

LX. "On the Final State of a System of Molecules in Motion Subject to Forces of any Kind," *Nature* 8 (1873): 537–538. MH document II-15, 143.

— "Atoms and Ether," *Nature* 8 (1873): 361. MMG document II-14, 131–132.

LXII. [A Discourse on] "Molecules." (A lecture delivered before the British Association at Bradford), *Nature* 8 (1873): 437–441; *Phil. Mag.* 46 (1873): 453–469; *Pharmaceutical Journal* 4 (1874): 404–405, 492–494, 511–513; *Popular Science Monthly* 4 (1874): 276–290. Reprinted in Knight (1968). MMG document II-16, 137–155.

— [Reply to Guthrie (1874)] *Nature* 10 (1874): 123.

LXVII. "Grove's 'Correlation of Physical Forces', " *Nature* 10 (1874): 302–304.

LXIX. "Van der Waals on the Continuity of the Gaseous and Liquid States," *Nature* 10 (1874): 477–480. MH document IV-3, 291–298.

LXXI. "On the Dynamical Evidence of the Molecular Constitution of Bodies." (A lecture delivered at the Chemical Society) *Journal of the Chemical Society* 13 (1875): 493–508; *Nature* 11 (1875): 357–359, 374–377. Report by Pfaundler (1875). MMG document II-24, 216–237.

— "Cavendish Laboratory," *Cambridge University Reporter*, 27 April, 1875: 352–354; 20 May, 1876: 496–498; 15 May, 1877: 434–435; 2 April, 1878: 420. (Lists of apparatus, etc.; e.g., in 1876 the Laboratory has "Model in plaster of Willard Gibbs' thermodynamic surface.")

LXXIII. "Atom," *Encyclopedia Britannica*, 9th ed., 3 (1875): 36–49. Reprinted in Knight (1968). MMG document II-23, 175–215.

LXXVI. "On the Equilibrium of Heterogeneous Substances," *Proc.*

Cambridge Phil. Soc. 2 (1876): 427–430. (Summary.) MH III-39, 262–265.

— "On the Equilibrium of Heterogeneous Substances." (Address on 24 May, 1876 at the South Kensington Conferences in connection with the Special Loan Collection of scientific apparatus. Reprinted from pp. 144–150 of the official volume of reports of the Conferences) *Phil. Mag.* 16 (1908): 818–824 (with prefatory note by J. Larmor). This is a longer version of the preceding item. MH document III-38, 257–262.

LXXVII. "Diffusion of Gases through Absorbing Substances," *Nature* 14 (1876): 24–25. Review of *Über die Diffusion der Gase durch absorbirende Substanzen.* Habilitationsschrift der Mathematischen und Naturwissenschaften Facultät der Universität Strassburg, vorgelet von Dr. Sigmund v. Wroblewski, ersten Assistaten am physikalischen Institute. Strassburg: G. Fischbach, 1876. MMG document III-46, 510–514.

LXXIX. "Instruments Connected with Fluids." From the Kensington Museum Hand Book. Discussion of low–pressure experiments.

LXXXIII. "Capillary Action," *Encyclopedia Britannica*, 9th ed. (1878) 5: 56–71.

LXXXVIII. "Constitution of Bodies," *Encyclopedia Britannica*, 9th ed. (1878) 6: 310–313. MMG II-30, 245–254.

LXXXIX. "Diffusion," *Encyclopedia Britannica*, 9th ed. (1878) 7: 214–221. MMG document III-52, 524–546.

— "Heat," *Encyclopedia Britannica*, 9th ed. (1878) 11: 554–589.

— "Physical Sciences," *Encyclopedia Britannica*, 9th ed. 19 (1878): 1–3.

XCI. "Tait's 'Thermodynamics', " *Nature* 17 (1878): 257–259, 278–280. MH document III-47.

XCIII. "On Stresses in Rarified Gases Arising from Inequalities of Temperature," *Proc. R. Soc. London* 27 (1878): 304–308; *Phil. Trans. R. Soc. London* 170 (1880): 231–256. Comments by Reynolds (1880, 1881, 1883). MH documents VI-11 and VI-31, 408–412, 462–494 respectively.

XCIV. "On Boltzmann's Theorem on the Average Distribution of Energy in a System of Material Points," *Trans. Cambridge Phil. Soc.* 12 (1879): 547–570. Report by Boltzmann

(1881). Comments by Rayleigh (1892). MH document V-7, 357–358.

— *The Scientific Papers of James Clerk Maxwell*, ed. W. D. Niven. Cambridge: Cambridge University Press, 1890. Reprint Dover Publications, New York, 1952–1965. Reviewed by Rayleigh (1890).

Secondary Sources

The entries are limited to publications on Maxwell and his work on Saturn's rings, atoms, kinetic theory, thermodynamics and statistical mechanics, and mechanical and thermal properties of matter. Articles on "Maxwell's Demon," "Maxwell's velocity distribution," etc. are not included unless they pertain to the nineteenth-century discussion of these issues.

Achinstein, Peter. "Theoretical Derivations," *Stud. Hist. Phil. Sci.* 17 (1986): 375–414. On Maxwell's derivation of his velocity distribution law.

Achinstein, Peter. "Scientific Discovery and Maxwell's Kinetic Theory," *Philosophy of Science* 54 (1987): 409–434.

Achinstein, Peter. Review of Garber, Brush, and Everitt, *Maxwell on Molecules and Gases. Foundations of Physics* 17 (1987): 425–433.

Adams, Henry, *Mont–Saint–Michel and Chartres.* Boston: Houghton Mifflin, 1905. Page 375 contains a reference to [Maxwell's] demon. (See Blackmur [1980], 234.)

[Anonymous] Review of *Theory of Heat Phil. Mag.* 43 (1872): 149–151.

[Anonymous] "Scientific Worthies. XVIII James Clerk Maxwell," *Nature* 24 (1881): 601. With facing portrait engraved from photograph.

Ariew, Roger, and Peter Barker, "Duhem on Maxwell: A Case-Study in the Interrelations of the History of Science and Philosophy of Science," *PSA 1986*, 1 (1986): 145–156. Mostly on electromagnetic theory.

Barus, Carl. "Maxwell's Theory of the Viscosity of Solids and Its Physical Verification," *Phil. Mag.* 26 (1888): 183–217.

Barus, Carl. "The Viscous Effect of Strains Mechanically Applied, as Interpreted by Maxwell's Theory," *Phil. Mag.* 27 (1889) : 155–177.

Baucia, Giovanni. "Microfenomeni e macrofenomeni secondo Maxwell in relazi one alla 'Teori a dinamica del calore'," *Physis* 15 (1973): 333– 350.

Bellone, Enrico. *Aspetti dell'approccio statistico alla meccanica: 1849–1905.* Firenze: G. Barbera Editore, 1972. (Publicazioni di Storia della Scienza della Domus Galileaeana, Sezione IV, Volume 1.)

Belopolskij, A. A. "Issledovanie smetseniya linii v spektr Saturna i ego kol'tsa" ("Recherches sur les deplacements des raises dans le spectre de Saturne et de son anneau"), *Izvestiya Imperatorsko i Akademii Nauk* [*Bulletin de l'Academie Imperiale des Sciences de St.-Pétersbourg*] 3 (1895): 379–403. Confirmed Maxwell's theory of Saturn's rings. See P. G. Kulikovsky, "Belopolsky, Aristarkh Apollonovich," *Dictionary of Scientific Biography* 1 (1970): 597–598.

Berger, M. S., ed. *J. C. Maxwell, The Sesquicentennial Symposium. New Vistas in Mathematics, Science, and Technology.* New York: North-Holland, 1984. Proceedings of a Symposium at the University of Massachusetts, Amherst, 16–18 October, 1981. Includes: quotations from Maxwell, 1–9; Einstein's quotations about Maxwell, 11– 14; short articles on various aspects of Maxwell's influence on physics and mathematics.

Bernhardt, H. "Über die Entwicklung und Bedeutung der Ergodenhypothese in den Anfängen der statistischen Mechanik," *N T M* 8 (1971): 13–25.

Bernstein, H. T. "J. Clerk Maxwell on the History of the Kinetic Theory of Gases, 1871," *Isis* 54 (1963): 206– 216.

Bertram, [Theodor?] [Report on Maxwell's Paper XXVII and on the abstract of Paper XXVIII in *Proc. R. Soc. London*], *Fort. Physik* 22 (1866): 544–555.

Blackmur, R. P. *Henry Adams.* New York: Harcourt Brace Jovanovich, 1980. Page 234–35, references to Maxwell's demon by Adams and Karl Pearson.

Bolton, H. C., and W. C. Price. "The Date of Birth of James Clerk Maxwell," *Notes Rec. R. Soc. London* 32 (1978): 213–214.

Boltzmann, Ludwig. "J. C. Maxwell. Über Boltzmanns Theorem, betreffend die mittlere Vertheilung der lebendigen Kraft in einem System materieller Punkte," *Ann. Phys. Beiblätter* 5 (1881): 403–417. English translation in *Phil. Mag.* 14 (1882): 299–312. Comments on Paper XCIV.

Boltzmann, Ludwig. *Gustav Robert Kirchhoff: Festrede zur Feier des 301. Gründungstages der Karl-Franzens-Universität zu Graz.* Leipzig: Barth, 1888. Includes the passage comparing Maxwell's method of deriving a result in kinetic theory to a style in music.

Boltzmann, Ludwig. "III Teil der Studien über Gleichgewicht der lebendigen Kraft." *Sitz. Math.-Physik. Cl. K. Bayerischen Akad. Wissen. München* 22 (1892): 329–358. English translation in *Phil. Mag.* 35 (1893): 153–173. Comments on Maxwell's proof of thermal equilibrium and his attempt to generalize it (Paper XXVIII) where he made a mistake which can be easily corrected.

Boltzmann, Ludwig. "On Maxwell's Method of Deriving the Equations of Hydrodynamics from the Kinetic Theory of Gases, " *Rep. 64th Meeting BAAS* (1894): 579.

Boltzmann, Ludwig "Zur Erinnerung an Josef Loschmidt" (Gedankrede, 29 Oktober 1895). In Boltzmann, *Populäre Schriften* Ausgewählt von E. Broda, 150–59. Braunschweig: Vieweg, 1979. Refers to Maxwell's Demon, which he says (incorrectly) was anticipated by Loschmidt (cf. Daub, 1970).

Boltzmann, Ludwig. *Vorlesungen über Gastheorie.* Leipzig: Barth, 1896–1898. English translation, *Lectures on Gas Theory,* introduction, notes, and bibliography by Stephen G. Brush, Berkeley: University of California Press, 1964. The reprint of *Vorlesungen* as Band 1 of *Ludwig Boltzmann, Gesamtausgabe*, hrsg. R. U. Sexl (Graz: Akademische Druck- u. Verlagsanstalt, 1981), includes a German translation of Brush's introduction, notes, and bibliography.

Boorse, H. A., and L. Motz, eds. *The World of the Atom.* New York: Basic Books, 1966. Includes a brief biography of Maxwell, and extracts from Paper XX.

Brock, W. H., and D. M. Knight. "The Atomic Debates: 'Memorable and Interesting Evenings in the Life of the Chemical Society'," *Isis* 56 (1965): 5–25. On lectures by the chemists Benjamin Brodie (1867) and Alexander Williamson (1869), and the discussion of atomic theory in which Maxwell played a small role.

Brock, W. H., ed. *The Atomic Debates: Brodie and the Rejection of the Atomic Theory.* New York: Humanities Press, 1967. Includes the above article by Brock and Knight. There are References to Maxwell in the papers by D. M. Dallas, "The Chemical Calculus of Sir Benjamin Brodie," and by W. H. Brock, "Some Correspondence Connected with Sir Benjamin Brodie's Calculus of Chemical Operations."

Brush, Stephen G. "The Development of the Kinetic Theory of Gases. IV. Maxwell," *Ann. Sci.* 14 (1958): 243–255. Reprinted in Brush, *The Kind of Motion We Call Heat.* Russian translation in *Dshems Klerk Maksvell, Stat'i i Rechi*, 288–304 (Moscow: Izdatel'stvo "Nauka," 1968).

Brush, Stephen G. "Development of the Kinetic Theory of Gases. VI. Viscosity," *Amer. J. Phys.* 30 (1962): 269–281.

Brush, Stephen G. *Kinetic Theory.* Volume 1. *The Nature of Gases and of Heat.* New York: Pergamon Press, 1965. Includes excerpts from Maxwell's Paper XX. German translation, *Kinetische Theorie*, Band I, *Die Natur der Gase und der Wärme* (Berlin: Akademie-Verlag, 1970).

Brush, Stephen G. *Kinetic Theory.* Volume 2. *Irreversible Processes.* New York: Pergamon Press, 1966. Includes Maxwell's paper XXVIII. German translation, *Kinetische Theorie*, Band II, *Irreversible Prozesse* (Berlin: Akademie-Verlag, 1970).

Brush, Stephen G. "Foundations of Statistical Mechanics 1845–1915," *Arch. Hist. Exact Sci.* 5 (1967): 145–183. Reprinted in Brush, *The Kind of Motion We Call Heat.*

Brush, Stephen G., and C. W. F. Everitt. "Maxwell, Osborne Reynolds, and the Radiometer," *Hist. Stud. Phys. Sci.* 1 (1969): 105–125. Reprinted in Brush, *The Kind of Motion We Call Heat.*

Brush, Stephen G. "Interatomic Forces and Gas Theory from Newton to Lennard-Jones, " *Arch. for Rational Mechanics and Analysis* 39 (1970) : 1–29.

Brush, Stephen G. "James Clerk Maxwell and the Kinetic Theory of Gases: A Review Based on Recent Historical Studies," *Amer. J. Phys.* 39 (1971): 631–640.

Brush, Stephen G. "The Development of the Kinetic Theory of Gases, VII. Heat Conduction and the Stefan-Boltzmann Law," *Arch. Hist. Exact Sci.* 11 (1973): 38–96. Reprinted in Brush, *The Kind of Motion We Call Heat.*

Brush, Stephen G. "The Development of the Kinetic Theory of Gases, VIII, Randomness and Irreversibility," *Arch. Hist. Exact Sci.* 12 (1974): 1–88. Reprinted in Brush, *The Kind of Motion We Call Heat.*

Brush, Stephen G. *The Kind of Motion We Call Heat: A History of the Kinetic Theory of Gases in the 19th Century.* Amsterdam: North-Holland Pub. Co., 1976. Reprinted, 1986.

Brush, Stephen G. "Irreversibility and Indeterminism: Fourier to Heisenberg," *J. Hist. Ideas* 37 (1976): 603–630. Reprinted in Brush, *Statistical Physics.*

Brush, Stephen G. "Statistical Mechanics and the Philosophy of Science," *PSA 1976* vol. 2, ed. F. Suppe and P. Asquith, 551–584. East Lansing, Michigan: Philosophy of Science Association, 1977. Reprinted in Brush, *Statistical Physics.*

Brush, Stephen G. *Statistical Physics and the Atomic Theory of Matter from Boyle and Newton to Landau and Onsager.* Princeton, N. J.: Princeton University Press, 1983.

Brush, Stephen G. "Gaseous Heat Conduction and Radiation in 19th Century Physics," in *History of Heat Transfer*, ed. E. T. Layton, Jr., and J. H. Lienard, 25–51. New York: American Society of Mechanical Engineers, 1988. Revised version of Brush, "Development," Part VII.

Butcher, J. G. "On Viscous Fluids in Motion," *Proc. London Math. Soc.* 8 (1871) : 103–135. On Maxwell's viscoelastic theory.

Campbell, Lewis, and William Garnett. *The Life of James Clerk Maxwell.* London: Macmillan, 1882. Second edition, 1884. First edition reprinted with a new preface and appendix with letters by R. H. Kargon; New York: Johnson Reprint Corp., 1969. (The Sources of Science Series, No. 85.) First edition reviewed by G. C[hrystal?] in *Nature* 27 (1882): 26–29.

Cannizzaro, S. "Considerations on Some Points of the Theoretic Teaching of Chemistry," *J. Chem. Soc. London* 10 (1872): 941–967. On the relation of the kinetic theory of gases to the chemical atomic theory.

Cardwell, D. S. *From Watt to Clausius: The Rise of Thermodynamics in the Early Industrial Age.* Ithaca: Cornell University Press, 1971. Reprinted, Ames: Iowa State University Press, 1989.

Cater, Harold Dean, ed. *Henry Adams and His Friends: A Collection of his Unpublished Letters.* Boston: Houghton Mifflin, 1947, 545–46. Letter from Henry Adams to Brooks Adams, 2 May 1903, mentioning Maxwell's Demon.

Chalmers, A. F. "On Learning from our Mistakes," *Brit. J. Phil. Sci.* 24 (1972): 164–173. Maxwell and Watson on the problem of specific heats in the kinetic theory.

Chapman, Sydney. "The Kinetic Theory of Gases Fifty Years Ago," *Lectures in Theoretical Physics*, vol. IX–C, ed. W. E. Brittin, 1–13. New York: Gordon and Breach, 1967. Reprinted in Brush, *Kinetic Theory.* Volume 3 and entitled "The Chapman-Enskog Solution of the Transport Equation for Moderately Dense Gases," 260–271. (New York: Pergamon Press, 1972).

Chapman, Sydney, and T. G. Cowling. *The Mathematical Theory of Non-Uniform Gases.* Cambridge: Cambridge University Press, 1939. 2d ed., 1952. Includes "Historical Summary," 380–388.

Clausius, R. "On the Dynamical Theory of Gases" *Phil. Mag.* 19 (1860): 434–436. On his method of finding the mean relative velocity, see W. D. Niven's note in Maxwell's *Scientific Papers*, vol. 1, 387.

Clausius, R. "Über die Wärmeleitung gasförmiger Körper," *Ann. Phys.* 115 (1862): 1–56. English translation in *Phil. Mag.* 23 (1862): 417–435, 512–524. Includes critical remarks on Maxwell's treatment of heat conduction in Paper XX.

Clausius, R. *Die Mechanische Wärmetheorie*, Bd. II., pt. 2. Aufl. Braunschweig: Vieweg, 1879, 315–16. Comments on Maxwell's Demon.

Clausius, R. "Über einige neue Untersuchungen über die mittlere Weglänge der Gasmolecüle," *Ann. Phys.* 10 (1880): 92–103. Further remarks on the average relative velocity of molecules in a gas.

Clausius, R. *Die Kinetische Theorie der Gase*, ed. by Dr. Max Planck und Dr. Carl Pulfrich. (Third volume of *Die Mechanische Wärmetheorie*, containing Clausius's published papers and ideas on the kind of motion that is heat.) Braunschweig : Vieweg, 1889–1891.

Clifford, W. K. "Atoms," in *Lectures and Essays*, ed. by L. Stephen and F. Pollock, vol. 1, 158–190. London: Macmillan, 1879. Based on lectures given in 1872. Discusses Maxwell's results on rigidity of a gas of noninteracting particles, and his hypothesis on atomic repulsion. See also other essays in this book.

Cohen, I. B. "Maxwell's Poetry," *Scientific American* 186 (3) (1952): 62–63.

Collier, John D. "Two Faces of Maxwell's Demon Reveal the Nature of Irreversibility." *Stud. Hist. Phil. Sci.* 21 (1990) : 257–268.

Cook, A. F., and F. A. Franklin. "Rediscussion of Maxwell's Adams Prize Essay on the Stability of Saturn's Rings," *Astronomical Journal* 69 (1964): 173–200.

Coulson, C. A. "Interatomic Forces: Maxwell to Schrödinger," *Nature* 195 (1962): 744–749.

Coulson, C. A. "Interatomic Forces," in Domb *Clerk Maxwell and Modern Science,* 43–69.

Crow, Christine M. *Paul Valery and Maxwell's Demon: Natural Order and Human Possibility.* Hull, England: University of Hull Publications, 1972.

Crowther, J. G. *British Scientists of the Nineteenth Century.* London: Kegan Paul, Trench, Trubner and Co., 1935. Chapter 7: Maxwell.

Daub, Edward E. "Atomism and Thermodynamics," *Isis* 58 (1967): 293–303. Maxwell on Rankine.

Daub, E. E. "Entropy and Dissipation,"*Hist. Stud. Phys. Sci.* 2 (1970) : 321– 354. On Tait, Maxwell, Clausius, and Kelvin.

Daub, E. E. "Maxwell's Demon," *Stud. Hist. Phil. Sci.* 1 (1970): 213–227.

Davies, A. M. *The Meteoric Theory of Saturn's Rings, Considered with Reference to the Solar Motion in Space.* London: Longmans, 1871. Anonymous review in *Nature* 4 (1871): 159, summarizes Maxwell's theory and says Davies "appears not to have seen Prof. Maxwell's work, as he ascribes to the perusal of a derived exposition of it the enlistment of his interest in favour of the satellite theory of the rings." See also R. A. Proctor, *Nature* 4 (1871): 346.

Deakin, Michael A. B. "Nineteenth Century Anticipations of Modern Theory of Dynamical Systems." *Arch. Hist. Exact Sci.* 39 (1988): 183–194. Maxwell's views on determinism and free will in LX. Newtonian physics need not lead to a fully determined universe.

DeMarzo, Carlo, *Maxwell e la Fisica Classica.* Roma Bari: Laterza, 1978. Chapter II: "La Teoria Cinetica dei Gas." Includes Italian translations of excerpts from Papers XVIII, XX, LXII, LXXI.

Deslandres, Henri. "Recherches spectrales sur les anneaux de Saturne," *Comptes Rendus* 120 (1895): 1155–1158. Confirmed the results of Keeler supporting Maxwell's theory of Saturn's rings.

Domb, C., ed. *Clerk Maxwell and Modern Science. Six Commemorative Lectures by Sir Edward V. Appleton, E. G. Bowen, C. A. Coulson, R. E. Peierls, Sir John Randall, R. A. Smith.* London: Athlone Press/New York: Oxford University Press, 1963. Includes Coulson (1963), Randall (1963). Reviewed by A. M. Bork *Science* 144 (1964): 724; by I. S. in *Cslky Cas. Hist.* 14 (1966): 634.

Domb, C. "James Clerk Maxwell—100 Years Later," *Nature* 282 (1979): 235–239.

Domb, C. "James Clerk Maxwell in London, 1860–1865," *Notes Records R. Soc. London* 35 (1980): 67–103.

Dorling, Jon, "Maxwell's Attempt to Arrive at Non-Speculative Foundations for the Kinetic Theory," *Stud. Hist. Phil. Sci.* 1 (1970): 229–248.

Duhem, P. See Ariew and Barker, "Duhem on Maxwell."

Dukov, V. M. "Atomisticheskie predstavleniia v rabotakh M. Faradeia i Dz. Maksvella," *Voprosy Istorii Estestvoznaniia i Tekhniki* 12 (1962): 178–184. ("Atomic concepts in the work of M. Faraday and J. Maxwell.")

Duncan, David. *Life and Letters of Herbert Spencer.* New York: Appleton, 1908. Includes correspondence with Maxwell.

Ehrenberg, W. "Maxwell's Demon," *Scientific American* 215 (November 1967): 103–110.

Ehrenfest, Paul, and Tatiana Ehrenfest, "Begriffliche Grundlagen der statistischen Auffassung in der Mechanik," *Encyklopadie der Mathematischen Wissenschaften* Bd. IV, Teil IV, Art. 32 (1911). English translation, *The Conceptual Foundations of the Statistical Approach in Mechanics*, translated by M. J. Moravcsik. Ithaca: Cornell University Press, 1959.

Elkana, Y. *The Discovery of the Conservation of Energy.* Cambridge, Mass.: Harvard University Press, 1974. Maxwell has a "tendency to attribute to great scientists [e.g., Helmholtz, Faraday] only 'correct' opinions" (p. 132).

El'yashevich, M. A. and T. S. Prot'ko. "Maxwell's Contribution to the Development of Molecular Physics and Statistical Methods," *Soviet Physics Uspekhi* 24 (1981): 876–903. Translated from *Uspekhi Fizicheskikh Nauk.* 135 (1981): 381–423.

Everitt, C. W. F. "Maxwell's Scientific Papers," *Applied Optics* 6 (1967): 639–646.

Everitt, C. W. F. "Maxwell, James Clerk," *Dictionary of Scientific Biography.* New York: Charles Scribner, 1974, vol. 9: 198–230.

Everitt, C. W. F. *James Clerk Maxwell: Physicist and Natural Philosopher.* New York: Scribner, 1975. Reviewed by P. M. Heimann, *Technology and Culture* 18 (1977): 264–265; P. Morrison, *Scientific American* 234 (5) (1976): 127.

Everitt, C. W. F. "Maxwell, the Man," *Physics Teacher* 22 (1984): 264–266. Review of Tolstoy's biography; corrects error about Maxwell's height .

Faye, H. "Note relative à un mémoire de M. Clerk–Maxwell, sur la stabilité des anneaux de Saturne," *Comptes Rendus* 75 (1872): 793–794.

Ferguson, Allan. "The Clerk Maxwell Centenary Celebrations," *Nature* 128 (1931): 604–608.

FitzGerald, G. F. "On Professor Osborne Reynolds's Paper On Certain Dimensional Properties of Matter in the Gaseous State," *Phil. Mag.* 11 (1881): 103–109. Includes a discussion of some of Maxwell's objections to Reynolds's work.

Fleck, G. M. "Atomism in late Nineteenth Century Physical Chemistry," *J. Hist. Ideas* 24 (1963): 106–114. Quotes Maxwell on vortex atoms.

Forbes, Eric G. *James Clerk Maxwell, FRSE, FRS, 1831–77.* Edinburgh: History of Medicine and Science Unit, 1982. (In the series "Scotland's Cultural Heritage,") 6 pp. + illus.

Ford, W. C., ed. *Letters of Henry Adams (1892–1918)*. Boston: Houghton Mifflin, 1938; New York: Kraus Reprint, 1969, 135–136, letter from Adams to Cecil Spring Rice, 11 November 1897, mentions Maxwell's Demon.

Fuller, A. T. "Clerk Maxwell's London Notebooks: Extracts Relating to Control and Stability, " *International Journal of Control* 30 (1979): 729–744.

Fuller, A. T. "James Clerk Maxwell's Cambridge Manuscripts: Extracts Relating to Control and Stability," *International Journal of Control* 35 (1982): 785–805; 36 (1982): 547–574; 37 (1983): 1197–1238; 39 (1984): 619–656; 43 (1986): 805–818; 43 (1986): 1135–1168. See esp. Part IV which deals with Maxwell's Saturn's Rings manuscript.

Fuller, A. T. "James Clerk Maxwell's Glasgow Manuscripts: Extracts Relating to Control and Stability," *International Journal of Control* 43 (1986): 1593–1612.

Garber, Elizabeth. Maxwell, Clausius and Gibbs: Aspects of the Development of Kinetic Theory and Thermodynamics. Ph. D. diss., Case Institute of Technology, 1966.

Garber, Elizabeth, "James Clerk Maxwell and Thermodynamics," *Amer. J. Phys.* 37 (1969): 146–155.

Garber, Elizabeth, "Clausius and Maxwell's Kinetic Theory of Gases," *Hist. Stud. Phys. Sci.* 2 (1970): 299–319.

Garber, Elizabeth, "Aspects of the Introduction of Probability into Phys ics," *Centaurus* 17 (1972): 11–39.

Garber, Elizabeth, "Molecular Science in Late-Nineteenth-Century Britain," *Hist. Stud. Phys. Sci.* 9 (1976): 265–297.

Garnett, William, "James Clerk Maxwell, F. R. S.," *Nature* 21 (1879): 43–46.

Garnett, William. *Heroes of Science: Physicists.* London: Society for Promoting Christian Knowledge, 1885. See, 278–308.

Garnett, William. "Energy," *Encyclopaedia Britannica*, New Werner Edition, vol. 8, 205–211. New York: Werner Co., 1903. On Maxwell's Demon in the works of Rayleigh and Preston.

Gavroglu, Kostas. "The Reaction of the British Physicists and Chemists to van der Waals' Early Work and to the Law of Corresponding States," *Hist. Stud. Phys. Biol. Sci.* 20 (1990): 199–237. Includes comments on Maxwell's Paper LXIX.

Gibbs, J. Willard. "A Method of Geometrical Representation of the Thermodynamic Properties of Substances by Means of Surfaces," *Trans. Acad. Sci. Connecticut* 2 (1873): 382–404. Reprinted in Gibbs *Collected Works*, vol. I (New Haven: Yale University Press, 1948). Remarks on usage of term "entropy" by Maxwell and others.

Gillispie, C. C. "Intellectual Factors in the Background of Analysis by Probabilities," in *Scientific Change*, ed. A. C. Crombie, 431–453. New York: Basic Books, 1963. Suggests that Maxwell was influenced in his original derivation of the velocity distribution law by reading John Herschel's review of books by Quetelet. See also comments by M. B. Hesse, 471–477.

Gillispie, C. C., "The Social and Intellectual Background of Statistical Mechanics," *Voice of America Forum Lectures, History of Science Series*, no. 7 (1964). Condensed version of above article.

Glazebrook, R. T. *James Clerk Maxwell and Modern Physics.* New York: Macmillan, 1896.

Goldman, Martin. *The Demon in the Aether: The Story of James Clerk Maxwell.* Edinburgh: Paul Harris, 1983. Reviewed by J. Calado, *Times Literary Supplement*, (20 April, 1984): 442; D. Gooding, *Isis*, 76 (1985): 281; P. M. Harman, *Ann. Sci.* 41 (1984): 400; J. O. Marsh, *Nature* 308 (1984): 800; J.-P. Mathieu, *Revue d'Histoire des Sciences* 38 (1985): 173; D. Siegel, *Physics Today* 38, no. 8 (1985): 66–67; C. Smith, *Brit. J. Hist. Sci.* 18 (1985): 120–121.

Goldsbrough, G. R. "The Stability of Saturn's Rings," *Phil. Trans. R. Soc. London* 244 (1951): 1–17. Comments on possible improvement of Maxwell's theory.

Grossmann [?]. [Report on abstract of Maxwell's Paper XXVIII in *Phil. Mag.* 32: 390–393,] *Fort. Phys.* 22 (1866): 261–263.

Guillemin, Amédee. "Les anneaux de Saturne: Conditions d'équilibre et constitution physique," *La Nature* 5 (1) (1877): 211–214; (2) (1877): 1–3, 20–23. Review of observations and theories.

Guthrie, J. [i.e. F.] "Kinetic Theory of Gases," *Nature* 8 (1873): 67.

Guthrie, F. "On the Equilibrium of Temperature of a Gaseous Column Subject to Gravity," *Nature* 10 (1874): 123.

Hansemann, G. "On the Equilibrium of Temperature of a Gaseous Column Subject to Gravity," *Nature* 9 (1873): 27. Supports Guthrie against Maxwell.

Hansemann, G. "Ueber den Einfluss der Anziehung auf die Temperatur der Weltkörper," *Ann. Phys. Ergänz.* (1874): 417–440.

Harig, G. "J. C. Maxwell, Versuch einer wissenschaftlichen Biographie," *Archives d'Histoire des Sciences et des Techniques, Academie des Sciences de l'U. R. S. S.* 6 (1935): 33–61. In Russian with German summary.

Harman [see also Heimann], P. M. "Edinburgh Philosophy and Cambridge Physics: The Natural Philosophy of James Clerk Maxwell," in *Wranglers and Physicists*, ed. P. M. Harman 202–224. Dover, N. H.: Manchester University Press, 1985.

Harman, P. M. "Mathematics and Reality in Maxwell's Dynamical Physics," in *Kelvin's Baltimore Lectures and Modern Theoretical Physics*, ed. R. Kargon and P. Achinstein, 267–297. Cambridge, Mass.: MIT Press, 1987.

Harman, P. M. "Newton to Maxwell: The *Principia* and British Physics", *Notes Records R. Soc. London* 42 (1988): 75–96.

Harman, P. M., ed. *The Scientific Letters and Papers of James Clerk Maxwell*, vol. 1: 1846–1862. New York: Cambridge University Press, 1990. General Introduction, xviii–xxii. Introduction, 1–32.

Harman, P. M. "Maxwell and Saturn's Rings: Problems of Stability and Calculability," in *An Investigation of Difficult Things: Essays on Newton and the History of the Exact Sciences,* ed. P. M. Harman and Alan Shapiro, 470–502. New York: Cambridge University Press, 1992.

Hayles, N. Katherine. *Chaos Bound: Orderly Disorder in Contemporary Literature and Science.* Ithaca: Cornell University Press, 1990. Chapter 2: "Self-Reflexive Metaphors in Maxwell's Demon and Shannon's Choice: Finding the Passages." Also in *Literature and Science: Theory and Practice* ed. Stuart Peterfreund, 209–370. Boston: Northeastern University Press, 1990.

Heimann, P. M. "Maxwell and the Modes of Consistent Representation," *Arch. Hist. Exact Sci.* 6 (1970): 171–213.

Heimann, P. M. "Molecular Forces, Statistical Representation and Maxwell's Demon," *Stud. Hist. Phil. Sci.* 1 (1970): 189–211.

Herschel, A. S. "On the Use of the Virial in Thermodynamics," *Nature* 18 (1878): 39–40. Comments on paper by Preston (*Nature* 17 (1877): 31), comparison with Maxwell's Demon.

Hirn, G. A. *Mémoire sur les Conditions d'Équilibre et sur la Nature Probable des Anneaux de Saturne.* Paris: Gauthier-Villars, 1872. He missed Maxwell's "beau travail" despite a careful search of the literature and learned of it only after completing his own work.

Hopley, Ian D. Clerk Maxwell's Contributions to Physics. Unpublished Ph. D. diss., University of London, 1956.

Howard, James L. Review of Maxwell's *Theory of Heat,* 10th ed., with corrections and additions by Lord Rayleigh. *Phil. Mag.* 33 (1892): 231

Ikenberry, E., and C. Truesdell, "On the Pressures and the Flux of Energy in a Gas According to Maxwell's Kinetic Theory, I," *Journal of Rational Mechanics and Analysis* 5 (1956): 1–54. For Part II see Truesdell (1956).

Jeans, J. H. "Clerk Maxwell's Method," in J. J. Thomson et. al. *James Clerk Maxwell: A Commemoration Volume, 1831–1931* (1931).

Jochmann, [Emil]. [Report on Maxwell's Paper XX], *Forts. Phys.* 15 (1859): 314–320; 16 (1860): 322–325.

Jochmann, [Emil]. [Report on Clausius, "On the dynamical ..." *Phil. Mag.* 19 (1860): 434–436], *Forts. Phys.* 16 (1860): 321–322. Maxwell's factor 2 for the average relative velocity is incorrect.

Jones, J. V., "James Clerk Maxwell at Aberdeen 1856–1860," *Notes Records R. Soc. London* 28 (1973): 57–81. Includes Maxwell's *Inaugural Lecture*, 3 November 1856 at Marischal College Aberdeen.

Jones, R. V. "The Complete Physicist: James Clerk Maxwell, 1831–1879," *Yearbook, Royal Society of Edinburgh*, session 1978–79 (pub. 1980): 5–23.

Kargon, Robert. "Model and Analogy in Victorian Science: Maxwell and the French Physicists," *J. Hist. Ideas* 30 (1969): 423–436.

Kartsev, Vladimir P. *Maksvell.* Moscow: Molodaia Gvardia, 1974.

Kartsev, V. P. "Edinburgskii Period Tvorchestva Maksvella," *Voprosy Istorii Estestvoznaniia i Tekhniki* 1, no. 58 (1977): 105–108.

Kartsev, V. P. "Maksvell i Chuvstevennyi Obraz Fizicheskogo Mira" ["Maxwell and a Sensual Image of the Physical World,"] *Voprosy Istorii Estestvoznaniia i Tekhniki* (1980): 95–101.

Kartsev, V. "The Mach–Boltzmann Controversy and Maxwell's Views on Physical Reality," in *Probabilistic Thinking, Thermodynamics and the Interaction of the History and Philosophy of Science*, Proceedings of the 1978 Pisa Conference on the History and Philosophy of Science, Vol. II, 199–205, ed. J. Hintakka et. al. Dordrecht and Boston: D. Reidel Pub. Co., 1981.

Keeler, J. E. "A Spectroscopic Proof of the Meteoric Constitution of Saturn's Rings," *Astrophysics Journal* 1 (1895): 416–427.

[Kelvin] Thomson, William. "Voltaic Potential Differences and Atomic Sizes," *Proc. Manchester Lit. Phil. Soc.* 9 (1870): 136–41. Also in *Nature* 2 (1870): 56–57. Mentions "Clausius's and

Maxwell's magnificent working out of the Kinetic Theory of Gases" as providing one of four ways to limit the size of atoms or molecules.

[Kelvin] Thomson, William. "Kinetic Theory of the Dissipation of Energy," *Nature* 9 (1874): 441–444; *Proc. R. Soc. Edinburgh* 8 (1874): 325–334. Reprinted in Kelvin, *Mathematical and Physical Papers*, vol. V. Cambridge: Cambridge University Press, 1911, 11–20. Uses Maxwell Demons to counteract dissipation.

[Kelvin] Thomson, William. "The Sorting Demon of Maxwell," *Proc. R. Inst.* 9 (1879): 113–114; abstract in *Nature* 20 (1879): 126. Reprinted in Kelvin, *Popular Lectures and Addresses* vol. 1. London: Macmillan, 1891, 144–148, also in his *Papers*, vol. V. London: Cambridge University Press, 1911, 21–23.

[Kelvin] Thomson, William. "Steps toward a Kinetic Theory of Matter," (Address to Section A of the British Association), *Rep. 54th Meeting BAAS* (1884): 613–622; *Nature* 30 (1884): 417–421. Discusses Maxwell's work and uses "an engineer corps of Maxwell's army of sorting demons."

[Kelvin] Thomson, William. "On a Decisive Test-Case Disproving the Maxwell-Boltzmann Doctrine Regarding Distribution of Kinetic Energy," *Proc. R. Soc. London* 51 (1892): 397–399.

Kelvin. "Nineteenth-Century Clouds over the Dynamical Theory of Heat and Light," *Phil. Mag.* 2 (1901): 1–40; *J. R. Inst.* 16 (1902): 363–397. Reprinted with additions as Appendix B to his *Baltimore Lectures on Molecular Dynamics and the Wave Theory of Light.* London: Clay, 1904, 486–527.

Klein, Martin, J. "Maxwell, His Demon, and the Second Law of Thermodynamics," *Amer. Sci.* 58 (1970): 84–97.

Klein, Martin, J. "The Maxwell-Boltzmann Relationship." In *Transport Phenomena 1973*, ed. J. Kestin, 297–308. New York: American Institute of Physics, 1973.

Knight, D. M. *Atoms and Elements, a Study of Theories of Matter in England in the Nineteenth Century.* London: Hutchinson, 1967.

Knight, D. M., ed. *Classical Scientific Papers: Chemistry.* London: Mills and Boon/New York: American Elsevier, 1968. Includes Maxwell's papers LXII and LXXIII.

Knott, C. G. *Life and Scientific Work of Peter Guthrie Tait.* Cambridge: Cambridge University Press, 1911. Includes important correspondence with Maxwell.

Larmor, Joseph, ed. *Memoir and Scientific Correspondence of the late Sir George Gabriel Stokes, Bart.* Cambridge: Cambridge University Press, 1907. 2 vols. Reprinted, New York: Johnson Reprint Corp., 1971. Includes important correspondence with Maxwell.

Larmor, J. "The Scientific Environment of Clerk Maxwell," in J. J. Thomson *James Clerk Maxwell: A Commemoration Volume, 1831–1931.*

Larmor, J., ed. "Origins of Clerk Maxwell's Electric Ideas as Described in Familiar Letters to William Thomson," *Proc. Cambridge Phil. Soc.* 32 (1936): 695–750.

Leff, Harvey S. "Maxwell's Demon, Power, and Time," *Amer. J. Phys.* 58 (1990): 135–142.

Leff, Harvey S., and Andrew F. Rex. "Resource Letter MD-1: Maxwell's Demon," *Amer. J. Phys.* 58 (1990): 201–209.

Leff, Harvey S., and Andrew F. Rex. *Maxwell's Demon: Entropy, Information, Computing.* Princeton, N. J.: Princeton University Press, 1990. Reprints papers of W. Thomson (1974), E. E. Daub (1970), P. M. Heimann (1970), M. J. Klein (1970) and others.

Locqueneux, Robert, Bernard Maitte, and Bernard Pourprix. "Les Statuts Epistémologiques des Modèles de la Théorie des Gaz dans les Oeuvres de Maxwell et Boltzmann," *Fundamenta Scientiae* 4 (1983): 29–54.

Lodge, O. J. Review of *The Scientific Papers of James Clerk Maxwell, Phil. Mag.* 31 (1891): 141–142. Regrets that "a few of his less laborious contributions to science are not included" as well as some of his letters.

Lowery, H., "The Joule Collection in the College of Technology, Manchester," *J. Scient. Instrum.* 7 (1930): 369–378; 8 (1931): 1–6. Includes letter from Maxwell to Balfour Stewart.

MacDonald, D. K. C. *Faraday, Maxwell, and Kelvin.* Garden City, N. Y.: Doubleday/Anchor, 1964.

Macfarlane, Alexander. *Lectures on Ten British Physicists of the Nineteenth Century.* New York: Wiley, 1919. Includes Maxwell among others.

Mendoza, E., "A Sketch for a History of the Kinetic Theory of Gases," *Physics Today* 14 (March 1961): 36–39.

Mendoza, E., "The Surprising History of the Kinetic Theory of Gases," *Mem. Proc. Manchester Lit. Phil. Soc.* 105 (2) (1962–63): 1–14.

Meyer, O. E., "Über die innere Reibung der Gase. 3. Über Maxwell's Methode zur Bestimmung der Luftreibung," *Ann. Phys.* 143 (1871): 14–26.

Meyer, O. E., "Ueber einen Beweis des Maxwell'schen Gesetzes für das Gleichgewicht von Gasmolecülen," *Ann. Phys.* 7 (1879): 317–321.

Meyer, O. E., "Ueber eine veränderte Form meines Beweises für Maxwell'sche Gesetz der Energievertheilung," *Ann. Phys.* 10 (1880): 296–304.

Muir, M. M. Pattison, "Chemical Equilibrium," *Nature* 21 (1880): 516–518. Maxwell's views on the thermodynamics of equilibrium and his popularization of the work of Josiah Willard Gibbs.

Myers, Greg. "Nineteenth-Century Popularization of Thermodynamics and Social Prophecy," *Victorian Studies* 28 (1985): 35–66. On Maxwell, molecules and Spencer's social theories as well as those of Henry Adams.

Neesen. [Report on W. Thomson's paper [Kelvin 1874].] *Fort. Phys.* 30 (1874): 673–674. Mentions Maxwell's Demon.

Neidhardt, W. Jim. "Biblical Humanism: The Tacit Grounding of James Clerk Maxwell's Creativity," *Perspectives on Science and Christian Faith* 41 (1989): 137–142.

Newman, James R. "James Clerk Maxwell," *Scientific American* 192 (June 1955): 58–71.

Osterbrock, Donald. "The Nature of Saturn's Rings: James E. Keeler's 'Prettiest Application of Doppler's Principle', " *Mercury* 14, no. 2 (March/April 1985): 46–47, 50–53, 62.

Pearson, Karl. *The Grammar of Science.* London: Scott, 1892, 100–1. Maxwell's Demon.

Pearson, Karl. *The Grammar of Science*, 2d ed. London: Black, 1900. Note VII (new in this ed.) uses Maxwell Demon to see into the past.

Pearson, Karl. "Old Tripos Days at Cambridge, as Seen from Another Viewpoint, " *Math. Gazette* 20 (1936): 27–36. Sketches of Stokes, Maxwell, Todhunter and other Cambridge mathematicians.

Pearson, Karl. *The History of Statistics in the 17th and 18th Centuries Against the changing Background of Intellectual, Scientific and Religious Thought*, edited by E. S. Pearson. New York: Macmillan, 1978. Page 360 refers to Maxwell's letter on singular solutions and free will.

Pendse, C. G. "Note on the Stability of Saturn's Rings," *Phil. Mag.* 16 (1933): 575–580. Maxwell's theory does not imply stability for uniform rigid circular ring though this can be proved by a slightly different method.

Pendse, C. G. "The Theory of Saturn's Rings," *Phil. Trans. R. Soc. London* A234 (1935): 145–176. Critique and revision of Maxwell's theory.

Pfaundler, L. [Report on Maxwell's Paper LXXI], *Fort. Phys.* 31 (1875): 47–49.

Planck, Max. "Maxwell's Influence on Theoretical Physics in Germany" in J. J. Thomson et. al. *James Clerk Maxwell: A Commemoration Volume, 1831–1931.*

Poincaré, H. "Note sur la stabilité de l'anneau de Saturne," *Bull. Astron.* 2 (1885): 507–508. Discusses Maxwell's theory of Saturn's rings. This article was itself the subject of an anonymous report, "The density of Saturn's Ring," *Nature* 33 (1886): 303. The reviewer corrected Poincaré on Maxwell. "For a ring of very small satellites (not for a fluid ring, as M. Poincaré erroneously states), Maxwell has shown the condition to be that the density should not exceed 1/300 part of that of Saturn."

Poincaré, H. "Le mécanisme et l'éxperience," *Revue de Metaphysique et de Morale* 1 (1893): 534–537. Mentions Maxwell's Demon.

Porter, Theodore M. "A Statistical Survey of Gases: Maxwell's Social Physics," *Hist. Stud. Phys. Sci.* 12 (1981): 77–116. Claims Maxwell's understanding of probability in his kinetic theory came from Quetelet's social statistics.

Porter, Theodore M. *The Rise of Statistical Thinking 1820–1900.* Princeton, N. J.: Princeton University Press, 1986.

Preston, S. Tolver. "Mode of the Propagation of Sound, and the Physical Condition Determining its Velocity on the Basis of the Kinetic Theory of Gases," *Phil. Mag.* 3 (1877): 441–453. Maxwell, to whom the paper was communicated, derived the ratio of the velocity of sound to molecular velocity.

Preston, S. Tolver. "On the Equilibrium of Pressure in Gases," *Phil. Mag.* 4 (1877): 77. Recognizes Maxwell's priority in deducing that molecules move equally in all directions.

Preston, S. Tolver. "On a Means for Converting the Heat-Motion Possessed by Matter at Normal Temperatures into Work," *Nature* 17 (1878): 202–204. Remark on Maxwell's Demon.

Preston, S. Tolver. "Temperature Equilibrium in the Universe in Relation to the Kinetic Theory," *Nature* 20 (1879): 28. Remark on Maxwell's Demon.

Preston, S. Tolver. "A Psychological Aspect of the Vortex-Atom Theory," *Nature* 21 (1880): 323. Quotes Maxwell's "Atom" (Paper LXXIII).

Preston, S. Tolver. "Some Remarks on the Kinetic Theory of Gases," *Phil. Mag.* 31 (1891): 441–443. Quotes a letter from Maxwell to himself on high-velocity molecules; discusses Maxwell's estimates of molecular sizes.

Price, Derek, J. "The Cavendish Laboratory Archives. 3 Facsimiles," *Notes Records R. Soc. London* 10 (1953): 139–147. Includes an undated letter on a possible "molecular aether" from Maxwell to Tait.

Proctor, Richard. *Saturn and its System.* London: Chatto and Windus, 1865. 2d ed., 1882. See, 133–134.

Proctor, Richard. "The Planet Saturn in July 1869," *Popular Science Review* 8 (1869): 252–60. Doesn't know why he has been credited with Maxwell's conclusion that rings are solid particles since he gave Maxwell full credit in his book on Saturn [cited above].

Randall, J. "Aspects of the Life and Work of James Clerk Maxwell," *Nature* 195 (1962): 427–434. On Maxwell's association with King's College.

Randall, J. "Aspects of the Life and Work of James Clerk Maxwell," in Domb, *Clerk Maxwell and Modern Science,* 1–25.

Rankine, W. J. Macquorn. "Actual Energy," *Phil. Mag.* 43 (1872): 160. Comment on Maxwell's statement in *Theory of Heat* about kinetic and potential energy.

Rayleigh [John William Strutt, Third Baron]. "Inaugural Address" *Rep. 54th Meeting BAAS* (1884): 1–23; *Nature* 30 (1884): 410–417. Reprinted in Rayleigh *Scientific Papers*, vol. II, 333–354. Cambridge: Cambridge University Press, 1902. Includes remarks on Maxwell's gas theory and the viscosity of gases at low densities.

Rayleigh [John William Strutt, Third Baron]. "Clerk–Maxwell's Papers," *Nature* 43 (1890): 26–27. Reprinted in Rayleigh *Scientific Papers*, vol. III, 426–428. Review of the edition of Maxwell's collected papers.

Rayleigh [John William Strutt, Third Baron]. "Remarks on Maxwell's Investigation Respecting Boltzmann's Theorem," *Phil. Mag.*

(1892): 356–359. Reprinted in Rayleigh *Scientific Papers*, vol. III, 554–557. On Maxwell's paper XCIV.

Reynolds, Osborne. "Note on Thermal Transpiration," *Proc. R. Soc. London* 30 (1880): 300. Comments on Maxwell's theory of the radiometer. See also Stokes (1880).

Reynolds, Osborne. "Certain Dimensional Properties of Matter in the Gaseous State. An Answer to Mr. G. F. FitzGerald," *Phil. Mag.* (1881): 335–342. Reply to Maxwell's comments on his previous paper.

Reynolds, Osborne. "Vortex Rings," *Nature* 29 (1883): 193–195. On Maxwell's and Thomson's views regarding "thermal effusion" and "thermal transpiration" of gases.

Roche, E. "Essai sur la constitution et l'origine du système solaire," *Mem. Acad. Sci. Montpellier*, 8 (1873): 235–324. Part VII, "De l'Anneau de Saturn," 289–301. He announced his results on impossibility of a solid satellite existing at the distance of the rings in 1849, "bien avant les travaux de M. Maxwell et de M. Hirn."

Rossis, Georges. "Sur la discussion par Brillouin de l'Expérience de Pensée de Maxwell," *Fundamenta Scientiae* 2 (1981): 37–44.

Roth. [Report on Maxwell's Paper XCIII.] *Fort. Phys.* 37 (1881): 197–199.

Schneider Ivo. Review of Garber, Brush, and Everitt, *Maxwell on Molecules and Gases. Isis* 80 (1989): 535–536.

Schuster, Arthur. "The Clerk-Maxwell Period," in *A History of the Cavendish Laboratory 1871–1910.* London: Longmans, Green, 1910.

Schuster, Arthur. *The Progress of Physics during 33 Years (1875–1908).* Cambridge: Cambridge University Press, 1911.

Schuster, Arthur. *Biographical Fragments.* London: Macmillan, 1932.

Schweber, Silvan S. "Demons, Angels and Probability: Some Aspects of British Science in the Nineteenth Century," in *Physics as Natural Philosophy*, ed. Abner Shimony and Hermann

Feshbach, 319–63. Cambridge Mass.: MIT Press, 1982. On Darwin, Buckle, Quetelet, and Maxwell.

Scott, Wilson. *The Conflict between Atomism and Conservation Theory 1644 to 1860.* New York: Elsevier, 1970.

Sheynin, O. B. "On the History of the Statistical Method in Physics," *Arch. Hist. Exact Sci.* 33 (1985): 351–382.

Shortland, Michael. "Maxwell, James Clerk (1831–1879)," in *Victorian Britain: An Encyclopedia*, ed. Sally Mitchell, 487. New York: Garland Publishing, 1988.

Smith, Crosbie. "From Design to Dissolution: Thomas Chalmers' Debt to John Robison," *Brit. J. Hist. Sci.* 12 (1979): 59–70. Discusses Maxwell's views on theology of nature and the Stewart-Tait *Unseen Universe.*

Smith, Crosbie, and M. Norton Wise. *Energy and Empire: A Biographical Study of Lord Kelvin.* New York: Cambridge University Press, 1989.

Smith-Rose, R. L. *James Clerk Maxwell, F. R. S., 1831–1879.* London: Longmans, Green, 1948.

Stallo, J. B. *The Concepts and Theories of Modern Physics.* New York: Appleton, 1881. Reprinted, Cambridge, Mass.: Harvard University Press, 1960. Chapter VIII is a critique of the kinetic theory of gases.

Stefan, Josef. "Untersuchungen über die Wärmeleitung in Gasen," *Sitz. Akad. Wiss. Wien* 65 (1872): 45–69. Comments on Maxwell's kinetic theory and its confirmation by experiments.

Stefan, Josef. "Über das Gleichgewicht und die Bewegung insbesondere die Diffusion von Gasgemengen," *Sitz. Akad. Wiss. Wien* 63 (1871): 63–124. Comments on Maxwell's theory of diffusion.

Stefan, Josef. "Untersuchungen über die Wärmeleitung in Gasen. Zweite Abhandlung. Relative Bestimmung der Wärmeleitermögen verschiedener Gase." *Sitz. Akad. Wiss. Wien* 72 (1876): 69–101. On the discrepancy between Maxwell's theoretical value for the heat conduction coefficient (as corrected by Boltzmann) and experimental values.

Stefan, Josef. "Über die Beziehung zwischen der Wärmestrahlung und der Temperatur," *Sitz. Akad. Wiss. Wien* 79 (1879): 391–428. Since the heat conduction coefficient is independent of pressure down to very low pressures, according to Maxwell's theory, data on heat radiation can be interpreted to show that the rate of heat loss from a hot body is proportional to the fourth power of its absolute temperature.

Stewart, Balfour. "Maxwell on Heat," *Nature* 5 (1872): 319–320. Review of *Theory of Heat.*

Stokes, G. G. "Note by the Communicator," *Proc. R. Soc. London* 30 (1880): 301. On Reynolds and Maxwell.

Stokes, G. G. "On the Bearings of the Study of Natural Science, and of the Contemplation of the Discoveries to Which That Study Leads, on Our Religious Ideas," *Journal of the Transactions of the Victoria Institute* (1880): 227–247. Remarks on Maxwell's Demon.

Stoney, G. Johnstone. "The Internal Motions of Gases Compared with the Motions of Waves of Light," *Phil. Mag.* 36 (1868): 132–141. Estimate of atomic magnitudes based on Maxwell's estimate of mean free path of molecules in gases.

Strutt, R. J. *Life of John William Strutt, Third Baron Rayleigh.* London: Arnold, 1924. Reprinted with additions by J. N. Howard, Madison: University of Wisconsin Press, 1968.

Tait, P. G. *Sketch of Thermodynamics.* Edinburgh: Edmonston and Douglas, 1868. Comments on kinetic theory of gases. Maxwell's three results, independent of hypotheses about intermolecular forces, are: equipartition of energy among molecules of unequal mass: Dalton's law of partial pressures: temperature is constant in vertical column of gas.

Tait, P. G. *Lectures on Some Recent Advances in Physical Science.* London: Macmillan, 1876. Remarks on Maxwell's Demon, 188–220 and on Maxwell's kinetic theory, 324–327.

Tait, P. G. "Zöllner's Scientific Papers," *Nature* 17 (1878): 420–422. Comments on the chapter on Thomson's (i.e. Maxwell's) Demon in Zöllner (1878).

Tait, P. G. "Clerk-Maxwell's Scientific Work," *Nature* 21 (1880): 317–21.

Tait, P. G. "Note on a Theorem of Clerk-Maxwell," *Proc. R. Soc. Edinburgh* 13 (1884, pub. 1886): 21–23. On Maxwell's proof of his velocity distribution law. See Tait's later papers on this topic, listed in Brush (1976), 764–765.

Tait, P. G. *Heat.* London: Macmillan, 1884. Comments on Maxwell's explanation of gaseous viscosity (p. 355), and his Demon (p. 366).

Taylor, William B. "Note on the Rings of Saturn," *Bull. Wash. Phil. Soc.* 6 (1884): 41–45 *Smithsonian Miscellaneous Collection* 33 (1888), art. 1. Maxwell's work on the meteoric constitution of Saturn's rings has been accepted by all physical astronomers.

Theerman, Paul Harold. James Clerk Maxwell: Physicist and Intellectual in Victorian Britain. Unpublished Ph. D. diss. University of Chicago, 1980.

Theerman, Paul. "James Clerk Maxwell and Religion," *Amer. J. Phys.* 34 (1986): 312–317.

Thompson, Silvanus P., *The Life of William Thomson, Baron Kelvin of Largs.* London: Macmillan, 1910.

Thomson, J. J. et al. *James Clerk Maxwell: A Commemoration Volume, 1831–1931.* Cambridge: Cambridge University Press, 1931. Includes "James Clerk Maxwell," by J. J. Thomson, 1–44; essays by Planck (1931), Larmor (1931), Jeans (1931) Albert Einstein, William Garnett, Ambrose Fleming, Oliver Lodge, R. T. Glazebrook, and Horace Lamb.

Tolstoy, Ivan, *James Clerk Maxwell: A Biography.* Chicago: University of Chicago Press, 1981. Reviewed by C. W. F. Everitt, *Physics Teacher* 22 (1984): 264–266; R. V. Jones, *Nature* 294 (1981): 294; N. Mott, *London Review of Books* (19 Nov. 1981): 18; R. O'Hanlon, *Times Literary Supplement*

(22 Jan. 1982): 83–84; L. Ohlon, *Lychnos* (1983): 240–241; D. Siegel, *Physics Today* 38, no. 8 (1985): 66–67; C. Smith, *Isis* 73 (1982): 480; P. Theerman, *Technology and Culture* 25 (1984): 136–138; D. B. Wilson, *Victorian Studies*, 28 (1985): 539–540.

Tremaine, Scott. Review of Brush, Everitt, and Garber, *Maxwell on Saturn's Rings*, *Journal of the Royal Astronomical Society of Canada*, 79 (1985): 35–37. Mentions several errors made by Maxwell and the relation of his work to modern research on accretion disks.

Truesdell, C. "A New Definition of a Fluid. II. The Maxwellian Fluid," *J. Math. Pures Appl.* 30 (1951): 111–158. Includes a review of non-linear viscous flow theories of Maxwell and others.

Truesdell, C. "On the Pressures and the Flux of Energy in a Gas According to Maxwell's Kinetic Theory, II," *Journal of Rational Mechanics and Analysis* 5 (1956): 55–128. For Part I see Ikenberry and Truesdell (1956).

Truesdell, C. "Une solution exacte des équations de Maxwell," *J. Math. Pures Appl.* 37 (1958): 119–133.

Turner, Joseph. The Methodology of James Clerk Maxwell. Unpublished Ph. D. diss., Columbia University, 1953.

Turner, Joseph. "A Note on Maxwell's Interpretation of some Attempts at Dynamical Explanation," *Ann. Sci.* 11 (1955): 238–245.

Turner, Joseph. "Maxwell on the Method of Physical Analogy," *Brit. J. Phil. Sci.* 6 (1955): 226–238.

Watson, H. W. *A Treatise on the Kinetic Theory of Gases.* Oxford: Clarendon Press, 1876; 2d ed. 1893.

Webb, Thos. Wm. "The Planet Saturn," *Intellectual Observer* 9 (1866): 247–267, 366–381, 466–469; 10 (1867): 49–59, 142–149, 194–202. Refers to Maxwell's theory, recently publicized by Richard Proctor.

Wheeler, L. P. *Josiah Willard Gibbs: The History of a Great Mind.* New Haven: Yale University Press, 1951.

Whiting, Harold. "Maxwell's Demons," *Science* 6 (1885): 83.

Willerding, E. "James C. Maxwell und die Stabilität der Saturnringe," *Phys. Blätter* 48 (1992): 799–803.

Wilson, David B. *Kelvin and Stokes: A Comparative Study in Victorian Physics.* Bristol: Hilger, 1987.

Winkelmann, A. A. "Über die Wärmeleitung der Gase," *Ann. Phys.* 153 (1875): 497–531; 157 (1876): 497–555. First two parts of a long series of papers reporting experimental tests of the heat conduction theories of Clausius, Maxwell, and Boltzmann.

Wise, M. Norton. "The Maxwell Literature and British Dynamical Theory," *Hist. Stud. Phys. Sci.* 13 (1983): 175–201.

Woodruff, A. E. "The Radiometer and How It Does Not Work," *Physics Teacher* 6 (1968): 358–364.

Woods, L. C. "Maxwell's Models," *Bull. Inst. Math. Appl.* 16 (1980): 11–15.

Zöllner, Friedrich. "Thomson's Dämonen und die Schatten Plato's," in Zöllner, *Wissenschaftliche Abhandlungen,* Band I, 710–732. Leipzig: Staackman, 1878. On Maxwell's Demon and William Thomson's writings on LeSage's gravific corpuscles. See review by Tait (1878).

Chronological Index to Correspondence

This is a chronological listing of all the correspondence, including referee reports, published in the three volumes of Maxwell papers on Saturn's rings, gas theory, molecules and heat.

Abbreviations

M = James Clerk Maxwell
MSR = *Maxwell on Saturn's Rings*
MMG = *Maxwell on Molecules and Gases*
MH = *Maxwell on Heat & Statistical Mechanics*
[ ] = referee report on paper by [ ]

	Date	From	To [report on]	Document number
1855	May 15	M	Thomson, W.	MH III-1
1856	July 4	M	Litchfield, R. B.	MSR 1
1857	May 20	M	Munro, C. J.	MH III-2
	Aug. 1	M	Thomson, W.	MSR 2
	Aug. 24	M	Thomson, W.	MSR 3
	Aug. 28	M	Campbell, L.	MSR 4
	Oct. 15	M	Litchfield, R. B.	MSR 5
	Nov. 14	M	Droop, H. R.	MSR 6
	Nov. 14	M	Thomson, W.	MSR 7
	Nov. 21	M	Tait, P. G.	MSR 8
	Nov. 28	M	Cay, J.	MSR 9
	Dec. 22	M	Campbell, L.	MSR 10
1858	Jan. 30	M	Thomson, W.	MSR 11
	Feb. 7	M	Litchfield, R. B.	MSR 12
	Feb. 17	M	Campbell, L.	MSR 13
	Apr. 29	M	Monro, C. J.	MSR 14
	July 24	M	Monro, C. J.	MSR 15

	Sep. 7	M	Stokes, G. G.	MSR 16
1859	May 30	M	Stokes, G. G.	MMG III-1
	Oct. 8	M	Stokes, G. G.	MMG III-3
1860	Jan. 5	M	Campbell, L.	MMG III-7
1862	Jan. 28	M	Droop, H. R.	MMG III-10
	Apr. 21	M	Campbell, L.	MMG III-11
1863	Aug. 25	M	Bond, G.	MSR 19
1864	Feb. 23	Herapath, J.	M?	MMG II-3
1865	Jan. 5	M	Cay, C. H.	MMG III-13
	May 1	M	Graham, T.	MMG III-14
	June 17	M	Tait, P. G.	MMG III-15
	July 19	M	Droop, H. R.	MMG III-16
1866	Feb. 7	M	?	MMG II-4
	Apr. 4	M	Tait, P. G.	MMG III-27
	Apr. 6	Tait, P. G.	M	MMG III-28
	Oct. 13	Thomson, W.	Stokes, G. G.[M]	MH II-2
	Dec. 18	M	Stokes, G. G.	MH II-3
1867	Dec. 6	Tait, P. G.	M	MH III-3
	Dec. 11	M	Tait, P. G.	MH III-4
	Dec. 13	Tait, P. G.	M	MH III-5
undated		M	P. G. Tait	MH III-6
("Catechism on Demons")				
	Nov. 13	M	Tait, P. G.	MMG II-6
	?	M	Tait, P. G.	MMG II-7
	Dec. 23	M	Tait, P. G.	MH III-7
1868	Mar. 5	M	Tait, P. G.	MH III-8
	Mar. 12	M	Tait, P. G.	MMG III-31
	Apr. 7	M	Pattison, M. M.	MH III-9
	Apr. 13	M	Pattison, M. M.	MH III-10
	July	M	Tait, P. G.	MH III-11
	Aug. 3	M	Tait, P. G.	MH III-12
	Oct. 13	M	Huggins, W.	MH VI-1
	Dec. 7	M	Thomson, W.	MH III-13
1869	Nov. 16	M	Thomson, W.	MH III-14
1870	Apr. 14	M	Thomson, W.	MH III-15
	Dec. 6	M	Strutt, J. W.	MH III-16
1871	Feb. 15	M	Tait, P. G.	MH III-17
	May 2	M	Tait, P. G.	MH III-18
	July 13	M	Thomson, J.	MH III-19
	July 21	Thomson, J.	M	MH III-20
	July 24	M	Thomson, J.	MH III-21

	?	M	Thomson, W.	MMG II-10
1872	Feb. 3	M	Tait, P. G.	MH III-24
	Feb. 12	M	Tait, P. G.	MH III-25
	Oct. 19	M	Campbell, L.	MMG III-32
1873	Summer	M	Tait, P. G.	MMG III-40
	Summer?	M	Tait, P. G.	MMG III-41
	Aug. 28	M	Rayleigh	MMG III-43
	Aug.	M	Tait, P. G.	MH II-8
	Sep. 4	Brown, A.C.	M	MMG III-44
	Dec. 1	M	Tait, P. G.	MH III-26
	Dec. 4	Spencer, H.	M	MMG II-17
	Dec. 5	M	Spencer, H.	MMG II-18
	Dec. 17	M	Spencer, H.	MMG II-19
	Dec. 30	Spencer, H.	M	MMG II-20
	?	M	Tait, P. G.	MH VI-2
1874	Feb. 24	M	[Crookes, W.]	MH VI-3
	July 4	M	Tait, P. G.	MH IV-1
	Sept. 2	M	Tait, P. G.	MH IV-2
	Oct. 13	M	Tait, P. G.	MH III-27
	Nov. ?	M	Andrews, T.	MH III-28
	?	M	[Andrews, T.]	MH IV-4
1875	Mar. 19	M	Tait, P. G.	MMG II-25
	Mar. 27	M	Thomson, J.	MH III-30
	April ?	M	[Crookes, W.]	MH VI-4
	?	M	Thomson, J.	MH III-31
	July 8	M	Thomson, J.	MH III-32
	July 15	M	Andrews, T.	MH III-33
	July 25	Andrews, T.	M	MH III-34
	Aug. 3	M	Stokes, G. G.	MH III-35
	Sept. 25	M	Stokes, G. G.	MMG II-26
1876	?	M	[Crookes, W.]	MH VI-5
	Feb. 14	M	Tait, P. G.	MMG III-45
	Apr. 7	M	[Reynolds, O.]	MH VI-6
	Apr. ?	M	[Schuster, A.]	MH VI-7
	May 15	M	Cay, R. D.	MH VI-8
	July 29	M	Tait, P. G.	MH III-40
	Oct. 13	M	Tait, P. G.	MH III-41
	Nov. 21	Ellicott, C. J.	M	MMG II-27
	Nov. 22	M	Ellicott, C. J.	MMG II-28
	Dec. 5	Preston, S. T.	M	MMG II-29
	? Dec. 28	M	Tait, P. G.	MH III-42

1877	June 30	M	Garnett, W.	MMG III-47
	Nov. 8	Clausius, R.	M	MH III-43
	Dec. 12	M	Tait, P. G.	MH III-44
1878	Jan. 23	M	[Crookes, W.]	MH VI-9
	Feb. 28	M	Tait, P. G.	MH III-45
	March 7	M	Thomson, W.	MH VI-10
	June 15	Thomson, W.	[M]	MH VI-12
	June 26	Tait, P. G.	M	MH VI-13
	Aug. 7	M	Tait, P. G.	MMG II-31
	Oct. 23	M	[Crookes, W.]	MH VI-14
	?	M	Tait, P. G.	MH II-19
?	?	M	Tait, P. G.	MH III-46
1879	Mar. 28	M	[Reynolds, O.]	MH VI-15
	May 31	M	Newcomb, S.	MMG II-37
	June 2	M	Newcomb, S.	MMG II-37
	?	M	Stokes, G. G.	MH VI-16
	July 4[7?]	Reynolds, O.	M	MH VI-17
	Aug. 18	Stokes, G. G.	M	MH VI-18
	Aug. 21	M	Stokes, G. G.	MH VI-19
	Aug. 25	M	Thomson, W.	MH VI-20
	Sept. 1	M	Thomson, W.	MH VI-21
	Sept. 2	M	Stokes, G. G.	MH VI-22
	Sept. 7	Thomson, W.	M	MH VI-23

Index